Rainer Pöttgen, Thomas Jüstel, Cristian A. Strassert (Eds.)

Applied Inorganic Chemistry

Also of interest

Applied Inorganic Chemistry
Volume 1: From Construction Materials to Technical Gases
Rainer Pöttgen, Thomas Jüstel, Cristian A. Strassert (Eds.), 2023
ISBN 978-3-11-079878-4, e-ISBN 978-3-11-079889-0

Applied Inorganic Chemistry
Volume 3: From Magnetic to Bioactive Materials
Rainer Pöttgen, Thomas Jüstel, Cristian A. Strassert (Eds.), 2023
ISBN 978-3-11-073837-7, e-ISBN 978-3-11-073347-1

Intermetallics
Synthesis, Structure, Function
Rainer Pöttgen, Dirk Johrendt 2019
ISBN 978-3-11-063580-5, e-ISBN 978-3-11-063672-7

Rare Earth Chemistry
Rainer Pöttgen, Thomas Jüstel and Cristian A. Strassert (Eds.) 2020
ISBN 978-3-11-065360-1, e-ISBN 978-3-11-065492-9

Zeitschrift für Kristallographie - Crystalline Materials
Rainer Pöttgen (Editor-in-Chief)
ISSN 2194-4946, e-ISSN 2196-7105

Applied Inorganic Chemistry

Volume 2: From Energy Storage to Photofunctional Materials

Edited by
Rainer Pöttgen, Thomas Jüstel, Cristian A. Strassert

DE GRUYTER

Editors
Prof. Dr. Rainer Pöttgen
Institut für Anorganische und Analytische Chemie
Westfälische Wilhelms-Universität Münster
Corrensstraße 30
48149 Münster
Germany
E-mail: pottgen@uni-muenster.de

Prof. Dr. Thomas Jüstel
Fachbereich Chemieingenieurwesen
Fachhochschule Münster
Stegerwaldstraße 39
48565 Steinfurt
Germany
E-mail: tj@fh-muenster.de

Prof. Dr. Cristian A. Strassert
Institut für Anorganische und Analytische Chemie
CiMIC – CeNTech – SoN
Westfälische Wilhelms-Universität Münster
Corrensstraße 28/30
48149 Münster
Germany
E-mail: ca.s@wwu.de

This book was carefully produced. Nevertheless, the authors and the publisher do not warrant the information contained herein to be free of errors. Readers are advised to keep in mind that statements, data, illustrations, procedural detail or other items may inadvertently be inaccurate.

ISBN 978-3-11-079878-4
e-ISBN (PDF) 978-3-11-079889-0
e-ISBN (EPUB) 978-3-11-079900-2

Library of Congress Control Number: 2022935001

Bibliographic information published by the Deutsche Nationalbibliothek
The Deutsche Nationalbibliothek lists this publication in the Deutsche Nationalbibliografie; detailed bibliographic data are available on the Internet at http://dnb.dnb.de.

© 2023 Walter de Gruyter GmbH, Berlin/Boston
Cover image: $LiYF_4:Pr^{3+}$ laser crystal, Danuta Dutczak/Thomas Jüstel
Typesetting: Integra Software Services Pvt. Ltd.
Printing and binding: CPI books GmbH, Leck

www.degruyter.com

Preface

The Periodic Table meanwhile lists 118 chemical elements, which leads to a vast number of inorganic compounds. Many of them have well-defined physicochemical properties, which are exploited for the realization of functional materials we all comfortably use in daily life without even thinking about it, including magnetic and optical materials, construction materials, materials for energy storage and conversion – just to name a few remarkable examples. The impact of inorganic chemistry in human evolution cannot be overstated, and is proven by the designation of historical ages, such as stone, copper, bronze or iron age (even golden ages and gold rush), or by geographical locations (such as the Silicon Valley and Argentina). While carbon-based organic chemistry has provided incredible breakthroughs in medicinal chemistry and plastic materials, there is no doubt that the solution of the most urgent problems currently faced by humanity will stem from inorganic chemistry providing high-density/high-stability materials for construction, information technologies, energy storage and conversion.

Chemical sciences and industries are often demonized, but the many indispensable materials we use in daily life impressively show how significantly they influence our society. Ecosystems, metabolic and pathophysiological processes, food production, construction in its broadest sense, mobility and energy conversion are determined by chemistry – these facts cannot simply be ignored! The present book summarizes the many basic examples of inorganic materials we use on a large scale in everyday life, but also niche products with thoroughly optimized properties. Many subchapters are written by experts from academia and industry. We tried to ensure a proper balance of topics, even though it is simply impossible to cover all aspects of applied inorganic chemistry. Nonetheless, we hope that we made a good compromise – if any topic is missing, this was unintentional. The final chapter focusses on energy flows and resources, which constitutes one of the most urgent topics. As a kind of appetizer for the following 16 chapters, we briefly summarize some applications for the elements of the first four rows of the Periodic Table. Several of these topics are picked up again in the following chapters:

Hydrogen: energy source; **helium:** low-temperature refrigerant; ballon gas, **lithium:** anode materials for lithium-ion batteries; **beryllium:** hardening component for light-weight alloys, non-spark alloys, X-ray windows; **boron:** hardening component for intermetallics; **carbon:** electrode materials, black pigment; **nitrogen:** source for ammonia and nitrate fertilizers, protective gas, low-temperature cooling; **oxygen:** medical gas, liquid oxygen for the Linz-Donauwitzer process in steel refinement; **fluorine:** uranium hexafluoride production; **neon:** helium-neon lasers; **sodium:** reducing agent; **magnesium:** alloying component and sacrificial anodes; **aluminum:** light-weight alloys, construction material; **silicon:** semiconductors; **phosphorus:** synthesis of phosphoric acid; matches; **sulfur:** vulcanization of rubber; **chlorine:** disinfection of water; **argon:** protective gas in chemical synthesis

https://doi.org/10.1515/9783110798890-202

and arc-welding; **potassium:** liquid sodium-potassium alloys as coolants in nuclear reactors; **calcium:** reducing agent in metallurgy; **scandium:** additive for aluminum-based alloys, component of electron emitters; **titanium:** steel additive, corrosion resistant alloys; **vanadium:** high-speed tool steels; **chromium:** stainless steel and chromium plating; **manganese:** ferromanganese, activator in LED phosphors; **iron:** steel and cast iron; **cobalt:** superalloys and samarium-cobalt magnets; **nickel:** catalysis and anti-corrosion coatings; **copper:** cables and water tubes; **zinc:** facade cladding, corrosion protection; **gallium:** gallium nitride, phosphide or arsenide semiconductors; **germanium:** semiconductors and detection technology; **arsenic:** doping of semiconductors; **selenium:** II-VI semiconductors and alloy additive for free cutting steel; **bromine:** special disinfection products and synthesis of flame retardants; **krypton:** excimer lasers, KrCl excimer discharge lamps. The reader might notice that transition metals and lanthanides are not even mentioned here; there would not be sufficient space in a preface to list their impact!

Such a book project is not realizable without the help of numerous colleagues and co-workers. We thank Gudrun Lübbering for continuous help with literature search and text processing and Thomas Fickenscher for providing with many photos of materials and devices. We are especially grateful to our colleagues for their immediate agreements to write up a subchapter. It is always challenging to compile a concise Table of Contents and find the right co-authors. We are indebted to the editorial and production staff of De Gruyter. Our particular thanks go to Kristin Berber-Nerlinger, Dr. Vivien Schubert and Melanie Götz for their continuous support during conception, writing and producing the present book.

Münster, Steinfurt, June 2022
Thomas Jüstel, Rainer Pöttgen, Cristian A. Strassert

This book contains two different tokens, pointing to:

list of references

recommended literature for further reading; i.e. relevant text books, review articles or important original articles

Contents

Volume 2 (From Energy Storage to Photofunctional Materials)

Volume 1 (From Construction Materials to Technical Gases)

Volume 3 (From Magnetic to Bioactive Materials)

List of contributors

Ackermann, Dr. habil. Lothar
Deutsche Stiftung Edelsteinforschung (DSEF)
Professor-Schlossmacher-Straße 1
55743 Idar-Oberstein
Germany
E-mail: lackermann@outlook.de

Agne, Dr. Matthias
Forschungszentrum Jülich GmbH
Helmholtz-Institut Münster (HI MS, IEK-12)
Corrensstraße 46
48149 Münster
Germany
E-mail: m.agne@fz-juelich.de

Apaydin, Dr. Dogukan H.
Institut für Materialchemie
TU Wien
Getreidemarkt 9/165
1060 Wien
Austria
E-mail: dogukan.apaydin@tuwien.ac.at

Arnault, Dr. Jean-Charles
Laboratoire des Edifices Nanométriques
Université Paris-Saclay, CEA, CNRS, NIMBE
91191 Gif sur Yvette
France
E-mail: jean-charles.arnault@cea.fr

Banik, Dr. Ananya
Institut für Anorganische und Analytische Chemie
Westfälische Wilhelms-Universität Münster
Corrensstraße 30
48149 Münster
Germany
E-mail: banik@uni-muenster.de

Bauer, Dr. Thomas
Deutsches Zentrum für Luft- und Raumfahrt (DLR)
Institut für Technische Thermodynamik
Thermische Prozesstechnik
Linder Höhe, Gebäude 26
51147 Köln
Germany
E-mail: thomas.bauer@dlr.de

Baur, Dr. Florian
Fachbereich Chemieingenieurwesen
Fachhochschule Münster
Stegerwaldstraße 39
48565 Steinfurt
Germany
E-mail: florian.baur@fh-muenster.de

Bayer, Dr. Bernhard
Institut für Materialchemie
TU Wien
Getreidemarkt 9/165
1060 Wien
Austria
E-mail: bernhard.bayer-skoff@tuwien.ac.at

Behrend, Prof. Dr.-Ing. habil. Detlef
Lehrstuhl Werkstoffe für die Medizintechnik
Fachbereich Maschinenbau und Schiffstechnik
Universität Rostock
Friedrich-Barnewitz-Straße 4
18119 Rostock
Germany
E-mail: detlef.behrend@uni-rostock.de

Behrens, Dr. Rainer
VDM Metals International GmbH
Kleffstraße 23
58762 Altena
Germany
E-mail: rainer.behrens@vdm-metals.com

Bertau, Prof. Dr. rer. nat. habil. Martin
Institut für Technische Chemie
TU Bergakademie Freiberg
Leipziger Straße 29
09599 Freiberg
Germany
E-mail: Martin.Bertau@chemie.tu-freiberg.de
and
Fraunhofer Technology Center for High-Performance Materials THM
Fraunhofer Institut for Ceramic Technologies and Systems IKTS
Am St.-Niclas-Schacht 13
09599 Freiberg
Germany

https://doi.org/10.1515/9783110798890-204

Binnewies, Prof. Dr. Michael
Institut für Anorganische Chemie
Naturwissenschaftliche Fakultät
Leibnitz Universität Hannover
Callinstraße 3-9
30167 Hannover
Germany
E-mail: michael.binnewies@aca.uni-hannover.de

Boos, Dr. Markus
Remmers GmbH
Bernhard-Remmers-Straße 13
49624 Löningen
Germany
E-mail: mboos@remmers.de

Boos, Dr. Peter
HeidelbergCement AG
Zur Anneliese 11
59320 Ennigerloh
Germany
E-mail: Peter.Boos@heidelbergcement.com

Bredol, Prof. Dr. Michael
Fachbereich Chemieingenieurwesen
Fachhochschule Münster
Stegerwaldstraße 39
48565 Steinfurt
Germany
E-mail: bredol@fh-muenster.de

Broll, Dr. Sascha
Broll-Buntpigmente GmbH & Co. KG
Karl-Winnacker-Straße 2-4
36396 Steinau
Germany
E-mail: drsascha@broll-buntpigmente.de

Buchner, Dr. Magnus R.
Anorganische Chemie, Fluorchemie
Philipps-Universität Marburg
Hans-Meerwein-Straße 4
35032 Marburg
Germany
E-mail: magnus.buchner@chemie.
uni-marburg.de

Busch, Dr. Frank
Materialprüfungsamt Nordrhein-Westfalen
Marsbruchstraße 186
44287 Dortmund
Germany
E-mail: busch@mpanrw.de

Buttler, Dr.-Ing. Torben Alexander
-ISAF- Institut für Schweißtechnik und
Trennende Fertigungsverfahren
Technische Universität Clausthal
Agricolastraße 2
38678 Clausthal-Zellerfeld
Germany
E-mail: buttler@isaf.tu-clausthal.de

Dewalsky, Dr. Martin V.
Am Gemeindeholz 6
82205 Gilching
Germany
E-mail: martinvdew@gmail.com

Dorsch, Leonhard Yuuta
Institut für Anorganische Chemie
Universität Leipzig
Johannisallee 29
04103 Leipzig
Germany
E-mail: leonhard.dorsch@uni-leipzig.de

Dramicanin, Prof. Dr. Miroslav
Vinca Institute of Nuclear Sciences
University of Belgrade,
PO Box 522
11001 Belgrade
Serbia
E-mail: dramican@vinca.rs

Eckert, Prof. Dr. Hellmut
Institut für Physikalische Chemie
Westfälische Wilhelms-Universität Münster
Corrensstraße 30
48149 Münster
Germany
and
Instituto de Física de Sao Carlos
Universidade de Sao Paulo
Avenida Trabalhador Saocarlense 400
Sao Carlos, SP 13566-590
Brasil
E-mail: eckerth@uni-muenster.de

Eder, Prof. Dr. Dominik
Institut für Materialchemie
TU Wien
Getreidemarkt 9/165
1060 Wien
Austria
E-mail: dominik.eder@tuwien.ac.at

Engel, Stefan
Universität des Saarlandes
Anorganische Festkörperchemie
Campus C4 1
66123 Saarbrücken
Germany
E-mail: stefan.engel@uni-saarland.de

Engels, Ir. Marcel
Forschungsinstitut für Glas | Keramik (FGK)
Heinrich-Meister-Straße 2
56203 Höhr-Grenzhausen
Germany
E-mail: marcel.engels@fgk-keramik.de

Epple, Prof. Dr. Matthias
Anorganische Chemie
Fakultät für Chemie
Universität Duisburg-Essen
Universitätsstraße 7
45141 Essen
Germany
E-mail: matthias.epple@uni-due.de

Faust, PD Dr. Andreas
European Institute for Molecular Imaging (EIMI)
Waldeyerstraße 15
48149 Münster
Germany
E-Mail: faustan@uni-muenster.de

Feser, Prof. Dr.-Ing. Ralf
Fachbereich Informatik und
Naturwissenschaften
Fachhochschule Südwestfalen
Frauenstuhlweg 31
58644 Iserlohn
Germany
E-Mail: feser.ralf@fh-swf.de

Fickenscher, Thomas
Institut für Anorganische und Analytische
Chemie
Westfälische Wilhelms-Universität Münster
Corrensstraße 30
48149 Münster
Germany
E-mail: thomasfi@uni-muenster.de

Fröhlich, Dr. Peter
Institut für Technische Chemie
TU Bergakademie Freiberg
Leipziger Straße 29
09599 Freiberg
Germany
E-mail: peter.froehlich@chemie.tu-freiberg.de

Ghidiu, Dr. Michael
Institut für Anorganische und Analytische
Chemie
Westfälische Wilhelms-Universität Münster
Corrensstraße 30
48149 Münster
Germany
E-mail: ghidiu@uni-muenster.de

Glaum, Prof. Dr. Robert
Institut für Anorganische Chemie
Rheinische Friedrich-Wilhelms-Universität
Gerhard-Domagk-Straße 1
53121 Bonn
Germany
E-mail: rglaum@uni-bonn.de

Grönefeld, Dr. Martin
Magnetfabrik Bonn GmbH
Dorotheenstraße 215
53119 Bonn
Germany
E-Mail: Martin.Groenefeld@Magnetfabrik.de

Haberkamp, Prof. Dr.-Ing. Jens
Fachbereich Bauingenieurwesen
Fachhochschule Münster
Corrensstraße 25
48149 Münster
Germany
E-mail: haberkamp@fh-muenster.de

Haneklaus, Dr. Nils
Td Lab Sustainable Mineral Resources
Universität für Weiterbildung Krems
Dr.-Karl-Dorrek-Straße 30
3500 Krems an der Donau
Austria
E-mail: nils.haneklaus@donau-uni.ac.at
and
Institut für Technische Chemie
TU Bergakademie Freiberg
Leipziger Straße 29
09599 Freiberg
Germany

Hayen, Prof. Dr. Heiko
Institut für Anorganische und Analytische Chemie
Westfälische Wilhelms-Universität Münster
Corrensstraße 48
48149 Münster
Germany
E-mail: heiko.hayen@uni-muenster.de

Hendriks, Dr. Theodoor
Forschungszentrum Jülich GmbH
Helmholtz-Institut Münster (HI MS, IEK-12)
Corrensstraße 46
48149 Münster
Germany
E-mail: t.hendriks@fz-juelich.de

Hermes, Dr. Wilfried
trinamiX GmbH
Industriestraße 35
67063 Ludwigshafen
Germany
E-mail: wilfried.hermes@trinamix.de

Herrmann, Dr. Fabian
Institut für Pharmazeutische Biologie und Phytochemie
Westfälische Wilhelms-Universität Münster
Corrensstraße 48
48149 Münster
Germany
E-mail: fabian.herrmann@wwu.de

Hosono, Prof. Dr. Hideo
Materials Research Center for Element Strategy (MCES)
Tokyo Institute of Technology
SE-1, 4259 Nagatsuta-cho
Midori-ku, Yokohama, Kanagawa, 226-8503
Japan
E-mail: hosono@mces.titech.ac.jp

Huppertz, Prof. Dr. Hubert
Institut für Allgemeine, Anorganische und Theoretische Chemie
Universität Innsbruck
Innrain 80–82
6020 Innsbruck
Austria
E-mail: Hubert.Huppertz@uibk.ac.at

Janiak, Prof. Dr. Christoph
Institut für Anorganische Chemie und Strukturchemie
Lehrstuhl für nanoporöse und nanoskalierte Materialien
Heinrich-Heine-Universität Düsseldorf
Universitätsstraße 1
40225 Düsseldorf
Germany
E-mail: janiak@hhu.de

Janka, PD Dr. Oliver
Universität des Saarlandes
Anorganische Festkörperchemie
Campus C4 1
66123 Saarbrücken
Germany
E-mail: oliver.janka@uni-saarland.de

Jin, Wenqi
Xinjiang Technical Institute of Physics & Chemistry
Chinese Academy of Sciences
40-1 South Beijing Road
830011 Urumqi
China
E-Mail: jwqineni@qq.com

Johrendt, Prof. Dr. Dirk
Department Chemie
Ludwig-Maximilians-Universität München
Butenandtstraße 5-13 (Haus D)
81377 München
Germany
E-mail: johrendt@lmu.de

Jüstel, Prof. Dr. Thomas
Fachbereich Chemieingenieurwesen
Fachhochschule Münster
Stegerwaldstraße 39
48565 Steinfurt
Germany
E-mail: tj@fh-muenster.de

Klapötke, Prof. Dr. Thomas M.
Department Chemie
Ludwig-Maximilians-Universität München
Butenandtstraße 5-13 (Haus D)
81377 München
Germany
E-mail: tmk@cup.uni-muenchen.de

Kohlmann, Prof. Dr. Holger
Institut für Anorganische Chemie
Universität Leipzig
Johannisallee 29
04103 Leipzig
Germany
E-mail: holger.kohlmann@uni-leipzig.de

Koller, PD Dr. Hubert
Institut für Physikalische Chemie
Westfälische Wilhelms-Universität Münster
Correnssstraße 30
48149 Münster
Germany
E-mail: hkoller@uni-muenster.de

Kratz, Dr. Nadja
Forschungsinstitut für Glas | Keramik (FGK)
Heinrich-Meister-Straße 2
56203 Höhr-Grenzhausen
Germany
E-mail: nadja.kratz@fgk-keramik.de

Kränkel, PD Dr. Christian
Leibniz-Institut für Kristallzüchtung (IKZ)
Max-Born-Straße 2
12489 Berlin
Germany
E-mail: christian.kraenkel@ikz-berlin.de

Krzywinski, Jacek
Magnetfabrik Bonn GmbH
Dorotheenstraße 215
53119 Bonn
Germany
E-Mail: Jacek.Krzywinski@Magnetfabrik.de

Langner, Dr. Bernd E.
Glockenheide 11
21423 Winsen
Germany
E-mail: langner@understanding-copper.com

Letz, Dr. Martin
SCHOTT AG
Research and Technology Development
Hattenbergstraße 10
55122 Mainz
Germany
E-mail: martin.letz@schott.com

Lider, Konstantin
Diener & Rapp GmbH & Co. KG Eloxalbetrieb
Junkersstraße 39
78056 Villingen-Schwenningen
Germany
E-mail: konstantin.lider@dienerrapp.de

Lox, Prof. Dr. Ir. Egbert S. J.
Am Laerchentor 8
36355 Grebenhain-Hochwaldhausen
Germany
E-mail: Egbert.Lox@gmail.com

Lovrincic, Dr. Robert
trinamiX GmbH
Industriestraße 35
67063 Ludwigshafen
Germany
E-mail: robert.lovrincic@trinamix.de

Maletz, Prof. Dr. Reinhard
VOCO GmbH
Anton-Flettner-Straße 1-3
27472 Cuxhaven
Germany
E-mail: r.maletz@voco.de

Matschke, Dr. Christian
BERLIN-CHEMIE AG
Glienicker Weg 125
12489 Berlin
Germany
E-mail: cmatschke@berlin-chemie.de

Mertens, Prof. Dr. Konrad
Fachbereich Elektrotechnik und Informatik
Fachhochschule Münster
Stegerwaldstraße 39
48565 Steinfurt
Germany
E-mail: mertens@fh-muenster.de

Mudryk, Dr. Yaroslav
Ames Laboratory, U.S. Department of Energy
Iowa State University
254 Spedding
Ames, IA 50011-2416
USA
E-mail: slavkomk@ameslab.gov

Niehaus, Dr. Oliver
Umicore AG & Co. KG
Rodenbacher Chaussee 4
63457 Hanau
Germany
E-mail: Oliver.Niehaus@eu.umicore.com

Pan, Prof. Dr. Shilie
Xinjiang Technical Insitute of Physics & Chemistry
Chinese Academy of Sciences
40-1 South Beijing Road
830011 Urumqi
China
E-mail: slpan@ms.xjb.ac.cn

Pavón Regaña, Dr. Ing. Sandra
Fraunhofer-Institut für Keramische Technologien und Systeme IKTS
Fraunhofer-Technologiezentrum Hochleistungsmaterialien THM
Am St.-Niclas-Schacht 13
09599 Freiberg
Germany
and
Institut für Technische Chemie
TU Bergakademie Freiberg
Leipziger Straße 29
09599 Freiberg
Germany
E-mail: sandra.pavon.regana@ikts.fraunhofer.de

Pecharsky, Prof. Dr. Vitalij K.
Ames Laboratory
Iowa State University
Ames, IA 50011-2416
USA
E-mail: vitkp@ameslab.gov

Piribauer, Dipl.-Ing. Dr. rer. nat. Christoph
Forschungsinstitut für Glas | Keramik (FGK)
Heinrich-Meister-Straße 2
56203 Höhr-Grenzhausen
Germany
E-mail: christoph.piribauer@fgk-keramik.de

Pöttgen, Prof. Dr. Rainer
Institut für Anorganische und Analytische Chemie
Westfälische Wilhelms-Universität Münster
Corrensstraße 30
48149 Münster
Germany
E-mail: pottgen@uni-muenster.de

Quirmbach, Prof. Dr. rer. nat. Dr. h.c. Peter
Technische Chemie und
Korrosionswissenschaften
Universität Koblenz-Landau
Universitätsstraße 1
56070 Koblenz
Germany
E-mail: pquirmbach@uni-koblenz.de

Reiss, Prof. Dr. Günter
Physics Department
Center for Spinelectronic Materials and
Devices
Universitätsstraße 25
33615 Bielefeld
Germany
E-mail: guenter.reiss@uni-bielefeld.de

Riedel, Prof. Dr. Sebastian
Institut für Chemie und Biochemie
Anorganische Chemie
Freie Universität Berlin
Fabeckstraße 34/36
14195 Berlin
Germany
E-mail: s.riedel@fu-berlin.de

Rieger, Dr. Thorsten
VDM Metals International GmbH
Kleffstraße 23
58762 Altena
Germany
E-mail: Torsten.Rieger@vdm-metals.com

Salvermoser, Dr. Manfred
Riedel Filtertechnik GmbH
Westring 83
33818 Leopoldshöhe
Germany
E-mail: manfred.salvermoser@riedel-filtertechnik.com

Sax, Dr.-Ing. Almuth
Technische Chemie und
Korrosionswissenschaften
Universität Koblenz-Landau
Universitätsstraße 1
56070 Koblenz
Germany
E-mail: asax@uni-koblenz.de

Schäferling, Prof. Dr. Michael
Fachbereich Chemieingenieurwesen
Fachhochschule Münster
Stegerwaldstraße 39
48565 Steinfurt
Germany
E-mail: michael.schaeferling@fh-muenster.de

Schmid, Jonas R.
Institut für Chemie und Biochemie
Anorganische Chemie
Freie Universität Berlin
Fabeckstraße 34/36
14195 Berlin
Germany
E-mail: jonas.schmid@fu-berlin.de

Schramm, Dr. Stefan
Merck KGaA
Frankfurter Straße 250
64293 Darmstadt
Germany
E-mail: stefan.schramm@merckgroup.com

Schupp, Prof. Dr. Thomas
Fachbereich Chemieingenieurwesen
Fachhochschule Münster
Stegerwaldstraße 39
48565 Steinfurt
Germany
E-mail: thomas.schupp@fh-muenster.de

Seifert, Dr. Markus
TU Dresden
Walther-Hempel-Bau
Mommsenstraße 4
01069 Dresden
Germany
E-mail: markus.seifert1@tu-dresden.de

Slabon, Prof. Dr. Adam
Chair of Inorganic Chemistry
University of Wuppertal
Gaußstraße 20
42119 Wuppertal
Germany
E-mail: slabon@uni-wuppertal.de

Staffel, Prof. Dr. Thomas
Research & Development, Phosphate solutions
BK Giulini GmbH, ICL Group Ltd.
Dr.-Albert-Reimann-Straße 2
68526 Ladenburg
Germany
E-mail: thomas.staffel@icl-group.com

Stephan, Dr. Tom
Deutsche Gemmologische Gesellschaft e.V. (DGemG)
Prof.-Schlossmacher-Straße 1
55743 Idar-Oberstein
Germany
E-mail: t.stephan@dgemg.com

Stengel, Dr. Ilona
Merck KGaA
Frankfurter Straße 250
64293 Darmstadt
Germany
E-mail: ilona.stengel@merckgroup.com

Stöwe, Prof. Dr. Klaus
Faculty of Natural Sciences
Institute of Chemistry, Chemical Technology
Technische Universität Chemnitz
09107 Chemnitz
Germany
E-mail: klaus.stoewe@chemie.tu-chemnitz.de

Strassert, Prof. Dr. Cristian A.
Institut für Anorganische und Analytische Chemie
CiMIC – CeNTech – SoN
Westfälische Wilhelms-Universität Münster
Corrensstraße 28/30
48149 Münster
Germany
E-mail: ca.s@wwu.de

Teliban, Dr. Iulian
Magnetfabrik Bonn GmbH
Dorotheenstraße 215
53119 Bonn
Germany
E-Mail: Iulian.Teliban@Magnetfabrik.de

Termath, Dr. Andreas
Clariant Plastics & Coatings (Deutschland) GmbH
Chemiepark Knapsack
Industriestraße 149
Gebäude 2703, R. 128
50354 Hürth
Germany
E-mail: andreas.termath@clariant.com

Trodler, Dr. Jörg
Trodler-EAVT
Technische Beratung für die Aufbau und Verbindungstechnik in der Elektronik
Grüner Weg 18/19
15712 Königs Wusterhausen
Germany
E-mail: joerg.trodler@trodler-eavt.de

Voigt, Dominik
Fachbereich Chemieingenieurwesen
Fachhochschule Münster
Stegerwaldstraße 39
48565 Steinfurt
Germany
E-mail: dv009200@fh-muenster.de

Voigt, Prof. Dr. Ingolf
Fraunhofer-Institut für Keramische Technologien und Systeme IKTS
Michael-Faraday-Straße 1
07629 Hermsdorf
Germany
E-mail: ingolf.voigt@ikts.fraunhofer.de

Warkentin, apl. Prof. Dr.-Ing. habil. Dr. rer. nat. Mareike
Fakultät für Maschinenbau und Schiffstechnik
Universität Rostock
Friedrich-Barnewitz-Straße 4
18119 Rostock
Germany
E-mail: mareike.warkentin@uni-rostock.de

Weigand, Prof. Dr. Jan J.
TU Dresden
Walther-Hempel-Bau
Mommsenstraße 4
01069 Dresden
Germany
E-mail: jan.weigand@tu-dresden.de

Werner, Prof. Dr. Jan
Forschungsinstitut für Glas | Keramik (FGK)
Heinrich-Meister-Straße 2
56203 Höhr-Grenzhausen
Germany
E-mail: jan.werner@fgk-keramik.de

Wendel, Dr. Jörg
Wendel GmbH Email- und Glasurenfabrik
Am Güterbahnhof 30
35683 Dillenburg
Germany
E-mail: joerg.wendel@wendel-email.de

Wilhelm, Dr. Dominik
TYROLIT - Schleifmittelwerke Swarovski K.G.
Swarovskistraße 33
6130 Schwaz
Austria
E-mail: Dominik.Wilhelm@Tyrolit.com

Winter, Dr. Florian
Culimeta Textilglas-Technologie GmbH & Co. KG
Werner-von-Siemens-Straße 9
49593 Bersenbrück
Germany
E-mail: fwinter@culimeta.de

Yang, Prof. Dr. Zhihua
Xinjiang Technical Institute of Physics & Chemistry
Chinese Academy of Sciences
40-1 South Beijing Road
Urumqi 830011
China
E-Mail: zhyang@ms.xjb.ac.cn

Zeier, Prof. Dr. Wolfgang
Institut für Anorganische und Analytische Chemie
Westfälische Wilhelms-Universität Münster
Corrensstraße 30
48149 Münster
Germany
E-mail: wzeier@uni-muenster.de

Ziegler, Raimund
Institut für Allgemeine, Anorganische und Theoretische Chemie
Universität Innsbruck
Innrain 80–82
6020 Innsbruck
Austria
E-mail: Raimund.Ziegler@uibk.ac.at

Zumdick, Dr. Markus
H.C. Starck Tungsten GmbH
Im Schleeke 78-91
38642 Goslar
Germany
E-mail: markus.zumdick@hcstarck.com

5 Energy storage and conversion

5.1 Battery materials

Michael Ghidiu, Theodoor Hendriks, Wolfgang Zeier

Electrical energy is high-quality energy, as it can be easily converted to do many different types of work (e.g. heating, lighting, mechanical and electrochemical). Due to mismatches between the time of energy generation and usage, as well as portability for versatility, storing electrical energy is of high importance. Storage can be accomplished physically (through capacitors, flywheels, hydroelectric, thermal) or through electrochemical processes. This was first done to a useful degree in the early nineteenth century by Alessandro Volta by selecting materials with different redox potentials and building them into the voltaic pile based on copper and zinc [1]. Since then, the study of chemistry and specifically inorganic materials has led to a much deeper understanding in storage of electrical energy and, thereby, advancements in battery technology. There are many ways to store and release electrical energy, which can dictate how a device is constructed. Two performance indicators on how much energy can be stored and how quickly it can be released are, respectively, the energy density (in Watt-hours per unit volume or mass) and the power density (in Watt per unit volume or mass). A way to summarize this is with a *Ragone plot*, showing these two densities on respective axes (example in Figure 5.1.1). Capacitors can quickly charge and discharge, and thus have a high power density; however, their energy density is very limited. Batteries typically have a high energy density but limited ability to charge or discharge quickly. Supercapacitors, or electrochemical capacitors, started to bridge the gap between the two. However, the ultimate goal lies at the top right of the plot; a device that can store a lot of energy and access it quickly. It is clear from Figure 5.1.1 that different chemistries result in different distributions across this plot. The discussion here will be limited to batteries, which consist of positive (cathode) and negative (anode) electrodes, separated by an ion-transporting but electron-blocking electrolyte. Batteries can be *primary* (designed for a single discharge followed by disposal) or *secondary* (able to be reliably recharged). During discharge, electrons are moved through a circuit due to driving forces of different redox potentials between the electrodes, where electrochemical reactions take place, generally involving metals changing their oxidation state. Charging reverses these processes by an applied potential difference. Because battery use involves physically moving material around inside the cell (through electrolyte movement, plating reactions, intercalations, etc.), knowledge of chemical structures, phase transitions, as well as diffusion, and hence the intrinsic inorganic material properties, is essential for understanding real-world performance and commercial application.

https://doi.org/10.1515/9783110798890-001

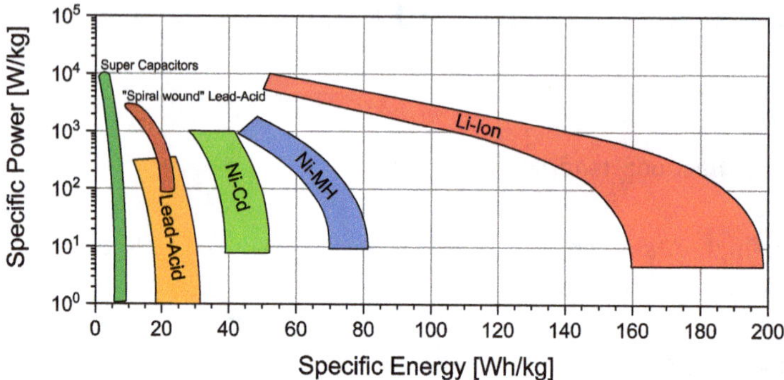

Figure 5.1.1: Battery technologies in terms of specific power density (W/kg) and energy density (Wh/kg). Adapted from [2].

5.1.1 The lead acid battery

The lead acid battery, invented in 1859, is still used in the automotive industry as the standard for energy storage for starting, lighting and ignition. Fully charged, the anode consists of Pb metal and the cathode of PbO_2, with an electrolyte of $H_2SO_{4(aq)}$ (concentrated sulfuric acid). During discharge, lead metal at the anode reacts with bisulfate anions (from the dissociation of one H^+ from H_2SO_4) to produce protons, electrons and lead sulfate (Pb(+**II**)):

$$Pb_{(s)} + HSO_4{}^-{}_{(aq)} \rightarrow PbSO_{4(s)} + H^+{}_{(aq)} + 2e^-$$

At the cathode, lead oxide (Pb(+**IV**)O_2) reacts with bisulfate and the produced protons, accepting the electrons through the circuit, to produce amorphous lead sulfate (Pb(+**II**)) and water:

$$PbO_{2(s)} + HSO_4{}^-{}_{(aq)} + 3H^+{}_{(aq)} + 2e^- \rightarrow PbSO_{4(s)} + 2H_2O_{(l)}$$

The end result is two plates of lead sulfate, effectively an electrochemical comproportionation reaction, separated by dilute sulfuric acid due to the participation and depletion of the electrolyte in the overall reaction. This means that the state of charge is directly proportional to the concentration of H_2SO_4, and hence the specific density of the electrolyte that can thus be easily measured. This was one of the first battery chemistries rechargeable by reversing the current; however, the capacity can be impeded by the crystallization of $PbSO_4$ rather than amorphous deposition; this is not reversible, but $BaSO_4$ can be used as an additive to intentionally promote a reversible crystallization of $PbSO_4$ that does not block the active material. Despite the toxicity of lead, this battery class has the benefits of ability to operate in a wide temperature range (−20 to 50 °C) with low self-discharge, as well as ease and widespread ability of recycling. However, energy density is fairly low, especially considering the materials (such as casing, and electronic battery management systems) that need to be added to go from the operating potential of a single cell to that of the battery pack with cells parallel and in series.

5.1.2 The alkaline battery

Ubiquitous alkaline batteries, reliant on zinc and manganese oxide electrodes, are good examples of primary batteries that normally cannot be recharged. At the anode, metallic zinc reacts with hydroxide in the electrolyte to produce zinc oxide (Zn(+**II**)):

$$Zn_{(s)} + 2OH^{-}{}_{(aq)} \rightarrow ZnO_{(s)} + H_2O_{(l)} + 2e^{-}$$

At the cathode, manganese oxide Mn(+**IV**) is reduced to manganese oxide Mn(+**III**), with water participating to form hydroxide with the lost oxygen of MnO_2 to replenish the electrolyte:

$$2MnO_{2(s)} + H_2O_{(l)} + 2e^{-} \rightarrow Mn_2O_{3(s)} + 2OH^{-}{}_{(aq)}$$

Alkaline batteries are advantageous in that they can be produced cheaply and in high volume for consumer applications (everything from toys, to cameras, to household appliances). However, their disadvantages are foremost that they cannot practically be recharged, in addition to their capacity being a strong function of the rate of discharge as well as being prone to leaking their caustic hydroxide electrolytes.

5.1.3 The nickel-cadmium and nickel-metal-hydride type

The nickel-cadmium (Ni-Cd) chemistry uses a similar electrolyte as the alkaline battery and is widely known for its use in earlier rechargeable cells and as battery packs in electric tools due to its ability to be readily recharged. These cells consist of a cadmium anode, NiOOH (nickel oxide-hydroxide) cathode and hydroxide electrolyte, typically $KOH_{(aq)}$. During discharge, at the anode (metallic cadmium), cadmium hydroxide (Cd(+**II**)) is formed:

$$Cd_{(s)} + 2OH^{-}{}_{(aq)} \rightarrow Cd(OH)_{2(s)} + 2e^{-}$$

At the cathode, nickel oxide-hydroxide (Ni(+**III**)) accepts electrons through the circuit and loses oxygen to become nickel hydroxide (Ni(+**II**)):

$$2NiO(OH)_{(s)} + 2H_2O_{(l)} + 2e^{-} \rightarrow 2Ni(OH)_{2(s)} + 2OH^{-}{}_{(aq)}$$

These cells can suffer from a “memory effect” where a partial discharge can lead to limited future discharge capacity, though this can be alleviated by deep discharge cycles making sure that all inorganic material is being converted. The cell can deliver up to around 1.3 V and the main benefits of this class of batteries are their cycle-life (over 1,000 cycles typically before their capacity reduces to below 50%, including the ability to handle deep discharge) and much higher energy density than the lead acid battery. However, cadmium, being a toxic heavy metal, is of concern for production and end-of-life.

The nickel-metal-hydride (NiMH) chemistry also relies on nickel oxide-hydroxide, but instead of cadmium, an intermetallic compound (M) is used, typically of composition AX_5, where A is a rare-earth metal and X is Ni, Co, Mn or Al; the electrolyte is typically aqueous KOH. During discharge, the intermetallic compound at the anode, having already been converted to a hydride, is reduced, moving an electron into the circuit and producing water:

$$MH_{(s)} + OH^-_{(aq)} \rightarrow M_{(s)} + H_2O_{(l)} + e^-$$

At the cathode, nickel oxide-hydroxide (Ni(+**III**)) reacts with water to regenerate OH^- and form nickel hydroxide (Ni(+**II**)):

$$NiO(OH)_{(s)} + H_2O_{(aq)} + e^- \rightarrow Ni(OH)_{2(s)} + OH^-_{(aq)}$$

This battery class does not suffer from the memory effect, has a high energy density, is very reliable and can be made mostly with inexpensive and less-toxic materials. However, the use of rare-earth metals is an economic concern. Nevertheless, these are needed to stabilize the anode against the hydroxide electrolyte for a long cycle life and much development was needed over decades to perfect the battery chemistry. NiMH batteries overtook Ni-Cd due to their advantages and enjoyed a large market share until Li-ion batteries became widespread. They are still often used for automotive applications. Similar to Ni-Cd batteries, the cell voltage is somewhat limited at around 1.25 V.

5.1.4 The lithium ion battery

Lithium-ion (Li-ion) batteries are relatively new, having been introduced commercially in the 1990s (although their chemistry was beginning to be understood much earlier) [3]. They have become the main power source for portable electronics and even automotive applications. Transitioning toward electric mobility, electric cars require lithium batteries with superior energy and power density, without compromising safety and environmental concerns [4, 5]. For stationary applications, lithium batteries are becoming more popular due to their high energy and power density [6]. At their introduction, these batteries worked on the "rocking-chair" principle, where Li^+ ions were shuttled across a Li^+-containing electrolyte from one intercalating electrode to the other, which correspondingly changed oxidation state. The common materials that were used for the anode and cathode were graphite and $LiCoO_2$ [3]. During discharge, at the anode, lithiated graphite (which has a distributed negative compensating charge) loses Li^+ from between its layers, moving electrons through the circuit:

$$Li_xC_6 \rightarrow C_6 + xLi^+ + xe^-$$

At the cathode, Li-deficient lithium cobalt oxide accepts Li^+ ions and electrons:

$$Li_{1-x}CoO_2 + xLi^+ + xe^- \rightarrow LiCoO_2$$

Li-ion cells of this type are assembled in the discharged state with pure graphite, so before these reactions can take place, the graphite must be lithiated by charging. The electrolyte is typically a mixture of organic liquids and ion-conducting Li salts. A comparison between this type and, previously discussed, Pb and Ni-based batteries is shown in Figure 5.1.2.

Due to high reactivity of the inorganic materials, unwanted reactions can happen at the interface where the liquid electrolyte contacts the anode and/or cathode. These form an *interphase* that consists of decomposition products of both the electrolyte and/or the electrode material; the so-called solid-electrolyte-interphase (SEI) may hamper conductivity of the Li^+ ions.

A broad spectrum of materials can be used for lithium-ion batteries, not only for the cathode and anode, but also for the electrolyte. The challenge for battery development is finding the right combination with the right additives to design the best battery for a given application, keeping in mind cycling stability, performance and possible degradation mechanisms. This has led to the common chemistry where a graphite anode is used with $LiPF_6$ as salt in organic liquid electrolytes and a cathode consisting of $LiCoO_2$ [3, 7]. While other anodes (such as Si, Li-metal and $Li_4Ti_5O_{12}$) have been extensively tested, graphite Li_xC_6 is most used commercially. Silicon suffers from large volume changes during cycling, causing loss of connection with the electrode and trapping of Li^+. Li-metal is very reactive, causing decomposition of the electrolyte and making the battery unsafe, as well as the formation of dendrites that can cause catastrophic cell failure when an electronically conducting bridge through the electrolyte is formed. $Li_4Ti_5O_{12}$ is a very stable anode, but due to its high electrical potential of about 1.5 V vs Li/Li^+, the voltage of the battery is lower, leading to a decreased energy density. For the electrolytes, the $LiPF_6$ salt offers the highest ion-mobility while the organic compounds (it is dissolved in mixtures of ethylene or dimethyl carbonate) offer a large enough stability versus the reactive conditions inside the cell, providing a reasonably safe battery. Therefore, these materials are the industry standard [7]. For the cathode, mostly insertion materials are used. In these materials, the Li^+ ion can travel within the material's framework. The three major commercial materials to date are the olivine $LiFePO_4$, the layered $LiCoO_2$ (and its derivatives) and the spinel $LiMn_2O_4$ (Figure 5.1.3).

For a cathode active material to be viable, certain criteria must be met: 1) ability of Li^+ to move freely inside the structure, 2) redox-active transition metals in the structure (typically limited to Mn, Ni and Co) and 3) a decent electronic conductivity

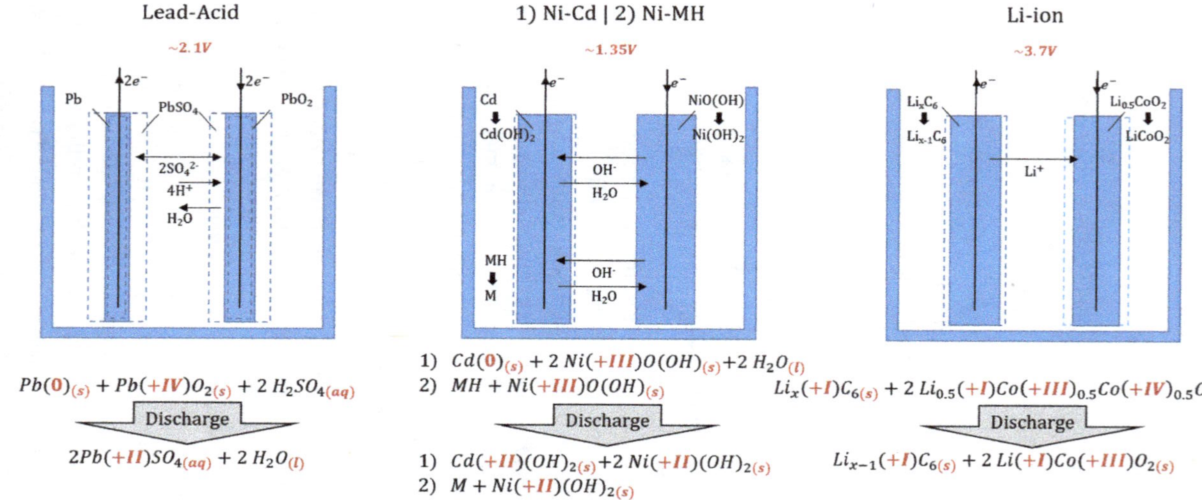

Figure 5.1.2: Working principles of rechargeable batteries in discharge with average potentials shown. Upon discharge, ions travel through the electrolyte while electrons travel through the circuit from anode to cathode. When charging, the ions are pushed in the reverse direction by an applied potential difference.

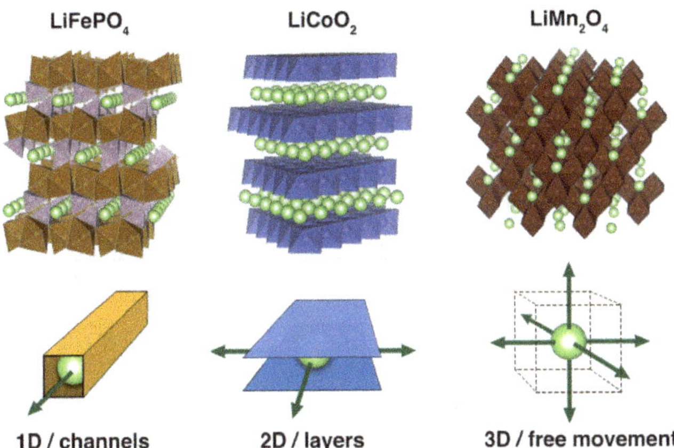

Figure 5.1.3: The structure and lithium-diffusion pathways for three major intercalation cathode materials; the olivine $LiFePO_4$, the layered $LiCoO_2$ and the spinel $LiMn_2O_4$ representing dimensionality of diffusion from 1D to 3D.

for transport of electrons toward the current collector. The structures in Figure 5.1.3 show that the currently used materials have 1-, 2- and 3-dimensional diffusion channels, respectively, through which Li ions must travel; this strongly influences diffusivity through the bulk materials. Hence, $LiFePO_4$ is actually a poor ionic and electronic conductor. This means that crystallographic orientation can be a concern, which can be important in some electrode geometries. Limited diffusion pathways can influence the cycle life and lifetime, and speed of charge and discharge.

In contrast to $LiFePO_4$, $LiCoO_2$ allows for fast Li^+ diffusion combined with a high energy density of 272 mAh/g with a voltage of 4 V. However, for structural stability reasons, only about half of that energy density can be used (140 mAh/g) as removing too much Li^+ from the framework leads to decomposition. This has led to chemistries where the Co is partly replaced by Ni, Mn or Al to yield the so-called NMC ($LiNi_xMn_yCo_zO_2$) or NCA ($LiNi_xCo_yAl_zO_2$) cathode active materials, all of which affect the average electrochemical insertion potentials. Furthermore, similar to lead and cadmium, cobalt is expensive, toxic, environmentally unfriendly, and mined under problematic conditions [8].

As an alternative to the two-dimensional active materials, the spinel $LiMn_2O_4$ has faster three-dimensional ionic diffusion, a higher operating voltage (4.1 V vs. Li/Li^+) and comparable practical energy density to $LiCoO_2$ (theoretically 148 mAh/g, typical 125 mAh/g) combined with lower costs of the material and absence of direct environmental or safety hazards. Substitutions, such as a quarter of the Mn with Ni to give $LiNi_{0.5}Mn_{1.5}O_4$, can raise the average potential and further increase the energy density. Table 5.1.1 below shows these three major cathode classes with their specific capacity (and practical capacity), average potential and (practical) energy density, serving to illustrate how these structural concerns come to play in real-world performance.

Table 5.1.1: Comparison of three major Li-ion cathode materials.

Framework	Compound	Specific capacity (mAh g^{-1})	Average potential (V vs. Li^0/Li^+)	Energy density (Wh kg^{-1})
Layered	$LiCoO_2$	272 (140)	4.0	1088 (560)
	$LiNi_{1/3}Mn_{1/3}Co_{1/3}O_2$	272 (200)	4.0	1088 (800)
Spinel	$LiMn_2O_4$	148 (125)	4.1	607 (513)
	$LiNi_{0,5}Mn_{1,5}O_4$	148 (125)	4.7	696 (588)
Olivine	$LiFePO_4$	170 (160)	3.45	587 (552)
	$LiFe_{1/2}Mn_{1/2}PO_4$	170 (160)	3.4/4.1	638 (600)

Adapted from [9]. Values in parentheses are typically achieved in practice.

Currently, the NMC/NCA materials offer the highest energy densities. Although NMC/NCA uses less cobalt, they are still costly. However, even with an excellent material for a certain performance in mind, there are still challenges for production. An example of this is the precipitation of NMC: current methods often employ careful control of parameters using hydroxides to precipitate the proper precursors for sintering the desired NMC [10]. Recently, there has been work to use oxalic acid to achieve the same effect with less stringent controls required, offering a better route to upscaling and control of morphology [11].

5.1.5 The solid-state battery

To maximize cell performance, there are multiple design choices for the electrolyte. To utilize Li-metal anodes and make safer cells avoiding flammable solvents in the electrolytes, solid electrolytes can be utilized to make all-solid-state cells. Indeed, one of the first to be realized was the lithium-iodine primary battery for pacemakers, utilizing an iodine-containing cathode that forms conducting LiI electrolyte when built onto a Li metal anode. At the anode, lithium gives an electron to the circuit and migrates to the cathode [12]:

$$2Li_{(s)} \rightarrow 2Li^+ + 2e^-$$

reacting with iodine:

$$X{\cdot}I_2 + 2e^- \rightarrow X + 2I^-$$

where X is a matrix that contains the iodine. The Li^+ and I^-, situated between the electrodes, form Li^+-conducting LiI in situ (although this electrolyte has a rather low conductivity and as such is not suited for high-power applications). The future of the lithium-ion battery is the evolution from liquid to solid electrolytes as this can allow for nonflammable systems and potentially higher energy densities when a solid electrolyte allows the utilization of Li-metal anodes. There are many families of inorganic materials that are very promising as fast ionic conductors to shuttle Li^+ between the electrodes. Although the ions must pass through a solid framework, the conductivities can actually approach or even surpass those of many liquid electrolytes. Thorough understanding of ionic transport mechanisms is required to carefully design the framework (via elemental substitutions, engineering of phases, etc.). Many of the same structural chemistry principles noted above in discussion of electrode materials apply here, as Li^+ ions must be quickly migrated through the structure, and pathway geometry is just one of the many deciding factors on performance. There is currently much research into making these types of cells viable for commercial products, such as the automotive sector [13].

5.1.6 Summary

Since the introduction of useful batteries for storing electrical energy by chemical means, inorganic materials have played a critical role. We still have a need for various classes of storage devices, as there is no single solution for every application. To better design within each type, a deeper understanding of inorganic chemistry is crucial. No battery satisfies all needs perfectly; there is a reason why conventional automobiles still use Pb-acid batteries and portable tools often still use NiCd where high power is needed. Table 5.1.2 summarizes the multidimensional aspects of a battery that must be considered when making a selection for an application.

Table 5.1.2: Summary of various properties of battery classes for comparison.

	Lead acid	Ni-Cd	Ni-MH	Li ion
Anode	Pb	Cd	MH ($LaNi_5H_6$)	LiC_6
Cathode	PbO_2	NiOOH	NiOOH	$LiCoO_2$
Electrolyte	Aqueous H_2SO_4	Aqueous KOH	Aqueous KOH	$LiPF_6$
Specific energy density (theory) (W h kg^{-1})	170	220	220	410
Specific energy density (practical) (W h kg^{-1})	30–40	40–60	75–100	120–150
Working voltage (V)	2.0–1.8	1.2–1.0	1.2–1.0	4.0–3.0

Table 5.1.2 (continued)

	Lead acid	Ni-Cd	Ni-MH	Li ion
Working temperature (°C)	−20 to + 50	−40 to + 45	−20 to + 45	−30 to + 80
Chemical overcharge potential	Yes	Yes	Yes	No
Chemical overdischarge protection	No	Yes	Yes	No
Cycle number, 100% DOD	>50	>1000	>1000	<1000
Relative toxicity	High	High	High	Medium
Safety	High	High	High	Medium
Estimated material costs	Medium	Medium	Medium-high	High
Estimated manufacture cost	Low	Medium	Medium	High

Adapted from [14].

References

[1] Volta A. On the electricity excited by the mere contact of conducting substances of different kinds. Philos Trans, 1800, 90, 403–31.

[2] IEA. Technology Roadmap – Electric and Plug-in Hybrid Electric Vehicles, IEA, Paris, 2011.

[3] Pistoia G. (Ed.) Lithium-Ion Batteries, 1st Edition, Springer, Berlin, 2013.

[4] Linden D, Reddy TB. Handbook of Batteries, 3rd Edition, 2002. DOI: https://doi.org/10.1002/9780470933886.ch1.

[5] Nitta N, Wu F, Lee JT, Yushin G. Li-ion battery materials: Present and future. Mater Today, 2015, 18, 252–64. DOI: https://doi.org/10.1016/j.mattod.2014.10.040.

[6] Xu T, Wang W, Gordin ML, Wang D, Choi D. Lithium-ion batteries for stationary energy storage. Jom, 2010, 62, 24–30. DOI: https://doi.org/10.1007/s11837-010-0131-6.

[7] Thompson DL, Hartley JM, Lambert SM, Shiref M, Harper GDJ, Kendrick E, Anderson P, Ryder KS, Gaines L, Abbott AP. The importance of design in lithium ion battery recycling-a critical review. Green Chem, 2020, 22, 7585–603. DOI: https://doi.org/10.1039/d0gc02745f.

[8] Banza Lubaba Nkulu C, Casas L, Haufroid V, De Putter T, Saenen ND, Kayembe-Kitenge T, Musa Obadia P, Kyanika Wa Mukoma D, Lunda Ilunga JM, Nawrot TS, Luboya Numbi O, Smolders E, Nemery B. Sustainability of artisanal mining of cobalt in DR congo. Nat Sustain, 2018, 19, 495–504. DOI: https://doi.org/10.1038/s41893-018-0139-4.

[9] Julien CM, Mauger A, Zaghib K, Groult H. Comparative issues of cathode materials for li-ion batteries. Inorganics, 2014, 2, 132–54. DOI: https://doi.org/10.3390/inorganics2010132.

[10] Ren D, Shen Y, Yang Y, Shen L, Levin BDA, Yu Y, Muller DA, Abruña HD. Systematic optimization of battery materials: Key parameter optimization for the scalable synthesis of uniform, high-energy, and high stability $LiNi_{0.6}Mn_{0.2}Co_{0.2}O_2$ cathode material for lithium-ion batteries. ACS Appl Mater Interfaces, 2017, 9, 35811–19. DOI: https://doi.org/10.1021/acsami.7b10155.

[11] Yao X, Xu Z, Yao Z, Cheng W, Gao H, Zhao Q, Li J, Zhou A. Oxalate co-precipitation synthesis of $LiNi_{0.6}Co_{0.2}Mn_{0.2}O_2$ for low-cost and high-energy lithium-ion batteries. Mater Today Commun, 2019, 19, 262–70. DOI: https://doi.org/10.1016/j.mtcomm.2019.02.001.

[12] Mathias W, Eldridge M, Moser JR, Schneider AA. The solid-state lithium battery : A new improved chemical pover source for implantable cardiac pacemakers. 1971, 5, 317–24.
[13] Ohno S, Banik A, Dewald GF, Kraft MA, Krauskopf T, Minafra N, Till P, Weiss M, Zeier WG. Materials design of ionic conductors for solid state batteries. Prog Energy, 2020, 2, 022001. DOI: https://doi.org/10.1088/2516-1083/ab73dd.
[14] Aravindan V, Gnanaraj J, Lee YS, Madhavi S. $LiMnPO_4$ – A next generation cathode material for lithium-ion batteries. J Mater Chem A, 2013, 1, 3518–39. DOI: https://doi.org/10.1039/c2ta01393b.

5.2 Magnetocaloric materials

Yaroslav Mudryk, Vitalij K. Pecharsky

Magnetocaloric cooling, also known as cooling with magnetocaloric effect or magnetic cooling, holds promise to become mainstream technology that in the foreseeable future could sustainably support most of the human society heat pumping needs including refrigeration and indoor climate control. Exploiting solids that exhibit large magnetocaloric effects (MCEs) makes heat pumping more efficient when compared to the ubiquitous vapor-compression approach and simultaneously eliminates volatile liquid refrigerants, majority of which have high global warming potentials and/or are ozone-depleting chemicals. Very generally, MCE quantifies reversible thermal responses of materials that contain magnetic moment-carrying species to changing external magnetic fields. The field-induced thermal effects in such materials stem from the ability of varying magnetic fields to either enhance or degrade magnetic order naturally established in magnetic solids at any given combination of temperature and pressure. Consequently, external magnetic fields reversibly manipulate magnetic, and under certain conditions, electronic and lattice entropies of those solids. The magnetocaloric effect, therefore, is one of the fundamental properties intrinsic to all magnetic materials but in the vast majority of them, the thermal effects achievable with magnetic fields that can be created with permanent magnets are too weak to be useful or even measurable at ambient conditions. In this chapter, we briefly review fundamentals of the magnetocaloric effect, explain which key properties make certain compounds useful for magnetocaloric cooling, and describe a few families of materials under consideration for applications in the cryogenic and near room temperature regimes. For in-depth coverage of the topic, we refer the interested reader to a number of available reviews [1–4].

5.2.1 What is the magnetocaloric effect and what makes some materials promising?

Consider a solid composed entirely of, or containing a fraction of, atoms or ions that carry magnetic moments in a certain initial magnetic field, H_I; typically $H_I = 0$. When the moments are non-interacting or interact with one another weakly, they are nearly completely disordered and the corresponding bulk magnetization, M_I, of such solid is nearly zero. *Isothermal* application of a final magnetic field $|H_F| > |H_I|$ at constant pressure forces the moments to align, at least partially, along the direction of H_F, thus enhancing magnetic order (that is, reducing magnetic disorder), increasing magnetization from M_I to M_F and, consequently, decreasing the magnetic component of the total entropy of the solid by $\Delta S_M = S_F - S_I < 0$. When the field-induced moment alignment (*magnetizing*) is performed without heat exchange with

https://doi.org/10.1515/9783110798890-002

the surroundings, i.e. *adiabatically*, lattice entropy (and electronic entropy if the solid is a metal) must increase by the equivalent of $|\Delta S_M|$ to maintain the total entropy of the solid constant. The result is enhanced lattice vibrations and rise of the global temperature of the solid; this simplified model assumes that changes in pressure, p, and volume, V, are negligible. When the magnetic field is removed (adiabatically as well), the material returns to its original disordered (low magnetization) state, so the magnetic entropy goes up by $\Delta S_M = S_I - S_F > 0$, the lattice and electronic entropies combined go down by $|\Delta S_M|$, and the solid cools.

The result of a single adiabatic $H_F \rightarrow H_I$ cycle equivalent to that described at the end of the previous paragraph was predicted independently by Debye [5] and Giauque [6]. Experimentally, cooling with magnetocaloric effect was validated by Giauque and MacDougall in 1933 [7] who used about 60 g of paramagnetic $Gd_2(SO_4)_3{\cdot}8H_2O$ precooled with liquid helium to 1.5–3.4 K in a steady 8 kOe magnetic field. When they quickly, that is, nearly adiabatically reduced the magnetic field from $H_F = 8$ kOe to $H_I = 0$, the salt cooled down to a few hundred milliKelvin (a low temperature record at the time!) due to disordering of the localized, purely spin magnetic moments of Gd atoms. Obviously, such cooling method is restricted by the magnitude of the temperature change due to adiabatic (de)magnetization. Known as adiabatic demagnetization refrigeration, this approach remains in common use in research laboratories to reach temperatures close to the absolute zero using either or both electronic and nuclear magnetic moments. The door to continuous cooling with magnetocaloric effect near room temperature was opened by Brown [8] who coupled periodic magnetic field-induced up and down temperature changes in elemental gadolinium to regenerative heat exchange with a fluid (water-alcohol mixture), demonstrating in principle the concept of *active magnetocaloric regeneration*. The latter increases temperature span beyond a single magnetic field-induced adiabatic temperature change of a given material, allowing for the development of near room temperature magnetic cooling applications.

The adiabatic temperature change, ΔT_{ad}, and the isothermal magnetic entropy change, ΔS_M, are two most important characteristics of magnetocaloric compounds. The former can be quantified directly, by measuring temperature change during rapid application and removal of magnetic field to a thermally insulated material. It is more common, however, to determine these quantities indirectly, using calorimetric and/or bulk magnetization measurements. For example, after measuring temperature dependence of material's magnetization in various applied magnetic fields, one can calculate the isothermal magnetic entropy change for a given $\Delta H = H_F - H_I$ by numerical integration using one of the Maxwell relations [9, 10]:

$$\Delta S_M(T)_{\Delta H} = \int_{H_I}^{H_F} \left(\frac{\partial M(T,H)}{\partial T} \right)_H \mathrm{d}H \tag{5.2.1}$$

Both ΔS_M and ΔT_{ad} can be calculated when the heat capacity of a material is known as a function of temperature and magnetic field [11]. The adiabatic temperature change can also be calculated from combined magnetization and heat capacity measurements:

$$\Delta T_{ad}(T)_{\Delta H} = -\int_{H_I}^{H_F} \left(\frac{T}{C(T,H)}\right)_H \left(\frac{\partial M(T,H)}{\partial T}\right)_H \mathrm{d}H \tag{5.2.2}$$

From eqs. (5.2.1) and (5.2.2), the rate at which bulk magnetization changes with temperature in the presence of constant magnetic field, $(\mathrm{d}M/\mathrm{d}T)_H$, defines both the sign and the extent of the magnetocaloric effect of a solid for a given ΔH. For cooperative magnetic ordering phenomena a $|(\mathrm{d}M/\mathrm{d}T)_H|$ plot exhibits a maximum at a magnetic transition temperature, which roughly corresponds to the maximum MCE. Not only $|(\mathrm{d}M/\mathrm{d}T)_H|$ must be large, which is common in weak magnetic fields for many materials in the immediate vicinities of their spontaneous ferro or ferrimagnetic ordering phase transformations, it must remain large as H_F increases. The latter is, however, uncommon because in conventional magnetic materials magnetic transitions broaden significantly when magnetic field increases, lowering $|(\mathrm{d}M/\mathrm{d}T)_H|$ values. Further, many magnetic materials, even the ones with large magnetization at room temperature, such as elemental iron, exhibit magnetic transitions, and correspondingly peak MCE values away from room temperature, which explains why we do not notice any significant thermal response in them from even the strongest permanent magnets. For this reason, only a handful of materials, discussed below are available for near room temperature magnetic cooling.

Beyond the critical large $|(\mathrm{d}M/\mathrm{d}T)_H|$ requirement, a number of additional materials characteristics, both related and not directly related to magnetism, must also be considered. Magnetocaloric effect should be fully reversible, meaning field and temperature hystereses must be negligible, otherwise unrecoverable energy losses during cyclic magnetizing and demagnetizing are inevitable. Creating strong magnetic fields on the Earth is difficult and rather expensive, hence large adiabatic temperature and isothermal entropy changes must be observed in the weakest possible magnetic fields. For example, magnetic field strengths on surfaces of the strongest known permanent magnets today is limited to about 15 kOe. This means that large magnetocaloric effects should be realized in fields lower than that, and preferably lower than 10 kOe, otherwise construction of expensive Halbach-type permanent magnet arrays is required. Considering eq. (5.2.2), the adiabatic temperature change of a material is inversely proportional to its heat capacity, so the latter should be minimized over the temperature range of interest. Efficient heat exchange with the heat transfer fluid requires that a good magnetocaloric solid has reasonably high thermal conductivity. Minimization of eddy current losses mandates low electrical conductivity. Other desirable characteristics of promising magnetocaloric materials include availability and affordability of its constituents, chemical and mechanical stability, easy synthesis and fabrication into shapes to maximize heat transfer, long-term durability and high gravimetric density (it is easier to create a large magnetic field in a small gap). At the time of writing this chapter, there were no materials that exhibited large MCE around 300 K in low magnetic fields and satisfied all of these requirements, which is one of the main reasons why magnetic refrigerators for near room temperature use are not yet commercially available. There are, however, several families of materials that either have been shown to work reasonably well in laboratory-scale magnetocaloric heat pumps or have been broadly accepted as potential candidates for both cryogenic and room temperature magnetocaloric cooling applications. These are briefly described below, focusing on inorganic compounds.

5.2.2 Magnetocaloric materials for cryogenics

Cooling with magnetocaloric effect was initially conceived as means to reach ultralow temperatures (milliKelvin [7]), and it performs very well in cryogenic applications. There are two main reasons. First, at low temperatures interactions between magnetic moments are weakly affected by thermal energy and $|(dM/dT)_H|$ are generally greater than at ambient conditions. We also note that in an ideal paramagnet, $|(dM/dT)_H|$ approaches infinity as T approaches 0. Considering the magnetic moments themselves, large magnetic moments lead to large magnetic entropies that can be manipulated by magnetic field. The maximum molar magnetic entropy of a solid, S_M, is given [1, 12] as $S_M = R\ln(2J + 1)$, where R is the universal gas constant and J is the total angular momentum quantum number. The J quantum numbers are the largest for heavy lanthanide elements, where $J = L + S$, and L is the orbital angular momentum quantum number and S is the spin quantum number. At low temperatures, where lattice heat capacity is small, magnetic entropy associated with lanthanide moments may even exceed lattice entropy facilitating large cryogenic magnetic-field induced temperature changes. For this reason lanthanide-based compounds containing Gd, Tb, Dy, Ho and Er are materials of choice for cryogenic MCE applications due to large available magnetic entropies and wide range of magnetic ordering temperatures that are commonly below 300 K. Second, lattice heat capacity of a solid, C, increases rapidly as a function of $(T/\theta_D)^3$, where θ_D is the Debye temperature, which means that the adiabatic temperature change, inversely proportional to C (see eq. (5.2.2)), is maximized at a few K.

Until recently, the cryogenic cooling was a niche technology relevant for a number of specialized applications, for example, basic science and space exploration. The shift toward clean energy and decarbonization of economy puts cryocooling into focus for a broader audience due to its potential for efficient hydrogen liquefaction ($T = 20$ K). In addition, precooling of natural gas (NG) and its liquefaction ($T = 120$ K), can reduce the cost of its transportation, thus becoming another focus area for magnetic refrigeration.

There are two major classes of materials considered for cryogenic applications: a) ionic and covalent compounds, such as oxides, salts and salt-like substances; b) intermetallic compounds and alloys. Most, if not all of these compounds contain rare earths.

5.2.2.1 Non-metallic compounds

Due to rather low concentration of the moment-carrying ions and low thermal conductivity, $Gd_2(SO_4)_3{\cdot}8H_2O$ – the original compound employed to demonstrate the principle of magnetic cooling – finds no practical use today [13]. In the 1970s, gadolinium gallium garnet, $Gd_3Ga_5O_{12}$ (commonly known as GGG), was suggested and became the lead material for cryogenic magnetocaloric cooling [13], due to its much improved properties compared to gadolinium sulfate. Later studies revealed that chemical modifications of GGG, for example $Gd_3(Ga_{0.5}Fe_{0.5})_5O_{12}$ (GGIG) and $Dy_{2.4}Gd_{0.6}Al_5O_{12}$ (DGAG) compounds, are even better suited to perform as low-temperature MCE materials [14–16]. In addition to larger magnetic entropy change, they also have better mechanical properties and thermal conductivity. A number of other complex gadolinium oxides and salts, such as $Gd(HCOO)_3$ [17], have been proposed to achieve and maintain low temperatures both via adiabatic demagnetization refrigeration and continuous magnetocaloric heat pumping. These oxides and salts do not absorb hydrogen and are good options for hydrogen liquefaction at ~20 K, but their cooling performance significantly drops above 25 K [13, 18]. At the same time, rare-earth mononitrides, *RE*N, where *RE* = Gd, Tb, Dy, and/or Ho have a range of magnetic ordering temperatures spanning from ~20 to ~100 K [19]. These materials possess high magnetic moment density per unit volume, good thermal conductivity and are inert to hydrogen. Holmium nitride, HoN, has a peak MCE at 18 K, right around the boiling point of hydrogen, while others can be used to cool gaseous H_2 from 77 K.

5.2.2.2 Intermetallic compounds

In an intermetallic compound, a reaction of two or more metals or metalloids results in bonding and properties that are typical of a metal. Thermal conductivity of intermetallics is usually much higher compared to oxides and salts. Large magnetic moments of lanthanide atoms and the ability to precisely tailor magnetic interactions between them (also known as indirect Ruderman-Kittel-Kasuya-Yosida, or RKKY [20] interactions) by alloying with *s*, *p* and *d* elements, allow for large MCEs, located (on the temperature scale) exactly where they are needed for applications, as long as the temperatures are not too high (generally below 300 K). As a result, a large number of rare earth intermetallics could be considered as promising magnetocaloric materials for cryogenic applications. Among them, for example, are binary REM_2 compounds with cubic $MgCu_2$-type crystal structure, commonly known as C15-type or Laves phases, where *RE* is a rare earth and M could be a transition metal, such as Ni or Co, an *s*-element such as Mg or a *p*-element such as Al. For instance, $ErAl_2$ has a large $|\Delta S_M|$ with a maximum of 36 J Kg^{-1} K^{-1} at T = 13 K for ΔH = 50 kOe [21], and $HoCo_2$ shows a $|\Delta S_M|$ of 12.5 J Kg^{-1} K^{-1} at T = 78 K for ΔH = 20 kOe [22]. Further, many isostructural binary lanthanide intermetallics easily form continuous solid solutions. Thus, $(Er_{1-x}Dy_x)Al_2$, which adopts the same room temperature $MgCu_2$-type crystal structure for all x, demonstrates excellent magnetocaloric effects between 13 K (T_C of $ErAl_2$) and 64 K (T_C of $DyAl_2$) controlled by adjusting x between 0 and 1 [2, 21]. It is worth noting that below T_C the structures of these intermetallics often distort from the isotropic cubic symmetry, reflecting their anisotropic magnetic structures [22]. Recently, large magnetocaloric effects in relatively low magnetic fields were discovered in several RE_2In compounds. These materials, among them Eu_2In, Pr_2In, and Nd_2In possess large MCEs (Eu_2In has $|\Delta S_M|$ of 27 J Kg^{-1} K^{-1} at T = 55 K for ΔH = 20 kOe) between 50 and 120 K [23–25]. Nd_2In, having a T_c of ~120 K is promising for natural gas liquefaction [25]. Despite large MCE and no detrimental energy losses from hysteresis, RE_2In compounds are reactive with air, which may limit their practical use.

5.2.3 Magnetocaloric materials for near-room temperature applications

As noted above, cooling with magnetocaloric effect serves cryogenic applications well. Specific functional properties, namely large $|(dM/dT)_H|$ and low heat capacity that make MCE attractive at low temperatures are, however, harder to realize as the temperature increases. The lattice heat capacity increases proportionally to $(T/\theta_D)^3$, saturating at 3R J/mol K at and above θ_D, while thermal energy reduces the strength of magnetic field – magnetic moment coupling, impeding alignment of magnetic moments and leading to disruption of magnetic ordering and reduction of $|(dM/dT)_H|$.

Among conventional ferromagnets, only a handful of highly concentrated magnetic materials such as Gd (discussed in Section 5.2.3.1) and $Gd_{1-x}Y_x$ alloys with $x \ll 1$ and Tb and $Gd_{1-x}Tb_x$ alloys [1] are suitable near room-temperature magnetocaloric refrigerants. To realize the potential for high efficiency of magnetocaloric heat pumping near room temperature, magnetic fields above 20 kOe are highly desirable, but they require superconducting magnets that are unrealistic in consumer applications. An alternative is to employ a different class of magnetocaloric materials, in which conventional magnetic entropy changes described by eq. (5.2.1) may be enhanced by either or both lattice and electronic entropy contributions associated with coupled magnetostructural or magnetovolume transformations [3, 26, 27]. Such solids are commonly known as materials exhibiting giant magnetocaloric effect [28], and most important families of those materials are discussed in Sections 5.2.3.2–5.2.3.7.

5.2.3.1 Gd

Figure 5.2.1 shows temperature and magnetic field dependent magnetocaloric effect as both the magnetic entropy changes and the adiabatic temperature changes of elemental Gd – a benchmark magnetocaloric material. The Gd metal has many of the necessary properties – large MCE, negligible hysteresis, reasonably high thermal conductivity and good mechanical stability – so there is no surprise that the majority of magnetic cooling prototype devices employ Gd as the active material. The MCE of Gd is substantial over a wide range of temperatures but the effect is clearly maximized near the spontaneous ferromagnetic ordering transition temperature of the element that occurs at $T_C = 294$ K. At the same time, gadolinium (and other conventional second-order ferromagnets) has one important shortcoming – all of the magnetic field changes shown in Figure 5.2.1 require superconducting magnets for their generation. While Gd has measurable MCE in 10–12 kOe fields achievable with permanent magnets, the cooling power of a magnetic cooling device powered by permanent magnets falls off rapidly at practical temperature spans between the hot and cold sides of the heat pump. Hence, in the last few decades a number of materials exhibiting a much larger magnetocaloric effect, dubbed as "giant", were developed as potential replacements of Gd for near room temperature refrigeration.

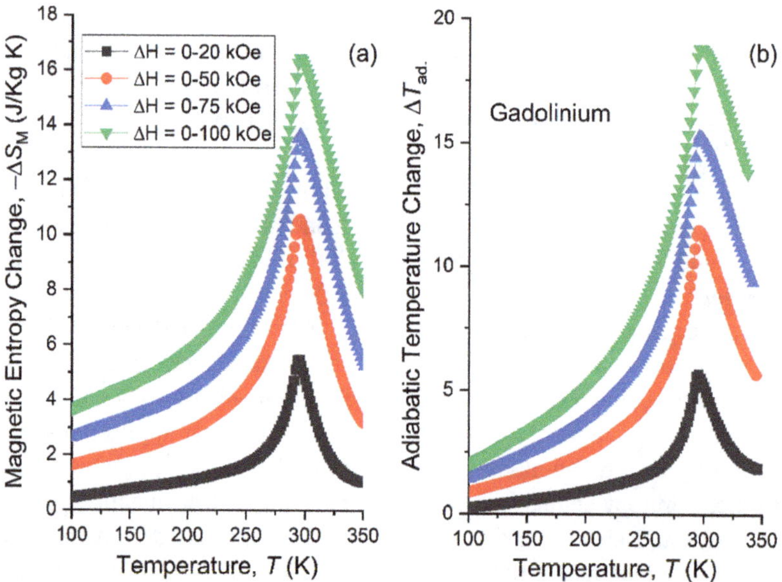

Figure 5.2.1: The magnetocaloric effect of elemental Gd as a) isothermal magnetic entropy changes and b) adiabatic temperature changes calculated from heat capacity data measured as functions of temperature in 0, 20, 50, 75 and 100 kOe applied magnetic fields [26].

5.2.3.2 $Gd_5Si_2Ge_2$

The term "giant magnetocaloric effect" was coined in 1997, when a large MCE was discovered near room temperature in a ternary intermetallic compound $Gd_5Si_2Ge_2$ and related $Gd_5Si_{4-x}Ge_x$ compounds [28, 29]. The maximum isothermal magnetic-field-induced entropy change of $Gd_5Si_2Ge_2$, $\Delta S_M = -14$ J Kg^{-1} K^{-1}, is 2.5 times higher than that of the Gd metal for the same magnetic field change of 20 kOe. The adiabatic temperature change, $\Delta T_{ad} = 7$ K, is greater as well, but the difference in $\Delta T_{ad.}$ is not as significant because of higher molar heat capacity of $Gd_5Si_2Ge_2$. The magnetocaloric effect in $Gd_5Si_2Ge_2$ is enhanced by the first order nature of the paramagnetic (PM) – ferromagnetic (FM) transition at T_C centered near 270 K. Both the magnetic (Figure 5.2.2a) and structural transformations (Figure 5.2.2b) occur in $Gd_5Si_2Ge_2$ simultaneously [30–34]. A specific magnetic state of $Gd_5Si_2Ge_2$ is always connected to a specific crystal structure: in the paramagnetic state, the compound retains monoclinic symmetry, but when it orders ferromagnetically, the crystallography changes from the monoclinic to a closely related orthorhombic. When the PM monoclinic $Gd_5Si_2Ge_2$ phase transforms into the FM orthorhombic phase, some of the interatomic bonds that connect pairs of quasi-two-dimensional atomic layers or slabs (Figure 5.2.2b) are created as the slabs shift with respect to one another. This bonding is associated with drastic changes in the magnetic exchange in the material, making it strongly ferromagnetic [27, 32–34]. Both the magnetic and structural transitions are reversible. The same transition may also be triggered by applied hydrostatic pressure and applied magnetic field. The first-order nature of the transition makes the change of magnetization with temperature nearly discontinuous (Figure 5.2.2a) and the resulting $|(dM/dT)_H|$ may reach very high values, remaining such in high magnetic fields. High $|(dM/dT)_H|$ enhances both the isothermal entropy change and the adiabatic temperature change, leading to giant MCE.

5.3.2.3 FeRh

Though the arrival of materials exhibiting giant magnetocaloric effects is rightfully traced to the discovery of $Gd_5Si_2Ge_2$ [28], a large MCE associated with first order magnetovolume phase transformation was earlier reported in nearly equiatomic FeRh compound [35, 36]. The magnetic transition leading to the giant MCE in FeRh is fundamentally different from both Gd and $Gd_5Si_2Ge_2$ because it is an order-order transition. Namely, in the low-temperature antiferromagnetic (AFM) phase magnetic moments are ordered antiparallel and the net magnetization is low. Above the transition the moments align parallel and the magnetic ordering state is ferromagnetic. The magnetization change at the AFM-FM transition is almost instantaneous and $|(dM/dT)_H|$ is similar to that of $Gd_5Si_2Ge_2$. This results in large value of $\Delta S_M =$ 10 J Kg^{-1} K^{-1} for a rather small $\Delta H = 10$ kOe, and the adiabatic temperature change for FeRh, $\Delta T_{ad} = 13$ K for $\Delta H = 20$ kOe, is highest reported among known intermetallic alloys and compounds. We note, however, that this high value of ΔT_{ad} is only

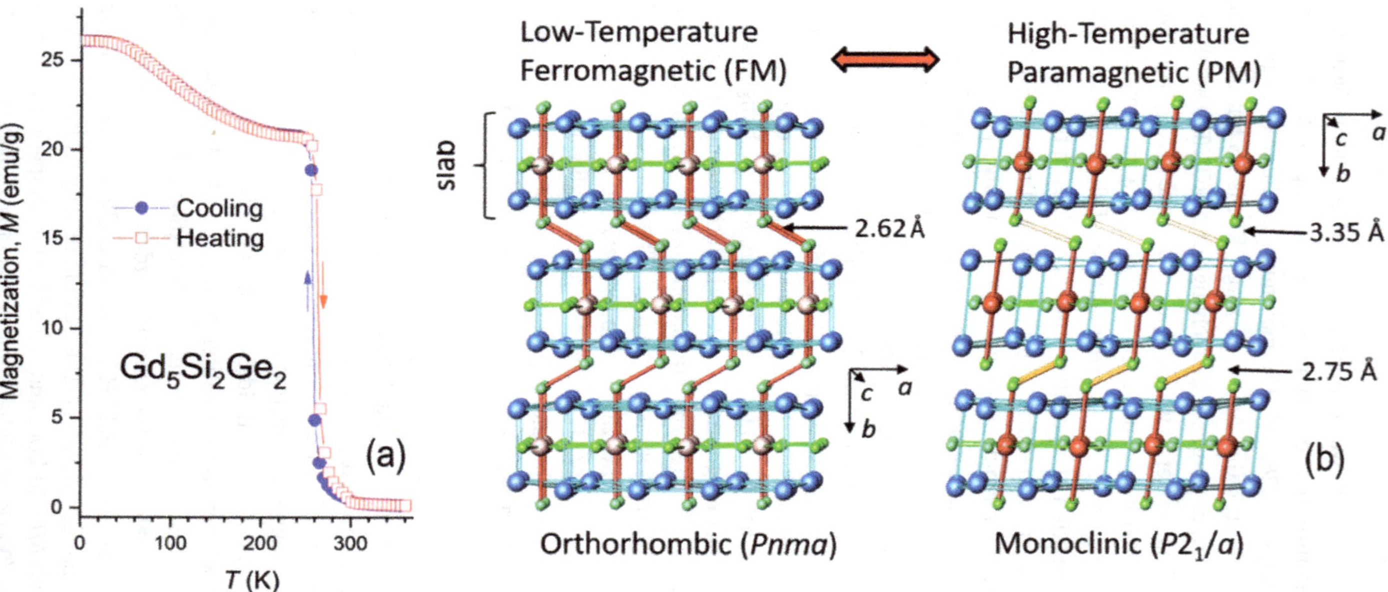

Figure 5.2.2: Magnetostructural transition in $Gd_5Si_2Ge_2$ – a) first-order magnetic phase transition in $Gd_5Si_2Ge_2$ at $T_C = 265$ K; b) structural change at the T_C between (left) low-temperature orthorhombic and (right) high-temperature monoclinic phases.

observed during first application of the magnetic field, and becomes substantially reduced during subsequent field changes due to rather broad hysteresis.

Due to a different nature of the underlying phase transition, the sign of MCE in FeRh is opposite to those of Gd and $Gd_5Si_2Ge_2$. Hence, the MCE in the latter two materials is called “direct”, but it is called “inverse” in the former. From the practical point of view this means that materials exhibiting “direct” magnetocaloric effect cool down when demagnetized (upon magnetic field removal), while those exhibiting “inverse” magnetocaloric effect cool down when magnetized (upon magnetic field application). This terminology has historic origin: all of the initial work on MCE going back to Debye and Giauque was done on materials that cool upon demagnetization, and therefore were considered standard, or having “direct” magnetocaloric effect. Another difference of FeRh from $Gd_5Si_2Ge_2$ is the absence of the structural change: even though FeRh exhibits a significant, 0.3% change of its lattice parameter a at the AFM-FM transition, its crystal structure remains the same cubic B2 CsCl-type above and below the transformation. It is important to note that the transition leading to giant MCE occurs in the narrow range of concentrations centered at $Fe_{0.49}Rh_{0.51}$ composition. Moving away from this nearly equiatomic chemistry toward either Fe or Rh destroys the order-order transition and converts the ordered CsCl-type structure into disordered body-centered cubic (bcc) or face-centered cubic (fcc) type. It is also worth noting that there is another, FM-PM order-disorder transition in nearly equiatomic FeRh at higher temperatures, but both $|(dM/dT)_H|$ and MCE in the vicinity of that high-temperature transition are small.

5.3.2.4 $LaFe_{13-x}Si_x$-based compounds and their hydrides

$LaFe_{13-x}Si_x$ compounds and their derivatives are considered by many the best practical MCE materials available [37] due to a combination of good performance and lack of expensive constituents, such as Rh or Ge. Because the binary $LaFe_{13}$ is not stable, Fe/Si substitution is needed to stabilize the cubic $NaZn_{13}$ crystal structure of $LaFe_{13-x}Si_x$ (the range of x is from 1.17 to approximately 3.5). When $x(Si) \geq 4$, the crystal structure becomes tetragonal and weakly magnetic; thus, hereafter we will discuss the cubic $LaFe_{13-x}Si_x$ compounds only. In cubic $LaFe_{13-x}Si_x$ both the magnetism and giant magnetocaloric effect are defined by $3d$ electrons of iron and the cubic $NaZn_{13}$-type crystal structure remains the same below and above the magnetic-ordering transition, similarly to FeRh. Depending on x, $LaFe_{13-x}Si_x$ exhibits either first or second-order transition at T_C. In compounds with the first-order transition there is a rather large phase volume change of 1.5%, and the total isothermal entropy change exceeds -20 J Kg^{-1} K^{-1} for a magnetic field change of $\Delta H = 20$ kOe [38–40]. The boundary between first and second-order transitions in $LaFe_{13-x}Si_x$ solid solutions is not clearly defined. The phases exhibit gradual weakening of the first-order nature of the transition at T_C as x increases from 1.2 to 1.8 [38], finally becoming a second-order phase transition when $x \geq 2$, where unit-cell volume change at the transition is reduced to 0. Materials that fall into the intermediate region may combine desirable properties, such as large MCE with much reduced hysteresis losses.

The $LaFe_{13-x}Si_x$ family of compounds exhibits excellent magnetocaloric properties when $1.17 \leq x \leq 1.8$, but their magnetic ordering temperatures, and, consequently, peak MCEs are located at and below 200 K. This makes them attractive candidates for certain cryogenic applications, such as propane gas liquefaction, but they are not suitable for use at ambient conditions. There are two known ways to bring their operating temperature range to room temperature: 1) partial substitution of Fe or Si with Co [39]; 2) expansion of the lattice with hydrogen [40]. In the first case, T_C increases because Co has a higher T_C compared to Fe, so the Co-substituted $LaFe_{10.97}Co_{1.03}Si$ has $T_C = \sim 293$ K [39]. However, the transition becomes second order, and its entropy change is less than a half when compared to the parent $LaFe_{11.8}Si_{1.2}$. The second approach, hydrogenation of $LaFe_{13-x}Si_x$ phases, preserves the first-order nature of the phase transformation and the giant MCE, while raising T_C (which depends on the amount of absorbed hydrogen) up to 340 K [40] by virtue of lattice expansion. The fully hydrogenated materials contain up to 1.65 hydrogen atoms per formula unit and, in principle, by varying hydrogen content one can control the T_C. However, compounds with less than full hydrogen content phase separate into hydrogen-rich and hydrogen-poor phases with two different T_C when held at room temperature from a few days to a few weeks [41]. This phenomenon severely reduces magnetic cooling performance of the material. Thus, these magnetocaloric materials must be fully hydrogenated for practical use. The required tuning of T_C is achieved by minor substitutions of Fe with Mn, which strongly reduces T_C of the parent $LaFe_{13-x}Si_x$. The Si content is adjusted as well, and the final hydrogenated alloys can be described by the formula $LaFe_xMn_ySi_zH_{1.65}$, where $11.22 \leq x \leq 11.76$, $0.06 \leq y \leq 0.46$ and $1.18 \leq z \leq 1.32$, with peak MCE tunable over 270–340 K temperature range [42].

5.3.2.5 Fe_2P-type materials

The Fe_2P structure type and its derivatives are commonly adopted by intermetallic compounds so it should not come as a surprise that another promising family of magnetocaloric phases was discovered in a series of transition metal compounds that crystallize in this structure. Namely, in 2002, a giant MCE was reported in the $MnFeP_{0.45}As_{0.55}$ compound with $\Delta S_M = -18$ J Kg^{-1} K^{-1} for $\Delta H = 20$ kOe just above 300 K [43]. Given that the presence of arsenic could be problematic in products intended for wide commercial applications, in these materials As was eventually replaced with Si. The $Mn_xFe_{1.95-x}(P_{1-y}Si_y)$ series of phases (with x ranging between 0.66 and 1.34 and y ranging between 0.33 and 0.54) was designed and optimized to show a large MCE in relatively weak magnetic fields (up to $\Delta S_M = -12.5$ J Kg^{-1} K^{-1} for $\Delta H = 10$ kOe) in a wide range of temperatures, from 210 to 380 K [44]. Similar to other giant MCE materials the effect was enhanced due to first-order transition, however, the volume change at T_C is minor, and the magnetoelastic transition manifests itself as a discontinuous change in the ratio of lattice parameter c over lattice parameter a [45]. It was also discovered that minor additions of boron reduce energy losses associated with hysteresis and improve mechanical stability, which are common problems of the undoped MnFe(P,Si) compounds. For example, $MnFe_{0.95}P_{0.595}B_{0.075}Si_{0.33}$ bulk material does not show signs of degradation after cycling 10,000 times through the magnetoelastic transition [45].

5.3.2.6 Heusler alloys

The Heusler alloys, a family of materials based on the NiMn parent composition, where Mn is partially substituted by a p-element, such as Al, Ga, In, Sn and Sb, are broadly researched due to a variety of interesting and potentially useful phenomena. For example, compounds with compositions close to Ni_2MnGa are known as ferromagnetic shape memory alloys. They have two transitions: the high-temperature magnetic ordering FM-PM transition and the low-temperature FM-AFM order-order transformation. The phenomenology is somewhat similar to FeRh (Section 5.3.2.3) but with one substantial distinction – the order-order transition here is coupled to a structural transformation between cubic high-temperature austenite phase and low-temperature low symmetry martensite. Because the order-order transition is commonly discontinuous, it may exhibit large magnetocaloric effect, which, similarly to FeRh, is inverse since the cooling occurs upon magnetization. Many of the Heusler alloys possess large MCE and have good mechanical properties but they are commonly plagued by large irreversibilities (hysteresis) [46–48]. A number of promising compositions, containing up to 4–5 elements, e.g. Ni-Mn-Co-In compounds, were identified, and concepts how to avoid the detriments of hysteresis were developed [48]. The Heusler alloy family remains promising for future research and development due to a large compositional space that allows manipulation and control of physical behaviors.

Another series of materials that are based on the NiMn composition includes $Ni_{50-x}Co_xMn_{50-y}Ti_y$ alloys, whose properties closely resemble those of the prototypical Ni_2MnX Heusler alloys. When $Ni_{50-x}Co_xMn_{35}Ti_{15}$ compounds ($10 \le x \le 15$) are prepared by rapid solidification (melt-spinning), the MCE of the produced ribbons reaches 23 J Kg^{-1} K^{-1} for $\Delta H = 15$ kOe [49].

5.3.2.7 Transition metal-based compounds with Ni_2In/TiNiSi structural transformation

Compounds with MM'X stoichiometry, where M and M' are transition metals (usually the compounds contain Mn) and X is a p-element can be compositionally-tuned to exhibit magnetostructural transitions from high-temperature PM state that adopts Ni_2In-type structure to low-temperature FM state adopting TiNiSi-type structure. This transition is associated with a discontinuous change in magnetization leading to large MCE at broad range of temperatures including those near the desired room temperature range [50, 51]. For example, a series of $Mn_{1-y}Fe_yNiGe_{1-x}Si_x$ alloys was shown to exhibit MCE as high as $\Delta S = -17$ J Kg^{-1} K^{-1} for $\Delta H = 20$ kOe at $T_C = 333$ K when $x = 0.3$ and $y = 0.26$ [50]. These compounds have some of the largest MCE reported but their functionality is diminished by extreme brittleness and the presence of large thermal hysteresis, which will result in energy losses during operation. It is also worth noting that Ge is non-abundant and expensive. Recently discovered $Mn_{0.5}Fe_{0.5}NiSi_{1-x}Al_x$ series of alloys ($x = 0.045$–0.07) does not contain Ge or any other expensive or toxic elements while maintaining the MCE functionality [52]. If issues with mechanical stability and hysteresis are solved, these compounds may become the much needed materials sought for near room temperature magnetic refrigeration.

References

[1] Tishin AM, Spichkin YI. The Magnetocaloric Effect and Its Applications, IoP Publishing, Bristol and Philadelphia, 2003.

[2] Gschneidner Jr KA, Pecharsky VK. Annu Rev Mater Sci, 2000, 30, 387–429.

[3] Gschneidner Jr KA, Pecharsky VK, Tsokol AO. Rep Prog Phys, 2005, 68, 1479–539.

[4] Franco V, Blázquez JS, Ipus JJ, Law JY, Moreno-Ramírez LM, Conde A. Prog Mater Sci, 2018, 93, 112–232.

[5] Debye P. Ann Phys, 1926, 81, 1154.

[6] Giauque WF. J Am Chem Soc, 1927, 49, 1864.

[7] Giauque WF, MacDougall DP. Phys Rev, 1933, 43, 768.

[8] Brown GV. J Appl Phys, 1976, 47, 3673–80.

[9] Foldeaki M, Chahine R, Bose TK. J Appl Phys, 1995, 77, 3528.

[10] Nevez Bez H, Yibole H, Pathak A, Mudryk Y, Pecharsky VK. J Magn Magn Mater, 2018, 458, 301–09.

[11] Pecharsky VK, Gschneidner Jr KA. J Appl Phys, 1999, 86, 565–75.

[12] Kittel C. Introduction to Solid State Physics, 2nd edition, John Wiley & Sons, New York, 1956.

[13] Barclay JA, Steyert WA. Cryogenics, 1982, 73–80.

[14] McMichael RD, Ritter JJ, Shull RD. J Appl Phys, 1993, 73, 6946–48.

[15] Brodale GE, Hornung EW, Fischer RA, Giauque WF. J Chem Phys, 1975, 62, 4041–49.

[16] Matsumoto K, Matsuzaki A, Kamiya K, Numazawa T. Jpn J Appl Phys, 2009, 48, 113002.

[17] Lorusso G, Sharples JW, Palacios E, Roubeau O, Brechin EK, Sessoli R, Rossin A, Tuna F, McInnes EJL, Collison D, Evangelisti M. Adv Mater, 2013, 25, 4653–56.

[18] Numazawa T, Kamiya K, Utaki T, Matsumoto K. Cryogenics, 2014, 62, 185–92.

[19] Yamamoto TA, Nakagawa T, Sako K, Arakawa T, Nitani H. J Alloys Compd, 2004, 376, 17–22.

[20] a) Ruderman MA, Kittel C. Phys Rev, 1954, 96, 99. b) Kasuya T. Prog Theor Phys, 1956,16,45; c) Yosida K. Phys Rev, 1957,106,893.

[21] a) Gschneidner Jr KA, Pecharsky VK, Gailloux MJ, Takeya H. Adv Cryog Eng, 1996, 42A, 465. b) Gschneidner KA, Pecharsky VK, Malik SK. Adv Cryog Eng, 1996,42A,475.

[22] Mudryk Y, Pecharsky VK, Gschneidner Jr KA. In: Proceedings of the 3rd IIF-IIR International Conference on Magnetic Refrigeration at Room Temperature, Des Moines, USA, 2009, 127–31.

[23] Guillou F, Pathak AK, Paudyal D, Mudryk Y, Wilhelm F, Rogalev A, Pecharsky VK. Nature Commun, 2018, 9, 2925.

[24] Biswas A, Zarkevich NA, Pathak AK, Dolotko O, Hlova IZ, Smirnov AV, Mudryk Y, Johnson DD, Pecharsky VK. Phys Rev B, 2020, 101, 224402.
[25] Liu W, Scheibel F, Gottschall T, Bykov E, Dirba I, Skokov K, Gutfleisch O. Appl Phys Lett, 2021, 119, 022408.
[26] Gschneidner Jr KA, Mudryk Y, Pecharsky VK. Scr Mater, 2012, 67, 572.
[27] Pecharsky VK, Cui J, Johnson DD. Phil Trans A, 2016, 374, 20150305.
[28] Pecharsky VK, Gschneidner Jr KA. Phys Rev Lett, 1997, 78, 4494.
[29] Pecharsky VK, Gschneidner Jr KA. Appl Phys Lett, 1997, 70, 3299–301.
[30] Gschneidner Jr KA, Pecharsky VK. Int J Refrig, 2008, 31, 945–61.
[31] Pecharsky VK, Holm AP, Gschneidner Jr KA, Rink R. Phys Rev Lett, 2003, 91, 197204.
[32] Morellon L, Algarabel PA, Ibarra MR, Blasco J, García-Landa B, Arnold Z, Albertini F. Phys Rev B, 1998, 58, R14721.
[33] Samolyuk GD, Antropov VP. J Appl Phys, 2003, 93, 6882–84.
[34] Pecharsky VK, Gschneidner Jr KA. Adv Mater, 2001, 13, 683–86.
[35] Nikitin SA, Tishin AM, Redko SV. Phys Met Metallogr, 1988, 66, 77.
[36] Nikitin SA, Myalikgulyev G, Tishin AM, Annaorazov MP, Asatryan KA, Tyurin AL. Phys Lett A, 1990, 148, 363–66.
[37] Jacobs S, Auringer J, Boeder A, Chell J, Komorowski L, Leonard J, Russek S, Zimm C. Inter J Magn Refrig, 2014, 37, 84–91.
[38] Shen BG, Sun JR, Hu FX, Zhang HW, Cheng ZH. Adv Mater, 2009, 21, 4545–64.
[39] Moore JD, Morrison K, Sandeman KG, Katter M, Cohen LF. Appl Phys Lett, 2009, 95, 252504.
[40] Fujita A, Fujieda S, Hasegawa Y, Fukamichi K. Phys Rev B, 2003, 67, 104416.
[41] Zimm CB, Jacobs SA. J Appl Phys, 2013, 113, 17A908.
[42] Basso V, Küpferling M, Curcio C, Bennati C, Barzca A, Katter M, Bratko M, Lovell E, Turcaud J, Cohen LF. J Appl Phys, 2015, 118, 053907.
[43] Tegus O, Brück E, Buschow KHJ, de Boer FR. Nature, 2002, 45, 150.
[44] Dung NH, Ou ZQ, Caron L, Zhang L, Cam Thanh DT, de Wijs GA, de Groot RA, Buschow KHJ, Brück E. Adv Energy Mater, 2011, 1, 1215.
[45] Guillou F, Porcari G, Yibole H, van Dijk N, Brück E. Adv Mater, 2014, 26, 2671.
[46] Krenke T, Duman E, Acet M, Wassermann EF, Moya X, Manosa L, Planes A. Nature Mater, 2005, 4, 450.
[47] Liu J, Gottschall T, Skokov KP, Moore JD, Gutfleisch O. Nature Mater, 2012, 11, 620.
[48] Gottschall T, Gracia-Condal A, Fries M, Taubel A, Pfeuffer L, Manosa L, Planes A, Skokov KP, Gutfleisch O. Nature Mater, 2018, 17, 929.
[49] Bez HN, Pathak AK, Biswas A, Zarkevich N, Balema V, Mudryk Y, Johnson DD, Pecharsky VK. Acta Mater, 2019, 173, 225.
[50] Wei Z-Y, Liu E-K, Li Y, Xu G-Z, Zhang X-M, Liu G-D, Xi X-K, Zhang H-W, Wang W-H, Wu G-H, Zhang X-X. Adv Electron Mater, 2015, 1, 1500076.
[51] Samanta T, Lloveras P, Saleheen AU, Lepkowski DL, Kramer E, Dubenko I, Adams PW, Young DP, Barrio M, Tamarit JL, Ali N, Stadler S. Appl Phys Lett, 2018, 112, 021907.
[52] Biswas A, Pathak AK, Zarkevich NA, Liu X, Mudryk Y, Balema V, Johnson DD, Pecharsky VK. Acta Mater, 2019, 180, 341.

5.3 Materials for thermoelectric devices

Ananya Banik, Matthias Agne, Wolfgang Zeier

All materials develop an internal voltage difference ΔV when they are subjected to a temperature gradient (defined by the temperature difference ΔT) across them, an effect described in solids by the Seebeck coefficient ($S = \Delta V / \Delta T$) [1]. This effect arises because the distribution of excited electrons is different at different temperatures, so a material that is not in thermal equilibrium (i.e. $\Delta T \neq 0$) has a driving force for electron transport. Thus, it is possible to use the Seebeck effect to drive an electric current. As a complementary effect, it is also possible to use an electric current to pump heat from one side of a material to the other, leading to an effective cooling. This is known as the Peltier effect. The Seebeck and Peltier effects are the so-called thermoelectric effects.

Although every material has a Seebeck and Peltier coefficient, not every material has the right combination of electronic and thermal properties to usefully harness thermoelectric effects. In fact, the vast majority of materials have negligible utility for thermoelectric applications. To understand why, we can look at the thermoelectric material figure-of-merit zT first proposed by Ioffe [2]

$$zT = \frac{\sigma S^2}{\kappa} T,$$

which is defined by the electronic conductivity σ, the Seebeck coefficient S, the total thermal conductivity κ and the absolute temperature T. Here, high electronic conductivity makes it possible to have electrons and holes move, a large Seebeck coefficient provides a high voltage, and low thermal conductivities are needed to maintain a temperature difference across the device. Since it turns out that the material properties are all interrelated, zT is difficult to optimize. For instance, σ and S tend to be inversely proportional to each other. Additionally, since electrons carry heat in addition to their electric charge, the thermal conductivity κ will increase with increasing σ. Ioffe realized early on that narrow band gap (<1 eV) semiconductors would be the best candidates for optimizing zT if the number of electronic carriers could be effectively tuned (Figure 5.3.1). To date, the best-performing thermoelectric materials have achieved $zT \approx 2$. Nevertheless, a material may be considered as a "good" thermoelectric if it has $zT \approx 1$, which is already a rarity among all known solids.

An intuitive understanding of zT is that it is a ratio of electronic power conversion to thermal losses of the material. While zT itself is not a definition of thermoelectric efficiency, it is related to the energy conversion efficiency of the material η as

$$\eta = \frac{\Delta T}{T_H} \frac{\sqrt{1+zT}-1}{\sqrt{1+zT}+1},$$

https://doi.org/10.1515/9783110798890-003

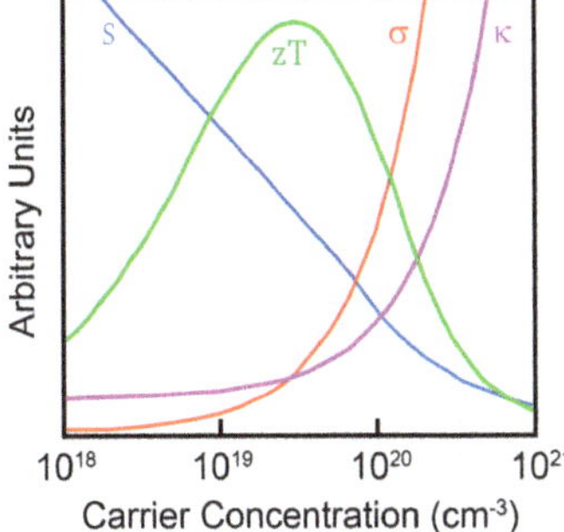

Figure 5.3.1: Optimizing the thermoelectric figure-of-merit zT requires the precise tuning of electronic carrier concentration for the desired operational temperature range.

which cannot exceed the Carnot efficiency ($\eta = \Delta T/T_H$) [1]. Clearly, zT should be as large as possible to maximize the conversion efficiency.

Although crystallinity is needed to have good electronic transport, one goal is to reduce the thermal conductivity mediated by atomic vibrations (phonons) toward the amorphous limit. In materials where phonons transport heat in a manner analogous to an ideal gas, then thermal transport can be significantly reduced if phonons of all wavelengths can be effectively scattered [3]. Long and mid-wavelength phonons are mostly scattered by grain boundaries and microstructural defects, whereas short-wavelength phonons are more affected by atomic-scale defects such as vacancies and impurity atoms. In addition to phonon scattering, phonon speed can be reduced through various methods [4, 5]. Further control over thermal transport in complex and disordered crystals is at the forefront of current research efforts [6]. General strategies to maximize zT have revolved around the "phonon-glass electron-crystal" concept [7] that aims to find materials with low thermal conductivities and then optimize their electronic transport.

Improvement of the electronic properties of materials can be guided by an understanding of atomic orbitals in conjunction with the band theory of solids [8]. Strategies for increasing the product σS^2 generally revolve around increasing the electronic density of states, including: increasing the band degeneracy by trying to push the valence band maximum (and/or conduction band minimum) to a low symmetry position that has higher multiplicity; converging electronic bands at different symmetry positions so that there is a higher effective band degeneracy; or, by adding so-called resonant dopants. Lower temperature electronic properties can be improved by mitigating grain boundary effects that are prevalent in some materials. In practice, these strategies can be implemented by doping or alloying the material to manipulate the orbital overlaps and, therefore, the electronic band structure [9]. Optimizing electronic properties by changing the dominant scattering mechanism of the electronic carriers may also be possible, likely through tuning the structural disorder of crystals.

Although optimizing the thermoelectric performance of a material is already a great undertaking, the implementation of a material into a working device is equally, if not arguably more, challenging. A typical device requires n-type (i.e. electron transport occurs in the conduction band) and p-type (i.e. electron transport occurs in the valence band) materials connected electrically in series and thermally in parallel [10] (Figure 5.3.2). Metal interconnects are used to connect the p and n-type thermoelectric "legs" and heat exchangers are used to direct heat through the thermoelectric module. Thermal expansion of the materials, as well as residual stresses at soldering/brazing joints, makes device construction a complex engineering problem. When all factors are considered, the best performing devices operate at ~10–20% thermal efficiency.

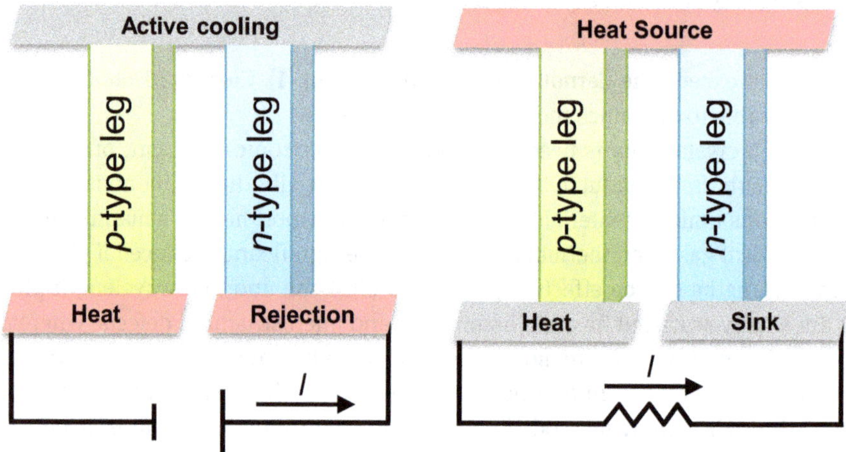

Figure 5.3.2: Schematic illustration of a thermoelectric cooler (left) and generator (right). In both cases, heat flow moves through the p and n-type materials in parallel, but they are connected electrically in series.

Nevertheless, thermoelectric generators have been used in space exploration since the 1960s, including the Apollo lunar missions. The Voyager space probes, which are the first man-made objects to leave the solar system, have been transmitting data continuously for 40+ years thanks to thermoelectric power. In these thermoelectric generators, electric power is produced from the heat of a small radioactive source, such as plutonium-238. For this reason, they could be (and have been) referred to as atomic batteries, although they are certainly not batteries in the usual sense. Now they are commonly referred to as radioisotope thermoelectric generators, or RTGs. Since there are no moving parts, RTGs are incredibly reliable – usually limited by the half-life of the radioactive source material. For situations where solar power is not an option, like when you are 22 billion kilometers from the sun (e.g. Voyager I), then RTGs are the only viable power source. Additionally, they have provided auxiliary power for the Curiosity and Perseverance Mars rovers since they can generate power at night and are unaffected should the solar panels be obstructed.

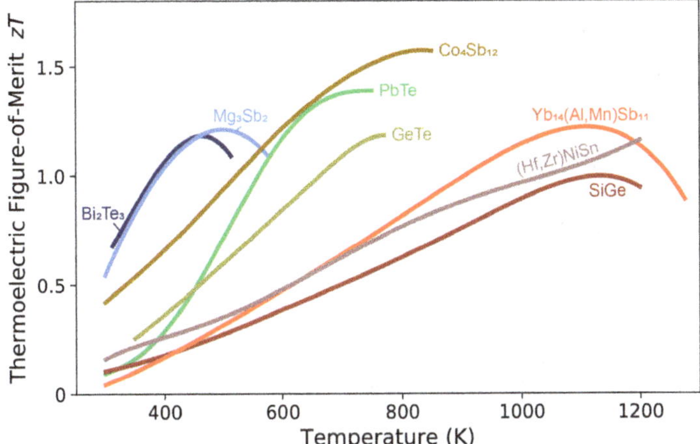

Figure 5.3.3: Thermoelectric figure-of-merit for materials designated for specified operational ranges. Data taken from ref [11–18].

Terrestrial applications of thermoelectric materials remain to be fully explored. As solid-state cooling devices they are more efficient than compressor-based cycles for chilling small volumes. The added benefit that they are silent makes them practical for personal refrigeration. One such use is for wine storage. As a power source, thermoelectric generators have potential for use in low-power internet-of-things devices and remote-sensing applications. On a larger scale, there is a strong desire to use thermoelectric generators for low-grade waste heat recovery. Nearly 2/3 of energy production is wasted due to distribution and use inefficiencies; and a high percentage of that waste is in the form of heat. Thermoelectric technologies could thus play a role in global energy sustainability.

Based on the working temperature regime, thermoelectric inorganic compounds can be categorized in three subclasses. Materials having peak zT in 25–200 °C are mostly used for cooling purpose. Mid-temperature (400–900 °C) and high-temperature ($T > 1000$ °C) thermoelectric materials are utilized for industrial waste heat recovery. The following sections summarize the state-of-art thermoelectric materials already used in commercial applications. An overview of the temperature dependence of zT for several of these compounds can be seen in Figure 5.3.3.

5.3.1 Low temperature and refrigeration materials

Narrow-gap semiconductor bismuth telluride (Bi_2Te_3) was the first material used for thermoelectric refrigeration purposes [19]. It crystallizes in a layered structure in which layers of Bi and Te (Te1-Bi-Te2-Bi-Te1) exist that are stacked by van der Waals interactions along the c-axis in the unit cell (Figure 5.3.4a). The unique

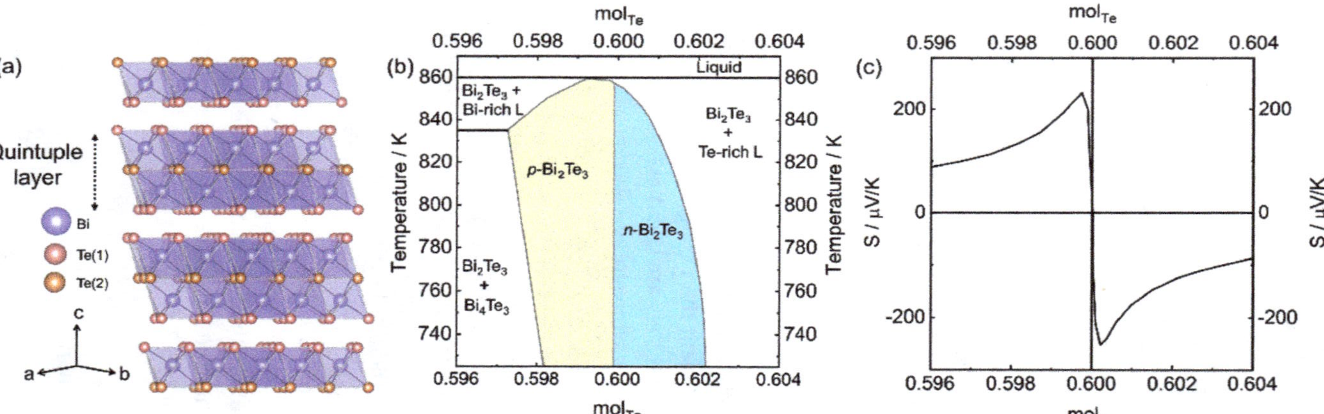

Figure 5.3.4: Thermoelectric properties of Bi_2Te_3 depending on stoichiometry (a) Crystal structure of Bi_2Te_3 showing Te1-Bi-Te2-Bi-Te1 stacks connected via van der Waals gap, (b) phase diagram showing phase width and accessible p and n-type regions depending on defect chemistry and (c) a schematic of the expected change in the Seebeck coefficient as the composition is subtly tuned across the phase width.

thermoelectric properties originate from its layered structure, large constituent atoms with relatively weak bonds and associated anharmonicity. As the congruent melting point of bismuth telluride is slightly off toward the Bi side of the Bi_2Te_3 composition (Figure 5.3.4b), one can prepare both p and n-type Bi_2Te_3 by appropriate choice of the synthesis route and stoichiometry [20] (Figure 5.3.4c). Approaches to enhance TE performance includes the compositional deviation from its stoichiometry, grain size manipulation, introduction of structural defects and solid-solution alloying. When substituting Bi_2Te_3 with other V–VI materials like Sb_2Te_3, Bi_2Se_3 shows the best result. Here, alloying reduces the lattice thermal conductivity via point defects scattering induced by mass and size contrast. Additionally, it also tunes the carrier concentration because of the formation of antisite defects. Nanostructured bulk Bi_2Te_3-M_2X_3 (M = Sb; X = S/Se/Te) alloys are typical thermoelectric materials used in commercial refrigerators [1, 21]. Both substitution and reducing grain size helps to lower the lattice thermal conductivity, all of which benefits the figure-of-merit.

Over the last 50 years, Bi_2Te_3-based phases have been the only practically applicable low temperature thermoelectric material, which limits the device design space. Mg_3Sb_2 is the recent addition to this subclass [22]. Despite its low density and simple structure, Mg_3Sb_2 exhibits lattice thermal conductivity as low as Bi_2Te_3 or PbTe, which is associated with the soft shear modulus and highly anharmonic acoustic phonons [23]. Additionally, the presence of multiple bands near the conduction band minima make this material promising as a high-performance thermoelectric. For example, n-type Mg_3Sb_2-Mg_3Bi_2 alloys show an exceptional thermoelectric performance (zT of 1.0–1.2 at 400–500 K) resulting from band convergence and an improved electronic conductivity [24,25]. The obtained zT value is higher than commercial Bi_2Te_3 phases, which makes this material promising for further optimization.

5.3.2 Mid-temperature thermoelectrics

For mid-temperature power generation, group IV-chalcogenides, such as PbTe and GeTe show best efficiency. Lead chalcogenides have a rich history since their discovery as thermoelectric materials. The cubic crystal structure and presence of multiple degenerate valleys near band edges made PbTe special for thermoelectric application (Figure 5.3.5a and b). Additionally, PbTe has a bandgap of 0.3 eV, which is large enough for operation at these higher temperatures. Dopants and intrinsic defects in PbTe allow for the carrier concentration to be optimally tuned in both n and p-type material [13, 26]. The maximum thermoelectric efficiency in PbTe was reached via the synergy of band engineering (resonance levels formation [27], band convergence [28]), carrier concentration optimization and microstructure manipulation [28]. For instance, nanoheterostructured PbTe/TAGS (TAGS: Tellurium-Antimony-Germanium-Silver) was used as the *p*-type leg in radio-isotope thermoelectric generators on the

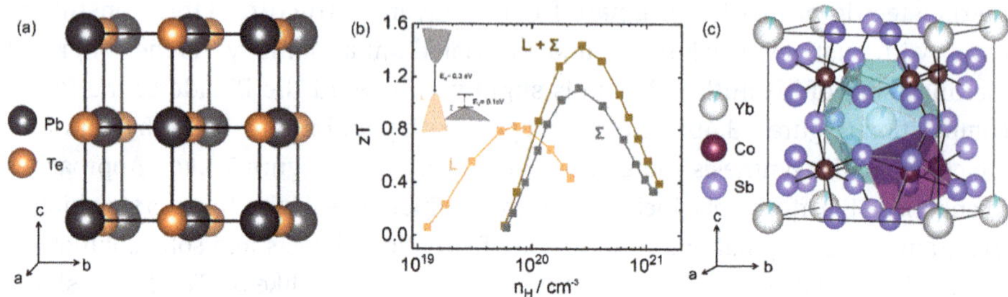

Figure 5.3.5: (a) The rock salt crystal structure typical of the chalcogenide thermoelectric materials such as PbTe. (b) A schematic band structure of PbTe showing the band gap and valence band offset that can be tuned by doping and substitution. Convergence of valence bands can result in significant rise in peak *zT* which is higher than carrier concentration optimization. (c) The crystal structure of skutterudite Co_4Sb_{12} where corner shared $CoSb_6$ octahedra form a cage-like structure with large void to contain so-called rattling atoms (e.g. ytterbium) that can greatly reduce the thermal conductivity.

NASA Mars rovers, Curiosity and Perseverance. The successful utilization of PbTe-based material has drawn the attention of the thermoelectric community toward SnTe, an environment friendly, cost-effective homologue of PbTe [29].

Another interesting inorganic materials class for mid-temperature power generation application is skutterudites [1]. These materials crystallize in a cubic $CoAs_3$-type structure type. The typical structure is composed of corner-shared XY_6 (X = Co, Rh, Ir; Y = P, As, Sb) octahedra forming voids in the center of the cubic unit cell. These voids represent large cages and if these void are filled, they can act as effective phonon-scattering centers reducing the lattice thermal conductivity by having a large degree of vibrational freedom in the cages (often called “rattling”) (Figure 5.3.5c) [30]. Since these atoms can also act as electron donors or acceptors, partially filling the void space of skutterudites can lead to an optimum electron concentration. Because of the “rattling” effect of these filler atoms, “void-filled” skutterudites show promising thermoelectric properties. In addition, long term high temperature stability makes the skutterudite class be the next choice of material for enhanced multimission radioisotope thermoelectric generator as a substitute of PbTe/TAGS.

5.3.3 High-temperature thermoelectric materials

The historical and continued use of thermoelectric materials in radioisotope thermoelectric generators, RTGs, makes the highest operational temperatures (>700 °C) of particular interest. Since low-thermal-conductivity materials typically also have weaker atomic bonding and lower melting temperatures, it is difficult to also balance thermal materials stability with overall thermoelectric performance. Nevertheless, there are several materials that meet these requirements.

Silicon is almost always the go-to semiconductor material, so it is not surprising that it has electronic properties that make it a candidate thermoelectric material. By itself, however, silicon has very high thermal conductivity to be efficient enough for practical applications. As it works out, when silicon is alloyed with germanium (Figure 5.3.6a), the thermal conductivity can be drastically reduced. The addition of ~30% germanium can reduce the thermal conductivity by a factor of ~10 [31]. This is somewhat surprising since Si-Ge solid solutions still have strong covalent bonds and a low average atomic mass, but this shows the importance of point defects for phonon scattering [32]. In other words, the difference in mass between Si and Ge is enough to prevent strong thermal transport. The success of Si-Ge solid solutions has led to the study of diamond-like semiconductors for thermoelectric applications more generally.

In the same way as the random Si-Ge solid solution exhibits strong thermal conductivity reduction due to site disorder, a similar effect can be achieved in crystals that have >100 atoms per unit cell. In these complex crystals there is an argument to say that, at some length scale, they look more structurally amorphous than crystalline [33]. Correspondingly, they conduct heat in a manner more similar to amorphous materials. An excellent example of a structurally complex material is $Yb_{14}MSb_{11}$ (M = Mn, Mg, Al, etc.), affectionately referred to as the 14-1-11 family of materials [34]. Indeed, 14-1-11 has 104 atoms per primitive unit cell and performs well as a high-temperature thermoelectric (Figure 5.3.6b). Consequently, it was used as part of the p/n-type leg in the RTG equipped on the Mars rover, Curiosity. From a research perspective, this material has provided the first insights as to how thermal conductivity may be further reduced when heat is transported diffusively, as in glasses [6].

Half-Heusler materials, having the general formula XYZ (X = Ti, Zr; Y = Fe, Ni, Co; Z = Sn), are currently being developed as next-generation high-temperature thermoelectric materials (Figure 5.3.6c) [35]. So far, however, there are limitations in effectively doping the materials and sufficiently reducing the thermal conductivity. Nevertheless, it has been shown that defects (particularly vacancies) play a crucial role in determining the carrier concentration and electronic band structure [36], and novel alloying strategies have been proposed. One novel insight to the vacancy structures of half-Heusler phases has led to the discovery of a subclass of materials called double half-Heusler phases [37], which are expected to have lower thermal conductivities than the parent members.

5.3.4 Novel thermoelectric materials

The complex electronic and thermal transport already at play in thermoelectric materials can be further complicated if one or more of the ions are kinetically mobile. This is to say that some candidate thermoelectric materials are also ionic conductors. Typical examples include $Cu_{2-\delta}X$ (X = S, Se, Te) [38, 39], Ag_2Se [40], CuAgX (X = S/Se/Te)

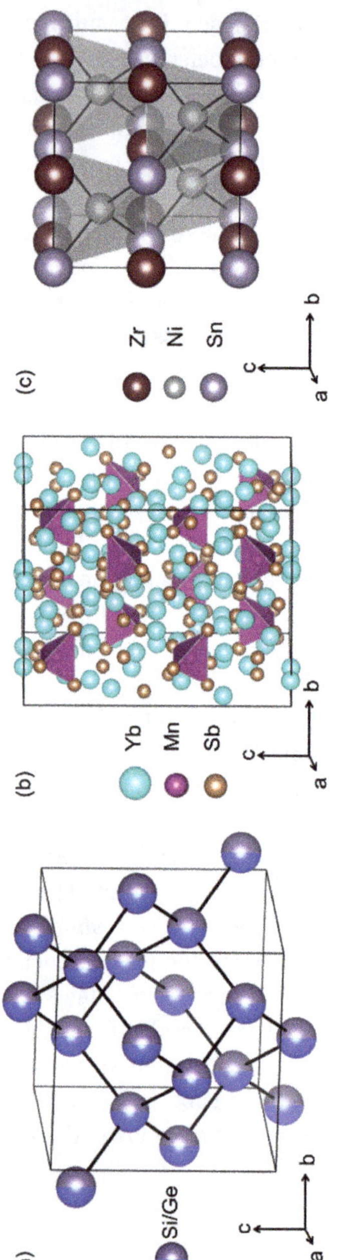

Figure 5.3.6: (a) Silicon-germanium solid solution where there is a random distribution of each atom on all crystallographic sites of the tetrahedral structure. (b) The many atoms unit cell of $Yb_{14}MnSb_{11}$ has structural complexity results in thermal transport glass-like at high temperatures. (c) The crystal structure of half-Heusler phases with general formula XYZ (e.g. ZrNiSn).

[41], Zn_4Sb_3 [42], Cu_5FeS_4 [43], Cu_7PSe_6 [44] and $Cu_{12}Sb_4S_{13}$ [45], where the cations (e.g. Cu, Ag, Zn ions) tend to be mobile at high enough temperatures. Although there have been reports of high zT in many of these materials, there have also been many cases of instability – with typical observations of metal whisker growth occurring during synthesis, measurement procedures and device operation. Consequently, there is an ongoing effort to find strategies that would enable the practical use of these thermoelectric materials with mobile ion substructures. One proposed strategy is to keep the voltage across the material below the thermodynamic stability limit [46, 47]. The otherwise optimistic electronic properties at mid-to-high temperatures, along with ultralow thermal conductivity values, makes these materials promising for future applications.

5.3.5 Summary

Thermoelectric materials epitomize the playing ground of solid-state chemistry: from complex crystal structures to complexities in bonding that give rise to unusual dynamic (vibrational) behaviors and/or unusual electronic structures. From an applications perspective, thermoelectric materials have helped mankind explore the solar system and may provide an opportunity to progress domestic energy efficiency. Such large goals should not overshadow other novel applications, for example that it is currently possible to wear a wristwatch that is thermoelectric powered and keep drinks cold in a car console using thermoelectric refrigeration. The future of wearable devices for personal cooling and power generation is also under investigation, but these technologies will likely encapsulate inorganic thermoelectric materials within a flexible polymeric support structure [48]. Such emerging markets as the "internet-of-things" further motivates the discovery of new complex thermoelectric materials, as well as the optimization of low and mid-temperature materials for practical devices.

References

[1] Snyder GJ, Toberer ES. Complex TE materials. Nat Mater, 2008, 7, 105–14.

[2] Ioffe AF. Semiconductor thermoelements and thermoelectric cooling. Phys Today, 1959, 12, 42. DOI: https://doi.org/10.1063/1.3060810.

[3] Vineis CJ, Shakouri A, Majumdar A, Kanatzidis MG. Nanostructured thermoelectrics: Big efficiency gains from small features. Adv Mater, 2010, 22, 3970–80. https://doi.org/https://doi.org/10.1002/adma.201000839.

[4] Slade TJ, Anand S, Wood M, Male JP, Imasato K, Cheikh D, Al Malki MM, Agne MT, Griffith KJ, Bux SK, Wolverton C, Kanatzidis MG, Snyder GJ. Charge-carrier-mediated lattice softening contributes to high *zT* in thermoelectric semiconductors. Joule, 2021, 5, 1168–82. DOI: https://doi.org/10.1016/j.joule.2021.03.009.

[5] Hanus R, Agne MT, Rettie AJE, Chen Z, Tan G, Chung DY, Kanatzidis MG, Pei Y, Voorhees PW, Snyder GJ. Lattice softening significantly reduces thermal conductivity and leads to high thermoelectric efficiency. Adv Mater, 2019, 31, 1900108. https://doi.org/https://doi.org/10.1002/adma.201900108.

[6] Hanus R, George J, Wood M, Bonkowski A, Cheng Y, Abernathy DL, Manley ME, Hautier G, Snyder GJ, Hermann RP. Uncovering design principles for amorphous-like heat conduction using two-channel lattice dynamics. Mater Today Phys, 2021, 18, 100344. https://doi.org/https://doi.org/10.1016/j.mtphys.2021.100344.

[7] Rowe DM. CRC Handbook of Thermoelectrics, CRC Press, Boca Raton, FL, 1995.

[8] Zeier WG, Zevalkink A, Gibbs ZM, Hautier G, Kanatzidis MG, Snyder GJ. Thinking like a chemist: intuition in thermoelectric materials. Angew Chem Int Ed, 2016, 55, 6826–41. https://doi.org/https://doi.org/10.1002/anie.201508381.

[9] Brod MK, Snyder GJ. Orbital chemistry of high valence band convergence and low-dimensional topology in PbTe. J Mater Chem A, 2021, 9, 12119–39. DOI: https://doi.org/10.1039/D1TA01273H.

[10] Tritt TM, Subramanian MA. Mater Res Soc Bull, 2006, 31, 188–198. https://doi.org/10.1557/mrs2006.44.

[11] Imasato K, Kang SD, Snyder GJ. Exceptional thermoelectric performance in $Mg_3Sb_{0.6}Bi_{1.4}$ for low-grade waste heat recovery. Energy Environ Sci, 2019, 12, 965–71. DOI: https://doi.org/10.1039/C8EE03374A.

[12] Xu B, Feng T, Agne MT, Zhou L, Ruan X, Snyder GJ, Wu Y. Highly porous thermoelectric nanocomposites with low thermal conductivity and high figure of merit from large-scale solution-synthesized $Bi_2Te_{2.5}Se_{0.5}$ hollow nanostructures. Angew Chem Int Ed, 2017, 56, 3546–51. DOI: https://doi.org/https://doi.org/10.1002/anie.201612041.

[13] Pei Y, LaLonde A, Iwanaga S, Snyder GJ. High thermoelectric figure of merit in heavy hole dominated PbTe. Energy Environ Sci, 2011, 4, 2085–89. DOI: https://doi.org/10.1039/C0EE00456A.

[14] Kumar A, Vermeulen PA, Kooi BJ, Rao J, van Eijck L, Schwarzmüller S, Oeckler O, Blake GR. Phase transitions of thermoelectric TAGS-85. Inorg Chem, 2017, 56, 15091–100. DOI: https://doi.org/10.1021/acs.inorgchem.7b02433.

[15] Zhou Z, Agne MT, Zhang Q, Wan S, Song Q, Xu Q, Lu X, Gu S, Fan Y, Jiang W, Snyder GJ, Wang L. Microstructure and composition engineering Yb single-filled $CoSb_3$ for high thermoelectric and mechanical performances. J Materiomics, 2019, 5, 702–10. https://doi.org/https://doi.org/10.1016/j.jmat.2019.04.008.

[16] Perez CJ, Wood M, Ricci F, Yu G, Vo T, Bux SK, Hautier G, Rignanese G-M, Snyder GJ, Kauzlarich SM. Discovery of multivalley Fermi surface responsible for the high thermoelectric performance in $Yb_{14}MnSb_{11}$ and $Yb_{14}MgSb_{11}$. Sci Adv, 2021, 7, eabe9439. DOI: https://doi.org/10.1126/sciadv.abe9439.

[17] Yu J, Xing Y, Hu C, Huang Z, Qiu Q, Wang C, Xia K, Wang Z, Bai S, Zhao X, Chen L, Zhu T. Half-heusler thermoelectric module with high conversion efficiency and high power density. Adv Energy Mater, 2020, 10, 2000888. https://doi.org/https://doi.org/10.1002/aenm.202000888.

[18] Bhandari CM, Rowe DM. Silicon–germanium alloys as high-temperature thermoelectric materials. Contemp Phys, 1980, 21, 219–42. DOI: https://doi.org/10.1080/00107518008210957.

[19] Goldsmid HJ, Douglas RW. The use of semiconductors in thermoelectric refrigeration. Br J Appl Phys, 1954, 5, 386–90. DOI: https://doi.org/10.1088/0508-3443/5/11/303.

[20] Witting IT, Chasapis TC, Ricci F., Peters M, Heinz NA, Hautier G, Snyder GJ. Adv Electron Mater, 2019, 5, 1800904. https://doi.org/10.1002/aelm.201800904.

[21] Poudel B, Hao Q, Ma Y, Lan Y, Minnich A, Yu B, Yan X, Wang D, Muto A, Vashaee D, Chen X, Liu J, Dresselhaus MS, Chen G, Ren Z. High-thermoelectric performance of nanostructured bismuth antimony telluride bulk alloys. Science, 2008, 320, 634–38. DOI: https://doi.org/10.1126/science.1156446.

[22] Wood M, Kuo JJ, Imasato K, Snyder GJ. Improvement of low-temperature zT in a Mg_3Sb_2–Mg_3Bi_2 solid solution via Mg-Vapor annealing. Adv Mater, 2019, 31, 1–5. DOI: https://doi.org/10.1002/adma.201902337.

[23] Peng W, Petretto G, Rignanese G-M, Hautier G, Zevalkink A. An unlikely route to low lattice thermal conductivity: Small atoms in a simple layered structure. Joule, 2018, 2, 1879–93. https://doi.org/https://doi.org/10.1016/j.joule.2018.06.014.

[24] Imasato K, Kang SD, Ohno S, Snyder GJ. Band engineering in Mg_3Sb_2 by alloying with Mg_3Bi_2 for enhanced thermoelectric performance. Mater Horizons, 2018, 5, 59–64. DOI: https://doi.org/10.1039/C7MH00865A.

[25] Kuo JJ, Yu Y, Kang SD, Cojocaru-Mirédin O, Wuttig M, Snyder GJ. Mg deficiency in grain boundaries of n-type Mg_3Sb_2 identified by atom probe tomography. Adv Mater Interfaces, 2019, 6, 1900429. DOI: https://doi.org/https://doi.org/10.1002/admi.201900429.

[26] Male J, Agne MT, Goyal A, Anand S, Witting IT, Stevanović V, Snyder GJ. The importance of phase equilibrium for doping efficiency: Iodine doped PbTe. Mater Horizons, 2019, 6, 1444–53. DOI: https://doi.org/10.1039/C9MH00294D.

[27] Heremans JP, Jovovic V, Toberer ES, Saramat A, Kurosaki K, Charoenphakdee A, Yamanaka S, Snyder GJ. Enhancement of thermoelectric of the electronic density of states. Science, 2008, 321, 1457–61. http://www.sciencemag.org/content/321/5888/554.abstract.

[28] Pei Y, Shi X, Lalonde A, Wang H, Chen L, Snyder GJ. Convergence of electronic bands for high performance bulk thermoelectrics. Nature, 2011, 473, 66–69. DOI: https://doi.org/10.1038/nature09996.

[29] Banik A, Roychowdhury S, Biswas K. The journey of tin chalcogenides towards high-performance thermoelectrics and topological materials. Chem Commun, 2018, 54, 6573–90. DOI: https://doi.org/10.1039/c8cc02230e.

[30] Hanus R, Guo X, Tang Y, Li G, Snyder GJ, Zeier WG. A chemical understanding of the band convergence in thermoelectric $CoSb_3$ skutterudites: Influence of electron population, local thermal expansion, and bonding interactions. Chem Mater, 2017, 29, 1156–64. DOI: https://doi.org/10.1021/acs.chemmater.6b04506.

[31] Vining CB, Laskow W, Hanson JO, Van Der Beck RR, Gorsuch PD. Thermoelectric properties of pressure-sintered $Si_{0.8}Ge_{0.2}$ thermoelectric alloys. J Appl Phys, 1991, 69, 4333. DOI: https://doi.org/10.1063/1.348408.

[32] Gurunathan R, Hanus R, Dylla M, Katre A, Snyder GJ. Analytical models of phonon–point-defect scattering. Phys Rev Appl, 2020, 13, 34011. DOI: https://doi.org/10.1103/PhysRevApplied.13.034011.

[33] Agne MT, Hanus R, Snyder GJ. Minimum thermal conductivity in the context of diffuson-mediated thermal transport. Energy Environ Sci, 2018, 11, 609–16. DOI: https://doi.org/10.1039/C7EE03256K.

[34] Brown SR, Kauzlarich SM, Gascoin F, Snyder GJ. $Yb_{14}MnSb_{11}$: New high efficiency thermoelectric material for power generation. Chem Mater, 2006, 18, 1873–77. DOI: https://doi.org/10.1021/cm060261t.

[35] Bos J-WG, Downie RA. Half-heusler thermoelectrics: A complex class of materials. J Phys: Condens Matter, 2014, 26, 433201. DOI: https://doi.org/10.1088/0953-8984/26/43/433201.

[36] Zeier WG, Schmitt J, Hautier G, Aydemir U, Gibbs ZM, Felser C, Snyder GJ. Engineering half-heusler thermoelectric materials using zintl chemistry. Nat Rev Mater, 2016, 1, 16032. DOI: https://doi.org/10.1038/natrevmats.2016.32.

[37] Anand S, Wood M, Xia Y, Wolverton C, Snyder GJ. Double half-heuslers. Joule, 2019, 3, 1226–38. DOI: https://doi.org/10.1016/j.joule.2019.04.003.
[38] Liu H, Shi X, Xu F, Zhang L, Zhang W, Chen L, Li Q, Uher C, Day T, Snyder GJ. Copper ion liquid-like thermoelectrics. Nat Mater, 2012, 11, 422–25. DOI: https://doi.org/10.1038/nmat3273.
[39] Di Liu W, Yang L, Chen ZG, Zou J. Promising and eco-friendly Cu_2X-based thermoelectric materials: progress and applications. Adv Mater, 2020, 32, 87–92. DOI: https://doi.org/10.1002/adma.201905703.
[40] Xiao C, Xu J, Li K, Feng J, Yang J, Xie Y. Superionic phase transition in silver chalcogenide nanocrystals realizing optimized thermoelectric performance. J Am Chem Soc, 2012, 134, 4287–93. DOI: https://doi.org/10.1021/ja2104476.
[41] Ishiwata S, Shiomi Y, Lee JS, Bahramy MS, Suzuki T, Uchida M, Arita R, Taguchi Y, Tokura Y. Extremely high electron mobility in a phonon-glass semimetal. Nat Mater, 2013, 12, 512–17. DOI: https://doi.org/10.1038/nmat3621.
[42] Snyder GJ, Christensen M, Nishibori E, Caillat T, Iversen BB. Disordered zinc in Zn_4Sb_3 with phonon-glass and electron-crystal thermoelectric properties. Nat Mater, 2004, 3, 458–63. DOI: https://doi.org/10.1038/nmat1154.
[43] Qiu P, Zhang T, Qiu Y, Shi X, Chen L. Sulfide bornite thermoelectric material: A natural mineral with ultralow thermal conductivity. Energy Environ Sci, 2014, 7, 4000–06. DOI: https://doi.org/10.1039/c4ee02428a.
[44] Weldert KS, Zeier WG, Day TW, Panthöfer M, Snyder GJ, Tremel W. Thermoelectric transport in Cu_7PSe_6 with high copper ionic mobility. J Am Chem Soc, 2014, 136, 12035–40. DOI: https://doi.org/10.1021/ja5056092.
[45] Lu X, Morelli DT, Wang Y, Lai W, Xia Y, Ozolins V. Phase stability, crystal structure, and thermoelectric properties of $Cu_{12}Sb_4S_{13-x}Se_x$ solid solutions. Chem Mater, 2016, 28, 1781–86. DOI: https://doi.org/10.1021/acs.chemmater.5b04796.
[46] Qiu P, Agne MT, Liu Y, Zhu Y, Chen H, Mao T, Yang J, Zhang W, Haile SM, Zeier WG, Janek J, Uher C, Shi X, Chen L, Snyder GJ. Suppression of atom motion and metal deposition in mixed ionic electronic conductors. Nat Commun, 2018, 9, 2910. DOI: https://doi.org/10.1038/s41467-018-05248-8.
[47] Qiu P, Mao T, Huang Z, Xia X, Liao J, Agne MT, Gu M, Zhang Q, Ren D, Bai S, Shi X, Snyder GJ, Chen L. High-efficiency and stable thermoelectric module based on liquid-like materials. Joule, 3, 2019, 1538–48. DOI: https://doi.org/10.1016/j.joule.2019.04.010.
[48] Nan K, Kang SD, Li K, Yu KJ, Zhu F, Wang J, Dunn AC, Zhou C, Xie Z, Agne MT, Wang H, Luan H, Zhang Y, Huang Y, Snyder GJ, Rogers JA. Compliant and stretchable thermoelectric coils for energy harvesting in miniature flexible devices. Sci Adv, 2018, 4, eaau5849. DOI: https://doi.org/10.1126/sciadv.aau5849.

5.4 Hydrogen storage materials

Holger Kohlmann, Leonhard Dorsch

The rising energy demand of modern societies in industrialized countries poses huge technical and social challenges. Key issues in this context are efficient energy storage and conversion. Hydrogen is a very interesting secondary energy source. Its energy content of 120.7 MJ kg^{-1} is higher than that of any hydrocarbon fuel. Hydrogen is the most abundant element in the universe, accounts for 17% of all atoms and 0.9% of the mass of the earth's lithosphere, hydrosphere, biosphere and atmosphere combined, i.e. hydrogen has almost limitless resources most of which are readily available (e.g. water). When burned in a combustion device or a fuel cell, it will produce only water as exhaust gas, making it environmentally friendly. On the road to CO_2 neutrality, many countries aim at increasing the use of hydrogen considerably (hydrogen roadmap of Europe [1], national hydrogen strategy of Germany [2], China Hydrogen Alliance [3]).

The so-called green, sustainable energy sources such as sunlight or wind for hydrogen production are key components of a hydrogen-based energy economy (Figure 5.4.1). The hydrogen economy has grown from a strategic dream to beginning commercial success nowadays. The decreasing cost of solar and wind technology and the growth of hydrogen infrastructure were important drivers of this development; major players in the field of hydrogen technology are China, Japan, South Korea and Germany [4]. One of the key challenges remains the storage of hydrogen. This contribution is about the chemistry behind hydrogen storage, gives the state-of-the-art for the year 2021 and points out major research directions for the near future and unresolved issues to be tackled. The focus is on *reversible* hydrogen storage.

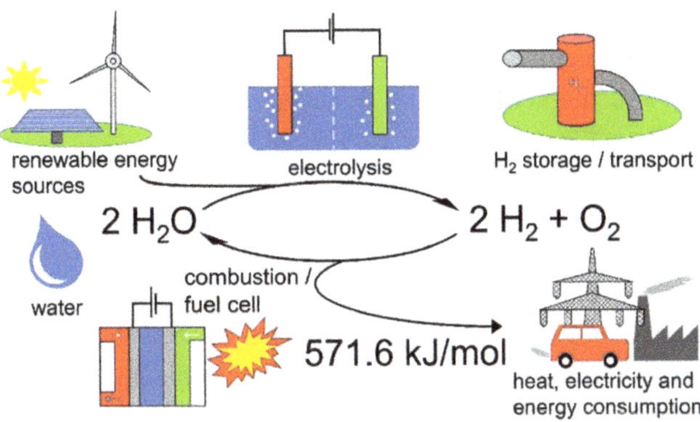

Figure 5.4.1: The hydrogen economy circle.

https://doi.org/10.1515/9783110798890-004

5.4.1 Principles of hydrogen storage and requirements

Hydrogen gas has a low density of 0.089 kg m^{-3} (= 0.089 g L^{-1}). The main purpose of storage methods is therefore to provide means of safely keeping hydrogen at higher densities and releasing it when needed. Hydrogen storage is based on different principles at an atomic scale. In the sequence of increasing interaction strength of hydrogen atoms or molecules with the storage material, these are:

- physical methods: no interaction; compressed H_2 gas or liquid; Chapters 5.4.2 and 5.4.3
- physicochemical methods: van der Waals interaction; H_2 adsorption (physisorption); Chapter 5.4.4
- chemical methods: chemical bonds; chemical compounds containing H atoms or hydride anions H^-; Chapter 5.4.5

The following requirements were set by the International Energy Agency for on-board hydrogen storage systems (DOE Technical Targets for Onboard Hydrogen Storage for Light-Duty Vehicles, ultimate technical system targets) [5]:

1) gravimetric storage capacity ≥ 6.5 mass% H_2
2) volumetric storage capacity ≥ 5% (≥ 0.05 kg L^{-1})
3) operating temperatures $233 \leq T \leq 333$ K
4) delivery pressure 0.5 MPa $\leq p \leq$ 1.2 MPa
5) cycleability ≥ 1500 cycles
6) system fill time ≤ 300 s
7) further issues: abundancy of elements, toxicity, environmental issues for production and waste management, recycling and second-life use.

Safety considerations for devices storing hydrogen also need to consider the broad range of explosive mixtures with air (4–76%), high diffusion rates through leaks and embrittlement of many metals and alloys.

Requirements 1 and 2 are determined by the chemical composition and the density of the storage material and the containment. Requirements 3 and 4 may be tuned by optimizing thermodynamic and kinetic properties of storage materials. For the uptake of hydrogen in a solid, the change in entropy is almost exclusively the standard entropy of hydrogen (−139.6 J K^{-1} mol^{-1}). The Gibbs free energy at ambient conditions may thus be easily calculated as $\Delta G = \Delta H - T\Delta S = \Delta H - (298\ \text{K}\ (-139.6\ \text{J K}^{-1}\ \text{mol}^{-1})) = \Delta H + 38.9\ \text{kJ mol}^{-1}$. This means that a metal hydride with a formation enthalpy of −38.9 kJ mol^{-1} is in perfect equilibrium with an atmosphere of 1 bar of hydrogen at 298 K, i.e. such a compound allows for hydrogen uptake and release very close to ambient conditions. This value of $\Delta H = -38.9$ kJ mol^{-1} is thus a thermodynamic benchmark for hydrogen storage materials.

Many of the above requirements may be tested by pressure-composition isotherms (p–c isotherms), which may be constructed from isochoric (increasing known doses of hydrogen gas while measuring pressure and hydrogen uptake) or isobaric measurements (keeping pressure constant during absorption or desorption). The hydrogen concentration at the end of the plateau defines the maximum hydrogen uptake; the length of the plateau region determines the capacity useful for reversible storage. For example, CaSi exhibits a hydrogen capacity of around 1.5%, while only about 1.1% (at 473 K) are accessible for reversible storage (Figure 5.4.2, left). From the slope of van't Hoff plots $\ln p_{eq}$ vs. $1/T$ (Figure 5.4.2, right) using the equation $\ln p_{eq} = \Delta H/RT - \Delta S/R$ (R: gas constant) the enthalpy of metal hydride formation may be determined. The *p–c* isotherms not only give information on the hydrogen capacity, but also on suitable operating temperatures and pressures and on reversibility by measuring absorption and desorption cycles.

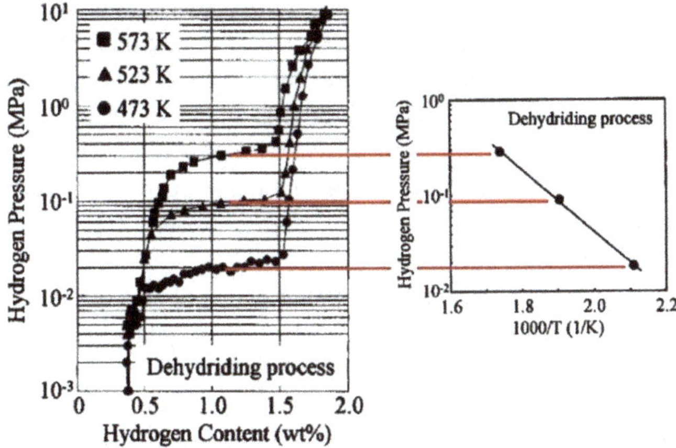

Figure 5.4.2: Hydrogen desorption isotherms (*p–c* isotherms) of CaSi (left) and according van't Hoff plot; reprinted from M. Aoki, N. Ohba, T. Noritake, S. Towata, Reversible hydriding and dehydriding properties of CaSi: Potential of metal silicides for hydrogen storage, *Appl. Phys. Lett.* **2004**, *85*, 387–388, with the permission of AIP Publishing; red lines added for reasons of clarity.

5.4.2 Storage as gaseous hydrogen: gas tanks, pipelines and caverns

Gas tanks up to 20 bar were already in use in the mid-nineteenth century, high pressure carbon dioxide tanks for beverages were first introduced in the 1870s. Today, gas tanks are usually divided into four types. Type I is a simple metal tube with round caps. It is the cheapest variant with pressures up to 300 bar and the *de facto* standard for industry and lab-scale applications. Type II introduces a fiber resin wrapping around the thick metallic hoop. It is preferred for stationary applications at higher pressures. Type III consists of a thin metal body fully wrapped in a fiber resin composite, whereas type IV tanks are made from a fiber resin wrapped polymeric liner with an integrated metallic port. Currently available hydrogen powered cars and busses are equipped with type III or IV tanks working at 350 or 700 bar. Type IV vessels allow non-cylindrical shapes as well [6–8].

For using the vastly existing natural gas hardware and infrastructure for hydrogen, some differences need to be considered, e.g. the higher gravimetric (33.3 kWh g^{-1}) but lower volumetric (3 Wh L^{-1}) energy density of hydrogen compared to natural gas. The low molecular weight of hydrogen also requires new compressors and drivers as conventional centrifugal compressors would need to spin three times as fast to achieve the same level of compression [9, 10].

Underground hydrogen storage in geological structures like natural and man-made salt caverns, aquifers and depleted oil and gas fields are ways of large-scale stationary storage. The experiences in the storage of hydrogen are scarce and currently limited to three salt caverns. The oldest one in Teesside (UK) has been in service since 1972 and consists of three caverns in a depth of 400 m storing up to 10^6 m^3 hydrogen at 50 bar. It supplies nearby ammonia and methanol plants. The other two are located at the US Gulf Coast in Texas, operating since 1983 and 2007 respectively, the former with a capacity of 3×10^7 m^3 in a cavern 850 m below the ground, serving the industry in Texas and Louisiana. In contrast to the very limited experience with underground hydrogen storage, natural gas underground storage is over 100 years old and currently performed at 680 facilities worldwide [11].

The facilities themselves are artificial accumulations of gas in depths of several hundred to over 1000 meters, consisting of the working gas that gets withdrawn and injected and the cushion gas which makes up around 30% of the injected volume and always resides there. The cushion gas prevents the inflow of water by providing the minimum pressure and also aids in the injection of the working gas. Advantages compared to surface facilities include safety, space concerns, low costs and the vast availability of suitable geological structures around the globe. Challenges involve the transformation of hydrogen into methane by bacteria as previously observed in coal gas (containing 50–60% H_2) facilities, reactions with the mineral matrix, reduced durability of metal hardware exposed to hydrogen and tightness against penetration. For the latter, aquifers and hydrocarbon deposits may be advantageous, as the pores of the rocks are filled with water, which shows a low solubility for hydrogen [11].

For long-distance hydrogen transport, pipelines are the most cost-effective method compared to ships, trucks or trains, with the exception of intercontinental transport. The typical diameter of a gas pipeline lies between 50–120 cm and the operating pressure ranges from 50 to 80 bar [9]. Like for the underground storage, the experience with transporting hydrogen compared to natural gas are limited. As of 2016, there were roughly 1600 km of hydrogen pipelines in Europe and around 5000 km worldwide [12].

5.4.3 Storage as liquid hydrogen: cryotanks

When the size of a storage device is decisive, liquid hydrogen with a three times higher energy density than hydrogen at 350 bar at room temperature is an interesting alternative. However, liquefaction at 20 K requires a lot of time and energy, which can be equivalent to 40% of the stored content, as compared to about 10% for compressed hydrogen [8]. Long-term storage of liquid hydrogen suffers from evaporation (boil off) by heat leakage through the tank, sloshing, flashing and ortho-para-conversion. Sloshing describes the motion of a liquid inside a moving vessel, leading to conversion into thermal energy. This can be tackled with anti-slosh baffles. Flashing can occur when liquid hydrogen under higher pressure is filled into a lower pressure vessel. When cooling hydrogen, the equilibrium of ortho and para-hydrogen changes drastically (298 K: 25% para, 0 K: 100% para). The thermodynamically driven conversion of ortho to para-hydrogen is exothermic and slow, and therefore a constant source of heat, causing enhanced boil off. For long-term storage, catalysts of the ortho to para-conversion like active charcoal, tungsten, or nickel paramagnetic oxides should be used [13].

The key to turn liquefied hydrogen into a viable energy storage option lies therefore not only in a strong insulation, but also in an appropriate boil off management. Storage vessels are composed of an inner (vacuum) insulated pressure vessel and an external protective jacket. The space in between can be filled with perlite, or other insulations to reduce thermal conductivity [8]. There are several options to enhance the efficiency like active cooling, buffering with metal hydrides and using the vented hydrogen to power engines or other systems. Furthermore, regasification offers vast amounts of cryogenic exergy that can be used to maximize the efficiency of the whole process [10].

Compared to the proposed use, cases of liquefied hydrogen like the hydrogen transport via ships between Australia and Japan and the ambitions to use it as the fuel for zero-emission commercial aircrafts, the usage of cryogenic hydrogen nowadays is limited. In space programs, it works in conjunction with liquid oxygen as rocket fuel and there are some prototypes and small-scale projects [8, 10].

Another option is the storage as cryocompressed hydrogen. The idea is to minimize the loss by evaporation while achieving a higher energy density than in room temperature compressed tanks. The vessels used are designed to withstand pressures up to 300 bar at ambient temperature and work at cryogenic temperatures of around 20 K. The higher operating pressures extend the time before hydrogen needs to be boiled off and can lead to higher storage densities than liquid hydrogen. The option to fill the tanks either with liquid hydrogen at 20 K, gaseous hydrogen at room temperature or anything in between is also an additional benefit [8].

5.4.4 Storage as adsorbed (physisorbed) hydrogen: porous materials

Physisorption of hydrogen is based on van der Waals interactions of H_2 molecules and an adsorbent surface. The process of hydrogen loading and release is usually fully reversible with an excellent cycle lifetime and it is fast – big advantages for reversible hydrogen storage. However, physisorption suffers from low adsorption enthalpies of typically <10 kJ mol^{-1}. This yields in low gravimetric densities of typically <3 mass%, which cannot compete with systems based on chemisorption (*vide infra*). Higher storage capacities may only be achieved at low temperature (and/or high gas pressure), i.e. the requirement for storage capacity and that for working conditions near ambient may not be fulfilled at the same time. For adsorption, high specific surface area is advantageous. According to Chahine's rule, around 1 mass% hydrogen are adsorbed on 500 m^2 g^{-1} surface area at 77 K and pressures above 20 bar. It is a good approximation for activated carbon and MOFs (see below) [14]. The drawback of high surface area is the usually accompanying low density, which results in low-volumetric hydrogen storage densities.

Zeolites (see also Chapter 7.3) are well-known, relatively low-cost, porous aluminosilicates, which may adsorb hydrogen in pores and channels. Storage capacities are small, typically <0.3 mass% at 273 K and <2% at 77 K [15]. Carbon nanomaterials typically have low-to-moderate storage capacity, e.g. for carbon nanofibers 0.7 mass % H_2 at 10 MPa, for single-walled carbon nanotubes <1 mass% at room temperature and <5 mass% at 77 K [15]. Unfortunately, the reproducibility of results, especially for hydrogen storage capacities, for carbon-based nanomaterials is rather poor [16]; this is mainly due to the intrinsic difficulty in characterizing these often complex mixtures or composites without a crystalline structure. Therefore, it is very difficult to judge their potential for reversible hydrogen storage. Recent strategies of research are increasing the micropore volume in activated carbon, optimizing size and packing of carbon nanotubes and the metal decoration of materials to increase coordination sites for hydrogen [14, 16, 17].

Metal-organic frameworks (MOFs) (see also Chapter 7.4) are coordination polymers built up from metal ions bridged by organic linkers. They often have porous structures, some with surface areas (Brunauer–Emmett–Teller (BET) areas) exceeding 6000 m^2 g^{-1}. High hydrogen capacities around 10% are only achieved far from ambient conditions (e.g. 9.95% at 77 K and 5.6 MPa H_2 pressure for NU-100 [18]). The ultramicroporous NU-1501-Al shows 14.0 mass% hydrogen capacity when changing between 77 K and 100 bar to 160 K and 5 bar, probably because of high BET areas of 7310 m^2 g^{-1} and 2060 m^2 L^{-1} [19]. The choice of metals is important, and usually a compromise between strong interaction to hydrogen and the weight has to be found. While the former favors highly charged transition or rare-earth metals, the latter favors lithium, sodium, magnesium and aluminum; a good compromise are often iron, zinc or heterometallic MOFs. The high capacities of Ni_2(*m*-dobdc) (*m*-dobd = 4,6-dioxido-1,3-benzenedicarboxylate) were explained by open metal cation sites, which show strong interaction with hydrogen molecules. Current research concentrates on the optimization of pore sizes (around 1 nm), e.g. by catenation, and increasing adsorption enthalpies (to ideally around 25 kJ mol^{-1}). Using flexible organic linkers may enable the framework structure to "breathe" upon exchange of guest molecules and show sponge-like behavior with contracted and expanded states. *M*-URJC-n (*M* = Co, Cu, Zn) is based on the 5,5′-thiodiisophthalic acid linker (H_4TBTC) and shows a special type of this feature, a gate-opening type adsorption mechanism at low pressures. This is of interest for gas separation since only hydrogen can introduce the "breathing" [19]. MOFs may also be used as scaffold for metal hydrides in nanoconfinement [19].

5.4.5 Storage as absorbed (chemisorbed) hydrogen: metal hydrides

In contrast to physisorption (Chapter 5.4.4), the absorption of hydrogen breaks the H–H bond in the H_2 molecule and incorporates H atoms or hydride anions (H^-) into the solid, thus forming a metal hydride. The interaction between hydrogen and metal atoms is much stronger than in the case of physisorbed H_2 molecules, which lifts the necessity for cooling or high pressures to achieve good storage capacities. Volumetric and gravimetric densities as well as reversibility of hydrogen uptake vary considerably, since metal hydrides are a vast and very diverse class of compounds. Binary metal hydrides are usually not suitable for hydrogen storage, some because of low gravimetric capacities (transition metals and lanthanides), those with high capacities for reasons of poor reversibility (AlH_3: 10.1 mass%, BeH_2: 18.2 mass%) or high desorption temperatures (LiH: 12.7 mass%). MgH_2 with 7.6 mass% is an exception and will be discussed in Chapter 5.4.5.2.

5.4.5.1 Transition metal-based intermetallic and complex hydrides

$LaNi_5$ and related compounds with partial substitution are well-known for electrochemical hydrogen storage in nickel metal-hydride (Ni-*M*H) rechargeable batteries (see also Chapter 5.1) with storage capacities of typically 80 Wh kg^{-1} and 250 Wh L^{-1}. The reversible intercalation (charging) and deintercalation (discharging) of hydrogen in $\mathit{Mm}Ni_{3.55}Mn_{0.4}Al_{0.3}Co_{0.75}$ (= *M*, *Mm* being Mischmetall, a mixture of rare-earth elements) occurs in an aqueous potassium hydroxide solution according to *M*H + NiOOH = *M* + $Ni(OH)_2$. Other materials have also been employed for *M*, e.g. *RE*-Mg-Ni compounds. Ni-*M*H batteries were widely used since the 1990s in portable electronic devices, but were largely replaced by lithium-ion batteries except for small appliances. Advantages compared to rechargeable lithium-ion batteries are the price, higher charge currents, ease of recycling, better energy throughput per lifetime and safety issues (overheating), however they have lower capacity. $LaNi_5$ is also used for hydrogen gas storage because of favorable thermodynamic values for hydrogen absorption (formation enthalpy of –30 kJ mol^{-1} H_2) upon formation of the hydride $LaNi_5H_6$ (Figure 5.4.3). Low desorption pressure slightly above 1 bar, fast absorption and desorption kinetics, easy activation and good cycling stability are almost ideal, but low gravimetric storage capacity (1.5%) limits its use to stationary applications [20]. Forklifts and other heavy duty vehicles are an exception, because they need heavy ballast, turning the disadvantage of low gravimetric storage capacity into an advantage.

Many ternary or quaternary transition metal hydrides contain homoleptic hydridometallate complexes, for example $[FeH_6]^{4-}$ in Mg_2FeH_6 or $[PdH_2]^{2-}$ in Na_2PdH_2. Mg_2Ni reacts to Mg_2NiH_4 with a reaction enthalpy of –64 kJ mol^{-1} H_2, which contains 18-electron complexes $[NiH_4]^{4-}$ [20]. This reaction is reversible and the basis for one of the early hydrogen storage materials with a gravimetric storage capacity of 3.7%. Unfortunately, many other complex hydrides, e.g. Mg_2FeH_6 with 5.5% gravimetric capacity, do not react reversibly upon hydrogen release. There are numerous further examples for intermetallic compounds with transition metals of AX, AX_2, AX_3 and AX_5 composition and their substitutional derivatives, which reversibly take up hydrogen [21]. They exhibit a very rich structural chemistry and sometimes further interesting properties (e.g. switchable mirrors, gas separation). For hydrogen storage, in general, they offer great volumetric capacity, suitable operating temperatures, delivery pressure, cycle lifetime and system fill time, but poor gravimetric density and often include less-abundant elements and critical resources.

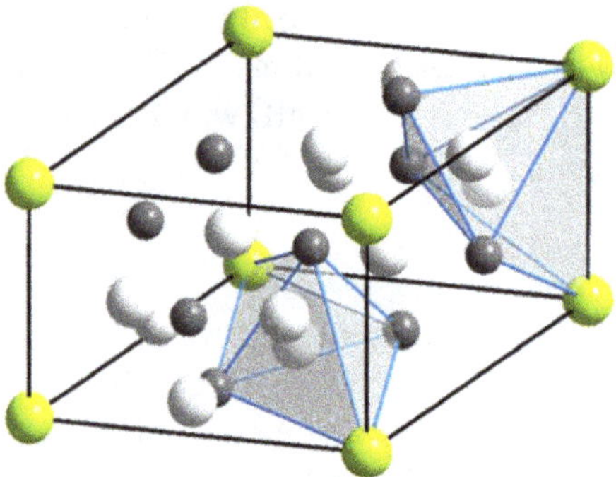

Figure 5.4.3: Crystal structure of $LaNi_5H_6$ (La atoms yellow, nickel atoms grey, hydrogen atoms white; *c* axis up) with hydrogen occupying distorted La_2Ni_3 trigonal bipyramids (simplified 2-site model for hydrogen [21]).

5.4.5.2 Magnesium-based hydrogen storage materials

Magnesium hydride is a white powder with the rutile-type structure, which can be synthesized by direct hydrogenation of magnesium powder. The reaction is rather sluggish, sensitive to the surface state and particle size of magnesium powder and reaction times vary between a few and some hundred hours; often successive regrinding between hydrogenation reaction steps is needed for complete reaction [21]. This is in part due to autopassivation, where a thin layer of MgH_2 on the surface of magnesium particles works as a huge diffusion barrier for hydrogen. Accordingly, ball milling is an efficient way to accelerate the hydrogenation of magnesium [22]. Because of the high thermodynamic stability (formation enthalpy of -74 kJ mol^{-1} H_2) the dehydrogenation requires rather high temperatures around 600 K and it is slow. Many magnesium-based intermetallic compound have been investigated with the purpose of improving hydrogen storage characteristics, e.g. MMg_2 (see above), MMg_3, M_5Mg_{41} and M_2Mg_{17}. Reduction of particle size can help in both lowering the desorption temperature and improving the kinetics (see Chapter 5.4.5.7). Agglomeration, however, destroys this effect after some cycles. Nanoconfinement of the magnesium nanoparticles by surrounding with an air-repelling, hydrogen-permeable polymer matrix [23] or in porous materials [24] prevents this detrimental effect. Other strategies are nanocomposites of magnesium with very small amounts of nickel or palladium and graphene or graphene oxide; while retaining a good gravimetric hydrogen density of 6%, hydrogenation was achieved around 370 K and dehydrogenation around 520 K, both within minutes and for tens or hundreds of cycles [22]. Many other additives were tried in order to improve the reaction kinetics, e.g. Ni, Co, Fe, Ti, TiF_3, Nb_2O_5; sometimes they are called catalysts although their role is not always clear. The formation of intermetallic phases such as $Mg_{0.9}Ti_{0.1}$, hydrides such as TiH_2 or Mg_2FeH_6, oxides such as $Mg_xNb_{1-x}O$ or hydride fluoride such as $MgH_{2-x}F_x$, were identified as intermediates of the hydrogenation and dehydrogenation reaction. In general, magnesium-based hydrides offer great gravimetric capacity, are based on abundant materials, have medium system fill time and volumetric capacity, but often poor operating temperature and delivery pressure.

5.4.5.3 Alanates and boranates

Group 13 elements show a very rich chemistry with hydrogen. Boron and aluminum form monovalent tetrahedral anions BH_4^- (tetrahydridoborate or boranate, sometimes called borohydride) and AlH_4^- (tetrahydridoaluminate or alanate), which may be combined with a broad variety of cations. Compounds with lithium and sodium are well-known and used as reducing and hydriding agents. Because of their high gravimetric hydrogen content, they are also of interest for hydrogen storage. $NaAlH_4$ (7.5 mass%) decomposes in three steps: $6\ NaAlH_4 \rightarrow 2\ Na_3AlH_6 + 4\ Al + 6\ H_2 \rightarrow 6\ NaH + 6\ Al + 9\ H_2 \rightarrow 6\ Na + 6\ Al + 12\ H_2$ around 500 K, 530 K and 710 K, respectively. Because these reactions suffer from sluggish kinetics of hydrogen release and uptake, catalysts such as metals, metal oxides, metal halides and carbon-based materials, are used [25]. Treatment of $NaAlH_4$ with 2 mol% $TiCl_3$ or $Ti(O\text{-}n\text{-}C_4H_9)_4$ in ether or toluene lowers the hydrogen desorption temperatures by about 80 K and considerably improves the reaction kinetics [25, 26]. Addition of carbon nanomaterials lowers the activation energies for all three hydrogenation steps considerably

(128→88, 247→99, 267→136 kJ mol^{-1} [25]). Issues to be tackled for these systems are the high working pressures of around 10 MPa and working temperatures well above 400 K.

Tetrahydridoborates show a very rich crystal chemistry and may be synthesized by metathesis reaction of readily available $LiBH_4$ or $NaBH_4$ with metal chlorides in a ball mill, e.g. $MnCl_2 + 2\ LiBH_4 \rightarrow Mn(BH_4)_2 + 2\ LiCl$ [27]. Alkali and alkaline metal tetrahydridoborates MBH_4 and $M'(BH_4)_2$ are either ionic solids or more covalent like $Be(BH_4)_2$ and $Al(BH_4)_3$. The latter has a very high gravimetric hydrogen content of 16.8%, but is not easy to handle due to pronounced air sensitivity, volatility, pyrophoric character and tendency for the release of diborane. One of the main problems of tetrahydridoborates as hydrogen storage materials is poor reversibility. Rehydrogenation often requires drastic conditions and has sluggish kinetics. Moreover, some materials like $LiBH_4$ are thermodynamically too stable and require high hydrogen desorption temperatures. These problems may be overcome by using so-called reactive hydride composites (simply a mixture of two phases and often not a composite), which allow tuning of thermodynamic and kinetic characteristics at the same time. The reaction $2\ LiBH_4 + MgH_2 \rightleftharpoons MgB_2 + 2\ LiH + 4\ H_2$ is reversible with 10% gravimetric storage capacity, at least at high temperatures. For mixtures of $LiBH_4$, $Mg(NH)_2$ and LiH, the working temperatures for reversible hydrogen storage are below 370 K [14]. The hydrolysis of sodium tetrahydridoborate, $NaBH_4$, readily produces hydrogen gas, but it is not reversible. The products, $NaBO_2$ hydrates, may be regenerated by reaction with Mg, MgH_2 or Mg_2Si [28].

5.4.5.4 Amides, imides, nitrides: hydrogen storage in metal-N-H systems

Hydrogen may react with nitrides, thereby forming imides and amides, e.g. $Li_3N + 2\ H_2 \rightleftarrows Li_2NH + LiH + H_2 \rightleftarrows LiNH_2 + 2\ LiH$ with a favorable capacity of 9.3 mass%. One of the peculiarities of this disproportionation is that positively polarized ($H^{\delta+}$ in Li_2NH and $LiNH_2$) and negatively polarized hydrogen ($H^{\delta-}$ in LiH) are produced. Hydrogen release often has high activation energy and poor kinetics. Side reactions release NH_3, especially for increased cycle life, causing safety issues and diminishing capacities [29]. By addition of small amounts of $TiCl_3$ and ball milling, these problems can be avoided. The addition of some silicon powder may reduce the desorption temperature by 200 K with respect to pure Li_3N by formation of Li_2Si [29]. Many other additions like Al, $AlCl_3$, C, $MgCl_2$, TiN, BN, MgH_2, CaH_2, TiH_2 (sometimes falsely called doping) were used with more or less success for lowering the desorption temperature and suppress unwanted side reactions.

Stoichiometric mixtures of lithium amide with magnesium hydride or lithium hydride with magnesium amide show favorable characteristics, according to $2\ LiNH_2 + MgH_2 \rightleftarrows Li_2Mg(NH)_2 + 2\ H_2 \rightleftarrows Mg(NH_2)_2 + 2\ LiH$ with 5.5 mass%. Although thermodynamics is favorable with $\Delta H = 41.6$ kJ mol^{-1} for the formation of the imide, kinetics is poor and requires temperatures above 480 K. Again, the addition of small amounts of hydrides (KH, $LiBH_4$, MgH_2) or halides ($CaCl_2$, $CaBr_2$, VCl_3) is helpful. The extension to mixtures with three main constituents offers even higher combinatorial potential, however, at the cost of complexity. The system $LiNH_2/MgH_2/LiBH_4$ produces a multitude of phases, e.g. $Li_4BN_3H_{10}$, $Li_2Mg(NH)_2$, $Li_4BH_4(NH_2)_3$, $Li_2Mg_2(NH)_3$, $Li_2BH_4NH_2$ just to name a few. While kinetics are improved for some compositions, hydrogen capacities are somewhat lowered. Further systems of interest include the respective ones with $NaNH_2$ instead of $LiNH_2$ and transition metal compounds like $K_2[Zn(NH_2)_4]/LiH$ [14, 22, 29]. None of the hydrogen-storage materials based on nitrides, imides and amides fully meet the requirements stated above (Chapter 5.4.1) with the biggest challenge in most cases being kinetics and cycle stability.

5.4.5.5 Zintl phases and their hydrides

Zintl phases consist of an alkali or alkaline earth metal combined with an electronegative p-block element and contain polyanions with structures similar to isoelectronic elements. As such, Zintl phases bridge the gap between ionic salts and intermetallic compounds. There are two options to incorporate hydrogen into a Zintl phase. In interstitial hydrides, hydrogen is exclusively coordinated to the metal atoms, while polyanionic hydrides feature covalent bonds between hydrogen atoms and the polyanionic substructure. Some Zintl phases like CaSi exhibit both types simultaneously. CaSi contains Si^{2-} chains (Figure 5.4.4, left) and incorporates hydrogen in Ca_4 tetrahedra and covalently bound to silicon atoms. In the hydride $CaSiH_{4/3}$ three silicon polyanion chains condense to form a ribbon-like hydrogen-decorated polyanion (Figure 5.4.4, right). The *p–c* isotherms show a promising plateau pressure for reversible hydrogen storage; however, storage capacities are low (Figure 5.4.2) [30–32].

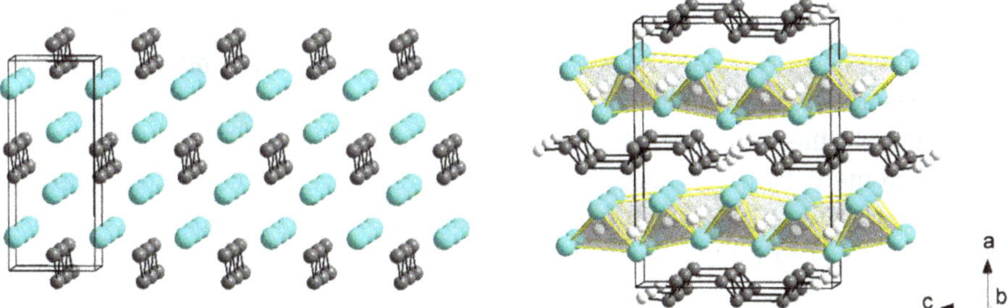

Figure 5.4.4: The crystal structures of CaSi (left) and γ-$CaSiH_{4/3}$ (right, H atoms bound covalently to Si atoms and H atoms on interstitial sites inside Ca_4 tetrahedra); Ca: turquoise, Si: grey, H: white.

The reaction from KSi to $KSiH_3$ takes place at 373 K and 50 bar H_2 while the complete dehydrogenation is achieved at 414 K in 1 bar H_2. With a comparatively high hydrogen content of 4.3 wt% and such mild reaction conditions it appears to be a good candidate for a Zintl phase hydrogen storage material [33]. The catch, however, is the extreme reactivity of KSi with moisture or air, rendering it unfit for any kind of safe application. Analogous compounds NaSi and LiSi are less reactive and lighter, but do not show reversible hydrogen uptake.

5.4.6 Storage in molecules

Hydrogen may be used to produce other basic chemicals and energy carriers such as methanol (power to gas) or ammonia. These processes are only considered here if hydrogen is released again, because otherwise this is not hydrogen storage. Ammonia has a high storage capacity (17.6 mass% H_2), well-known synthesis conditions and established infrastructure. Drawbacks are the efficiency of hydrogen release, which needs catalysts, e.g. transition metals, metal imides and amides, toxicity and environmental issues. For methanol (12.6 mass% H_2) and formic acid, there are similar problems.

Gravimetric hydrogen densities are huge in ammonia borane, NH_3BH_3 (19.6 mass%), being isoelectronic to ethane, and hydrazine borane, $N_2H_4BH_3$ (15.5 mass%); the decomposition temperature is low (around 370 K), but the release of hydrogen is accompanied by formation of diborane (B_2H_6), borazine ($B_3N_3H_6$) and ammonia (NH_3) for ammonia borane, and hydrazine (N_2H_4) and ammonia (NH_3) for hydrazine borane, causing severe safety problems due to emission of toxic gases. Moreover, reversibility is rather poor and both ammonia and hydrazine borane are air-sensitive solids, the latter an explosive by shock. Attempts to tame these hydrogen-rich compounds revived interest in the long known amidoboranes $M(NH_2BH_3)_n$ and hydrazinidoboranes $M(N_2H_3BH_3)_n$ with $n = 1$ or 2, M = alkali or alkaline earth metal [34]. They provide even lower desorption temperatures than ammonia or hydrazine borane and purer hydrogen evolved, however, at the cost of lower, but still rather high gravimetric densities (e.g. 12% for $Mg(NH_2BH_3)_2$ and 10.6% for $Mg(N_2H_3BH_3)_2$ [34]). The major drawback of all these H-B-N materials and their metal salts are the irreversibility of dehydrogenation. Rehydrogenation is not possible, which severely impairs any serious application in reversible hydrogen storage.

For aromatic hydrocarbons, the infrastructure is already present, there are no high gas pressures or temperatures, and storage capacities are good, e.g. cyclohexane – benzene with 7.1 mass%, methylcyclohexane – toluene with 6.1 mass%, naphthalene-decaline with 7.3 mass%. However, kinetics of hydrogenation and dehydrogenation are poor and catalysts are costly due to the presence of noble metals at elevated temperatures [15].

5.4.7 Conclusion: current and future challenges of hydrogen storage

Hydrogen storage may be realized as compressed gas (5.4.2), liquid hydrogen (5.4.3), physisorbed hydrogen molecules (5.4.5) or chemisorbed hydrogen atoms (5.4.5). Chemistry is mainly involved in the latter two, by the synthesis of new sorbent materials and in the control of hydrogen uptake and release in the storage medium. The examples in this non-comprehensive overview clearly show that up to date no hydrogen storage system meets all requirements (as stated in 5.4.1), or is fundamentally superior to all others; some, however, have severe principal limitations making it unlikely that they will succeed on a large scale. It is quite likely that in the near future a variety of storage methods (5.4.2–5.4.6) will further develop in parallel and the demand as well as the ability to produce such devices economically on a large scale will be decisive in the long run. It is also very important to put the seven requirements for hydrogen storage (Chapter 5.4.1) into perspective, because they were developed for the onboard hydrogen storage for light-duty vehicles. There are many other applications, for which such requirements will be different, sometimes to the opposite. An example is the forklift and other heavy-duty vehicles (5.4.5.1), where heavy instead of light storage materials are advantageous. Volume and gravimetric capacity also plays much less of a role for stationary hydrogen storage, e.g. for private households.

On the side of materials development, the continuous improvement of existing and exploration of new compounds and materials will certainly go on. In this context, the investigation of reaction pathways has become a very useful method and is expected to show a growing impact. In-situ methods for time-resolved studies are gaining importance both for the synthesis of storage materials and to study the basic steps of hydrogen uptake and release, revealing valuable insights into intermediate products, kinetics and reversibility [35]. On the applied side, the hydrogen infrastructure will have to develop further for a sustainable hydrogen economy. Pipelines, filling stations, large-scale stationary storage, rail or road based large containers and transport ships will get more and more attention. If the future brings larger demand for small-scale devices, e.g. in cars, the mass production will also put new issues at stake, e.g. recycling and second life of materials or the criticality of resources (Chapter 14).

After considering these topics of applied chemistry for hydrogen storage, we should not forget the fascinating basic chemistry of the materials involved. Hardly any other class of substances is as diverse as metal hydrides in terms of chemical bonds, structural and physical properties. The chemistry of hydrogen storage materials will certainly hold exciting chemistry in store for the future, both for fundamental and applied research.

References

[1] European Commission, A hydrogen strategy for a climate-neutral Europe, 2020; https://ec.europa.eu/energy/sites/ener/files/hydrogen_strategy.pdf, accessed 13 May 2021.
[2] Bundesministerium für Wirtschaft und Energie, The National Hydrogen Strategy, 2020; https://www.bmwi.de/Redaktion/EN/Publikationen/Energie/the-national-hydrogen-strategy.pdf?__blob=publicationFile&v=6, accessed 13 May 2021.
[3] China Hydrogen Alliance; http://www.h2cn.org.cn/en/about.html, accessed 13 May 2021.
[4] Sazali N. Int J Hydrogen Energy, 2020, 45, 18753–71.
[5] Office of Energy Efficiency & Renewable Energy, DOE Technical Targets for Onboard Hydrogen Storage for Light-Duty Vehicles; https://www.energy.gov/eere/fuelcells/doe-technical-targets-onboard-hydrogen-storage-light-duty-vehicles; accessed 13 May 2021.
[6] Barthelemy H. Int J Hydrogen Energy, 2012, 37, 17364–72.
[7] Robertson IM, Sofronis P, Nagao A, Martin ML, Wang S, Gross DW, Nygren KE. Metall Mater Trans, 2015, 46B, 1085–103.
[8] Barthelemy H, Weber M, Barbier F. Int J Hydrogen Energy, 2017, 42, 7254–62.
[9] Wang A, Jens J, Mavins D, Moultak M, Schimmel M, van der Leun K, Peters D, Buseman M. Gas for climate 2050: Analysing future demand, supply, and transport of hydrogen. 2021,70–82; https://gasforclimate2050.eu/wp-content/uploads/2021/06/EHB_Analysing-the-future-demand-supply-and-transport-of-hydrogen_June-2021_v3.pdf, accessed 12 August 2021.
[10] Wijayanta AT, Oda T, Purnomo CW, Kashiwagi T, Aziz M. Int J Hydrogen Energy, 2019, 44, 15026–44.
[11] Tarkowski R. Renew Sustain Energy Rev, 2019, 105, 86–94.
[12] H_2 Tools / Pacific Northwest National Laboratory, Hydrogen Pipelines, 2016 https://h2tools.org/hyarc/hydrogen-data/hydrogen-pipelines, accessed 12 August 2021.
[13] Sherif SA, Zeytinoglu N, Veziroglu TN. Int J Hydrogen Energy, 1997, 22, 683–88.
[14] Hirscher M, Yartys VA, Baricco M, von Colbe JB, Blanchard D, Bowman Jr RC, Broom DP, Buckley CE, Chang F, Chen P, Cho YW, Crivello J-C, Cuevas F, David WIF, de Jongh PE, Denys RV, Dornheim M, Felderhoff M, Filinchuk Y, Froudakis GE, Grant DM, MacA Gray E, Hauback BC, He T, Humphries TD, Jensen TR, Kim S, Kojima Y, Latroche M, Li H-W, Lototskyy MV, Makepeace JW, Møller KT, Naheed L, Ngene P, Noreus D, Nygård MM, Orimo S-I, Paskevicius M, Pasquini L, Ravnsbæk DB, Sofianos MV, Udovic TJ, Vegge T, Walker GS, Webb CJ, Weidenthaler C, Zlotea C. J Alloys Compd, 2020, 827, 153548.
[15] Kustov LM, Kalenchuk AN, Bogdan VI. Russ Chem Rev, 2020, 89, 897–916.
[16] Jain V, Kandasubramanian B. J Mater Sci, 2020, 55, 1865–903.
[17] Durbin DJ, Maladier-Jugroot C. Int J Hydrogen Energy, 2013, 38, 14595–617.
[18] Broom DP, Webb CJ, Fanourgakis GS, Froudakis GE, Trikalitis PN, Hirscher M. Int J Hydrogen Energy, 2019, 44, 7768–79.
[19] Zeleňák V, Saldan I. Nanomater, 2021, 11, 1638.
[20] Kohlmann H. Metal hydrides. In: Meyers RA. (Ed.) Encyclopedia of Physical Sciences and Technology, 3rd edition, vol. 9, Academic Press, 2002, 441–58.
[21] Schlapbach L. (Ed). Hydrogen in intermetallic compounds II: Surface and dynamic properties, applications. Top Appl Phys, 1992, 67.
[22] Yartys VA, Lototskyy MV, Akiba E, Albert R, Antonov VE, Ares JR, Baricco M, Bourgeois N, Buckley CE, Bellosta von Colbe JM, Crivello J-C, Cuevas F, Denys RV, Dornheim M, Felderhoff M, Grant DM, Hauback BC, Humphries TD, Jacob I, Jensen TR, de Jongh PE, Joubert J-M, Kuzovnikov MA, Latroche M, Paskevicius M, Pasquini L, Popilevsky L, Skripnyuk VM, Rabkin E, Sofianos MV, Stuart A, Walker G, Wang H, Webb CJ, Zhu M. Int J Hydrogen Energy, 2019, 44, 7809–59.

[23] Jeon K-J, Moon HR, Ruminski AM, Jiang B, Kisielowski C, Bardhan R, Urban JJ. Nat Mater, 2011, 10, 286–90.
[24] Nielsen TK, Besenbacher F, Jensen TR. Nanoscale, 2011, 3, 2086–98.
[25] Ali NA, Ismail M. Int J Hydrogen Energy, 2021, 46, 766–82.
[26] Bogdanović B, Schwickardi M. J Alloys Compd, 1997, 253–254, 1–9.
[27] Paskevicius M, Jepsen LH, Schouwink P, Černý R, Ravnsbaek DB, Filinchuk Y, Dornheim M, Besenbacher F, Jensen TR. Chem Soc Rev, 2017, 46, 15651634.
[28] Abdelhamid HN. Int J Hydrogen Energy, 2021, 46, 726–65.
[29] Garroni S, Santoru A, Cao H, Dornheim M, Klasen T, Milanese C, Gennari F, Pistidda C. Energies, 2018, 11, 1027. 
[30] Häussermann U, Kranak VF, Puhakainen K. Struct Bond, 2011, 139, 143–62.
[31] Aoki M, Ohba N, Noritake T, Towata S. Appl Phys Lett, 2004, 85, 387–88.
[32] Auer H, Yang F, Playford HY, Hansen TC, Franz A, Kohlmann H. Inorganics, 2019, 7, 106.
[33] Chotard J, Tang WS, Raybaud P, Janot R. Chem Eur J, 2011, 17, 12302–09.
[34] Castilla-Martinez CA, Moury R, Demirci UB. Int J Hydrogen Energy, 2020, 45, 30731–55.
[35] Kohlmann H. Eur J Inorg Chem, 2019, 4174–80.

5.5 Solar cell materials

Konrad Mertens

Since the 1970, the climate crisis has come more and more into the conscience of mankind. In those days, the change of the climate was relatively theoretical: a curve of CO_2 emissions suddenly rising steeply up the in the twentieth century (Figure 5.5.1).

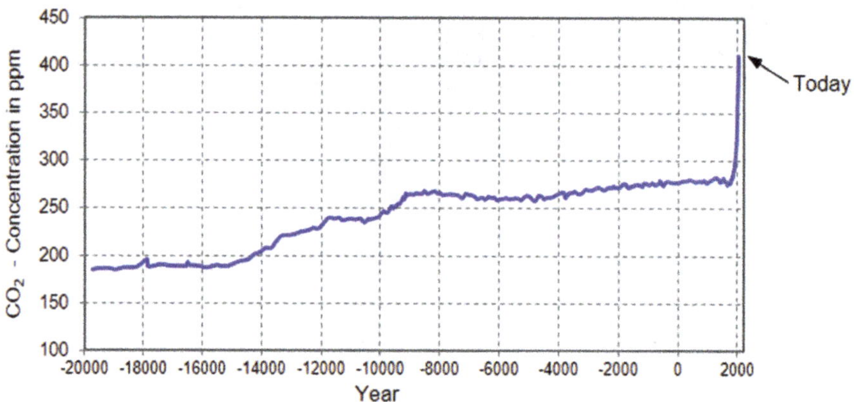

Figure 5.5.1: Development of the CO_2 content in the atmosphere in the last 22000 years: Noticeable is the steep rise since the start of industrialization [1–3].

Nowadays, the problems coming along with the climate change are more and more visible: a gradual heating of the earth's temperature, bush fires, coral bleaches, heavier storms, change of climate zones and so on. Mankind gets more and more aware that measures have to be taken to limit climate change. One decisive solution to the problems mentioned above is the use of regenerative energies like wind power, water power and photovoltaics. Particularly photovoltaics play a special role in the ranks of the different energies: It is scalable from watts to gigawatts, it can be used on and off grid and it has a relatively small impact on the environment. Therefore, it is worth to have a closer look on this technology.

5.5.1 A short history of crystalline silicon solar cells

In 1954, Daryl Chapin, Calvin Fuller and Gerald Pearson in the Bell Labs presented the first silicon solar cell with an area of 2 cm^2 and an efficiency of 6% [4]. This success was based on the coinventor of the transistor, the American Nobel laureate William B. Shockley. In 1950, he presented an explanation of the method of

https://doi.org/10.1515/9783110798890-005

functioning of the p-n junction and thus laid the theoretical foundations of the solar cells used today.

The Bell cell combined for the first time the concept of the p-n junction with the internal photo effect. In this, the p-n junction serves as a conveyor that removes the released electrons. Thus, this effect can be more accurately described as the "depletion layer photo effect" or also as the "photovoltaic effect". First applications of the new technology were satellites powered with solar cells as well as the terrestrial use of photovoltaics like transmitter stations, signal systems, remote mountain huts, etc. However, a change in thinking was brought about with the oil crisis in 1973. Suddenly, alternative sources of energy became the center of interest. In 1977, in the Sandia Laboratories in New Mexico, a solar module was developed with the aim of producing a standard product for economical mass production. The accident in the nuclear power plant in Harrisburg (1979) and especially the reactor catastrophe in Chernobyl (1986) increased the pressure on governments to find new solutions in energy supply. From the end of the 1980s, research in the field of photovoltaics intensified especially in the United States, Japan and Germany. In addition, research programs were started in the construction of grid-coupled photovoltaic plants that could be installed on single-family homes. In the beginning of the new millennium, several countries started feed-in tariffs for photovoltaic plants which nudged the production of solar modules. This mass production led to a drastic reduction of the prices of solar modules.

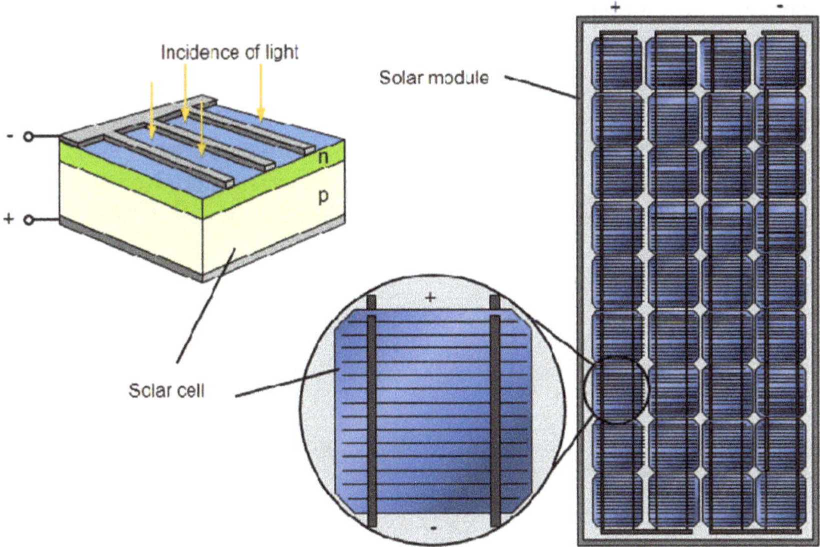

Figure 5.5.2: Solar cell and solar module as basic components of photovoltaics [5].

5.5.2 Technology of crystalline silicon solar cells

5.5.2.1 Principle of the structure

The concrete build-up of a silicon solar cell is shown in Figure 5.5.3.

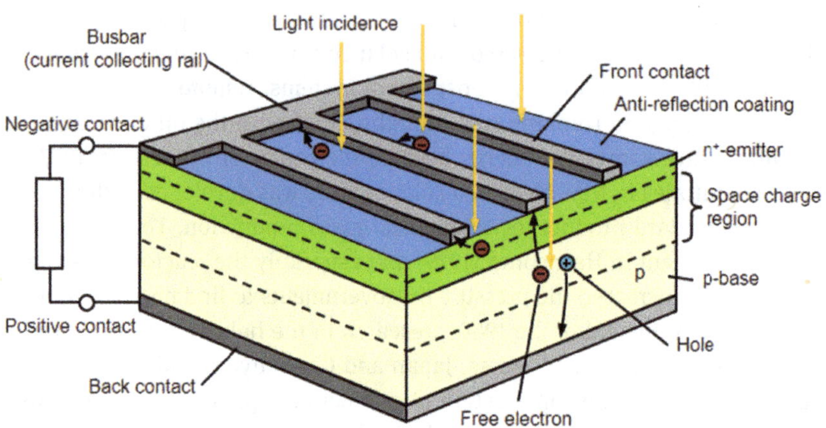

Figure 5.5.3: Typical build-up of a silicon solar cell.

Basically, it consists of a p-n junction. This is asymmetrically doped: At the bottom is the p-base and at the top a heavily doped n^+-emitter. The terms base and emitter come from the starting times of the bipolar transistor and have been adopted for solar cells. If light penetrates the cell, then every absorbed photon generates an electron-hole pair. The particles are separated by the field of the space charge region and moved to the contacts: The holes through the base to the bottom back contact and the electrons through the emitter to the front contacts. These are small metal strips that transport the generated electrons to the current collector rail (busbar). If a load is connected to the two poles of the solar cell, then this can draw off the generated electrical energy.

5.5.2.2 What happens in the individual cell regions?

The situation in the interior of the solar cells is shown in greater detail in Figure 5.5.4.

As sunlight consists of a broad range of wavelengths, the absorption of photons also is dependent on the concrete energy of the photons. For instance, blue light has the highest absorption coefficient with the result, that most of the light energy is already absorbed after less than 1 μm (the so-called penetration depth). Infrared radiation, in comparison, has penetration depths of more than 100 μm. For this reason, we will look more closely at the generation of photocurrent at different depths in the cell. Let us consider photon ①. It is absorbed in the highly doped emitter.

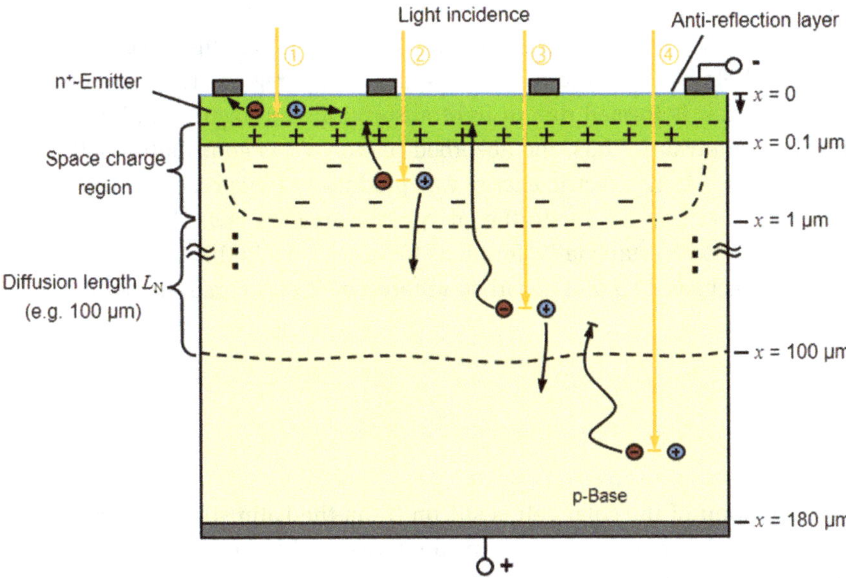

Figure 5.5.4: Cross-section of a solar cell: Each individually generated electron-hole pair has a different chance of contributing to the photocurrent.

Because of the high degree of doping, the diffusion length is extremely short so that the generated hole probably recombines before reaching the space charge region. The particularly highly doped upper edge of the emitter is also, occasionally, called the "dead layer" in order to emphasize that this is where the highest recombination probability is situated.

What happens to photon ②? Absorption takes place within the space charge region. The field prevailing in the space charge region separates the generated electron-hole pair and drives the two charge carriers into different directions. The electron is moved to the n-region and from there, further to the minus contact of the solar cell. The hole is moved in the opposite direction. It must travel a relatively long way through the base to the plus contact. As it is in the p-region during this movement, the probability of recombination is slight. Thus, practically all generated electron-hole pairs can be used for the photocurrent.

Photon ③ is absorbed only deep in the solar cell. The generated electron is not situated in an electrical field but diffuses as a minority charge carrier with little motivation throughout the crystal. If, by chance, it arrives at the edge of the space charge region, then it is pulled to the n-side by the prevailing field, where it can flow as a majority carrier to the contact. As the electron was still generated within the diffusion length, the probability that it can maintain itself up to the space charge region is relatively high.

In contrast to that, electron ④ is a true loser and is absorbed only in the lowest region of the solar cell. Although the electron diffuses through the p-base, it recombines with a hole before it can reach the space charge region. Thus, although an electron-hole pair is formed due to light absorption, an electron and a hole are "eliminated" afterward. Thus, the absorbed photon 4 has contributed nothing to the photocurrent. As no electric energy was produced in this case, the crystal has only become a bit warmer due to the energy conservation law. This confirms the importance of good crystal quality for high efficiency. Only in this way, a long diffusion length is achieved so that absorbed infrared light rays can even be used deep in the cell.

5.5.2.3 Production of crystalline silicon cells

5.5.2.3.1 From sand to silicon

The starting point of the solar cell is silicon (from the Latin silicia: gravel earth). First, the silicon is reduced in an electric arc furnace with the addition of coal and electrical energy at approximately 1800 °C:

$$SiO_2 + 2\,C \rightarrow Si + 2\,CO$$

Thus, we obtain so-called metallurgical silicon (MG-Si) with a purity of approximately 98%. The designation is because this type of silicon is also used in the production of steel. For use in solar cells, the MG-Si must still go through complex cleaning. In the so-called silane process, the finely ground silicon is mixed in a fluidized bed reactor with hydrochloric acid (hydrogen chloride, HCl). In an exothermic reaction, this results in trichlorosilane ($SiHCl_3$) and hydrogen:

$$Si + 3\,HCl \rightarrow SiHCl_3 + H_2$$

Now the trichlorosilane can be further cleaned by means of repeated distillation. Fortunately, the boiling point is only at 31.8 °C. The reclamation of the silicon takes place in a reactor (Siemens reactor) in which the gaseous trichlorosilane with hydrogen is fed past a 1350 °C hot thin silicon rod. The silicon separates out at the rod as highly purified polysilicon. This results in rods, for instance, of length 2 m with a diameter of approximately 30 cm (Figure 5.5.5). The polysilicon should have a purity of at least 99.999% (5 nines, designation 5 N) in order to be called solar-grade silicon (SG-Si). However, for a normal semiconductor technology used in the production of computer chips and so on, this degree of purity would be insufficient; here a purity of 99.9999999% (9 N, electronic grade: EG-Si) is typical. As the Siemens process is rather energy consuming, the search has been on for years for alternatives to cleaning silicon. One possibility is the use of fluidized bed reactors (FBR) in which the purest silicon is continually separated. This is achieved by blowing small dust-shaped silicon seed crystals into the reactor instead of using seed rods.

These then grow with the help of trichlorosilane and hydrogen to small silicon spheres. The FBR have higher production rates and a 70% lesser energy usage than the Siemens reactor [6]. However, processing is difficult and requires much skill and experience.

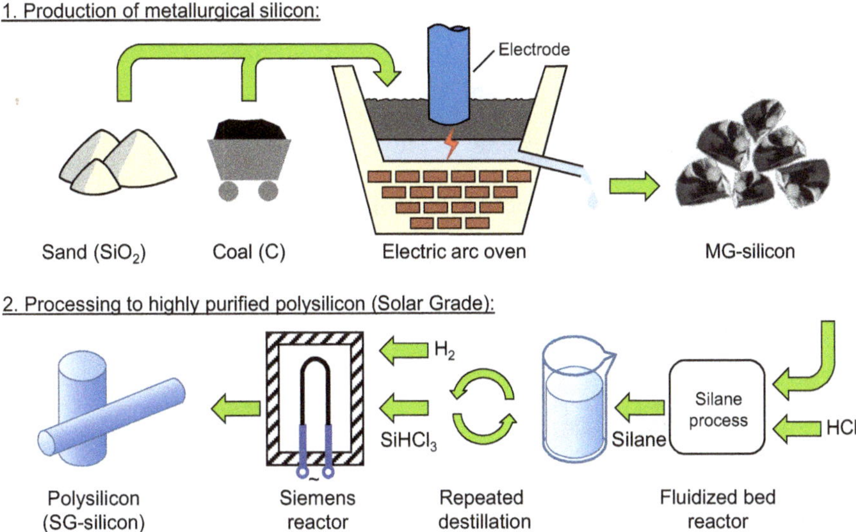

Figure 5.5.5: Production of polysilicon from quartz sand: (a) Production of metallurgical silicon. (b) Processing to highly purified polysilicon (solar grade).

5.5.2.3.2 Production of monocrystalline silicon

The Czochralski process (CZ process) is the process that has been used most for production of monocrystalline silicon. For this purpose, pieces of polysilicon are melted in a crucible at 1450 °C, and a seed crystal, fixed to a metal rod, is dipped into the melt from above. Then, with light rotation, it is slowly withdrawn upward whereby fluid silicon attaches to it and crystallizes (Figure 5.5.6).

Thus, eventually, a monocrystalline silicon rod (ingot) is formed, whose thickness can be adjusted by the variation of temperature and withdrawal speed. Rods with a diameter of up to 30 cm and a length of up to 2 m can be produced with this method. For photovoltaics, the diameter is typically 6 inch (15 cm) and larger.

5.5.2.3.3 Production of multicrystalline silicon

The production of multicrystalline silicon is much simpler. Figure 5.5.7 shows the principle: Pieces of polysilicon are poured into a graphite crucible and brought to a melt, for instance, using induction heating. Then the crucible is allowed to cool from the bottom by the heating ring slowly pulled upward. At various places on the

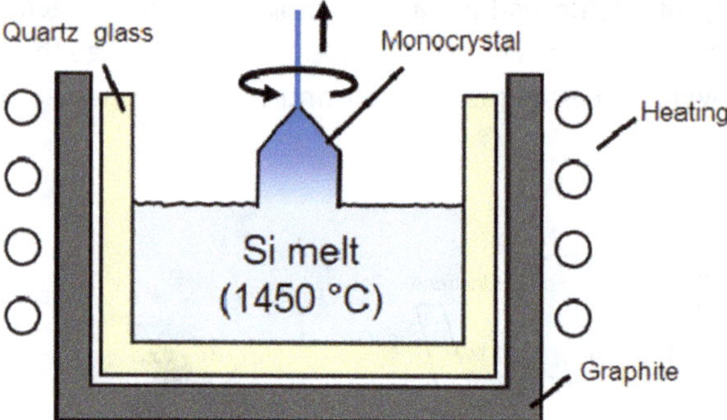

Figure 5.5.6: Production of monocrystalline silicon rods by means of the Czochralski process.

bottom of the crucible, small monocrystals are formed that grow sideways until they touch each other. With the vertical cooling process, the crystals grow upward in a column (columnar growth).

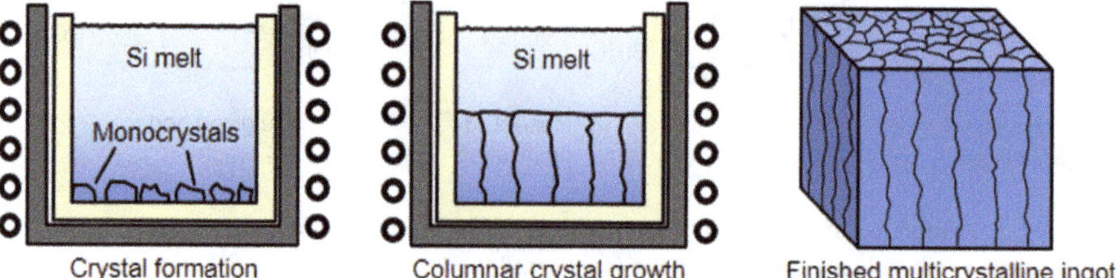

Figure 5.5.7: Production of multicrystalline silicon.

At the boundary layers, crystal displacements are formed that later become centers of recombination in the cell. For this reason, one tries to let the monocrystals become as large as possible. The column structure also has the advantage that minority carriers generated by light do not have to cross over a crystal boundary in the vertical direction. Because of the poorer material quality of multicrystalline silicon, the efficiency of solar cells made from this material is typically 2–3% below that of monocrystalline solar cells. After the crystallization of the whole melt, the silicon block (ingot) is divided into cubes (bricks) of 5 or 6 inches along the edges.

5.5.2.3.4 Wafer production

After production, the ingots must be sawed into individual sheets (wafers). This is mostly done with wire saws that remind one of an egg cutter (Figure 5.5.8). A wire with the thickness of 100–140 µm moves at high speed through a paste of glycol and extremely hard silicon-carbide particles and carries these with it into the saw gap of the silicon. With the current wafer thicknesses of 150–180 µm, there are saw losses that are almost as large as the used parts. More and more producers use saw wires encrusted with diamond particles in order to refrain from the use of silicon-carbide particles. [7].

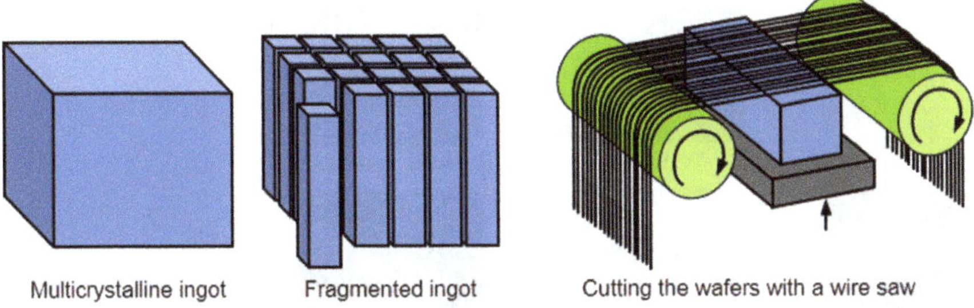

Figure 5.5.8: Production of multicrystalline wafers: After the fragmentation of the ingots into individual bricks, they are cut into wafers with the wire saw.

5.5.2.3.5 Solar cell production

Figure 5.5.9 shows the process steps for producing standard silicon cells.

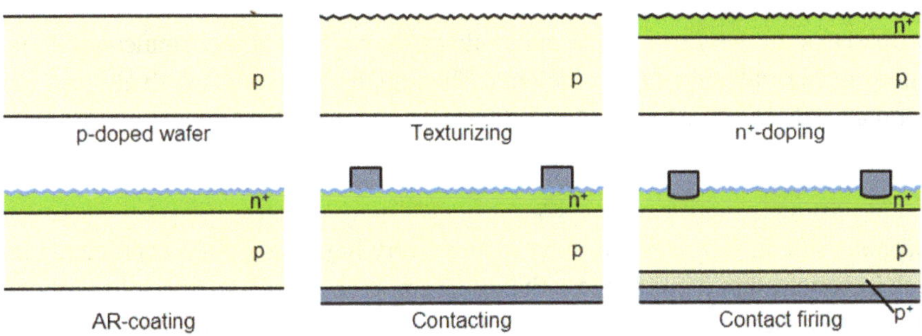

Figure 5.5.9: Process steps for producing standard cells.

Starting with the p-doped wafer, a texturizing step follows to remove contaminants or crystal damage of the surface. The formation of the p-n junction is then achieved by the formation of the n^+-emitter by means of phosphorous diffusion. In the next step, the deposition of the so-called antireflection coating with silicon nitride (Si_3N_4) is carried out, which causes a passivation of the surface at the same time. The application of the contacts occurs in the screen-printing process. For this purpose, a mask with slits is placed on the cell, and metal paste is brushed on. In this way, it is placed on the wafer only at particular positions. The formation of the rear-side contacts occurs in two steps. First, the soldering contact surfaces of silver paste are applied.

Then, the rest of the rear side is fully covered with aluminum. The front-side contacts are applied in the next step.

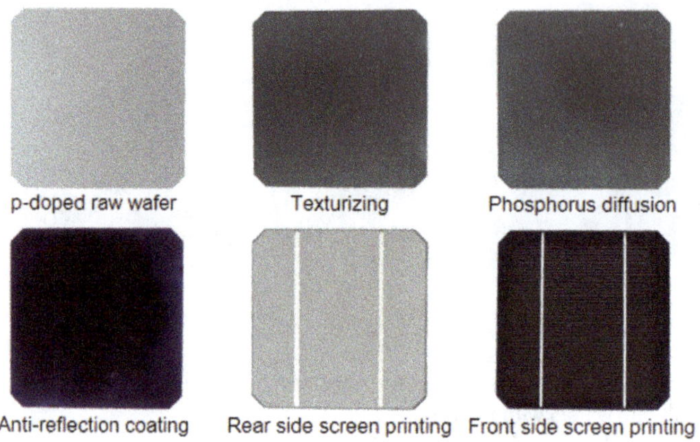

Figure 5.5.10: View of a monocrystalline cell after the respective production steps [5].

5.5.3 Thin-film cells

Thin-film cells are very promising materials as they all use direct semiconductors, resulting in high absorption coefficients. This means that only relatively thin absorber materials can be used to collect the electrons and holes.

5.5.3.1 Amorphous silicon solar cells

Amorphous silicon solar cells (a-Si cells) show very high absorption coefficients resulting in penetration depths of less than 0.25 μm. With respect to crystalline silicon solar cells this means, that only 1/10 to 1/100 of the thickness of the absorption material is needed! Figure 5.5.11 shows the buildup of an amorphous solar cell. The cell is pin-doped and uses a transparent conductive oxide (TCO) to contact the structure below the glass pane.

The absorbing layer only has a thickness in the range of 0.25 to 0.4 μm. However, the efficiency of the cell only lies in the order of 6%! One reason for this lies in the fact, that a degrading process occurs, when the cell is exposed to sunlight. This is caused by the so-called "Staebler-Wronski effect", named after the two scientists who first described it in detail [8]. The reason for this effect is in the strained Si-Si bonds that are caused by the irregular crystal structure. In the recombination of the electron-hole pairs generated by light, these weak bonds are "split open" and, as new dangling bonds, form new recombination centers for the minority carriers (see Figure 5.5.12). The split-open bonds also represent additional space charge

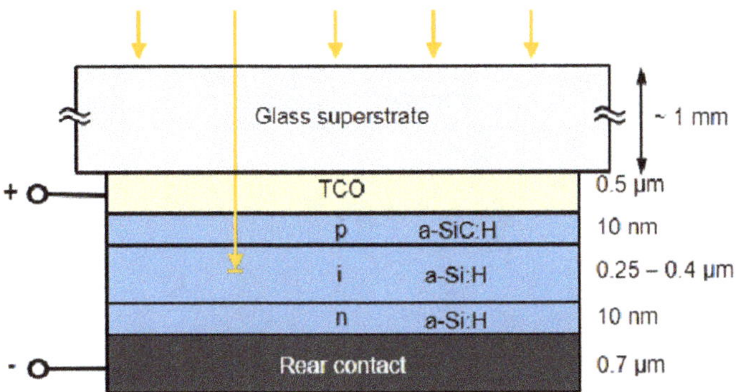

Figure 5.5.11: Structure of the a-Si thin-film cell: The overall thickness of the deposed material is less than 2 µm.

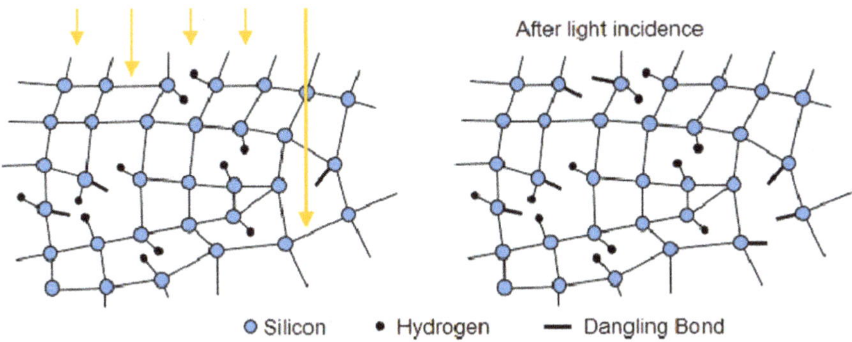

Figure 5.5.12: Depiction of the Staebler-Wronski effect: with incident light, the weak bonds in the crystal are split open [9].

regions that can weaken the built-in field in the pin cells. After a certain light radiation time, all the weak bonds are split open so that the efficiency of the cell stabilizes itself.

One measure to increase the cell efficiency is to use tandem cells (see Figure 5.5.13). Here, an additional a-SiGe layer is applied on top of the a-Si absorber layer. Depending on the portion of germanium atoms, this alloy can possess a band gap between 1.4 and 1.7 eV. Also depending on the wavelength, incident light is absorbed at different depths: Short-wavelength light ("blue") above 1.7 eV only manages to reach the first pin cell. The upper pin cell is transparent for long-wavelength ("red") light; it is, therefore, only absorbed in the lower cell. As both cells are switched in series, the weaker cell determines the overall current. For this reason, the thickness of the individual absorber layers must be selected so that the two cells achieve approximately the same current ("current matching"). Even the tandem cells show efficiencies of below 10% which is significantly lower than the ones of c-Si or other thin-film technologies. Therefore, a-Si cells nowadays are almost completely vanished from the power marked. However, special consumer applications like solar watches, solar pocket calculators or solar lamps are still using this technology.

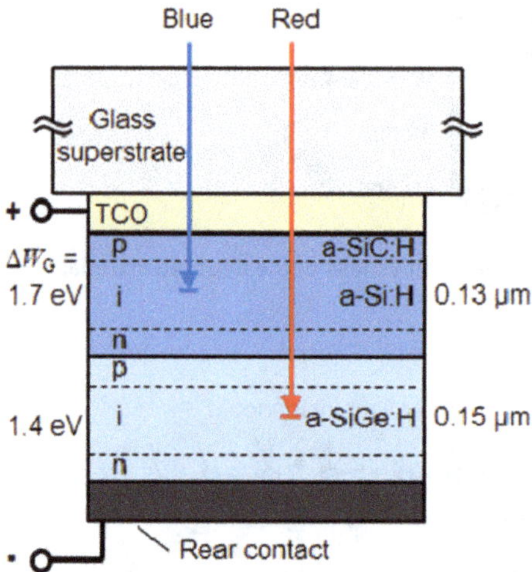

Figure 5.5.13: Example of a superstrate tandem cell [10].

5.5.3.2 CIS cells

CIS cells use materials of the chalcopyrite group that are generally summarized under the abbreviation of CIS or CIGS. What they have in common is that they crystallize in the structure of the mineral chalcopyrite (copper pyrites – $CuFeS_2$). Research on CIS cells has been carried out since the 1970s. In 1978, ARCO Solar was successful in the production of CIS cells with 14.1% [11]. But disappointment soon followed: The efficiency sank drastically in the transfer to larger areas. Already in the 1990s, efficiencies rose above 10%, rising again to 15% as the material knowledge had improved. In the last years, again a variety of improvements were introduced in the cells structure. Figure 5.5.14 shows the setup of such a modern CIGS cell. The color variation visualizes the "grading" of the band gap. To the bottom, the gallium content is enhanced, resulting in an increasing band gap of up to 1.1 eV. This leads to the generation of a "back surface field" which again reduces the recombination in the cell. Other means are an additional buffer layer of undoped zinc oxide to reduce leakage currents and a post-deposition treatment (PDT) to improve the crystal structure of the absorber.

The combination of the different improvements has drastically enhanced the efficiencies of CIGS cells in the last years to up 23% (lab cell).

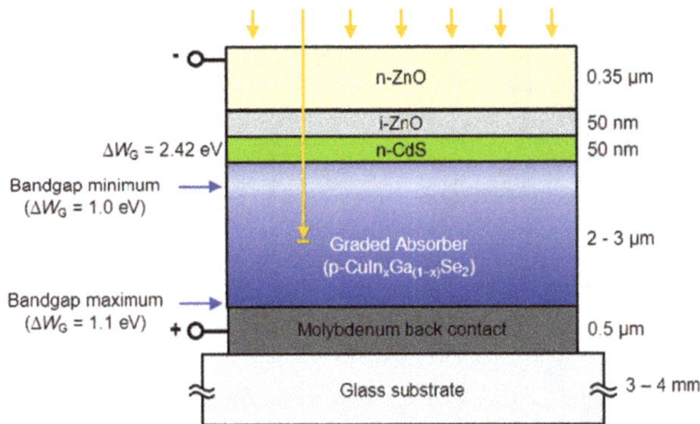

Figure 5.5.14: Setup of a modern CIGS cell: Significant efficiency gains compared to precedent cells result from the "graded absorber" and the additional i-ZnO intermediate layer, according to [12].

5.5.3.3 Cells from cadmium telluride

Cadmium telluride (CdTe) is a compound semiconductor of Group II and VI elements. It is a direct semiconductor with a band gap of $\Delta W_G = 1.45$ eV. A great advantage of this material is that it can be deposited in various ways with good quality as a thin film. The usual method is thermal evaporation over a short distance (CSS – close-spaced sublimation). In this process, the semiconductor sources are heated to approximately 500 °C. At this temperature, the semiconductors vaporize and deposit on the somewhat lower temperature substrate. Figure 5.5.15 shows a modern cell structure.

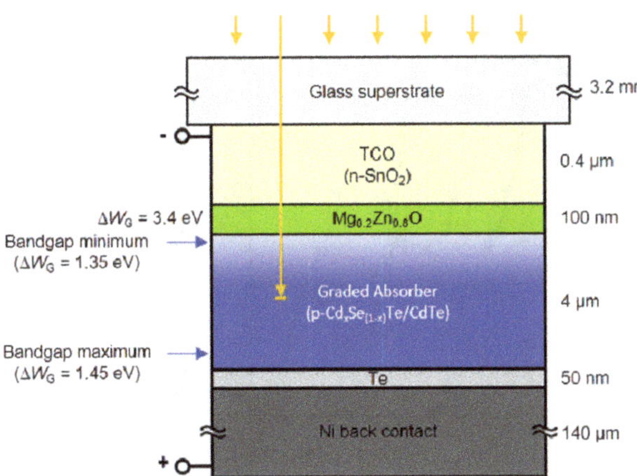

Figure 5.5.15: Setup of a modern CdTe cell: Significant efficiency gains compared to precedent cells result from the "graded absorber" and the additional $Mg_{0.2}Zn_{0.8}O$ intermediate layer, according to [13].

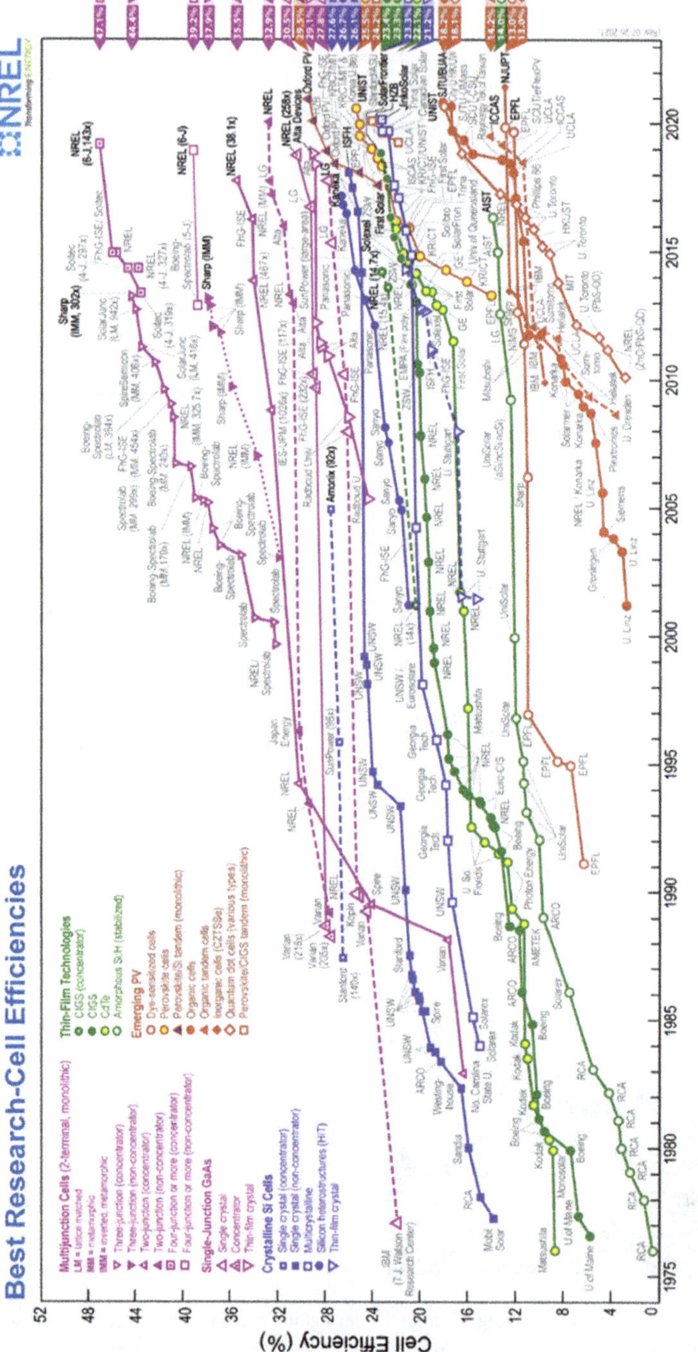

Figure 5.5.16: Development of record cell efficiencies over the last 45 years: Practically all technologies continue to show increasing efficiencies [15].

Also with CdTe, it is possible to vary the band gap of the absorber. This is done with the ternary material $CdSe_{(1-x)}Te_x$. In the upper part of the absorber the band gap can be reduced by the addition of selenium to 1.35 eV and then increases again to the 1.45 eV of the CdTe [13]. As in the case of CIGS, this leads to an improvement in absorption in the near infrared. In addition, the crystal quality rises resulting in the increase in the lifetime of the charge carriers. This in turn leads to an increase in the open-circuit voltage of the cell [14].

The record efficiency of CdTe cells is 22%; the company "First Solar" nowadays offers large modules (2.5 m^2) with an efficiency of 18%.

5.5.4 Conclusions

Since the invention of the first solar cell in the bell labs in 1954, solar cells have made tremendous progress. Starting with an efficiency of 6% on a cell with 2 cm^2, the technology has come of age. Figure 5.5.16 presents the efficiency of various cell types over the last 45 years. Modern solar cells show efficiencies clearly above 20% with the outlook for further increase.

A relatively new and promising technology is perovskite cells, which can drive the efficiencies to higher values. Here however, the stability is still a mayor question which has to be solved.

At the same time, the cost of solar cells and solar modules has been drastically reduced: Starting at a price of about 30 Euro/$Watt_P$ in the 1980s, the price has been reduced to 25 Ct/$Watt_P$, a reduction auf less than 1/100! Already nowadays, photovoltaic plants in many countries are the cheapest energy sources. This technology can therefore be a strong support to face the climate crisis.

References

[1] Monnin E, Steig EJ, Siegenthaler U, Kawamura K, Schwander J, Stauffer B, Stocker TF, Morse DL, Barnola J-M, Bellier B, Raynaud D, Fischer H. Evidence for substantial accumulation rate variability in Antarctica during the Holocene, through synchronization of CO_2 in the Taylor Dome, Dome C and DML ice cores. Earth Planetary Sci Lett, 2004, 224, 45–54. DOI: https://doi.org/10.1016/j.epsl.2004.05.007.

[2] Neftel A, Friedli H, Moor EH, Lötscher H, Oeschger H, Siegenthaler U, Stauffer B. Historical carbon dioxide record from the Siple Station ice core. 1994. http://cdiac.ornl.gov/trends/co2/siple.html. Accessed February 16 2022.

[3] www.esrl.noaa.gov/gmd/ccgg/trends. Accessed February 16 2022.

[4] Chapin D, Fuller CS, Pearson GL. A new silicon p–n junction photocell for converting solar radiation into electrical power. J Appl Phys, 1954, 25, 676–77.

[5] Mertens K. Photovoltaik - Lehrbuch zu Grundlagen, Technologie und Praxis. Hanser Verlag, 2020.

[6] Alsema EA, de Wild-scholten MJ, Fthenakis VM. Environmental impacts of PV electricity generation – A critical comparison of energy supply options. In: 21st European Photovoltaic Solar Energy Conference and Exhibition, September 4-8, 2006, Dresden, Germany.
[7] Podewils C. Diamantdraht zum Sägen. Photon, 2009, 77.
[8] Staebler D, Wronski C. Reversible conductivity changes in discharge produced amorphous Si. Appl Phys Lett, 1977, 31, 292.
[9] Repmann T. Stapelsolarzellen aus amorphem und mikrokristallinem Silizium, Dissertation. RWTH Aachen, Germany, 2003.
[10] Zeman M. Advanced amorphous silicon solar cell technologies. In: Poortmans J, Arkhipov V. (Eds). Thin Film Solar Cells – Fabrication, Characterization And Applications, Wiley-VCH, Weinheim, Germany, 2006.
[11] Mitchell K, Eberspacher C, Ermer J, Pier D. Single and tandem junction $CuInSe_2$ cell and module technology. In: 20th IEEE Photovoltaic Specialists Conference, Las Vegas, Conference Record, 1988, 2, 1384–89.
[12] Powalla M, Paetel S, Hariskos D, Wuerz R, Kessler F, Lechner P, Wischmann W, Friedlmeier TM. Advances in cost-efficient thin-film photovoltaics based on $Cu(In,Ga)Se_2$. Engineering, 2017, 3, 445–51. DOI: 10.1016/J.ENG.2017.04.015.
[13] Sites JR, Munshi A, Kephart J, Swanson D, Moore A, Song T, Sampath WS. Enhancements to cadmium telluride cell efficiency. In: 33rd European Photovoltaic Solar Energy Conference and Exhibition, 2017, 998–1000. DOI: 10.4229/EUPVSEC20172017-3CP.1.3.
[14] Exclusive: First Solar's CTO Discusses Record 18.6% Efficient Thin-Film Module, 2015, (accessed August 02, 2021, at http://www.greentechmedia.com/articles/read/exclusive-first-solars-ctodiscusses-record-18-6-efficient-thin-film-mod).
[15] Best Research-Cell Efficiency Chart, NREL, 2021. https://www.nrel.gov/pv/cell-efficiency.html, Accessed August 12 2021.

5.6 Thermal energy storage materials

Thomas Bauer

Heat or cold is a form of energy and can be stored in various ways and for many different applications. There are several ways to classify thermal energy storage (TES) materials and systems. In general, three types of TES systems are distinguished on the basis of the physical or physicochemical principle:

- **Sensible heat storage** results in an increase or decrease of the storage material temperature, stored energy is proportional to the temperature difference of the used materials. Either solids, liquids or a solid-liquid mixture is utilized (the storage density of gases is typically too low).
- **Latent heat storage** is connected with a phase transformation of the storage materials (phase change materials; PCM), typically changing their physical phase from solid to liquid and vice versa. Stored energy is equivalent to the heat (enthalpy) for melting and freezing.
- **Thermochemical storage** (TCS) is based on reversible thermochemical reactions. The energy is stored in the form of chemical compounds created by an endothermic reaction and it is recovered again by recombining the compounds in an exothermic reaction. The heat stored and released is equivalent to the heat (enthalpy) of reaction. Ad and absorption processes are also utilized. Sorption can be considered as a physical process. Nevertheless, it is common for sorption TES to be classified as thermochemical storage.

Further distinguishing features of thermal energy storage are discussed below. The most important property of a thermal energy storage device for material selection is its **temperature.** The temperature range extends from below 0 °C (e.g. ice as a latent heat storage medium) to over 1000 °C (e.g. regenerator in the high-temperature process industry). In the medium temperature range (0 to 120 °C), water is the dominant liquid storage medium (e.g. heating and cooling of buildings).

TES systems differ also in terms of the **heat transfer fluid** (HTF). HTFs may be of single-phase type (e.g. water, air, thermal oil) or two-phase type with a condensation and evaporation process in the system (e.g. water/steam). The storage material and HTF may be one single medium (direct storage concept, e.g. water as HTF and storage medium). Some HTFs cannot be stored directly; the energy must be transferred from the HTF to a separate storage medium (indirect storage concept, e.g. air as HTF and a ceramic as storage medium).

Another classification criterion is the **duration** of thermal energy storage between charging and discharging. The range extends from minutes (thermal buffering of processes), to hours and days (e.g. water storage for space heating; molten salt storage for electricity generation), to months (e.g. seasonal storage with water for space heating).

https://doi.org/10.1515/9783110798890-006

The following list contains further distinguishing features of TES [1–9]:

- Thermal capacity or size in kWh (e.g. pocket warmer 0.01 kWh, water storage in a single-family house for daily demand 2–60 kWh, molten salt storage in solar thermal power plants up to 4000000 kWh).
- Thermal power in kW for the charge and discharge process; multiplying the thermal power by the characteristic charging and discharging duration gives the thermal capacity.
- Properties of the storage material and other materials of the storage system (e.g. suitable thermal properties, availability, corrosivity, toxicity, environmental impact, flammability, durability, thermal stability, vapor pressure).
- The number of tanks or storage units. They are characterized by different zones with charged and discharged storage material.
- Efficiency and heat losses of the storage system.
- The associated energy conversion processes for charging and discharging (e.g. heat-to-heat, power-to-heat, heat-to-power generation).
- Stationary (e.g. hot water tank) and mobile (e.g. latent heat pocket warmer) use.
- The technological readiness level of thermal energy storage systems (e.g. research, precommercial, commercial).
- The energy density in terms of size (kWh/m^3) and weight (kWh/kg).
- Operational costs in €/(kWh ·year) and investment costs in €/kWh.

Table 5.6.1 lists typical TES materials with the associated storage principles. It also indicates the intended use of the material (see Table S, E, C, F). Low-cost inorganic materials are mainly used for TES. Hydrocarbons are sometimes used as TES media (e.g. paraffins in buildings as PCM) and as container materials, but are not considered further here. Furthermore, materials that are still in research and development or precommercial phase and metallic container materials are not discussed further (see gray text in the table).

Zeolites (see also Chapter 7.3) are crystalline, microporous, hydrated aluminosilicate compounds for thermochemical storage. The general formula is $M_w[Al_xSi_yO_z] \cdot nH_2O$, where *M* is an alkali metal or alkaline earth metal cation and the number of adsorbed water molecules. Zeolites have a high specific inner surface area of approximately 1000 m^2 g^{-1}. Water can be reversibly adsorbed on this surface. For thermochemical storage, their high heat of adsorption and their ability to hydrate and dehydrate while maintaining structural stability are used. Applications include, for example, self-cooling beer kegs, efficient dishwashers and temperature control in logistics (e.g. transport boxes) [8].

The following text refers to the commercial use of inorganic heat storage materials (see black text in the Table 5.6.1) with the following three subsections: solids, water and salts. A more comprehensive overview of the field of thermal energy storage can be found in the literature [1–9].

Table 5.6.1: Overview of the four thermal energy-storage principles with utilized materials. The letters have the following meaning: S = storage material; E = material to extended heat transfer surface, C = container material, F = filler material. Materials or letters marked gray are non-commercial, organic or not utilized as storage material. These materials are excluded in this work.

	Sensible heat storage in solids	Sensible heat storage in liquids	Latent heat storage (PCM)	Thermochemical heat storage and sorption
Natural stones (e.g. packed bed)	S	F		
Soil/ground	S			
Concrete	S			
Ceramics	S	F		
Metals	S, C	S, C	S, E, C	E, C
Graphite	S		E	E
Metal oxides, slag	S	F		S
Water		S	S	
Thermal oil		S		
Salt anhydrous		S	S	S
Salt hydrates			S, C	S
Salt solutions				S
Polymers (e.g. paraffins)		C	S, C	C
Metal hydrides				S
Zeolites				S
Silica gels				S

5.6.1 Solids

Geothermal or **underground thermal energy storage** (UTES) systems refer to systems that use buried devices designed to exchange heat with the surrounding ground. The storage medium is the surrounding ground, e.g. soil, rock or water containing layers. UTES systems are typically subdivided as follows [10, 11]:

- Aquifers thermal energy storage (ATES): storage in natural enclosed water-saturated soil structures.
- Borehole thermal energy storage (BTES): near-surface use of geothermal probe fields.

- Cavern thermal energy storage (CTES): storage in natural caverns or artificial caverns (e.g. reuse of gas caverns or mines).
- (Technical systems in basins such as tanks and pits are not considered as UTES here).

Geothermal probes use soil as storage medium (e.g. rock and also water-bearing strata). Locations with high groundwater flow are avoided in the application. Geothermal probe storage systems can be used as both heat and cold storage facilities. They are most efficient at low temperatures (e.g. 35 °C for underfloor heating). Compared to probe storage systems, geothermal ground collectors are installed near the surface in loose soil at depths ranging from about 1 to 2 meters. As a result, they are subject to seasonal temperature fluctuations [8].

In buildings, the thermal mass of the **building components** is used as TES. The thermal mass of the building can passively stabilize the temperature in the interior spaces by absorbing and releasing heat via the surfaces. In addition to passive temperature control, there are also active systems with integrated pipes in parts of the building. The thermal conductivity of building materials (e.g. concrete) is typically relatively low, which limits the heat transfer rates. Integrated pipes are therefore usually used for faster charging and discharging. A heat transfer fluid (e.g. water) flows in the pipes (e.g. plastic) to absorb and release heat. Systems with pipes integrated into the (structural) concrete are also used. Concrete foundation storage systems with integrated pipes in piles, base plates and walls, for example, are commercially available. Not only in the foundation area of buildings, but also in the buildings themselves, thermally active components can be used to store thermal energy. For example, underfloor heating with pipes is common, but ceiling and walls can also be used [8].

Concrete is an attractive sensible heat storage material, because of its low cost due to inexpensive aggregates, the availability throughout the world and the simple processing route via casting. Concrete is a composite material consisting of fine aggregates (e.g. sand) and coarse aggregates (e.g. gravel, crushed rocks) bonded with a cement paste (e.g. Portland cement). Typically, a water-based liquid cement paste is used, which hardens (cures) over time. In the hydraulic binding process, free water reacts with the cement paste minerals to form stable hydrated mineral phases. The cement paste "glues" the aggregate together. Special concretes with hydraulic binders up to 500 °C have been developed for TES in the high-temperature range. These concretes lose part of their hydrated phases and release steam at high-temperatures. Special care should be taken during initial heating when pressure builds up in the concrete block [12].

Storage heaters are used for space heating. They are directly charged with surplus electricity. In the past, this was done with base-load electricity at night (night storage heater) and today charging with surplus regenerative electricity from photovoltaics and wind is becoming attractive. The discharging process takes place either passively via an appropriately designed thermal insulation or actively via an air blower. **Iron oxide** bricks (e.g. magnetite Fe_3O_4) heated to about 600 °C are often used as storage media (sometimes called feolite bricks) [7, 8]. The gravimetric heat capacity values of iron oxides, given in J kg^{-1} K^{-1}, are moderate. On the other hand, iron oxides have high density values (e.g. given in kg m^{-3}). This results in a high volumetric heat capacity (e.g. given in MJ $m^{-3}K^{-1}$) compared to other inexpensive solid materials.

Regenerators, also called regenerative heat exchangers, comprise a storage inventory (e.g. ceramics) that is heated and cooled through an intermittent heat transfer between a hot and a cold fluid (e.g. air, flue gas). The storage is operated typically with a gas in counterflow: during charging, the hot gas enters the storage module at the "hot" end and during discharging the flow direction is reversed and the cold gas enters at the "cold" end. During discharging, heat is transferred from the solid to the gas and during charging, the gas releases the heat to the solid. Existing industrial examples include solutions in the steel industry (Cowper), the glass industry and industrial air purification systems (e.g. regenerative thermal oxidizer systems). This type of TES typically uses ceramic bricks as a storage media and allows the highest storage temperatures of up to 1600 °C. The maximum application temperature depends on the specific material. The surface-to-volume ratio of the bricks is adapted to the required charging and discharging power demand. For example, ceramic bricks with a large surface area (e.g. extruded ceramic honeycombs) can be used for short charging and discharging times, while bricks with a smaller surface area (e.g. ceramic checker bricks) are used for longer time periods. In practice, oxides (e.g. silica, alumina, magnesia and iron oxide), carbonates (e.g. magnesite) and their mixtures are usually used as **refractory materials** for regenerators [1, 6–8].

5.6.2 Water

The most widespread storage medium for liquid TES is water. Water has the following advantages [7, 8, 13]:
- It is abundant and inexpensive.
- It has a high heat capacity and latent heat of melting/solidification.
- It has a density that decreases with temperature and a relatively low thermal conductivity, which is an advantage for thermal stratification to separate the hot charged and cold discharged zone; or even several zones on top of each other.
- It can be easily stored in all kinds of containers.
- Experience with water and related components is common.
- It can be used also as HTF (no heat exchanger required).
- It is easy to handle (e.g. non-toxic, non-combustible).
- It is easily mixable with additives (antifreeze, anticorrosive).
- It is easy to dispose.

Water has also some disadvantages, which can be summarized as follows:
- It only provides a limited operation range from 0 °C (freezing) to 100 °C (boiling) without pressurizing.
- It can be corrosive.
- It has high vapor pressures (e.g. about 10 bar at 180 °C).

The type of water storage and the scope of application are very different. Application fields include human comfort (e.g. hot-water bottle), space heating and cooling, domestic hot water, local/district heating networks and industrial process steam. Table 5.6.2 shows the different areas of application for sensible heat storage in water [8, 9, 11, 13]. Below room temperature, cold storage is utilized for space cooling. If the heat source allows sufficiently high temperatures, it is often cost-optimal to select the maximum storage temperature around 100 °C. The maximum operation temperature is limited by the critical point of saturated water (374 °C, 221 bar). At high temperatures and pressures, the cost for the pressure vessel often becomes disproportionately high. Hence, typical systems operate at a maximum temperature of 250 °C with an upper pressure limit of 40 bar [7, 8]. Pressurized water-storage systems are used for high-temperature heating networks and industrial process steam supply. It can be seen that also the size (capacity) varies in a wide range. The storage duration between charging and discharging is mostly in the range of hours to days. Seasonal storage (e.g. with temporal shift from summer to winter) requires large dimensioned systems with a low surface-to-volume-ratio to avoid large heat losses. Water-storage system also differ widely in terms of the container materials. They include steel, concrete (e.g. reinforced concrete, prestressed concrete), plastic (e.g. fiber reinforced plastic) and earth basins (e.g. with liners).

Table 5.6.2: Overview of the applications for sensible heat storage in water with characteristic properties. They are sorted by max. temperature.

	Temperature	**Size**	**Duration**	**Application**
Cold storage tank	2–20 °C	10–6000 m^3	Hours (daily)	Space cooling, cooling networks
Hot water bottle	max. 60 °C	0.002 m^3	typ. <1 h	Human comfort
Water gravel (WGTES)	max. 90 °C	1000–8000 m^3	Months	Heating networks
Pit (PTES)	max. 80–95 °C	max. 200000 $m^{3\#}$	Months	Heating networks
Seasonal tank (TTES)	max. 95 °C§	1000–12000 m^3	Months	Heating networks

Table 5.6.2 (continued)

	Temperature	Size	Duration	Application
One-zone flat-bottom tank	max. 98 °C	max. 50000 m^3 *	Hours-days	Heating networks
Two-zone flat-bottom tank	max. 115 °C	max. 33000 m^3 *	Hours-days	Heating networks
Single-house tank	max. 90–110 °C	0.05–5 m^3	Hours-days	Space heating/domestic water
Medium size tanks	max. 90–160 °C	up to 200 m^3	Hours-days	Space heat & heating networks, process heat
Displacement pressure tank	max. 180 °C	max. 300 m^3 $	Hours-days	Heating networks
Sliding pressure (Ruths)	typ. max. 250 °C	max. 300 m^3 $	typ. <1 h	Process steam

*= in Germany in 2020; $ = approx. max. size per storage tank; # = in Denmark Vojens; § = higher values with steel liner feasible.

Figure 5.6.1 shows a large-sized pit storage in the construction phase to supply a heating network.

Figure 5.6.1: Pit storage in the construction phase in Toftlund Denmark in 2019. Visible is the plastic liner. The storage is filled with 80000 m^3 water and connected to a heating network (Source Ramboll).

In addition to sensible TES, water is also used as a phase change material (PCM), also known as **ice storage.** In the past, blocks of ice were cut to size from a frozen lake and stored in cellar vaults until summer to keep food cold. Today, there are only a few test facilities for seasonal storage with natural snow or ice. A typical application today is the cooling of large buildings using artificially produced ice (e.g. offices). Examples with smaller amounts of ice are the cold transport of food, drinks and medicines (e.g. transport boxes, cooling bags). The heat transfer from ice (solid) to the heat sink is typically limited. Therefore, there are different concepts to improve the heat transfer for ice storage. Concepts include pumpable ice slurries, ice-on-coil and macro-encapsulation with polymers [5].

5.6.3 Salts

For TES, several inorganic single salts and salt mixtures are utilized. A salt (e.g. NaCl) consist of a cation (e.g. Na^+) and anions (e.g. Cl^-), mostly with ionic binding. Typically, the salt chemistry is determined by the anion and therefore anions mark the salt classes (e.g. chlorides, nitrates, sulfates). For TES and HTF salt media, the water amount and type of binding plays a major role on the utilized temperature level and TES type. It can be distinguished between: salt solutions (1), salt hydrates (2) and anhydrous salts (3) and their mixtures as shown in Table 5.6.3 [14]. They are utilized in all three types of TES (sensible heat, latent heat with PCMs and thermo-chemical storage). Some approaches are in a research and development phase. The following discussion focuses on commercial applications.

Table 5.6.3: Overview of salts utilized in TES and HTF systems. They are sorted by max. temperature [14].

Temperature level	Salt type process	TES type, HTF usage	Development status
<0 °C	Water-salt mixtures (e.g. brines)	PCM slurry, HTF	Commercial
0–100 °C	Melting of salt hydrates in crystal water	PCM	Commercial
40–300 °C	Dehydration of salt hydrates	TCS	R&D
40–150 °C	Absorption in concentrated salt solutions	TCS	R&D
120–500 °C	Solid-liquid conversion in anhydrous salts	PCM	R&D
100–800 °C	Anhydrous molten salts	Sensible, HTF	Commercial

Table 5.6.3 (continued)

Temperature level	Salt type process	TES type, HTF usage	Development status
100–800 °C	Anhydrous solid salts	Sensible	Minor R&D
100–800 °C	Solid-solid conversion in anhydrous salts	PCM	Minor R&D
250–1000 °C	Dissociation of anhydrous solid salts	TCS	R&D

For refrigeration below 0°C, solution of **water-salt (brine)** and water-glycol can be utilized. Applications for brines are industrial processes and space cooling. For example, the mixture NaCl (22 wt%)-H_2O has an eutectic freezing point at −21 °C. The heat capacity can be increased in a narrow temperature window with the help of a PCM. These pumpable liquids with a PCM are called phase change slurries (PCSs). PCSs can be used both as HTF and as TES medium. Solutions with water itself as PCM are most commonly used. In such so-called ice slurries, part of the water in the water-salt solution freezes [5].

Salt hydrates are solid compounds which contain water. The water is integrated into the crystal structure. Characteristic for this crystal structure is a certain ratio of water molecules to the salt. The general formula for a salt hydrate is $M_xN_y{\cdot}nH_2O$, where *M* is the cation, *N* is the anion and the number of crystal water molecules. Due to this stable crystal structure of salt hydrates, the melting temperature is higher than that of water. The solid hydrate crystals release their water of crystallization at a certain temperature (or a temperature range) and the anhydrous salt dissolves in this water and forms a liquid (in some cases only part of the crystal water may be released). PCMs based on salt hydrates utilize the enthalpy of this "melting" process. The PCM is "charged" when heat is added and the salt dissolves in the released crystal water. This process is reversed when heat is withdrawn [5, 15]. Some typical materials with their phase change temperature and melting enthalpy are the following:

- Calcium chloride hexahydrate; $CaCl_2{\cdot}6H_2O$; 29 °C; ~170 kJ kg^{-1}
- Sodium sulfate decahydrate (Glauber's salt); $Na_2SO_4{\cdot}10H_2O$; 32 °C; ~250 kJ kg^{-1}
- Sodium acetate trihydrate; $Na(CH_3COO){\cdot}3H_2O$; 58 °C; ~230 kJ kg^{-1}

Typical advantages of salt hydrates compared to other PCMs include the low price, low toxicity, small changes in volume during phase change, high thermal conductivity and high melting enthalpy. As PCMs, salt hydrates can have disadvantages in terms of incongruent melting, vapor pressure and supercooling. For these disadvantages, measures have been developed to eliminate or minimize them which are described in detail in literature [5, 15]. Incongruent melting can lead to phase separation (precipitated solid) and thus reduced cycling stability. The vapor pressure of salt hydrates is usually slightly lower than that of water. Salts hydrates at or above 100 °C usually have a considerable vapor pressure and require closed containments. Supercooling occurs during cooling (heat discharge) when no sufficient nucleation sites are available. The liquid is cooled below the melting temperature, no crystallization occurs and the stored heat is not discharged. For most applications, subcooling is undesirable, but for pocket warmers, for example, subcooling is a desired phenomenon. Typically, sodium acetate trihydrate is heated in a plastic bag in hot water above the melting temperature of 58 °C. During cooling below the melting temperature to room temperature, the PCM remains liquid (supercooled melt). The pocket warmer is discharged by bending a metal clip that triggers crystallization and immediate heat release (Figure 5.6.2). In addition to the example of a pocket warmer shown, other applications of salt hydrates are in the temperature management of buildings. In addition, there are also other markets such as functional textiles, temperature stabilization of mechanical/electrical equipment and logistic temperature control (e.g. food and medical transport boxes) [5, 16, 17].

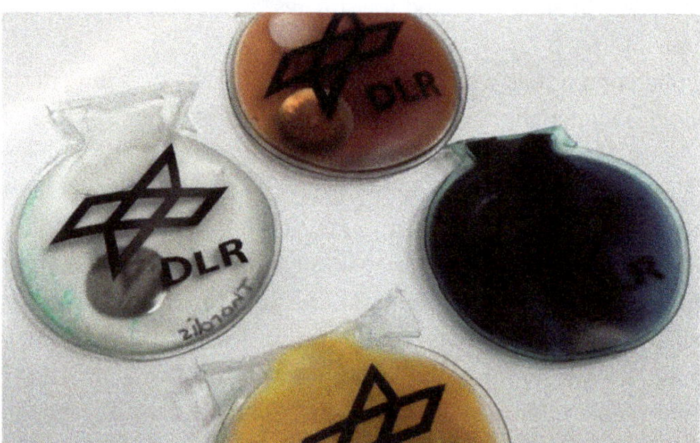

Figure 5.6.2: Pocket warmer with sodium acetate trihydrate in a plastic bag. Bending the metal clip triggers rapid crystallization and heat release.

For TES at high temperature, mainly **anhydrous salts** are used as liquid sensible heat storage medium, which is known as "molten salt" storage [14, 18]. Molten salt is attractive as a heat transfer and a heat storage medium above 100 °C. The major advantages of molten salts are high heat capacity, high density, high thermal stability, relatively low cost, high viscosity, non-flammability and low vapor pressure. The low vapor pressure results in storage designs without pressurized vessels. Molten salts have a characteristic lower and upper temperature operating limit. The lower limit is defined by the melting temperature (or liquidus). The minimum operating temperature is usually chosen a few tens of degrees higher than the melting temperature to safely avoid solidification of the salt during operation. The maximum operating temperature may be limited by different factors. Typical limitations are thermal salt decomposition, high salt vapor pressure and a high corrosion rate of construction materials. Mixtures of different salts have the advantage that they have lower melting temperatures compared to their single salts, but can have similar thermal stability limits. Therefore, salt mixtures can have a wider temperature range and a lower risk of freezing compared to single salts.

The most important salt class of commercial interest for TES are nitrates with a certain amount of nitrites. Chlorides and carbonate anhydrous salts are at the R&D stage. In commercial concentrating solar power (CSP) plants, almost exclusively a non-eutectic salt mixture of 60 wt% sodium nitrate and 40 wt% potassium nitrate is utilized. This mixture is commonly referred to as Solar Salt with an operation temperature range from 290 to 560 °C. For a temperature difference of 250 K, the volumetric heat capacity, or also called storage density, of the medium reaches a value of about 200 kWh m^{-3}. The commercial CSP plants utilize a two-tank storage system. This tank system works with two constant tank temperatures. One tank is "hot" (e.g. 385 or 560 °C) and the other is "cold" (e.g. 290 °C). During discharge, the salt is pumped out of the "hot tank", cools down in the steam generator and is returned to the cold tank. Therefore, the filling levels change during charging and discharging. Figure 5.6.3 shows a typical installation with 28500 tons of molten solar salt and 7.5 h storage duration for a 50 MW_{el} CSP plant. Several such storage systems are in commercial operation. By the end of 2019, the worldwide dispatchable power generation from molten salt storage in CSP plants was about 3 GW_{el} with an electrical storage capacity of 21 GWh_{el}. These storage facilities thus make an important contribution to grid stabilization.

Figure 5.6.3: Example of a 1000 MWh_{th} two-tank molten salt storage system of a concentrating solar power plant in Spain (Source: Andasol 3).

References

[1] Beckmann G, Gilli PV. Thermal Energy Storage: Basics, Design, Applications to Power Generation and Heat Supply, Springer, Wien and New York, 1984.

[2] Dincer I, Rosen MA. Thermal Energy Storage, John Wiley & Sons, New York, Chichester, 2002.

[3] Dinter F, Geyer M, Tamme R. Thermal Energy Storage for Commercial Applications, Springer, Berlin, 1991.

[4] Garg HP, Mullick SC, Bhargava AK. Solar Thermal Energy Storage, D. Reidel Publishing Company, Dordrecht, Boston and Lancaster, 1985.

[5] Mehling H, Cabeza LF. Heat and Cold Storage with PCM – An up to Date Introduction into Basics and Applications, Springer, Berlin and Heidelberg, 2008.

[6] Schmidt FW, Willmott AJ. Thermal Energy Storage and Regeneration, Hemisphere Publishing Corporation, Washington D.C., 1981.

[7] Bauer T, Steinmann W-D, Laing D, Tamme R. Annu Rev Heat Transfer, 2012, 15, 131–77. doi: 10.1615/AnnualRevHeatTransfer.2012004651.

[8] Stadler I, Hauer A, Bauer T. Chapter 10: Thermal energy storage. In: Sterner M, Stadler I. (Eds.) Handbook of Energy Storage: Demand, Technologies, Integration, Springer, Berlin and Heidelberg, 2019.

[9] Stadler I, Kraft A, Bauer T, Faatz R, Grabowski S, Harms G, Herrmann U, Kleimaier M, Maximini M, Ritterbach E, Roth T, Kühne J, Paschen I, Kähler L. Wärmespeicher in NRW – Thermische Speicher in Wärmenetzen sowie in Gewerbe- und Industrieanwendungen, Report EA615, Energieagentur.NRW, 2020.

[10] Homepage and Lexikon of Bundesverband Geothermie e.V., https://www.geothermie.de/bibliothek/lexikon-der-geothermie/u/utes.html

[11] Bott C, Dressel I, Bayer P. Ren Sust Energy Rev, 2019, 113, 109241. doi: 10.1016/j.rser.2019.06.048.

[12] Laing D, Bahl C, Bauer T, Fiss M, Breidenbach N, Hempel M. Proc IEEE, 2012, 100, 516–24. doi: 10.1109/jproc.2011.2154290.

[13] Urbaneck T. Short introduction of our interests // Thermal energy storages, 12. April 2013, Presentation in Stellenbosch (South Africa), http://sterg.sun.ac.za/wp-content/uploads/2013/04/Urbaneck-tse1.pdf

[14] Bauer T, Pfleger N, Laing D, Steinmann W-D, Eck M, Kaesche S. Chapter 20: High temperature molten salts for solar power application. In: Lantelme F, Groult H. (Eds.) Molten Salt Chemistry: From Lab to Applications, Elsevier, Burlington and San Diego, 2013.

[15] Tamme R, Bauer T, Hahne E. Heat Storage Media, in Ullmann's Energy: Resources, Processes, Products, Wiley-VCH Verlag GmbH & Co. KGaA, Weinheim, 2015.

[16] Homepage of Quality Association PCM, https://www.pcm-ral.org/pcm/en/

[17] Xie N, Huang Z, Luo Z, Gao X, Fang Y, Zhang Z. Appl Sci, 2017, 7, 1317. doi: 10.3390/app7121317.

[18] Bauer T, Odenthal C, Bonk A. Chemie Ingenieur Technik, 2021, 93, 534–46. doi: 10.1002/cite.202000137.

5.7 High-energy materials

Thomas M. Klapötke

Energetic materials are defined as being compounds or mixtures of substances which contain the fuel and oxidizer and which react readily with the release of energy and gas. However, although energetic materials release a lot of energy, it is still magnitudes less than that produced by a typical nuclear reaction in which a process occurs that changes the nucleus of an atom. In fact, for comparison, the combustion of ethane (producing CO_2 and H_2O) releases 1560 kJ/mol – which is almost twice the energy that CompB releases on detonation! Although the lower energy released on explosion of CompB is perhaps surprising, in energetic materials, the pertinent point is that the considerable amount of energy which is released on detonation, is released in an extremely short period of time. A comparison of the energies released in different types of processes is given in Figure 5.7.1.

High-energy materials can be classified according to their properties and uses as shown in Table 5.7.1 [1–3], and can be used either as pure substances, or, more commonly, as part of an explosive formulation. Explosive formulations, such as the commonly used CompB, consist of 59.5% RDX, 39.5% TNT and 1% wax, in addition to the high-energy (explosive) material.

The accidental discovery of black powder in China (~220 BC) signaled the beginning of energetic materials. The composition of blackpowder is in fact inorganic: 75% KNO_3, 10% sulfur and 15% charcoal. Despite this, it was only as late as the thirteenth and fourteenth centuries before the properties of black powder were investigated in Europe, and at the end of the thirteenth century, black powder was finally introduced into the military world. Finally, in 1425, as a result of greatly improved production methods, black powder (or gunpowder) was then introduced as a propellant charge for smaller and later also for large caliber guns. Despite this very long history, even today ca. 100,000 pounds of black powder are still used in the US military per year, for example, in "time blasting fuses". Another important landmark in the history of energetic materials was the commercialization of nitroglycerine (NG) in the form of dynamite in 1867 by Alfred Nobel. Since then, great efforts and strides have been made in developing energetic materials which are not just explosive, but which show increased performance, while exhibiting a marked lower sensitivity to external stimuli.

Although black powder is an inorganic mixture, nitroglycerine (NG) (like most traditional high-energy materials, especially high explosives, also called secondary explosives), is generally classed as belonging more to organic chemistry (see Table 5.7.2). However, a closer look at primary explosives and many rocket propellants shows that many can be considered as inorganic compounds (see next chapter).

https://doi.org/10.1515/9783110798890-007

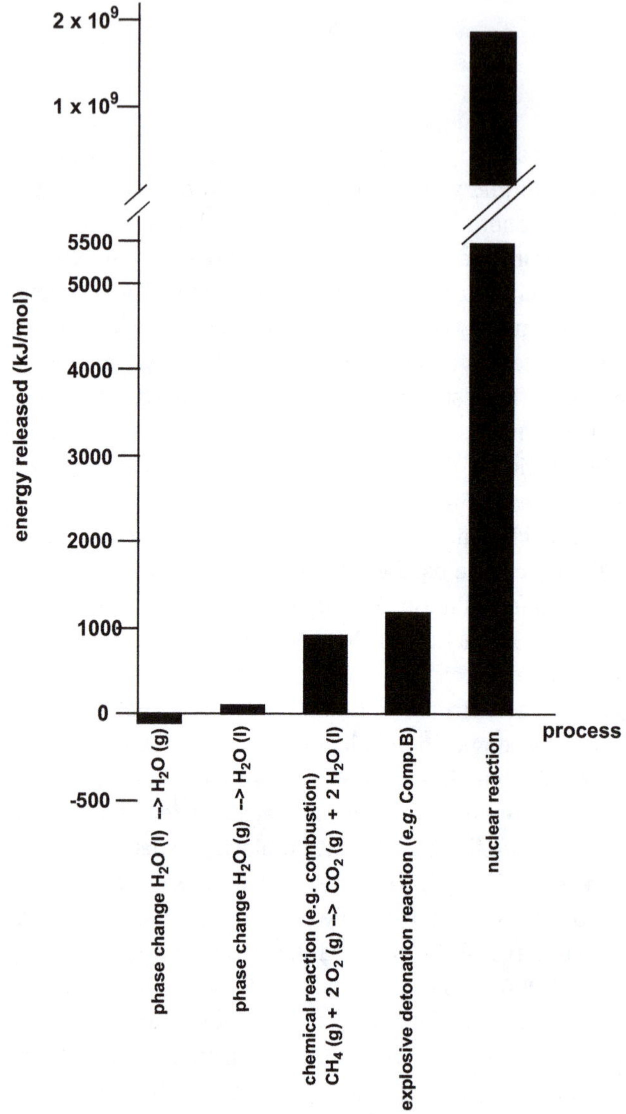

Figure 5.7.1: Comparison of the energies released in typical examples of different processes.

Table 5.7.1: Classification of high-energy materials.

Energetic materials			
High explosives (HEs) (secondary explosives)	**Primary explosives**	**Propellants**	**Pyrotechnics**
HEs cannot simply be initiated through heat or shock as primary explosives can be. Primary explosive must be used, which produces a shockwave that initiates the HE; usually higher performance than primary explosives	Show very rapid transition from combustion (or deflagration) to detonation; are much more sensitive toward friction, impact or heat than high explosives are; generally lower detonation velocities, detonation pressures and heat of explosion *c.f.* HEs; initiators of HEs, main charges, propellants	rocket propellants combust in controlled manner and don't detonate, however, propellant charge powders burn much quicker than rocket propellants which results in a higher pressure for gun propellants *c.f.* rocket propellants; can be solid or liquid; can be single, double or triple base, hypergolic	Substance or mixture of substances which is designed to produce an effect by heat, light, sound, gas or smoke or a combination of these effects as a result of a non-detonative self-sustaining exothermic chemical reaction(s) which doesn't rely on external oxygen to sustain the reaction
E.g. TNT, HMX (Octogen), RDX (Hexogen)	E.g. lead azide, mercury fulminate, lead styphnate, tetrazene	E.g. black powder (gunpowder),	*e.g.* $Sb_2S_3/KClO_3/P_{red}/$ glass
Applications	**Applications**	**Applications**	**Applications**
	Civil: initiator of HNS (hexanitrostilbene) in deep oil wells	Civil: charge in pyrotechnic munitions Military: charge in pyrotechnic munitions; from pistols to artillery weapons; large artillery and cannons, rockets	Civil: fireworks, signal flares, matches Military: smoke munition, signal flares, decoy munition

5.7.1 Inorganic explosives

5.7.1.1 Primary explosives

In contrast to secondary explosives (e.g. β-HMX, TNT, RDX), primary explosives are substances which show a very rapid transition from combustion (or deflagration) to detonation. In addition, primary explosives are also considerably more sensitive toward heat, impact or friction than secondary explosives [1–3]. Crucially, primary explosives generate either a large amount of heat or a shockwave enabling transfer of the detonation to a less sensitive secondary explosive. Consequently, they are used

Table 5.7.2: Historical overview of some important secondary explosives [4].

Substance	Acronym	Development	Application	Density (g cm^{-3})	Explosive power[a]
Black powder	BP	1250–1320	1425–today	ca. 0.89	29
Nitroglycerine 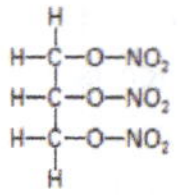	NG	1863	Propellant charges	1.6	170
Picric acid 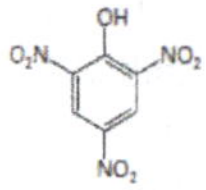	PA	1885–1888	WW I	1.77	100
nitroguanidine 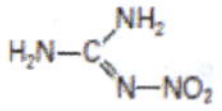	NQ	1877	Propellant charges, airbags	1.71	99
Trinitrotoluene 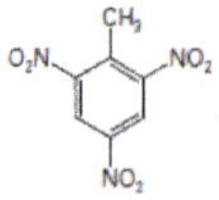	TNT	1880	WW I	1.64	116
Nitropenta 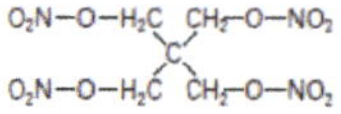	PETN	1894	WW II–today	1.77	167
Hexogen 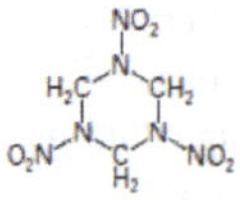	RDX	1920–1940	WW II–today	1.81	169
Octogen 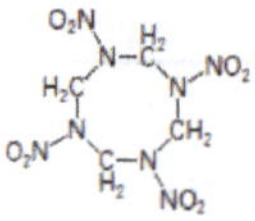	HMX	1943	WW II–today	1.91 (β-form)	169
Hexanitrostilbene 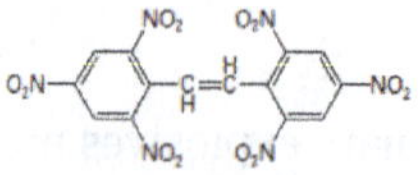	HNS	1913	Oil exploration	1.74	86

Table 5.7.2 (continued)

Substance	Acronym	Development	Application	Density ($g\ cm^{-3}$)	Explosive power[a]
Triaminotrinitrobenzene 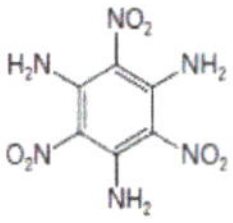	TATB	1888	Nuclear weapons	1.94	79
HNIW 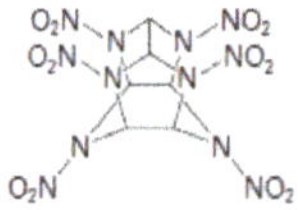	CL-20	1987	Under evaluation	2.04 (ε-form)	136
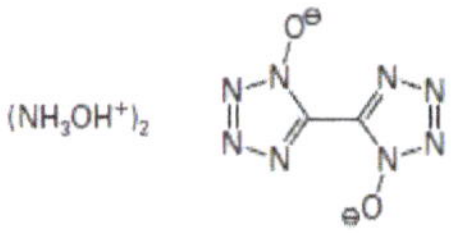 $(NH_3OH^+)_2$	TKX-50	2012	Under evaluation	1.877	160

[a]relative to picric acid.

as initiators for secondary booster charges (e.g. in detonators), main charges or propellants. It is important to note two defining characteristics of primary explosives (e.g. $Pb(N_3)_2$), namely, that despite the fact they are considerably more sensitive than secondary explosives (e.g. RDX), the detonation velocities typical of primary explosives are generally lower than those of secondary explosives. Two well-known primary explosives which are commonly used are lead azide (LA) and lead styphnate (LS) (see Table 5.7.3). Despite being less powerful than LA, LS has the advantage of being easier to initiate.

Mercury fulminate (MF, see Table 5.7.3) was initially used as the primary explosive in blasting caps, however, it was replaced by LA and LS by Alfred Nobel, since both LA and LS were found to be safer to handle than MF (Table 5.7.3). However, the long-term use of LA and LS has caused considerable lead contamination in military training grounds, resulting in the current worldwide activities searching for heavy metal-free replacements to LA and LS. These ongoing investigations into lead-free alternatives are important, since in 2016, the US military still used ca. 2000–3000 kg LA per year.

In the hunt for lead-free replacements for primary explosives, one compound, namely copper(I) 5-nitrotetrazolate has been the most extensively investigated and tested (Table 5.7.3). Copper(I) 5-nitrotetrazolate has been developed under the name of **DBX-1** by the company Pacific Scientific EMC, and has been shown to be thermally stable up to 325 °C (DSC). The impact sensitivity of DBX-1 is 0.04 J (ball-drop instrument) and therefore similar to the value of 0.05 J for LA. Other factors

Table 5.7.3: Overview of some important primary explosives and promising heavy metal-free alternatives [4].

Substance	Acronym and structure	VoD (m/s)	$P_{C\text{-}J}$† (kbar)	IS† (J)	FS† (N)	ESD† (J)	ρ (g/cm³)
Mercury fulminate	MF 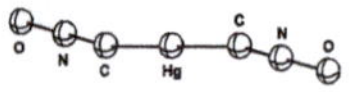	4976 (calcd.)	246 (calcd.)	0.62	5.3	0.025	4.43
Lead azide	LA $Pb(N_3)_2$	6077 (calcd.)	349 (calcd.)	0.05	0.1–1	0.007	4.8
Lead styphnate monohydrate	LS 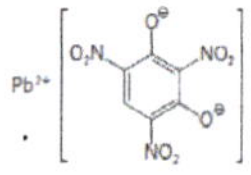 · H_2O	6098 (calcd.)	244 (calcd.)	2.5–5		0.0009	3.02
Copper(I) 5-Nitrotetrazolate	DBX-1 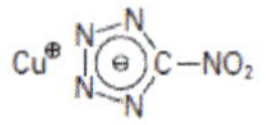	7000	367	0.04	0.098	0.000012	2.584
Calcium nitriminotetrazolate	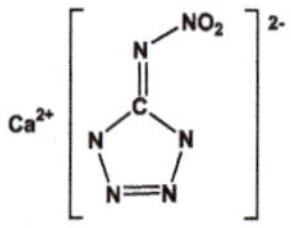	7970	271	50	112	0.15	2.0
$[NH_4]_2[Cu^{II}(NT)_4(H_2O)_2]$ (NT = 5-Nitrotetrazolate)	NH_4FeNT	7700		3		>0.36	2.2
Diazidodinitrophenol	DDNP 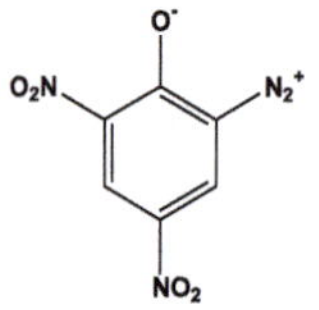	7331 (calcd.)	219 (calcd.)	1.5	20	0.15	1.63
Tetrazene	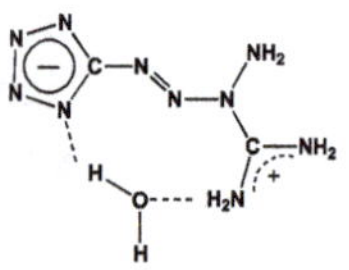	8820 (calcd.)	268 (calcd.)	1		0.01	1.7
(1-*H*-Tetrazol-5-yl) triaz-2-en-1-ylidene	MTX-1 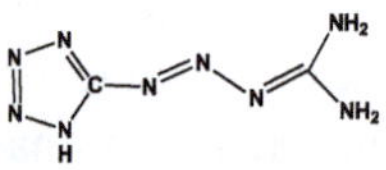	9729 (calcd.)	338 (calcd.)	0.02		3–4 mJ	2.351

Table 5.7.3 (continued)

Substance	Acronym and structure	VoD (m/s)	$P_{C\text{-}J}$† (kbar)	IS† (J)	FS† (N)	ESD† (J)	ρ (g/cm³)
Cyanuric azide	TTA, TAT	7866 (calcd.)	226 (calcd.)	0.18	0.1	1.2 mJ	1.73

† VoD corresponds to the detonation velocity, $P_{C\text{-}J}$ to the detonation pressure at the Chapman-Jouguet point, IS to the impact sensitivity, FS to the friction sensitivity and ESD to the electrostatic sensitivity

are also important for future primary explosives, such as a high thermal stability (DBX-1 is stable at 180 °C for 24 h in air and for 2 months at 70 °C) and simple preparation. DBX-1 can be obtained from sodium nitrotetrazolate (NaNT) and $CuCl_2$ in HCl/H_2O solution at a higher temperature, but the best preparation for DBX-1 in yields of 80–90% is shown in the following equation where sodium ascorbate, $NaC_6H_7O_6$, is used as the reducing agent:

$$CuCl_2 + NaCN_4NO_2 \xrightarrow{NaC_6H_7O_6 \text{ as reducing agent, } HCl/H_2O \text{ solution, 15 mins, elevated } T} DBX-1$$

5.7.1.2 Secondary explosives

The most common applications for energetic materials are either as high explosives (secondary explosives) or in propellant formulations. In terms of propellant formulations, there are certain factors which are important in determining the effectiveness of new energetic molecules in these formulations. For example, a high density (ρ), good oxygen balance (Ω_{CO2}, see below), high detonation/combustion temperatures and a high specific impulse (I_{sp}) is vital for rocket propellant formulations, while lower combustion temperatures combined with a high force and pressure, as well as a high N_2/CO ratio of the reaction gases is crucial for gun propellants.

Secondary explosives (HEs) differ from primary explosives since HEs cannot be initiated simply through heat or shock [1–3]. In fact, in order to initiate a secondary explosive, it is necessary to use a primary explosive, since the shockwave of the primary explosive initiates the secondary explosive. Although primary explosives are more sensitive to external stimuli, the performance of a secondary explosive is usually higher than that of a primary explosive. Typical secondary explosives which are currently used on a large scale are TNT, RDX, HMX, NQ and TATB (Table 5.7.2), as well as HNS and NG (e.g. in dynamite form) for civil applications.

Current trends in the research of modern secondary explosives can be arranged into three branches (Figure 5.7.2), with the overall emphasis being placed on insensitive munitions which demonstrate lower toxicity (also for the detonation and biological degradation products) while achieving higher explosive performance.

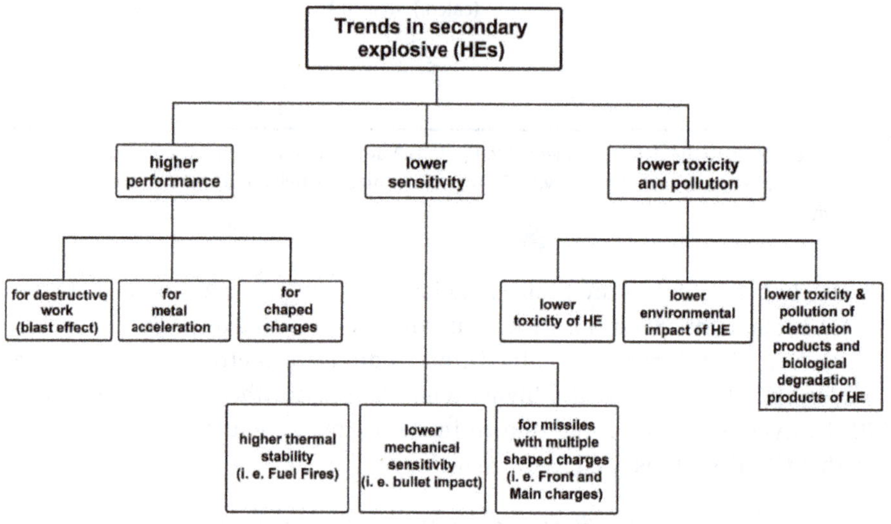

Figure 5.7.2: Current trends in the research and development of new, modern secondary explosives.

A higher performance for secondary explosives is of fundamental importance and always desired. The main performance criteria are: 1. heat of explosion Q (in kJ kg^{-1}), 2. detonation velocity VoD (m s^{-1}), 3. detonation pressure $P_{C\text{-}J}$ (in kbar), and less importantly, 4. explosion temperature T (K) and 5. volume of gas released V per kg explosive (in L kg^{-1}).

Although CL-20 (see Figure 5.7.2) has been investigated to an enormous extent over the recent past, it has the great disadvantage of a multistep, complicated and expensive synthesis, and in addition, it also has a very high impact sensitivity (4 J, ε-CL-20), which classifies it as sensitive under the UN classification for the transport of dangerous goods (UN classifications for IS: <4 J = very sensitive, 4–35 J = sensitive). The considerable cost of CL-20 production is highly problematic for its large-scale production, especially since established secondary explosives such as RDX are relatively cheap to produce. CL-20 is also a good illustrative example of the problems which can arise if phase transitions occur in a secondary explosive in relevant temperatures ranges, since several polymorphs of CL-20 are known (β, γ, ε, high-pressure polymorph ζ as well as the hydrate CL-20 · $^1/_2H_2O$ which is often incorrectly referred to as being the α-polymorph of CL-20 in the literature). Only the ε-polymorph of CL-20 is used for applications. Unfortunately, however, the γ-polymorph – and not the ε-polymorph – is the thermodynamically most stable one, and the ε to γ-phase transition occurs in the 157–176 °C temperature range. Other reports in the literature state that both the β and ε-phases slowly convert to the γ-phase at elevated temperatures, meaning that elevated temperatures have to be avoided in order to preserve phase purity of the ε-CL-20, otherwise the ε-CL-20 is contaminated with other phases. The issues of phase purity are further complicated by solvents and the method of crystallization having a considerable effect on causing phase changes. Ideally, a secondary explosive shows only one polymorph in the temperature range relevant for applications, since this would eliminate the considerable issues which can arise if phase purity isn't present.

In contrast to the concept of cage-strain which is the concept behind the increased energy of the CL-20 molecule, a new approach involving nitrogen-rich (also referred to as high-nitrogen) compounds is a particularly interesting modern method for synthesizing new energetic materials with improved performance. In addition to high nitrogen contents (>60%), oxidizing groups can also be included to improve the oxygen balance (Ω_{CO2}). The basis for adopting a nitrogen-rich approach is that these nitrogen-rich molecules are endothermic and it is possible to exploit the fact that the element nitrogen is unique among the period 2 elements in that the bond energy per two electron bond increases in the series single – double – triple bond. Since the $N\equiv N$ triple bond in elemental nitrogen (N_2) is so strong, the formation of $N\equiv N$ bonds (bond order of 3) generated from compounds containing N–N bonds with bond orders of 1–2 results in the release of a lot of energy. However, incorporating N–N bonds with low bond orders into energetic compounds can result in a reduction of the stability. One approach which has proven particularly successful in achieving compounds with nitrogen contents, but which are also energetic, has been the synthesis of compounds which incorporate triazole or tetrazole rings. In this context, tetrazole (five-membered rings containing only 1 carbon atom) compounds have been particularly well-investigated since they constitute a compromise between the usually less energetic triazole rings and usually kinetically too labile pentazole rings (five-membered rings consisting of only N atoms) (Figure 5.7.3).

Out of the numerous energetic tetrazole compounds that have been extensively investigated and reported in the literature in recent years in attempts to find new nitrogen-rich secondary explosives, TKX-50 is one of the most exciting, since it shows great promise for future application as a secondary explosive [5, 6]. TKX-50 combines low IS, FS and toxicity with high thermal stability, density and detonation velocity (Figure 5.7.4). In addition to these excellent properties, it can be prepared in a facile, inexpensive synthesis. TKX-50 is a future generation, high-performance explosive with increased safety without compromising in explosive performance (Figure 5.7.4).

The interest in TKX-50 has been considerable since its first report in only 2012, and many varied aspects of its chemical and physical properties have been reported by groups from all over the world. TKX-50 can be prepared on a >100 g scale starting

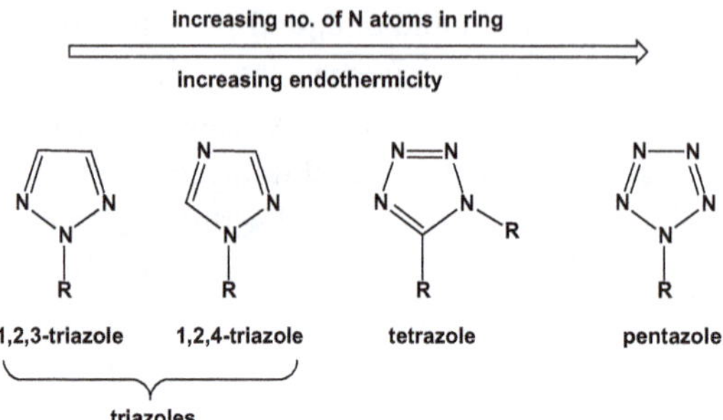

Figure 5.7.3: General connectivities of triazoles (1,2,3-triazole (left), 1,2,4-triazole (right)) (left), tetrazole (middle) and pentazole (right).

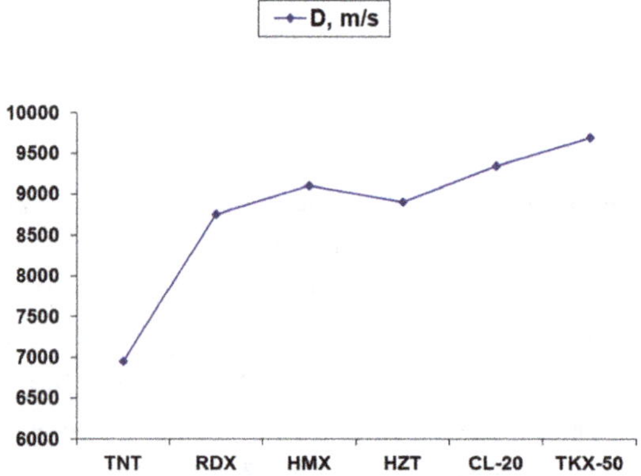

Figure 5.7.4: Performance of conventional explosives (D = VoD = velocity of detonation).

from the reaction of 5,5′-(1-hydroxytetrazole) with dimethyl amine, according to the pathway outlined in Figure 5.7.5.

The future of secondary energetic materials will undoubtedly involve further investigations into nitrogen-rich compounds – including triazoles and tetrazoles and their corresponding derivatives, for example, *N*-oxides such as TKX-50, as well as those in which nitrogen-rich rings and chains are combined within a molecule.

Future research will lead to even more powerful, high-nitrogen high-oxygen explosives with enhanced and superior detonation parameters to fulfill the lethality requirements of all forces. New energetics will provide vastly increased energy content over RDX, up to several times the energetic performance. In addition, these materials will have a high energy density with high activation energy.

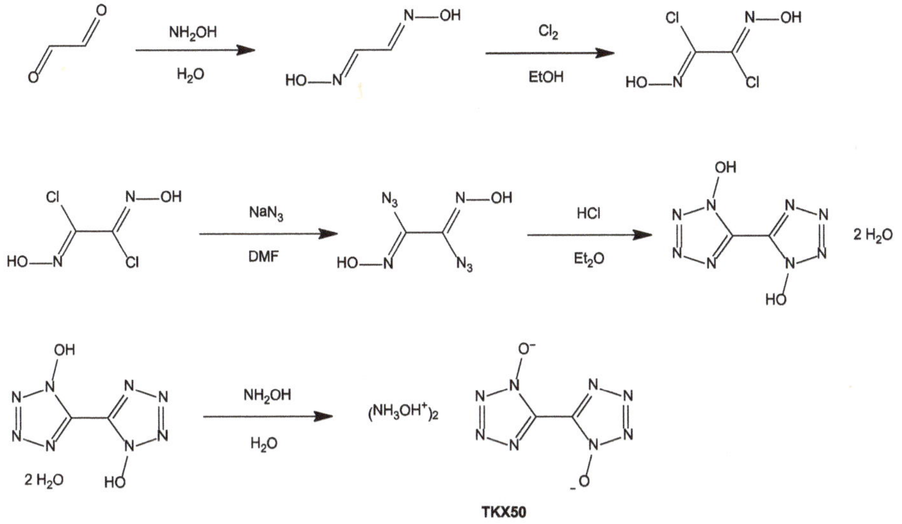

Figure 5.7.5: Synthesis of TKX-50.

5.7.2 Inorganic rocket fuels

Generally, rocket propellants can be categorized as being either for solid propellants or for liquid propellants (Figure 5.7.6). Solid propellants can be further categorized as being either double-base (homogeneous) or composite propellants (heterogeneous). Homogeneous propellants are generally based on NC, whereas heterogeneous propellants are generally based on AP. Double-base propellants are homogeneous NC/NG formulations, whereas composite propellants are mixtures of a crystalline oxidizer (e.g. ammonium perchlorate, AP) and a polymeric binder, which has been cured with isocyanate (e.g. hydroxy-terminated polybutadiene, HTPB/diisocyanate) and which also contains the fuel (e.g. Al). The purpose of the polymer matrix (the binder) is to form a solid, elastic body consisting of the propellant ingredients with suitable mechanical properties. The binder also acts as a fuel, since it contains mainly hydrogen and carbon [1–3].

5.7.2.1 Solid propellants

The solid propellants of almost all solid rocket boosters are based on a mixture of aluminum (Al) acting as the fuel and ammonium perchlorate (AP) acting as the oxidizer [1–3]. In addition to its applications in munitions – primarily as an oxidizer for solid rocket and missile propellants – AP is also used as an airbag inflator in the automotive industry, as well as being a component in fireworks. Unfortunately, a consequence of its large-scale use combined with its high aqueous solubility, chemical stability, and

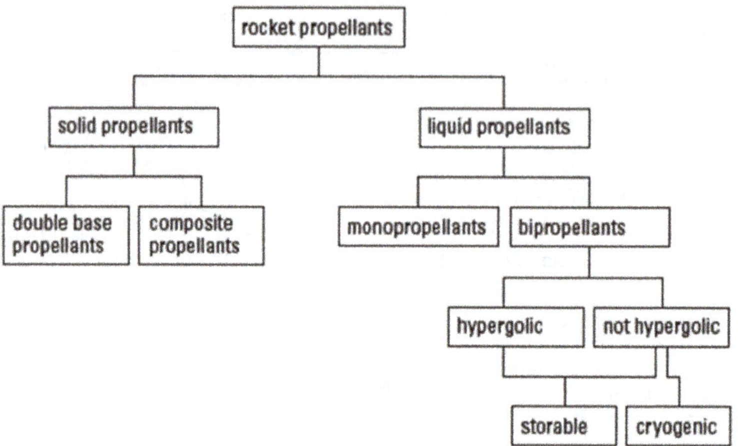

Figure 5.7.6: Classification of rocket propellants.

persistence is that it has become widely distributed in surface and ground water systems. Despite limited information being available about the effects of perchlorate on aquatic life, it is known that perchlorate disrupts the endocrine system and interferes with normal thyroid function, impacting both growth and development in vertebrates. In order to reduce the detrimental effects resulting from perchlorate competition for iodine binding sites in thyroids, investigations into adding iodine to culture water to mitigate perchlorate effects has been undertaken.

In attempts to move away from halogen-containing oxidizers, one of the current most promising chlorine-free oxidizers is ammonium dinitramide (ADN, Figure 5.7.7), first synthesized in 1971 in Russia (Oleg Lukyanov, Zelinsky Institute of Organic Chemistry) and currently being commercialized today by EURENCO.

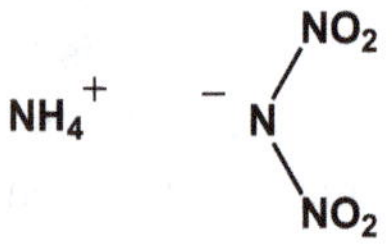

NH_4^+ $\quad ^-N(NO_2)_2$

Figure 5.7.7: Structure of ammonium dinitramide (ADN).

While ADN shows the best oxygen balance ($\Omega_{CO_2} = 25.8\%$, *cf.* AP 34.0%) of all presently discussed AP replacements, it still has some stability issues in terms of its compatibility with binders, as well as possessing only a low thermal stability ($T_{dec.} = 127$ °C). The oxygen balance Ω_{CO2} (in %) is defined as follows:

$$\Omega_{CO_2} = \frac{\left[d - (2a) - \left(\frac{b}{2}\right)\right] \times 1600}{M}$$

After undergoing melting at 91.5 °C, the thermal decomposition of ADN occurs at 127 °C. The main decomposition pathway is reported to be based on the formation of NH_4NO_3 and N_2O, followed by the subsequent thermal decomposition of NH_4NO_3 to N_2O and H_2O at higher temperatures. In addition, side reactions forming NO_2, NO, NH_3, N_2 and O_2 have also been described.

5.7.2.2 Liquid propellants

Liquid rocket propellants can be differentiated as being either monopropellants or bipropellants [1–3]. Monopropellants are endothermic liquids (e.g. hydrazine, N_2H_4), which show exothermic decomposition in the absence of oxygen. This usually involves a catalyst, such as e.g. Shell-405; Ir/Al_2O_3.

$$N_2H_4 \xrightarrow{catalyst} N_2 + 2H_2 \, \Delta H = -51 \, kj/mol$$

Since monopropellants have relatively small energy contents and low specific impulses, they are usually only used in small missiles and small satellites (for correcting orbits), where no large thrust is necessary.

In a bipropellant system, the oxidizer and the fuel are transported in two separate tanks and are injected into the combustion chamber. Bipropellants can be separated in two different ways. One approach is to categorize them as cryogenic bipropellants (which can be handled only at very low temperatures and are therefore unsuitable for military applications, e.g. H_2/O_2) or storable bipropellants (e.g. monomethylhydrazine/HNO_3). A second approach is to classify them in accordance with their ignition behavior as being either hypergolic or non-hypergolic mixtures.

The term "hypergolic" describes rocket propellants whose components react spontaneously with one another (in less than 20 ms) and ignite partly explosively when they are mixed or are in contact with each other. Hypergolic propellants are always combinations of oxidizers and reducing agents. The immediate reaction of the fuel on contact with the oxidizer by burning at the time of injection into the combustion chamber ensures that it is not possible for too much fuel to accumulate in the combustion chamber prior to ignition. This is important, since such a scenario could lead to an explosion and damage of the rocket engine. A further important advantage is that the ignition definitely occurs, which is important for weapon systems such as intercontinental rockets, pulsed engines and upper stages of launch vehicles (re-ignition in space) for example. (*Note:* Hypergolic systems are also used in incendiary devices e.g. for mine clearance.) The combination of a hydrazine derivative (hydrazine, MMH (Figure 5.7.8) and UDMH (Figure 5.7.8)) as the fuel with HNO_3 or NTO (dinitrogen tetroxide) as the oxidizer are practically the only hypergolic propellants used today. It is worthwhile to point out the presence of the weak N–N single bond in each of these molecules.

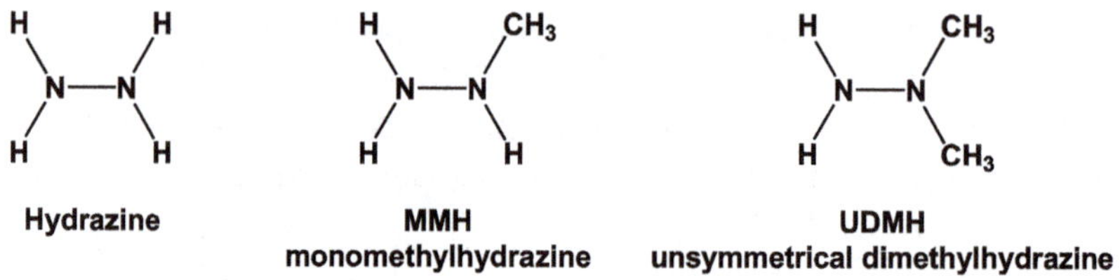

Hydrazine

MMH
monomethylhydrazine

UDMH
unsymmetrical dimethylhydrazine

Figure 5.7.8: Connectivities of hydrazine, MMH (monomethylhydrazine) and UDMH (unsymmetrical dimethylhydrazine) – examples of fuels which can be used in hypergolic propellant systems.

Table 5.7.4 shows a summary of different possible liquid bipropellant mixtures and the corresponding relevant performance parameters. As can be noted from Table 5.7.4, in addition to oxygen-containing oxidizers such as dinitrogen tetroxide (N_2O_4, dimer of NO_2), hydrogen peroxide (H_2O_2) or liquid oxygen (LOX), elemental fluorine or a mixture of elemental fluorine and elemental oxygen (FLOX) can also be used as oxidizers. However due to the technical difficulties in handling liquid fluorine and the formation of highly corrosive and toxic HF as a combustion product, it is usually not the oxidizer of choice for such systems.

Table 5.7.4: Examples of bipropellant systems. T_c = temperature in the combustion chamber and I_{sp} = specific impulse.

Oxidizer	Fuel	T_c (°C) [a]	I_{sp} (s) [b]
LOX (liquid O_2)	H_2	2740	389
LOX	CH_4	3260	310
LOX	N_2H_4	3132	312
F_2 (l)	H_2	3689	412
	N_2H_4	4461	365
FLOX (mixture of F_2 (l) and O_2 (l), 30/70)	H_2	2954	395
N_2O_4	MMH, $CH_3HN\text{-}NH_2$	3122	289
H_2O_2	N_2H_4	2651	287
H_2O_2	MMH, $CH_3HN\text{-}NH_2$	2781	285

References

[1] Klapötke TM. Chemistry of High Energy Materials, 6th edition, Walter de Gruyter, Berlin/Boston, 2022. 

[2] Keshavarz MH, Klapötke TM. Energetic Compounds – Methods for the Prediction of their Performance, 2nd edition, De Gruyter, Berlin, Boston, 2020.

[3] Keshavarz MH, Klapötke TM. The Properties of Energetic Materials, De Gruyter, Berlin, Boston, 2018.

[4] Klapötke TM. Energetic Materials Encyclopedia, 2nd edition, De Gruyter, Berlin/Boston, 2021. 

[5] Klapötke TM. TKX-50: A highly promising secondary explosive. In: Klapötke TM, Trache D. (Eds.) 8th Chemistry Days, Ecole Militaire Polytechnique, Springer, Singapore, 2021.

[6] Klapötke TM. Ein Neuer Sekundär-Sprengstoff: TKX-50. Chemie unserer Zeit, 2020, 54, 234–41. DOI: 10.1002/ciuz.201900068.

5.8 Nuclear materials

Nils Haneklaus

Nuclear material is a term commonly used when referring to uranium, plutonium and thorium. Natural uranium including uranium ore concentrate and depleted uranium are further characterized as "source material" and uranium enriched in ^{235}U, ^{233}U and ^{239}Pu is called "special fissionable material" [1]. Uranium that is the most relevant nuclear material is a naturally occurring element (99.284 wt% ^{238}U, 0.711 wt% ^{235}U and 0.0055 wt% ^{234}U) with an average concentration of 2.8 ppm in the Earth's crust, making it more abundant than gold, silver or mercury and slightly less abundant than cobalt, lead or molybdenum [2]. The present chapter provides a brief overview of nuclear fuels, the most common application of nuclear materials as well as nuclear batteries and nuclear waste, two topics that recently received increased attention.

5.8.1 Nuclear fuels

Nuclear reactors require special fissionable material or in fewer cases natural uranium as nuclear fuel for their operation. Today and for the foreseeable future (at least until 2050), uranium enriched in ^{235}U is the most important nuclear fuel. All of the roughly 450 commercial nuclear reactors operating worldwide today use uranium fuel [3]. Low-enriched uranium (usually 3–5 wt% ^{235}U) as uranium dioxide (UO_2) is the predominant fuel for nuclear power reactors used today. Uranium dioxide (UO_2) to be used as nuclear reactor fuel can be produced from enriched uranium hexafluoride (UF_6) through reaction with ammonia to form ammonium diuranate $(NH_4)_2U_2O_7$ that is calcined to form UO_3 and U_3O_8. Heating with hydrogen or ammonia results in the desired uranium dioxide (UO_2) that is mixed with an organic binder, pressed into pellets and sintered to form the characteristic fuel pellets used in nuclear fuel rods that ultimately provide the heat to warm water used to directly (boiling water reactor) or indirectly (pressurized water reactor) produce steam that turns a turbine connected to an electric generator that ultimately converts this mechanical energy into electrical power. The process from uranium conversion to nuclear waste storage is referred to as the nuclear fuel cycle, although it can hardly be considered a cycle. Uranium exploration, mining and milling are important processes that are needed prior to uranium conversion. Uranium conversion from ore concentrate to gaseous uranium hexafluoride (UF_6) is required for enrichment. Some nuclear reactors with excellent neutron efficiency such as the heavy-water reactors primarily used in Canada called CANDU (Canada Deuterium Uranium) reactors can directly use natural uranium skipping the enrichment process. Spent nuclear fuel still contains slightly more ^{235}U than

https://doi.org/10.1515/9783110798890-008

natural uranium, so that reprocessing to separate the available ^{235}U from spent light water reactor fuel is done by some countries and the recycled uranium is again used for fuel fabrication (Figure 5.8.1).

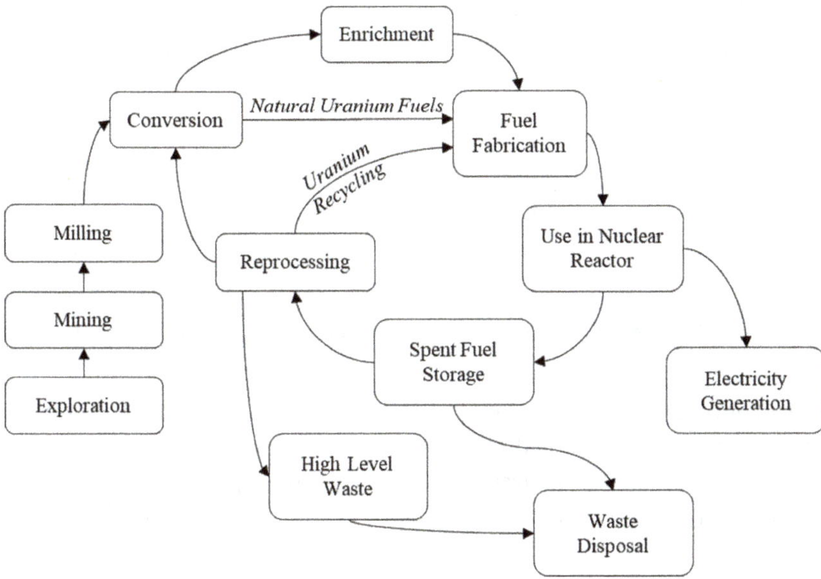

Figure 5.8.1: Overview of the nuclear fuel cycle from exploration to waste disposal.

5.8.2 Nuclear batteries

Nuclear batteries, sometimes referred to as atomic- or radioisotope batteries, are devices that use the energy from radioactive decay to generate electricity. Unlike nuclear reactors that rely on a nuclear chain reaction and are thus critical, nuclear batteries operate in subcritical mode without the occurrence of a nuclear chain reaction. Nuclear batteries are thus much simpler and the idea was first demonstrated by Henry Moseley in 1912 when he showed that he could generate a current by charged particles. In comparison the first (artificial) nuclear reactor, Enrico Fermi's Chicago Pile-1, reached criticality in 1942 and it would take another six years until a first nuclear reactor, Oak Ridge's X-10 Pile, produced enough electricity to power a light bulb. From these humble beginnings nuclear reactors provided some 10% of globally produced electricity in 2020. Nuclear batteries on the other hand turned out to be rather expensive and find application in niche markets for power sources that must be able to operate unattended for longer periods of time as they are found in spacecrafts, pacemakers, deep underwater systems and automated scientific stations in remote parts of the world [4]. These nuclear batteries can be further categorized in thermal and non-thermal converters (Figure 5.8.2).

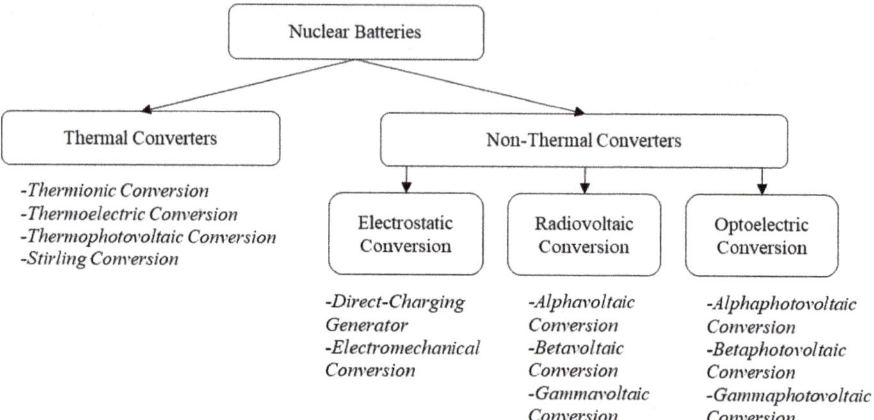

Figure 5.8.2: Basic differentiation of nuclear batteries in thermal and non-thermal converters.

Thermal converters use decay heat to produce electricity while non-thermal converters extract energy from emitted radiation before it is degraded into heat. Non-thermal generators can further be classified according to the type of particle used and by the mechanism by which the particles energy is converted [4]. The device first tested by Henry Moseley in 1913 for instance can be characterized as a direct-charging generator. A charged radium particle created electricity here. Recently, radiovoltaic conversion and here particularly betavoltaics, devices that use betavoltaic conversion, are of increased interest in the scientific community [5, 6]. A betavoltaic battery generates electric current from beta particles emitted from a radioactive source, such as tritium or promethium. In a non-thermal conversion process nuclear radiation generates electricity by producing electron-hole pairs. They are formed by the ionization trail of the beta particles traversing a semiconductor. Betavoltaic power sources are useful in low-power electrical applications where a durable energy source is needed.

5.8.3 Nuclear waste

Nuclear waste or radioactive waste is a hazardous waste material that contains radioactive material. Although radioactive waste is largely associated with nuclear power production, it is actually accumulated by a number of activities outside nuclear power production such as nuclear medicine and particularly mining such as rare-earth element mining that probably generates the largest quantities of nuclear waste by weight and volume today. Classifying nuclear waste is notoriously difficult as national regulations vary from country to country and can further vary from state to state. In this context, the International Atomic Energy Agency (IAEA) differentiates between six categories of materials: (1) exempt waste (EW), (2) very low-level waste (VLLW), (3) very short-lived waste (VSLW), (4) low-level waste (LLW), (5) intermediate-level waste (ILW) and (6) high-level waste (HLW). Figure 5.8.3 provides a conceptual illustration of the waste classification scheme proposed by IAEA [7].

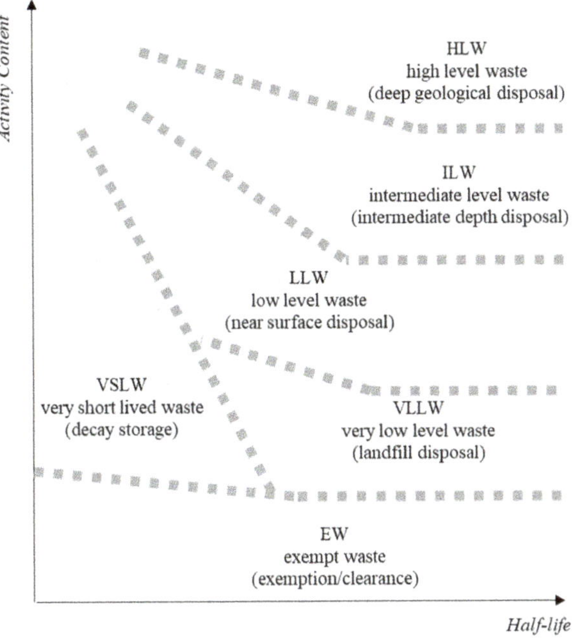

Figure 5.8.3: Conceptual illustration of the nuclear waste classification scheme proposed by IAEA.

Besides the six material categorizes offered by IAEA, two other classifications are important and particularly relevant in mining: naturally occurring radioactive material (NORM) and technically enhanced naturally radioactive material (TENORM). Both NORM and TENORM show low levels of radioactive materials. A common example would be natural uranium or thorium that can be associated with phosphate rock (relevant for mineral fertilizer production) or monazite (relevant for rare-earth element mining). When NORM materials are mined their tailings are considered TENORM if the radioactive constituents have not been isolated and could be become very low-level waste (VLLW), very short-lived waste (VSLW), low-level waste (LLW) or even intermediate-level waste (ILW) that require special disposal as indicated in Figure 5.8.3.

References

[1] IAEA, IAEA Safeguards Glossary, 2001. ISBN 92-0-111902-x.
[2] WNA, Uranium Mining Overview, 2020. https://world-nuclear.org/information-library/nuclear-fuel-cycle/mining-of-uranium/uranium-mining-overview.aspx
[3] WNA, Nuclear Power in the World Today, 2021. https://www.world-nuclear.org/information-library/current-and-future-generation/nuclear-power-in-the-world-today.aspx
[4] Prelas MA, Weaver CL, Watermann ML, Lukosi ED, Schott RJ, Wisniewski DA. Progr Nucl Energy, 2014, 75, 117–48.
[5] Zhou C, Zhang J, Wang X, Yang Y, Xu P, Li P, Zhang L, Chen Z, Feng H, Wu W. ECS J Solid State Sci Techn, 2021, 10, 027005.
[6] Spencer MG, Alam T. Appl Phys Rev, 2019, 6, 031305.
[7] IAEA, Classification of radioactive waste: safety guide, ISBN 978-92-0-109209-0.

6 Ionic solids

6.1 Resources: minerals, recycling and urban mining

Martin Bertau, Peter Fröhlich, Sandra Pavón

Raw material supply can be safeguarded by different means. The most reliable one is mining. Though regarded critically, mining products offer raw materials in primary product quality and as such their specifications conform to the needs of industry. It has to be emphasized here that mining is not done for the sake of digging into the earth's crust. Mining is indispensable for securing raw material supply of an economic area, the European Union for instance. Particularly in view of the rapidly growing world population along with the rapidly growing needs for food and consumer goods, raw material cycles today need to be fed with additional material, the only source of which is mining. In order to meet the challenges of a safe future, any industrial production has to obey the rules of being

1. climate friendly
2. zero waste
3. low carbon footprint
4. energy and resource efficient and
5. water saving

As pointed out in Chapter 2.1, one reasonably does not differentiate between raw materials of primary and secondary origin (circular resources chemistry). Moreover, one has to recognize that all the challenges of a secure raw material supply can be solved only through actors free of ideological mindsets. One such is the fairy tale of modern mining being dirty, another one is recycling being environmentally benign. Recycling means restoring primary product quality. In contrast to what is being encountered in ores, i.e. primary raw materials, there is a plethora of differently composed EoL and EoU products with each of which constituting a complex polymetallic mixture where the respective elements are present in *minimal* amounts, i.e. in *maximal* dilution. Ores in contrast display a limited set of elements to be treated where the respective elements are present in *maximal* amounts, i.e. in *minimal* dilution. It was for that reason why the term urban mining was coined for anthropogenic or technical deposits in which certain elements are concentrated locally. Yet, when considering a mobile phone, it is still in economically unattractive amounts. Consequently, recycling in itself is the most energy consuming way to recover elements. But it is part of the truth, too, that secondary raw materials are an important brick stone for raw material supply, although it is in reality downcycling what commonly is perceived of as recycling as mentioned in Chapter 2.1. On the longer run, though, we will not be in the situation to afford parallel industries that produce either primary product quality which

https://doi.org/10.1515/9783110798890-009

meets the specifications of producing industry or secondary products of whatever quality for which in the worst case a market needs to be created. Frankly spoken, the time of this early era of secondary raw material treatment is over.

It is for all those reasons stated above that industrial raw material production has to look ahead. Politics, stakeholders and decision-makers need to be open to entirely new processes, even when running danger of completely ignoring present and past technological solutions. To put it into one word, future challenges are solved best with future approaches. Any solution to push raw material supply forward inevitably has to bear in mind that solving the problems of tomorrow will not succeed with the toolbox of yesterday. What sounds like a truism is a serious issue, since on global scale national economies appear not prepared for the challenges of the forthcoming thirty years. It is obvious that Europe relies on and depends on raw material imports from producing countries. And it is well known, too, that Europe is struggling hard for securing its raw material base. In a nutshell, the story is about getting access to the required amounts and qualities of raw materials. The way is not the goal, but the goal determines the way.

Future proof approaches will have to work differently from traditional ones. In fact, one solution lies in the circular resources chemistry approach according to which there is no more distinction between primary, secondary or regenerative origin of raw materials. Future processes will be origin independent, i.e. they are robust enough to tolerate a broad input material variability (Chapter 2.1, Figure 2.1.6). One can compare this with a funnel, where the upper part, the opening, collects and pretreats the input streams until on a very early stage a common intermediate is produced which then is processed along one single process (the funnel's chimney). As a consequence, the result cannot be worse than primary product quality. The following chapters present such technologies that address the future simply adhering to the five-point agenda from above. The door to true recycling is open, and the message is clear: we already do have these processes, and these processes do work economically.

6.1.1 Lithium recovery from mineral ore and lithium batteries: a selective approach via the COOL process

The lithium issue is clearly a textbook example how the topics raised in Chapter 2.1 and above can be realized.

The COOL process (**CO_2-leaching**) is an example for such a future-directed process. The core step (and name giving) is leaching of the lithium-bearing material with supercritical CO_2 (*sc*-CO_2). In fact, it is suitable for both, primary feedstock (mineral ores) and secondary raw materials (lithium batteries). It consumes no chemicals in contrast to established routes, works with regenerative energies and produces no waste. At the same time, it is the most cost-efficient production process for *bg*-Li_2CO_3 so far [1]. This fact is important as much as it shows that novel approaches may well combine superior technology and economic efficiency.

From Figure 6.1.1, it is easily conceivable that lithium demand cannot be met by classic mine production. Lithium recycling is de facto not in operation worldwide [2]. This situation is aggravated through competing markets such as lightweight alloys [3]. As a consequence, there has been a significant price push to 61500 USD/t Li_2CO_3 in mid-June 2022 (Figure 6.1.1) [4]. The price for $LiOH{\cdot}H_2O$ was even higher with 75000 USD/t [5].

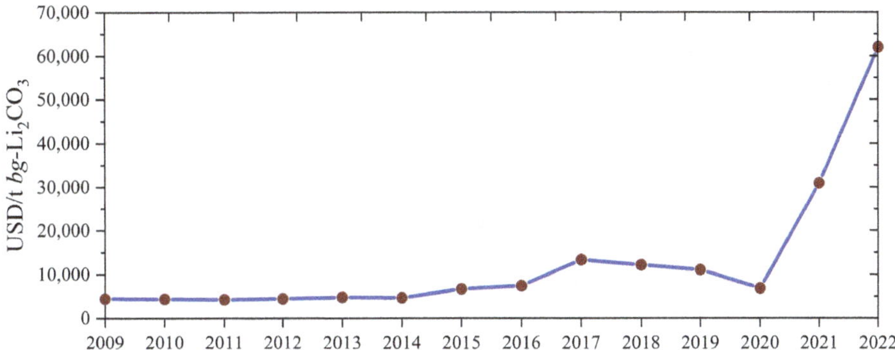

Figure 6.1.1: Lithium price chart 2009–2021.

6.1.1.1 Lithium deposits and extraction processes

Li_2CO_3, which is the trade form of lithium, was first produced on an industrial scale in 1923/1924 at the Hans-Heinrich smelter in Langelsheim (Harz) [6, 7]. The starting material was the lithium-iron mica zinnwaldite, which was mined in the Zinnwald/Cínovec deposit on the German-Czech border. Later, deposits were developed in the USA, Australia, Africa and South America.

The fact that lithium is not a scarce raw material is shown by the worldwide reserves of ~21 million t Li and resources (approx. 86 million t Li). With a global production volume in 2020 of ~82000 t Li, this corresponds to a static range of 256 (1198) years [2]. Figure 6.1.2 shows the geographic distribution of global lithium resources.

About 53% of the mineable lithium resources are located in South America, particularly in Chile, Bolivia and Argentina. Another 1.3 million t Li reserves are in North America (USA, Canada) and 1.5 million t Li in China, while Europe possesses only ~4% of the world's lithium reserves, while 4.7 million t Li reserves (22%) are located in Australian pegmatites.

The primary lithium resources are distributed with ~70% predominantly in salt lakes (which in South America are called salars) and lithium-bearing minerals. Currently, the most significant salar lithium deposit is Salar de Atacama in Chile

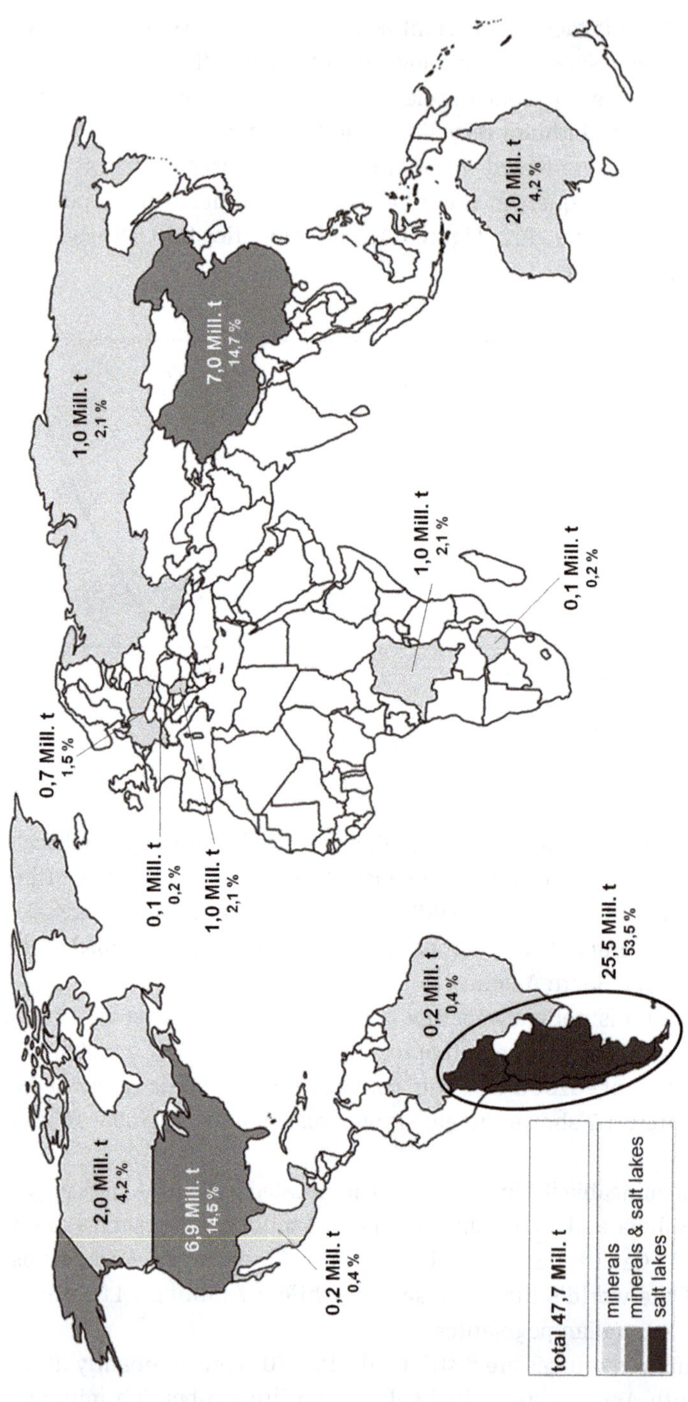

Figure 6.1.2: Geographic distribution of global lithium resources [4, 8].

(6.3 million t Li) which is situated in 2300 m altitude and encompasses 3200 km^2, where lithium concentrations of more than 4000 ppm are found. An even larger lithium deposit of 10.2 million t Li is described for the Salar de Uyuni in Bolivia, although the altitude (3650 m), the lower lithium content of 320 ppm on average, a much less favorable Li/Mg ratio compared to the Salar de Atacama, and poorer climatic conditions for concentration by evaporation have so far resulted in no large-scale lithium production [9–11]. Other large lithium deposits in the form of salt lakes are located in Argentina and in China [11, 12].

Due to low production costs and low technological complexity, lithium compounds are currently mainly extracted from salt lakes or salt pans. The economically most important salt lake is again the Salar de Atacama in Chile.

The basic process steps for the extraction of lithium derivatives from salt lakes always include the concentration of the lithium-containing solution in artificially created evaporation basins (solar ponds) with separation of crystallizing salts, precipitation of by-products and impurities and the extraction of the lithium compounds. In the following, the process steps are explained in more detail using the example of large-scale lithium carbonate extraction at the Salar de Atacama (Figure 6.1.3) [7].

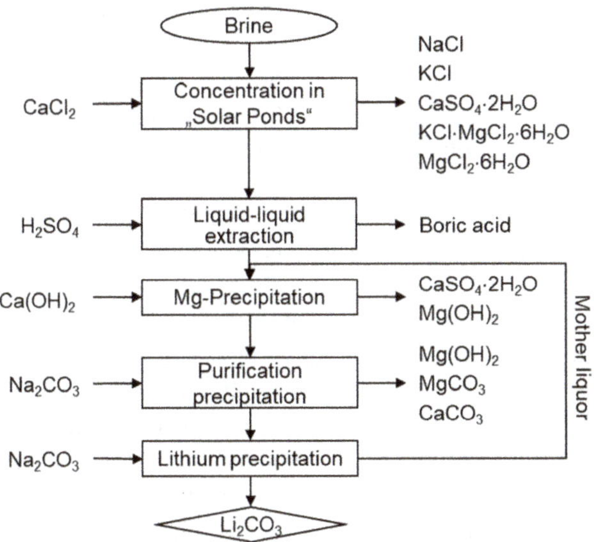

Figure 6.1.3: Process steps for the extraction of lithium carbonate from the Salar de Atacama (Chile). The time required is 12–18 months during which the Li concentration increases from <0.2% to 4–6%. The process consumes 10–18 t chemicals per t Li_2CO_3 and produces 38–130 t waste per t Li_2CO_3, while causing severe environmental impact through extensive groundwater use in the Atacama Desert.

First, the brine is concentrated from ~0.15% Li to ~6.0% lithiated brine using solar energy in shallow evaporation ponds. During concentration, a series of crystallizations begins due to increasing salt concentrations. First, precipitation of gypsum ($CaSO_4 \cdot H_2O$) and halite (NaCl) occurs, followed by sylvinite (NaCl·KCl) and sylvite (KCl). Potassium chloride represents a marketable product in this process and is recovered from the deposits by flotation.

At ~4.5% Li, crystallization of carnallite ($KCl \cdot MgCl_2 \cdot 6H_2O$) and then bishopite ($MgCl_2 \cdot 6H_2O$) sets in. When the brine is further concentrated to about 5.5–6.0% Li, lithium carnallite ($LiCl \cdot MgCl_2 \cdot 6H_2O$) may crystallize under certain circumstances, which reduces the lithium yield in the brine. In those cases, the crystallized lithium carnallite is mixed with fresh brine and brought back into solution. In addition to lithium chloride (35–40%), the concentrated lithium brine also contains magnesium (1.0–4.0%) and boron (0.5–1.5%) in the form of magnesium and lithium borates.

The separation of boron is essential, as it interferes with electrolysis processes (e.g. for the production of metallic lithium), particularly due to its accumulation in the electrolyte. For this purpose, a mineral acid, such as hydrochloric acid or sulfuric acid is added to crystallize boron as boric acid in the pH range 0–4. The separated boric acid crystals are washed with cold water, dried and represent a marketable byproduct.

The remaining boron (<0.5%) is separated by liquid-liquid extraction to below 5 ppm. To the boron-free crude solution, the mother liquor of the lithium carbonate precipitate (0.1–0.3% Li) from the previous batch, is added and thereby diluted from approx. 6% Li to a lithium content of 0.9–1.5% Li. The lower lithium content minimizes lithium losses during the subsequent two-stage separation of magnesium and calcium.

At this point, it also becomes apparent why the Li/Mg ratio plays an important role, because due to the diagonal relationship in the periodic table, both metals have similar properties. It becomes particularly critical when, as here, magnesium is separated from lithium by carbonate precipitation; in this case, the process parameters must be selected in such a way that the target metal lithium is not affected. On a technical scale, the problem is solved by precipitating and separating poorly soluble magnesium carbonate at pH = 7–9 by adding sodium carbonate. This separates up to 95% of the magnesium contained. To obtain lithium carbonate in high purities (>99%), milk of lime and sodium carbonate are then added to the filtrate and magnesium hydroxide and calcium carbonate are precipitated. After separation again, the lithium-containing solution (0.8–1.2% Li) is heated to 70–90 °C and 80–90% of the lithium contained is precipitated as lithium carbonate by adding sodium carbonate. By filtration, product washing with hot distilled water and final drying, lithium carbonate is obtained in purities >99%. The liquid residue (mother liquor) is recycled into the process and is used to dilute the subsequent charge.

In addition to extraction from salt lakes, the processing of primary resources of mineral origin is also of great economic importance. Lithium-bearing minerals can be divided into pegmatitic-pneumatolytic deposits, pegmatites and greisen and hectorite and jadarite deposits. Table 2.1.1 in Chapter 2.1 summarizes the main representatives of lithium-bearing minerals.

The most important representative is the lithium alumosilicate spodumene. The largest deposits are located in Greenbushes, Australia (0.85 million t Li) and Jiajika, China (0.48 million t Li) [11, 12]. Spodumene is particularly characterized by the high theoretical lithium content of 8.0 wt% Li_2O. The world's largest mine is the Greenbushes mine in Western Australia.

Lepidolite is also mined commercially. There are deposits in China (Jiangxi Province), Portugal (Guarda) and Vietnam (Quang Ngai). In addition, lepidolite often occurs together in pegmatites. Of importance are the associations with amblygonite (Karibib, Namibia, 0.15 mt Li), with spodumene (Bikita, Zimbabwe, 0.06 mt Li) and muscovite (Gonçalo, Portugal) [11].

Similar to lithium extraction from salt lakes, there is also a whole range of industrially used processes for processing lithium-bearing minerals. The basic process steps for spodumene, petalite, lepidolite and zinnwaldite are shown in Figure 6.1.4 [13].

Typically, all mined and crushed ores are first subjected to mechanical, physical or chemical beneficiation processes. The primary objective here is to obtain an ore concentrate with higher lithium content. This is achieved by separating the gangue of quartz (SiO_2), albite ($Na[AlSi_3O_8]$) or muscovite ($KAl_2[(OH,F)_2AlSi_3O_{10}]$), among others, by means of flotation processes, magnetic separation or grain size classification [14].

The most commonly used process for lithium recovery from silicate lithium ores is pyrometallurgical roast digestion with sulfuric acid. This process is particularly used for lithium recovery from spodumene concentrate and is therefore also referred to as the spodumene process (Figure 6.1.4). In the sulfuric acid digestion of the naturally occurring α-spodumene, only a low lithium mobilization is achieved. Therefore, the aluminosilicate is first converted to the acid-labile β-spodumene in a calcination process at temperatures ≥1000 °C. The resulting improved reactivity to mineral acids is used to transfer >90% of the contained lithium as lithium sulfate into the liquid phase by roasting the calcined starting material under sulfuric acid addition at temperatures of 250–300 °C and subsequent water leaching [15].

A base (e.g. sodium hydroxide, slaked lime) is added to separate the digested minor components that form poorly soluble hydroxides, in particular aluminum, but also magnesium and iron. Calcium separation is carried out after addition of sodium carbonate (soda ash) by precipitation as calcium carbonate at pH ~10. The purified solution containing lithium sulfate is then concentrated to lithium contents of ~11 g/L by evaporation or via an ion exchanger [16].

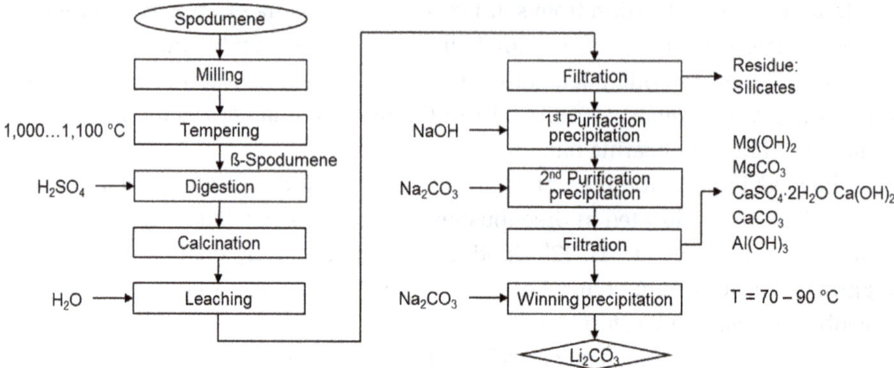

Figure 6.1.4: The spodumene process. 25–30 t chemicals are consumed per t Li_2CO_3 and 80–120 t waste are produced per t Li_2CO_3.

The final product precipitation is performed analogously to lithium carbonate recovery from salt lakes. At temperatures close to the boiling point of water (90–100 °C), sodium carbonate is added. This takes advantage of the solution anomaly of the target compound and precipitates the lithium carbonate, which is poorly soluble in heat. After filtration, washing with hot water and drying, technical lithium carbonate (*tg*-Li_2CO_3, >99.0%) is obtained.

An industrially established alternative to sulfuric acid digestion is the so-called limestone burning. In this process, not only is the phase transformation from α- to β-spodumene brought about at temperatures of 800–1000 °C with the addition of limestone ($CaCO_3$), but also the deacidification of limestone to burnt lime (CaO) brings about an in-situ reaction with the ore. Subsequent water leaching yields an alkaline solution containing lithium hydroxide [17]. Subsequently, contained calcium carbonate is first precipitated with the addition of CO_2 or sodium carbonate before lithium carbonate extraction is carried out by the spodumene process.

In addition to the two technically relevant processes, sulfuric acid digestion and limestone burning, a whole series of other processes has been developed for the mobilization of lithium from spodumene (or petalite). However, what all the listed processes have in common is that they have not yet been scaled up to commercial scale [14].

Currently, the Zinnwald/Cínovec deposit is attracting new attention after recent exploration data showed that it is the world's second largest siliceous lithium deposit after Greenbushes (Australia) [8]. The use of domestic resources also has the advantage that the supply of raw materials can be at least partially decoupled from geopolitical imponderables; and last but not least, mining is carried out in accordance with the high European standards for nature conservation and environmental protection.

With ~652000 t Li (1.40 mill. t Li_2O), the deposit in Zinnwald/Cínovec is the largest silicate lithium deposit in Europe. Of this, 96000 t Li (207000 t Li_2O) is distributed among the German and 556000 t (1.20 mill. t Li_2O) among the Czech deposits [8, 18].

However, zinnwaldite exhibits some properties that significantly complicate its extraction compared to the other technically important minerals, especially spodumene or lepidolite. The most significant European lithium ore contains Fe, Al and F, a deadly mixture, which prevents Li from being recovered under economically viable conditions.

Therefore, a novel strategy for lithium extraction from enriched zinnwaldite concentrate had to be developed. The COOL process is a textbook example of how future-directed processes may work. It is a hybrid approach, which means that it works origin-independently and as such adheres to the principle of circular resources chemistry. There is a broad substrate variability (cf. Chapter 2.1, Figure 2.1.6) where no distinction is made between different ore minerals from different origin, and it is suited for input material from both primary and secondary origin.

The key step for lithium minerals is a thermal treatment. This step is useful in two aspects: (i) Fluoride is released as SiF_4, which can be recovered in a scrubber with aqueous HF. The resulting hexafluoro siliceous acid (H_2SiF_6) is then sold to the market where it serves as a useful fluorine intermediate in fluorine chemical industry. To put it in one word, the extremely harmful and toxic environmental pollutant fluoride has been inverted from a costly disposal situation to an income situation which by the way provides the fluoride for lithium battery conducting salts, such as lithium hexafluorophosphate ($LiPF_6$), without the need for further interference with nature. (ii) At temperature levels >900 °C, lithium silicates become unstable and crystal structures break down. As a consequence, the lattice bricks of lithium silicates rearrange upon thermal decomposition, by which action the original mineral gets completely lost in favor of a species, which is the most stable entity under these conditions. This is β-spodumene. All other siliceous lithium minerals react to give β-spodumene ($LiAl[Si_2O_6]$), silicon tetrafluoride (SiF_4) and hematite (Fe_2O_3), corundum (Al_2O_3) and/or leucite ($K[AlSi_2O_6]$). Thus, a new structure is formed, the high-temperature modification, which is stable under elevated temperature. It is a property of the latter to be less dense and mechanically more brittle. However, under high-temperature conditions this structure type is the most stable one. The process profits from the fact that β-spodumene is highly susceptible to acid attack. The reason is that acid digestion is done at a far lower temperature level, where the high-temperature modification is thermodynamically unstable. Consequently, acid attack helps to liberate the intrinsic energy stored inside the crystal body as a result of thermal treatment. At the same time, the exergonic lattice decomposition profits from a considerable entropic contribution upon dissolving the ions in the digestion solution. The same applies for sulfate attack. This is why β-spodumene is so easily digestible in acids, and why it can be attacked so easily even by weak acids such as carbonic acid [19]. Figure 6.1.5 displays the COOL process. As can be seen there, the lithium from spent batteries is recycled, too. However, here, in its truest meaning, restoring primary product quality. Since both material streams are processed along one single, identical process chain, the secondary input material cannot be recovered in a quality inferior to that of the primary material stream. As the crude products is obtained already in battery-grade quality, i.e. a purity of ≥99.5% Li_2CO_3, battery recycling, too, cannot be worse that primary product quality, with the additional benefit that no costly and energy intensive refinery to battery grade is necessary.

The solid residue obtained after the leaching step is rich in leucite, an aluminum-silicate. In addition, it contains iron oxide and aluminum oxide, which altogether build a perfect base to make geopolymers (see Excursus Geopolymers).

Excursus Geopolymers

Geopolymers are inorganic and calcium-free polymers based on silicon and aluminum oxide. The prefix "geo" symbolizes the use of mineral inorganic materials and "polymer" the inner structure. As inorganic polymers, they are free of plastic and benefit from the properties of real polymers. Their production is CO_2-poor or even CO_2-free. Geopolymers resemble, to some extent, classical binder cement, which is a mixture of different calcium silicates, i.e. salts. These unconventional binders, however, are rapid hardening, shrink proof, have high compressive strengths and are stable against leaching. Furthermore, they are non-flammable, temperature resistant, dimensionally stable and resistant to all inorganic and organic acids (except hydrofluoric acid). As a binding agent ("cold cement"), they are therefore superior to concrete. Geopolymers can be foamed and have competitive insulation properties in regard to polystyrene.

Their fabrication is easy as it mostly follows the established route to cement, mortar and concrete through stirring a two-component mixture of silicon and aluminum-containing raw materials with an aqueous alkaline solution. The mixing process takes place at temperatures between 20 and 80 °C.

Geopolymers are of utmost interest for the energy and raw material shift since inorganic residues no longer need to be disposed of. As potential construction material, they are particularly interesting for industry. In this way, the cycle of recyclable materials can be closed, and the production of waste can be avoided. Geopolymers based on residual materials are no more expensive than conventional concrete but offer all the advantages of an inorganic polymer. And there are evident benefits compared to classic cement or concrete. Worldwide, 4.1 bill. t of cement is produced each year. The downside: with ~8% of global anthropogenic CO_2 emissions, the cement industry is one of the most important single emitters of climate-relevant CO_2. Geopolymers are a climate-friendly alternative to cement. They can be produced easily and cost-effectively at ambient conditions with little effort. The use of secondary raw materials avoids expensive primary raw materials and mining interventions in nature. They also avoid energy-intensive manufacturing processes such as the high-temperature thermal activation of cement clinker at 1450 °C. As a result, there is a positive CO_2 balance (savings potential of up to 80%) and an effective contribution to achieving climate-protection goals. The use of residual materials that would otherwise have to be landfilled avoids costs for expensive primary raw materials, and there are no costs for landfilling. Instead, residual materials are reintegrated into the value chain. Rather than costs, profits are made.

Carbon footprint of geopolymers vs. Portland cement.

Technical advantages are their significantly higher temperature stability, which is why they tolerate much wider temperature ranges in use. And they offer safety: in case of fire, there is no cracking or spalling. They do not contain lime, which makes them highly resistant to chemicals and corrosive conditions. They can develop compressive strengths comparable to high-strength concrete after just one day. Their rapid setting behavior makes them suitable for mass production of precast elements. Leaching-sensitive ingredients, such as those contained in commercially available cement, are firmly chemically bound in the geopolymer and are thus stable against leaching.

Last but not the least, geopolymers are fully recyclable.

6.1.1.2 Lithium batteries [20]

The global market for lithium-ion batteries (LIB) has been continuously rising since their inception in 1991. Due to the growing demand for electric mobility, the global LIB market was valued at $ 36.7 billion in 2019 and is predicted to reach $ 129.3 billion by 2027 [21]. Rechargeable LIB in cordless electronic devices is another rapidly increasing industry. Although lithium is abundant in the Earth's crust, the accessible quantities are insufficient to meet market demands, necessitating efficient recycling methods. The fact that LIB chemistry and technology are continuing in development exacerbates the dilemma, resulting in a large range of battery types. On the other hand, the European Union directive 2006/66/EC mandates that at least 50% of all used LIB be recycled [22].

Despite their structural variation, all LIBs have a similar underlying structure [23]. The anode is usually a copper foil with a graphite coating, and the cathode is usually an aluminum foil with an intercalated Li compound. A porous polyolefin separates the anode and cathode compartments, while the electrolyte is a combination of an organic solvent and a lithium salt. A sealed container constructed of aluminum, steel, special polymers or highly refined aluminum composite foils encloses these

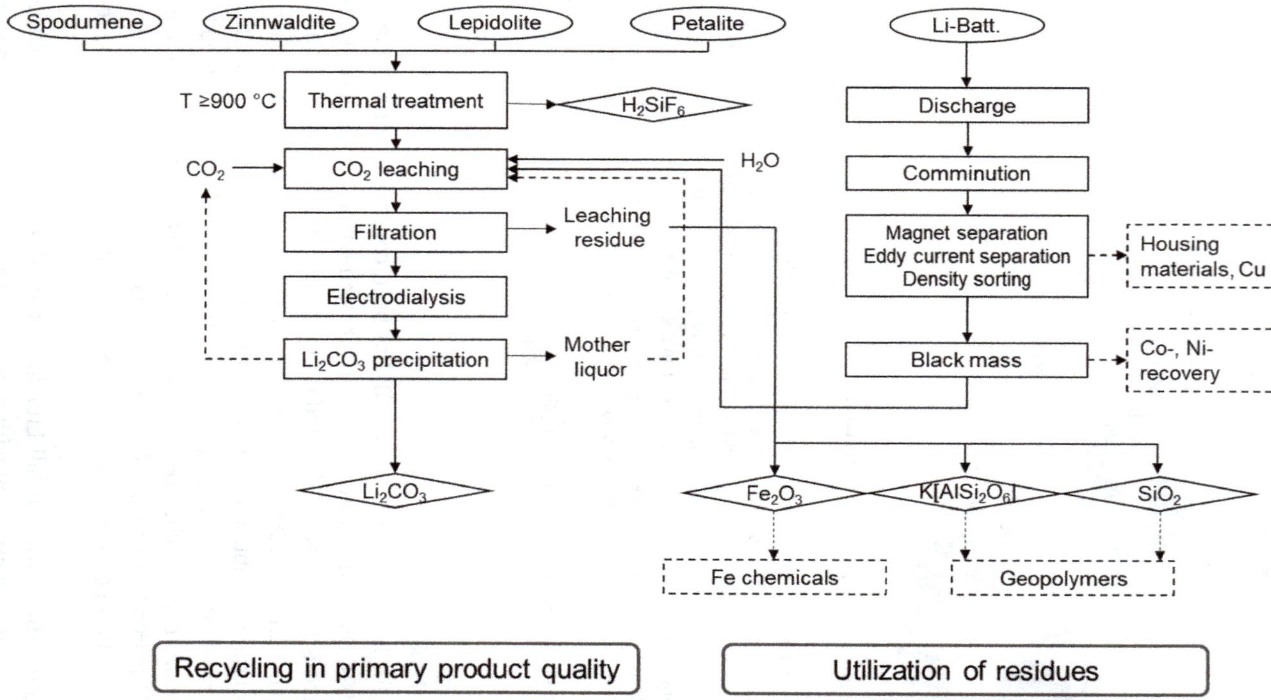

Figure 6.1.5: The COOL process. A novel, holistic process that is capable of processing both primary and secondary raw materials. It fully exhibits the properties of modern processes: 1. climate friendly, 2. zero waste, 3. low carbon footprint, 4. energy and resource efficient and 5. water saving. No lithium gets lost through recycling the mother liquor, which in turn renders the process water saving. There is no chemical input apart from CO_2, which can be generated from thermally treating waste black mass from spent lithium batteries. Inorganic residues are converted to market established construction material.

cells [23]. Current commercial LIBs can be classified into five categories based on the lithium salts utilized (Table 6.1.1):

Table 6.1.1: Lithium battery types and chemical compositions of the cathode materials [24].

Nr.	Cathode material	Composition
1	LCO	$LiCoO_2$
2	NCM	$LiNi_xCo_yMn_zO_2$ | $x + y + z = 1$
3	LMO	$LiMn_2O_4$
4	NCA	$LiNi_xCo_yAl_zO_2$ | $x + y + z = 1$
5	LFP	$LiFePO_4$

Evidently, any LIB recycling process must be independent of the type of the spent LIB and necessarily needs to be applicable for mixtures of different LIB from (i) different waste streams and (ii) different generations in past and future.

Pyrometallurgical and/or hydrometallurgical processes are used in most of the current recycling procedures. Pyrometallurgical approaches are linked to high energy consumption, high capital expenditures, the possibility of toxic gas emissions and complicated extraction techniques. Besides that, targeted lithium recovery is at best challenging and, in fact, either impossible or commercially unviable [25]. Furthermore, plastics and electrolyte cannot be recycled. It is difficult to meet the specified recycling rate (50 wt%) because both components account for 40–50 wt% of the wasted battery [26]. Leaching procedures with inorganic or organic acids followed by precipitation and/or solvent extraction provide the target metals with high efficiencies, yet with an unclear process economy on a large scale if lithium is addressed, too. The established lithium battery recycling processes aim at the more profitable Co and Ni [17, 27]. However, the high recycling rates can only be achieved by using high quantities of acid which in turn produce large amounts of wastewater [28]. Already the costs of the chemicals for acid digestion and subsequent neutralization exceed the intrinsic metal value by far. Furthermore, the low leaching selectivity, especially in the case of inorganic acids, necessitates extensive purification steps, which render the entire process complex and economically not viable.

The COOL process is the third player in the game. It combines moderate thermal treatment for primary feedstock with smooth lithium-leaching conditions for secondary feedstock. In fact, the CO_2 partial pressure of sparkling water is sufficient to extract the lithium from the black mass, while leaving cobalt and nickel unaltered. In a nutshell, the COOL process offers more degrees of freedom, because the delithiated black mass can be treated conventionally both by pyro and hydrometallurgical methods. This way the COOL process is the door opener to a full recycling of waste LIB [20].

6.1.2 Phosphate recovery from phosphate rock and sewage: a selective approach via the PARFORCE process

Phosphorus (P) is an essential nutrient for all living organisms. With a concentration of 1200 mg P/kg in the Earth's crust and 1,050 mg P/kg in the continental crust, its abundancy is on the fifteenth place [29]. More than 90% of the mined phosphate ore is used for the production of mineral fertilizers [30]. There are three types of phosphate ores: (i) sedimentary ore, phosphate rock, which are the most important P source, (ii) igneous phosphate ore from magmatic origin and (iii) biogenic ore, which are known as guano [16]. In phosphate rock, it is apatite, $Ca_5(PO_4)_3(OH,F,Cl)$, and francolite, $Ca_5(PO_4,CO_3)_3F$, are the actual phosphate bearing minerals [31].

The main producing countries of phosphate rock are China, Morocco and USA. The production of phosphate rock in Morocco reached a value of 38 million tons in 2021 (Figure 6.1.6) [2, 32].

For mineral fertilizer production, phosphate rock is processed to phosphoric acid through the wet phosphoric acid (WPA) process [33].

Fertilizer production is the main application of phosphoric acid next to feed and food production as shown in Figure 6.1.7.

For the production of phosphoric acid, H_3PO_4, from phosphate rock, the ore is subjected to a series of beneficiation steps, among which there is flotation and calcination for organic matter removal. The enriched material is then digested with sulfuric acid (Figure 6.1.8). This final process step before P-fertilizer production can begin, but must solve all the problems that could not be solved during beneficiation, such as increasing P grade, elimination of contaminants, etc. It is the successful operation of this process stage that determines whether the entire process from mining to phosphoric acid is economical.

The coproduct, gypsum dihydrate ($CaSO_4 \cdot 2H_2O$), results from the reaction of both the calcium portion of apatite and francolite and the remaining limestone fraction with the sulfate of the sulfuric acid (eq. 6.1.1). There are high procedural requirements for the digestion process in order to obtain phosphate-free gypsum. Only then can the gypsum be marketed. Otherwise, phosphate-rich gypsum is obtained, the disposal of which is costly (70 USD/t). Between 4 and 6 t of low-radioactive phosphogypsum per ton P_2O_5 is produced as relevant by-product/waste. About 85% of the 5.6–7.0 billion t phosphogypsum produced globally over the lifetime of the phosphate industry are disposed of in wet- or dry stacks in 52 countries worldwide. The digestion process also leads to the mobilization of transition metals, which are removed from the crude phosphoric acid in post-treatment steps [16]. The solubility of gypsum and metal salts in both water and phosphoric acid must be considered when purifying the acid.

Crude phosphoric acid is very impure and requires cost-intensive purification. When concentrated and allowed to stand, a considerable part of the insoluble

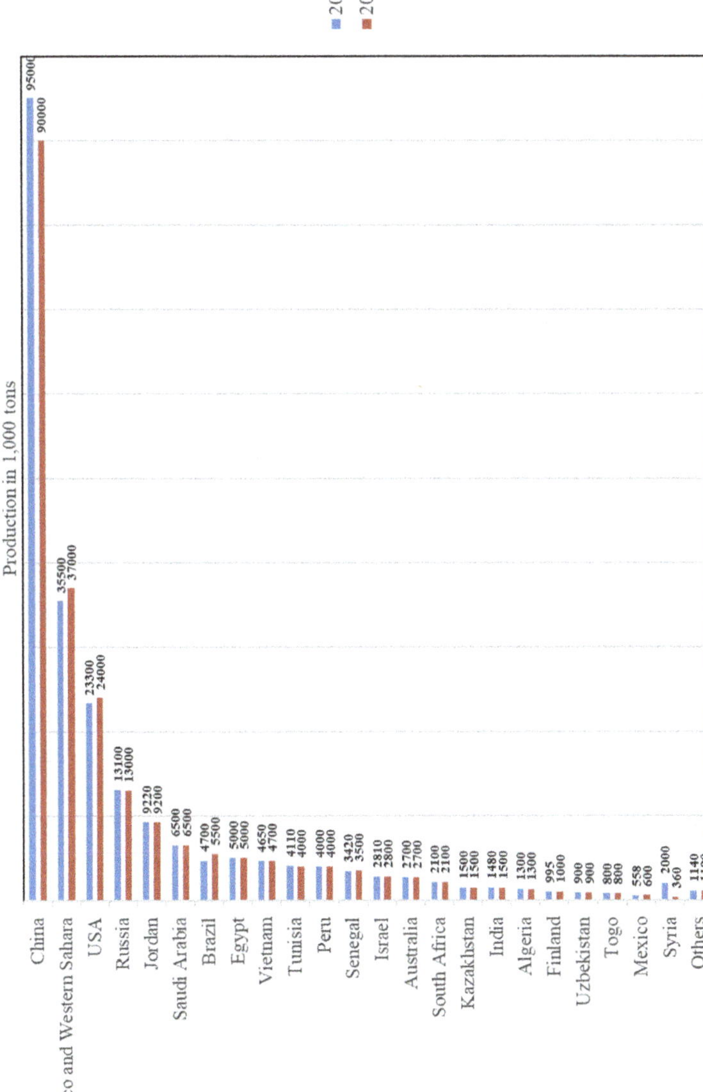

Figure 6.1.6: Mining production of phosphate rock by countries in 2019 and 2020. Data from [2, 32].

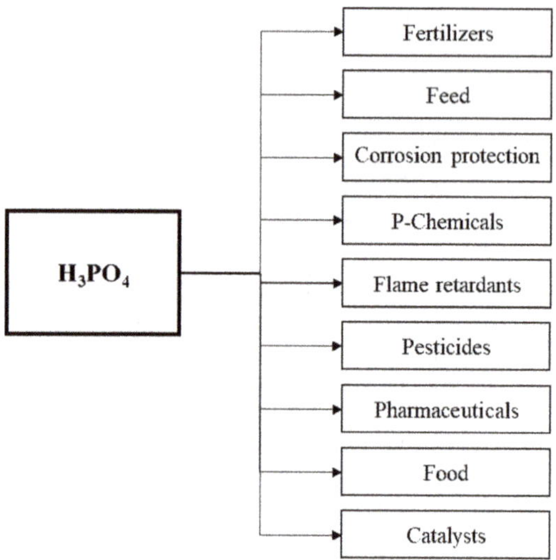

Figure 6.1.7: Application of phosphoric acid, H_3PO_4.

impurities settles in sludge form, which is separated and disposed of. Depending on its use, the acid must be concentrated. For fertilizer production, a H_3PO_4 with 40 to 54% P_2O_5 is required, while for shipping the concentration is 52 to 72% P_2O_5. The evaporation process is technically complex, not only because of the high corrosiveness of the acid and the precipitation of salts and gypsum but also owed to the acidic exhaust gases containing, released which contain fluorine compounds and phosphoric acid mist (eq. 6.1.1). The released fluorine reacts with the siliceous components from apatite to produce hexafluorosilicic acid as by-product (eq. 6.1.2). However, technical use has been low so far.

$$Ca_5(PO_4)_3(OH,F) + 5\,H_2SO_4 + 10\,H_2O \rightarrow 5\,CaSO_4 \cdot 2H_2O + 3\,H_3PO_4 + HF \quad (6.1.1)$$

$$SiO_2 + 6\,HF \rightarrow H_2SiF_6 + 2\,H_2O \quad (6.1.2)$$

Since phosphoric acid is strongly hygroscopic, the removal of water is highly energy intensive. A much more extensive cleaning is possible through precipitation of the interfering ions followed by multistage liquid-liquid countercurrent extraction of H_3PO_4 [34–37].

Apart from the gypsum issue there is another major drawback inherent to sulfuric acid. This is the unfavorable chemical property of the sulfates formed upon digesting P-ores of primary or secondary origin that contain trivalent ions, specifically Fe and Al. The resulting sulfates interact non-specifically with different chemical entities in a very broad pH range between 3 and 11, forming slimes and precipitates the removal of which is extremely tedious. In most cases, the technical effort exceeds the

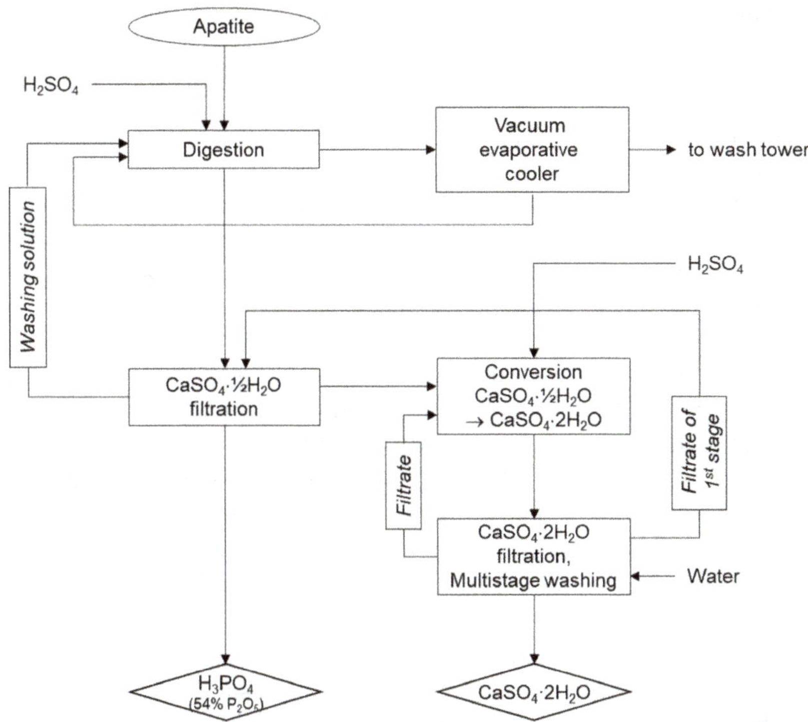

Figure 6.1.8: The wet-chemical process for phosphoric acid production along the hemihydrate route. The initially formed gypsum (hemihydrate) is rich in phosphate and as such not marketable. It is therefore recrystallized in the presence of excess sulfuric acid to well-marketable, almost phosphate-free gypsum dihydrate, $CaSO_4·2H_2O$. Excess sulfuric acid, H_2SO_4, from the recrystallization step is refunnelled into the digestion process, thus reducing acid losses to a minimum. Only the hemihydrate route produces marketable gypsum, but not all processes are operated this way.

value of the contained acid by far. Therefore, any alternative for the wet chemical process will have to utilize an acid different from sulfuric acid.

In the classical wet chemical process, these undesired chemical interactions of trivalent metal cations are superimposed by the far more abundant divalent calcium, for which reason they appear less prominent. But they contribute to the effort that has to be undertaken during purifying crude phosphoric acid, so-called green acid. However, alternative processes for extracting impurities from H_3PO_4 still lack technical maturity.

The problems associated with gypsum formation do not occur if nitric or hydrochloric acid is used instead of sulfuric acid. This can be done analogous to the Odda process (Nitrophoska) with nitric acid (eq. 6.1.3) or hydrochloric acid (eq. 6.1.4). Both by-products calcium nitrate and calcium chloride are processed to fertilizer or road salt.

$$Ca_5(PO_4)_3(OH, F) + 10\,HNO_3 \rightarrow 5\,Ca(NO_3)_2 + 3\,H_3PO_4 + HF \tag{6.1.3}$$

$$Ca_5(PO_4)_3(OH, F) + 10\,HCl \rightarrow 5\,CaCl_2 + 3\,H_3PO_4 + HF \tag{6.1.4}$$

The purification of the crude phosphoric acid can be done according to established techniques [38].

In order to substitute the established WPA for a process that is characterized by low carbon emission, zero waste and energy and resource efficiency, a completely different technology is required. One example for such a technology is the PARFORCE process ("Phosphoric Acid Recovery from Organic Residues and Chemicals by Electrochemistry"). Once developed for phosphate recycling from sewage sludge ashes (SSA), it became evident soon that this is a generally applicable approach to recover phosphate from a broad variety of input materials, such as SSA, phosphate rock or magnesium ammonium phosphate (MAP), $MgNH_4PO_4$, which is also known as struvite. In addition, PARFORCE is suited for a series of phosphate-containing secondary raw materials, e.g. production waste, calcium phosphates or bone meal ashes. It only produces to market-established products, while the use of chemicals, likewise the generation of waste is kept to a minimum. Unlike other phosphoric acid production processes, the PARFORCE technology works with electrical energy and thus can be operated fully on the basis of regenerative energies. This is of particular interest, where sufficient solar energy as is abundant, such as in Africa [39].

Another highly important issue is acid consumption. Up to now, as soon as a process becomes wet chemical, there is the need for excess acid to digest the ore (or secondary raw material) to complete the process within reasonable, i.e. economically viable timelines. Typically, this excess acid has to be removed, i.e. the spent chemical is not only lost, it is even overconsumed by the alkaline substances needed to adjust the pH to neutrality. This is not only questionable in terms of resource efficiency, but also typically the economical death blow. In the WPA, this issue has been solved by reintegrating excess acid in the digestion reactor when sulfuric acid is used as the digestion medium (Figure 6.1.8).

In the case of hydrochloric acid, the situation becomes even more favorable and economically more interesting once there is solar energy at hand. In that case, the formed intermediary, $CaCl_2$, which is of low value and more or less disposed of as road salt, can be split by means of electrochemistry. One product is $Ca(OH)_2$ (slaked lime), for which there is a market all over the world (80–90 USD/t). The other product is HCl, which is reintroduced into the digestion vessel, thus losing no excess acid. The only amount of acid consumed is the hydrogen chloride required for the digestion step in stoichiometric amounts according to P-ore composition (Figure 6.1.9).

Another crucial benefit is that no freshwater is required. Instead, seawater can be used. As the digestion process is conducted with hydrochloric acid, HCl, there is no disturbing interference by the sodium chloride, NaCl, in seawater ($c \approx 35$ g/L).

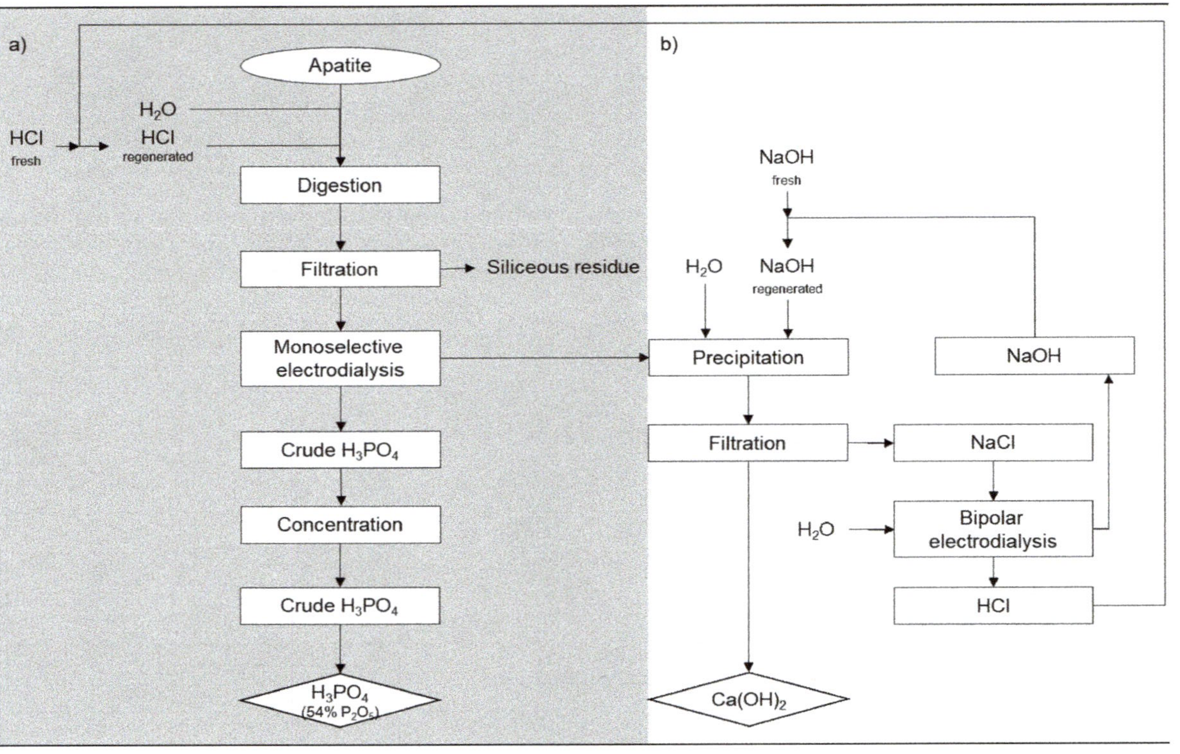

Figure 6.1.9: The PARFORCE process. a) The core process (grey) is an electrochemical process unit, which can be run with HCl or HNO_3. b) The regeneration (white) unit serves to regenerate HCl and NaOH by means of bipolar electrodialysis, so that the only amount of acid that has to be added is the amount that been drawn off in the form of H_3PO_4. Where HCl is used, filtered seawater is sufficient as the water source. The entire process can be run using regenerative energies.

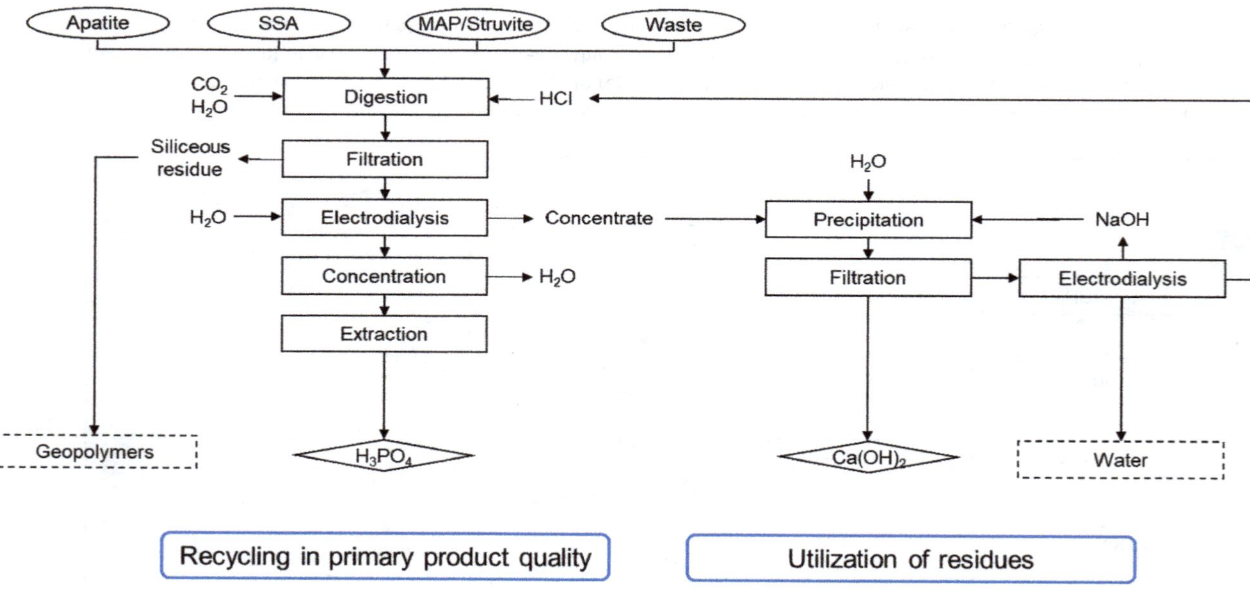

Figure 6.1.10: The PARFORCE process is a circular resources chemical approach for both processing of phosphate rock (apatite) and recycling of phosphate from sewage sludge ash (SSA), calcined magnesium ammonia phosphate (MAP), what is struvite, and phosphate-containing waste streams from industrial processes or water treatment (waste).

As mentioned above, other secondary resources like sewage sludge ash, struvite and production residue are equivalent for phosphoric acid production. In order to develop such processes, the legal requirements for obtaining marketable products from waste must be established which is already done in some European countries. One of these sources is sewage sludge which contains between 20 and 55 g P/kg dry matter. After extracting phosphorus from solid sludge phase into the aqueous phase, it can be precipitated as struvite ($MgNH_4PO_4$). An advantage besides the separation of phosphorus is the binding of nitrogen as ammonia. In the PARFORCE process struvite from sewage sludge is calcined to separate ammonia for fertilizer production. Phosphorus is extracted by an acid leaching step using HCl. In addition to ammonia, a magnesium chloride solution is produced which is recirculated and used again for struvite precipitation onside the waste water treatment plant. The process impresses with its diversity of feedstocks and phosphoric acid is also extracted from SSA or production waste in this way (Figure 6.1.10).

The leaching residue consists of different silicates and heavy metals from the waste water which are incorporated in geopolymers (see Section 6.1.1). The next step involves the separation of Fe^{3+} and Al^{3+} by ion exchange or solvent extraction as chlorides to recycle them as precipitation agent for phosphate in the waste water treatment process. Dissolved impurities like heavy metals ions (Ni^{2+}, Pb^{2+}, Cu^{2+} etc.) are separated from the phosphoric acid by electrodialysis as an electrochemical membrane process. The impurities remain together with $CaCl_2$ or $MgCl_2$ in stream from where they are precipitated. Calcium and magnesium chloride are reused either as road salt or precipitated salt for struvite. This means that no residual materials remain that have to be disposed of and the material cycle is closed (zero waste concept). The phosphoric acid is very pure and becomes concentrated by vacuum evaporation of water to a commercial concentration of 54% P_2O_5.

6.1.3 Rare earth recovery from end-of-life products

Rare-earth metals (REM) are a group of elements comprising 14 lanthanides, yttrium and scandium, which have become indispensable in the development of high-tech technologies in recent years. Because of their growing demand in these advanced applications and their small and opaque market, they have become a critical commodity [40, 41]. About 40% of the total worldwide consumption of REM is in battery alloys and permanent magnets as high-growth markets (Figure 6.1.11). Nevertheless, the use of these valuable metals seems to be the center of attention in modern and clean technologies in the upcoming years.

Considering that >75% of global reserves of REM are placed in China, the United States, Vietnam and Brazil, along with the fact that currently ~71% of the annual production of these metals is Chinese [2], there is a real risk of supply causing price volatility and geopolitical problems. In fact, China set maximum quotas, introducing

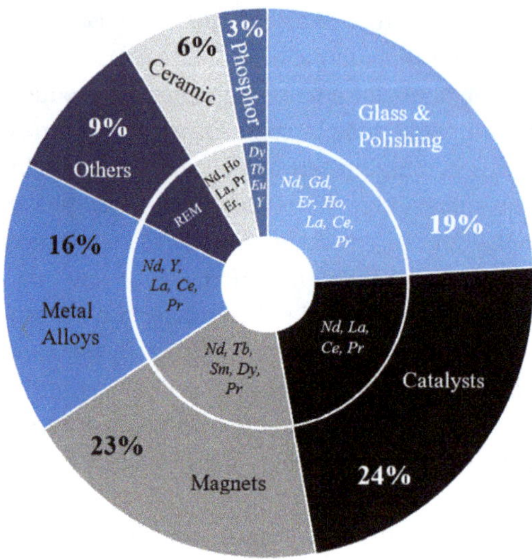

Figure 6.1.11: Breakdown of estimated REM consumption by sector with the REM utilized in each application. Data from [42].

license allocation, imposed taxes and regulated REM exports due to increased demand and the dominant position of REM production. Trade policy in this country caused a worldwide price peak for REM in 2011. To secure the world's RE demand in Europe reducing import dependencies, many researchers focused their investigations on the development of recycling processes from preconsumer scrap, industrial residues, end-of-life (EoL) products and their waste streams. Furthermore, the annual growth rate for electrical and electronic waste equipment (WEEE) in Europe amount of 7% with a forecast of reaching 52.2 Mt in 2021 [43]. As a consequence, a WEEE directive has been established with the purpose of promoting recycling and reuse instead of disposal [44]. Dealing with RE recovery from WEEE can not only protect the environment but also promote the sustainable development of rare earth resources.

Although all REM are necessary and present in such advanced devices, which are indispensable for a continuing development of modern life, five of them are considered critical metals (CRM) based on the importance of clean energy and supply risk. Thus, yttrium, dysprosium, neodymium, terbium and europium are the main target REM for recovery, contributing to a circular economy. Due to the high REM content, especially the CRM ones, into permanent magnets (NdFeB and SmCo), fluorescent lamp scraps and NiMH batteries, the REM recovery is mainly focused on these EoL products. Despite the efforts by the community to develop REM recycling processes, only a few have been developed to sufficient maturity to operate them on an industrial scale. Due to the drop-off in prices in recent years along with the nonuniform waste streams, became these recycling processes cost-inefficient with technology readiness levels (TRL) are <7 [45].

Over the years, hydrometallurgical and pyrometallurgical processes have been the main routes of REM recovery. A wet chemical EoL recycling for fluorescent lamps (2008) and NiMH batteries (2011) have already been developed by Solvay Rare Earths and Umicore on an industrial scale [46]. REMs have been also recovered from fluorescent lamp scraps by Osram process, which was patented in 2012 [47]. Hydrometallurgical approaches include the common acid digestion step, which is cost-intensive. Furthermore, a wet treatment requires working with an acid excess, becoming a high acidic waste stream, which makes inevitable the neutralization step. Thus, the chemical consumption by this recycling route is considerably high. Regarding pyrometallurgical processes, the cost-energy became a problem since the temperature required for REM recovery is >800 °C. To tackle these obstacles from both traditional recycling routes, an unconventional approach, the so-called solid-state chlorination (SSC) is emerging for REM extraction and processing since this alternative has been proven to be more efficient and cheaper [48]. The chlorination process consists of the conversion of the metal content of the raw material into water-soluble metal chlorides. Although different chloride agents can be used, $NH_4Cl_{(s)}$ is a promising chloride salt because the REM recovery is reached using a temperature range of 250–300 °C, thus reducing the energy consumption process compared to the use of $CaCl_{2(s)}$, $NaCl_{(s)}$ or $Cl_{2(g)}$. Therefore, SSC offers several advantages compared to acid leaching: (i) reduction of chemicals consumption, (ii) reduction of chemicals costs, (iii) avoidance of neutralization step, (iv) no regeneration of acidic waste water streams, (v) a circular economy step since the unreacted $HCl_{(g)}$ and $NH_{3(g)}$ in the exhaust gasses can recombine upon cooling, obtaining $NH_4Cl_{(s)}$ for reuse in next runs and (vi) earning revenue by selling a marketable coproduct like $NH_{3(g)}$ (purity >99.9998% and [Cl^-] <5 ppm[49]) obtained by its excess in the exhaust gas.

According to the circular resources chemistry principle, two recycling processes have been developed. REM, mainly Y and Eu, were recovered from waste phosphor by the SSC process and subsequent leaching with water. This route, known as the Sepselsa process, went into industrial application in 2015 processing 27 tons of phosphor since then [46]. Following the same chlorination process, REM from NdFeB and SmCo magnets were extracted within the MagnetoRec project. A first continuously operated reactor with a performance of a half-ton-per-day capacity is in operation in Freiberg (Germany) since July 2021.

6.1.3.1 Fluorescent lamps (Sepselsa process)

Since the content of REM in phosphors of lamps can reach 27.9% [50], the processing of their waste streams and therefore, the use of this EoL product as feasible metal raw materials provides a promising alternative to recover and recycle REM. In order to reduce the disposal in landfills of such lamps, the development of recycling processes is mandatory. 0.9 Mt of lamp wastes were produced worldwide in 2019 of which EoL phosphors (0.2 Mt containing 23% of REM) till now are not recycled and were disposed of in landfills in Europe. Although the manufacture of fluorescent lamps had a constant growth in the last decade, EU regulations regarding energy labeling and eco-design requirements established that most halogen and traditional fluorescent lamps will be phased out as of September 2023 [51]. They will be replaced by modern lutetium aluminum garnet (LAG) LED lamps. This change will generate a considerable waste stream that will reach recycling markets and therefore largely lack adequate REM recovery processes. Another important issue to be aware of is the mercury content into the phosphor powder reaching, without which fluorescent lamps do not work. For instance, ~300 t of Hg-contaminated fluorescent lamps containing 10% of REM are yearly deposed downhole in Germany [47].

Modern fluorescent lamps, known as tri-band lamps, use a mix of three different phosphors (Table 6.1.2) [13, 52].

Table 6.1.2: Emission colors and composition of rare earth phosphors [13].

Phosphor	Name	Emission color
$Y_2O_3:Eu^{3+}$	YOE	red
$La(PO_4):Ce^{3+}, Tb^{3+}$	LAP	green
$(Ce^{3+}, Tb^{3+})MgAl_{11}O_{19}$	CAT	green
$(Gd^{3+}, Ce^{3+}, Tb^{3+})MgB_5O_{10}$	CBT	green
$BaMgAl_{10}O_{17}:Eu^{2+}$	BAM	blue
$(Ca, Sr, Ba)_5(PO_4)_3Cl:Eu^{2+}$	ScAp	blue

Although up to six different REM can be found in fluorescent lamps, namely cerium, gadolinium, lanthanum, terbium, europium and yttrium, the last two are the ones with the largest amount, up to 55% in the red phosphors [52]. In this sense, many researchers are focused on recovering both REM.

Mechanical separation is the most common route to remove components such as glass, plastic and aluminum end caps. Afterward, a physical treatment consisting of crushing the EoL product is carried out to facilitate the collection of phosphors fractions. Unfortunately, these fractions often contain large amounts of impurities, mainly glass cullet and metallic parts like the remaining electrodes. The processing of the phosphor powder fractions has been performed in the aforementioned traditional hydrometallurgical mining technologies, being Osram or Solvay processes the most well-known due to their industrial application. However, it has been demonstrated that the cost-efficiency for the REM recovery process is increased using the SSC instead of wet treatments. The Sepselsa process converts the metals contained in the phosphor powder into chloride compounds by reaction with dry $HCl_{(g)}$. These metal chlorides are easily dissolved with water reaching a REM recovery of 88.6% [46], which can be a marketable product after precipitation in oxalate form and further calcination stage to achieve rare earth oxides (REO). Although this approach has been successfully applied in a real environment by FNE Entsorgungsdienste Freiberg GmbH (Germany), the product obtained was a mixture of REO instead of individual oxides. Therefore, an optimization of the chlorination route, Sepselsa process 2.0, was achieved by combining the SSC with the solvent extraction process (Figure 6.1.12). This new approach not only is able to recover REM from phosphors but also obtain single REO (Y_2O_3 and Eu_2O_3) as products with purities >99% [48].

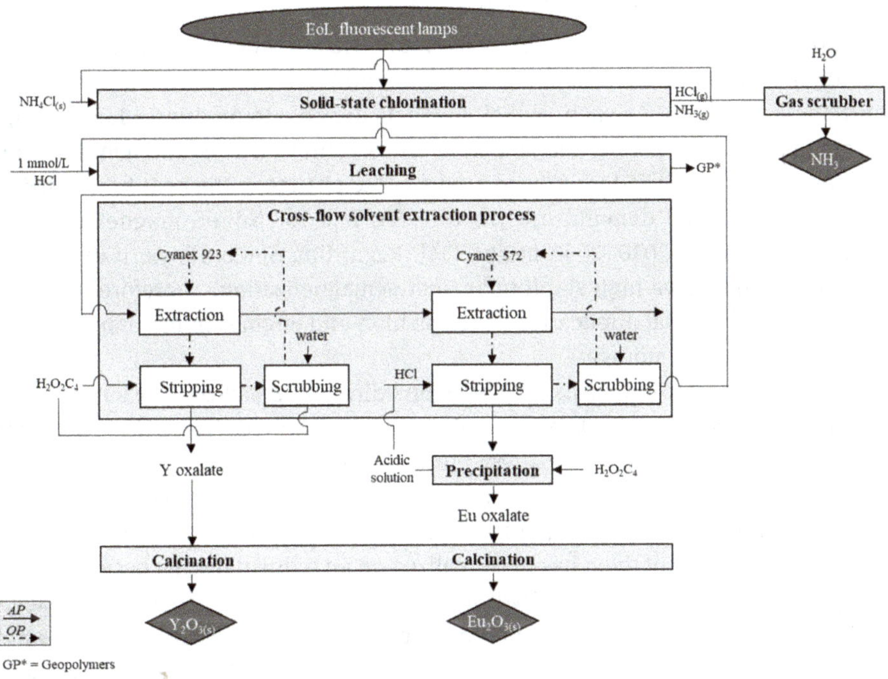

Figure 6.1.12: Industrial flowsheet for fluorescent lamp waste processing to recover RE as oxides [48].

The Sepselsa process 2.0 considers the Circular Resources Chemistry concept, too, and implements the zero-waste principle as far as possible. For instance, recirculating $NH_4Cl_{(s)}$ is beneficial from circularity (reduced chemicals consumption) and costs point of view. The solid residue obtained after the leaching step is rich in aluminum silicates, which is a perfect base to make geopolymers to be used as paving stones, paving slabs, canal and tunnel construction, etc. A particular environmental benefit is the recovery of fertilizer-grade ammonia (NH_3), which is the off-product of NH_4Cl decomposition once HCl has been consumed in the digestion step. It can be marketed directly and thus contributed considerably to covering process costs.

6.1.3.2 Permanent magnets (MagnetoRec process)

Permanent magnets demand has been increasing in recent years because of their strongest magnetic alloys content currently available with magnetic energy densities ≤450 kJ/m^3 [53]. With the focus on the development of clean energies, it is expected that this sector holds the first place in the priority of REM recycling because of the difficulty of finding substitutes. Neodymium, samarium, praseodymium, dysprosium and terbium are the common REM used in such magnets.

Neodymium-iron-boron (NdFeB) and samarium-cobalt (SmCo) are the two main permanent magnets. The former are much more frequently used in far higher amounts because of their magnetic strength and their non-restricted size due to brittleness problems. As such, NdFeB magnets are able to produce a higher magnetomotive force in smaller magnet sizes. Considering their requirement on wind turbine generators (WTG), this trend seems not to change in the near future, being the average annual demand growth rate forecast for NdFeB magnet in these WTGs, +10% in the 2020–2030 period [54]. Regarding SmCo alloys, it can be affirmed that they have high stability against demagnetization. Therefore, they are mainly used in special applications such as high-end engines in motorsports or in hard disk drive data storage.

Since 2011, when the peak of REM commodity prices had been reached, many attempts have been undertaken to develop processes for recovering these valuable metals. In fact, 1 kg Nd_2O_3 with ≥99% purity achieved a price of 250 USD/t, which however has dropped to ~66 USD/t today. Despite efforts to develop an efficient and economical process for recovering REM from permanent magnets to the best of our knowledge none of them has been realized on an industrial scale because of the sharp fall in raw material prices. Hydrometallurgical routes have been almost exclusively preferred for NdFeB and SmCo magnets. Pyrometallurgical approaches are avoided because in these EoL products there is corrosion, plastics, adhesives and nickel, cobalt and zinc as accompanying elements. As mentioned above for fluorescent lamps, hydrometallurgical processes involve an initial acid digestion step, which is intrinsically cost-intensive. Once again, the SSC process is an alternative to tackle all the drawbacks of the current hydrometallurgical approaches. In this sense, a cost-efficient process was developed within the MagnetoRec project, which allows the REM recovery from both, NdFeB and SmCo magnets.

85% of REM was recovered from EoL NdFeB magnet by using the SSC process with subsequent leaching with acetate buffer (pH = 3) [49]. The solution obtained after both steps not only contained REM but also iron as the main component of such magnets. REF_3 and $FeCl_2$ as products were obtained after fluoride precipitation with $HF_{(aq)}$ (Figure 6.1.13a). Regarding the recovery of REM from SmCo magnets, selective separation was achieved [55]. Thus, Sm was recovered whereas Co remained in the solid residue (Figure 6.1.13b).

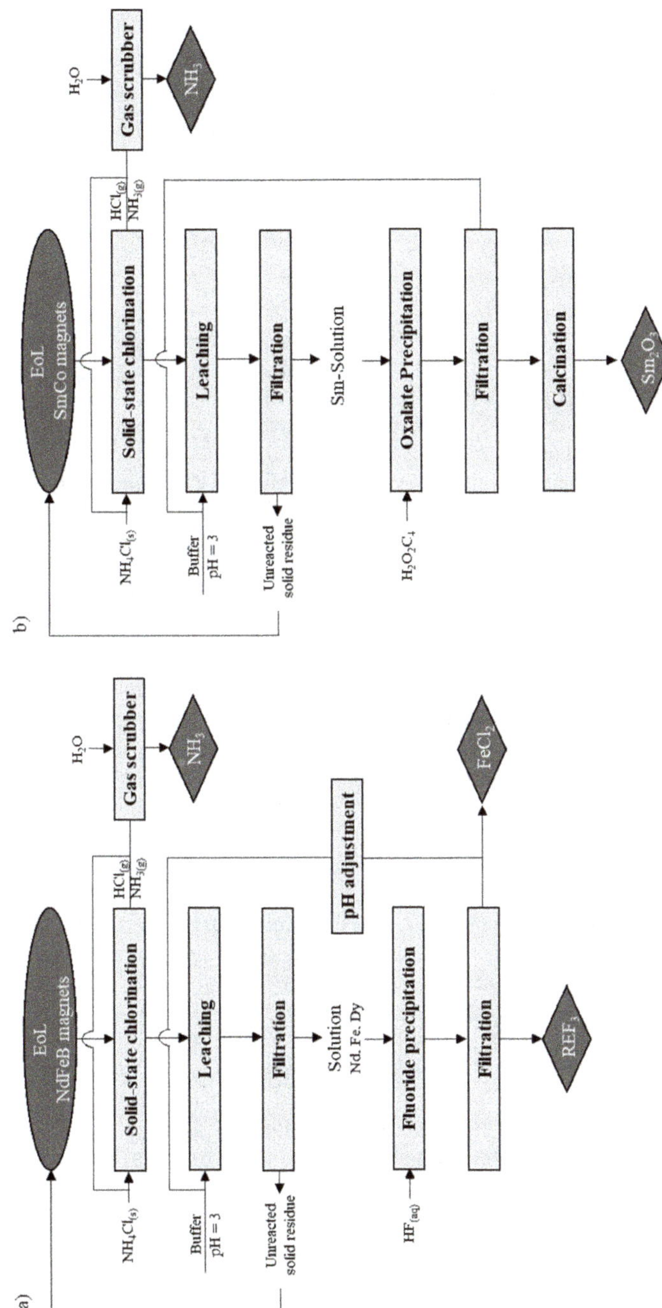

Figure 6.1.13: REM recovery from EoL permanent magnets. a) NdFeB, b) SmCo.

When compared with the literature, the MagnetoRec process for REM recovery from EoL permanent magnets works at less than half the costs of published alternatives. Such has been the success of both processes that in 2021 an industrial-scale reactor has been installed for the recovery of REM from permanent magnets, regardless of whether the magnet is NdFeB or SmCo.

References

[1] Rentsch L, Martin G, Bertau M, Höck M. Chem Eng Technol 2018, 41, 975–82.

[2] U.S. Geological Survey. Mineral Commodity Summaries 2021, 2021.

[3] Martin G, Rentsch L, Höck M, Bertau M. Energy Storage Mater 2017, 6, 71–79.

[4] Trading Economics. "Lithium Carbonate", available at https://tradingeconomics.com/commodity/lithium, 2022. accessed june 14th 2022

[5] LME. "Lithium at the LME," available at https://www.lme.com/Metals/EV/Lithium-prices, 2022, accessed june 14th 2022

[6] von Girselwad C, Weidmann H. Verfahren zur Gewinnung des Lithiumgehaltes von lithiumhaltigen Mineralien, DE000000616379A, 1933.

[7] Voigt W. In: Kausch P, Bertau M, Gutzmer J. (Eds.). Strategische Rohstoffe – Risikovorsorge, Springer Spektrum, Berlin, Heidelberg, 2014.

[8] Schmidt M. Rohstoffrisikobewertung – Lithium. DERA Rohstoffinformationen 33, Berlin, Germany, 2017, 138. www.deutsche-rohstoffagentur.de, accessed december 12th 2021

[9] Neukirchen GRF. Die Welt Der Rohstoffe: Lagerstätten, Förderung und wirtschaftliche Aspekte, Springer, Berlin, Heidelberg, 2016.

[10] An JW, Kang DJ, Tran KT, Kim MJ, Lim T, Tran T. Hydrometallurgy 2012, 117–118, 64–70.

[11] Kesler SE, Gruber PW, Medina PA, Keoleian GA, Everson ME, Wallington TJ. Ore Geol Rev 2012, 48, 55–69.

[12] ResearchInChina. "China Lithium Carbonate Industry Report," 2009. accessed december 11th 2021.

[13] Fröhlich P, Lorenz T, Martin G, Brett B, Bertau M. Angew Chem Int Ed 2017, 56, 2544–80.

[14] Chagnes A, Swiatowska J. Lithium Process Chemistry: Resources, Extraction, Batteries, and Recycling, Elsevier Science, Amsterdam, 2015.

[15] Brödermann P, Krüger G, Heng R. Verfahren zur Gewinnung von Lithiumcarbonat, DE 3622105 A1, 1988.

[16] Bertau M, Müller A, Fröhlich P, Katzberg M. Industrielle Anorganische Chemie, Wiley-VCH, Weinheim, 2013.

[17] Chagnes A, Pospiech B. J Chem Technol Biotechnol 2013, 88, 1191–99.

[18] Neßler J, Seifert T, Gutzmer J, Müller A. The Zinnwaldit-Lithium Project 2015. 10.5281/zenodo.3938862

[19] Martin G, Schneider A, Voigt W, Bertau M. Miner Eng 2017, 110, 75–81.

[20] Pavón S, Kaiser D, Mende R, Bertau M. Metals 2021, 11, 1–14.

[21] Research AM. "Global Lithium-ion Battery Market. Opportunities and Forecast 2020–2027," available at https://www.alliedmarketresearch.com/lithium-ion-battery-market. accessed december 11th 2021

[22] "Directive 2006/99/EC of the European parliament and of the council of 6 September 2006 on batteries and accumulators and waste batteries and accumulators and repealing Directive

91/157/EEC. 2006," available at https://www.legislation.gov.uk/%0Aeudr/2006/99/contents %0A, accessed december 11th 2021
[23] Velázquez-Martínez R, Valio O, Santasalo-Aarnio J, Reuter A, Serna-Guerrero M. Batteries 2019, 5, 68.
[24] Fan F, Li E, Wang L, Lin Z, Huang J, Yao Y, Chen Y, Wu R. Chem Rev 2020, 120, 7020–63.
[25] Gao W, Zhang X, Zheng X, Lin X, Cao H, Zhang Y, Sun Z. Environ Sci Technol 2017, 51, 1662–69.
[26] Harper S, Sommerville R, Kendrick E, Driscoll L, Slater P, Stolkin R, Walton A, Christensen P, Heidrich O, Lambert G. Nature 2019, 575, 76–86.
[27] Lv W, Wang Z, Cao H, Sun Y, Zhang Y, Sun Z. ACS Sustain Chem Eng 2018, 6, 1504–21.
[28] Fan E, Li L, Wang Z, Lin J, Huang Y, Yao Y, Chen R, Wu F. Chem Rev 2020, 120, 7020–63.
[29] Tiessen H, White PJ, Hammond JP. (Eds.) The Ecophysiology of Plant-Phosphorus Interactions. Plant Ecophysiology, Vol. 7, Springer, Dordrecht, 2008. https://doi.org/10.1007/978-1-4020-8435-5_1
[30] Liu X, Ruan Y, Li C, Cheng R. Int J Miner Process, 2017, 167, 95–102.
[31] Kaba OB, Filippov LO, Filippova IV, Badawi M. Powder Technol 2021, 382, 368–77.
[32] Jasinski S, U.S. Geological Survey, Mineral Commodity Summaries, 2022, accessed june 14th, 2022
[33] Abdennebi M, Benhabib N, Goutaudier K, Bagane C. J Mater Environ Sci 2017, 8, 557–65.
[34] Zou D, Jin Y, Li J, Cao Y, Li X. Sep Purif Techno 2017, 172, 242–50.
[35] Assuncao MC, Cote G, Andre M, Halleux H, Chagnes A. RSC Adv 2017, 7, 6922–30.
[36] Chen M, Li J, Jin Y, Luo J, Zhu X, Yu D. J Chem Technol Biotechnol 2018, 93, 467–75.
[37] Kahn MA, Muhammad A, Din I-D, Khan KM, Ambreen N. J Chem Soc Pakistan 2013, 35, 144–46.
[38] Fröhlich P, Eschment J, Martin G, Lohmeier R, Bertau M. The PARFORCE-Technology (Germany). In: Schaum C. (Hrsg.) Phosphorus Polluter Resour. Futur. Remov. Recover. From Wastewater, IWA Publishing, London, 2018.
[39] Steiner G, Wellmer F, Scholz R, Zenk L, Schernhammer E, Haneklaus N, Bertau M. Resour Conserv Recycl 2022. submitted for publication.
[40] Favot M, Massarutto A. J Environ Managem 2019, 240, 504–10.
[41] Li J, Li M, Zhang D, Gao K, Xu W, Wang H, Geng J, Huang L. Chem Eng J 2020, 382, 122790.
[42] Kooroshy J, Tiess G, Tukker A, Walton A. (Eds.) ERECON, Strengthening the Supply European Rare Earths Supply-Chain: Challenges and Policy Options, 2015.
[43] Alvarado-Hernández L, Lapidus GT, González F. Waste Manag 2019, 95, 53–58.
[44] European Commission, Comission Regulation (EU) 2019/2020 – Laying down Ecodesign Requirements for Light Sources and Separate Control Gears Pursuant to Directive 2009/125/EC and Commission Regulations (EC) No 244/2009, (EC) No 245/2009 and (EC) No 1194/2012, Vol. 2020, 2020.
[45] Pavón S, Lorenz T, Fortuny A, Sastre AM, Bertau M. Waste Manag 2021, 122, 55–63.
[46] Lorenz T, Bertau M. Recycling of Rare Earth Elements. Phys Sci Rev 2017, 2, 20160067.
[47] Huckenbeck T, Otto R, Haucke E. Verfahren zur Rückgewinnung Seltener Erden aus Leuchtstofflampen sowie zugehörige Leuchtstoffe und Lichtquellen, WO2012143240 A2, 2012.
[48] Pavón S, Lapo B, Fortuny A, Sastre AM, Bertau M. Sep Purif Technol 2021, 272, 118879.
[49] Lorenz T, Bertau M. J Clean Prod 2019, 215, 131–43.
[50] Ippolito NM, Innocenzi V, De Michelis I, Medici F, Vegliò F. J Clean Prod 2017, 153, 287–98.
[51] European Commission, Commission Regulation (EC) No 244/2009 – Implementing Directive 2005/32/EC of the European Parliament and of the Council with Regard to Ecodesign Requirements for Non-Directional Household Lamps, 2009.

[52] Tunsu C, Petranikova M, Ekberg C, Retegan T. Sep Purif Technol 2016, 161, 172–86.
[53] Yang Y, Walton A, Sheridan R, Güth K, Gauß R, Gutfleisch O, Buchert M, Steenari B-M, Van Gerven T, Jones PT, Binnemans K. J Sustain Metall 2017, 3, 122–49.
[54] Schulze R, Buchert M. Resour Conserv Recycl 2016, 113, 12–27.
[55] Lorenz T, Bertau M. J Clean Prod 2020, 246, 118980.

6.2 Phosphates

Robert Glaum, Thomas Staffel

Hydroxylapatite $Ca_5(PO_4)_3(OH)$ and related mineral resources form the basis for all phosphate and phosphoric acid production which amounted in the year 2020 worldwide to an equivalent of approx. 47 Mio. tons P_2O_5 (forecast FAO [1]). The majority (approx. 95%) of phosphate/phosphoric acid is used for fertilizer production (see Chapter 6.6). Yet, there are many further applications of inorganic phosphates making use of their particular acid/base, complexing, and redox properties. These applications range from phosphates as detergents, anti-corrosives, food additives (see Chapter 6.5), flame retardants (see Chapter 1.6), construction materials, and coatings (see Chapter 1.7). Applications include also use in dental and health care products. Eventually, special applications of transition metal phosphates in lithium ion secondary batteries, in heterogeneous catalysis and as color pigments will be outlined.

6.2.1 Phosphate reserves/resources, need and recovery

Natural phosphate minerals ("phosphorite" or "phosphate rock" [2]) are either of igneous (magmatic) origin or sedimentary deposits. The former have lower P_2O_5 content (approx. 5 wt%) than the latter (up to 30 wt%) [3]. In most cases phosphate rock has to be enriched in P_2O_5 content to above 20% ("benefication") to be suitable for phosphate/phosphoric acid production. Phosphate rock consists mainly of hydroxylapatite $Ca_5(PO_4)_3(OH)$ [4] and related minerals of the apatite family (e.g. $Ca_5(PO_4)_3X$ (X: F [5], Cl, Br, O^{2-}, CO_3^{2-}). Carbonate-apatite like $Ca_{9.75}[(PO_4)_{5.5}(CO_3)_{0.5}]CO_3$ [6] with heterovalent substitution of OH^- (A-type) and PO_4^{3-} (B-type) by CO_3^{2-} is also frequently occurring and particularly interesting with respect to its crystal chemistry. Apatite-based minerals can contain traces of uranium and thorium (predominantly as U^{4+} /Th^{4+} on Ca^{2+} sites [7]). The uranium/thorium content generally is low, nevertheless its recovery has been discussed repeatedly [8]. Enrichment of U/Th during mineral workup creates, however, an environmental risk. Ecologically it is also important that for beneficiation of one ton of phosphate rock 8 to 15 tons of water are required [3]. Sufficient water supply is essential for exploiting phosphate rock deposits.

Besides apatite-type minerals, aluminum-containing crandallite (ideal formula: $CaAl_3(HPO_4)(PO_4)(OH)_6$, related to alunite $KAl_3(SO_4)_2(OH)_6$ [9]) and wavellite $Al_3(PO_4)_2(OH, F)_3 \cdot 5H_2O$ [10] are of some importance. The well-known minerals vivianite $Fe_3(PO_4)_2 \cdot 8\ H_2O$ and monazite $RE^{III}PO_4$ (RE = rare earth element) are of no economic

Note: In preparation of this Section (6.2) consultation of and cross-referencing to the chapter "Phosphoric Acid and Phosphates" co-authored by one of us (T.S.) in Ullmann's Encyclopedia of Industrial Chemistry (6th ed., Wiley-VCH, 2008) is acknowledged.

https://doi.org/10.1515/9783110798890-010

relevance for phosphate/phosphoric acid production. In order for a phosphate rock to be suitable for a wide range of processing options, the M_2O_3/P_2O_5 ratio (M_2O_3: total content of $Fe_2O_3 + Al_2O_3$) must generally be below approximately 0.10. The ratio M_2O_3/P_2O_5 has to be below 0.12, if MgO is included in the total metal oxide content, to avoid manufacturing problems, in particular while synthesizing ammonium phosphates ($NH_4H_2PO_4$, $(NH_4)_2HPO_4$) [3].

World phosphate rock production since 1975 is shown in Figure 6.2.1. The production is forecast to reach 260 million tons (corresponding to approx. 60 Mio. tons P_2O_5) in the year 2024 [11]. Currently (2021), global phosphate rock resources are estimated at about $71{\cdot}10^9$ t. Peak phosphorus with the depletion of the phosphate mines in China and the USA (worlds 1st and 2nd largest producers 2020) is expected within the next 60 years. While the peak phosphorus hypothesis is disputed [12], it is generally accepted that 70% of global phosphate rock resources are located in Morocco [13, 14].

While there will be no immediate shortage in phosphate rock supply it has to be kept in mind that phosphate fertilizer is essential for global food production and without substitute. In future this might lead to an increasing need for recovering and recycling phosphate from sewage water.

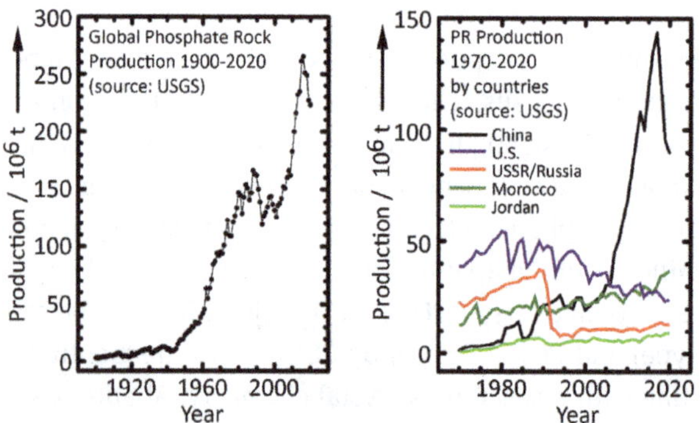

Figure 6.2.1: Phosphate rock production (global 1900–2020 and by main countries 1970–2020 [1, 15]).

6.2.2 Phosphoric acid

Synthesis of phosphoric acid H_3PO_4 (monophosphoric or orthophosphoric acid; approx. $83 \cdot 10^6$ t/a world production in 2015 with est. P_2O_5 content of slightly above 55%) is accomplished by two different processes which are summarized in reactions 6.2.1 and 6.2.2 [16]. The wet process with digestion of apatite by sulfuric acid leads to crude phosphoric acid with a rather low P_2O_5 content of approx. 30 wt% and gypsum $CaSO_4 \cdot 2\,H_2O$ or hemihydrate $CaSO_4 \cdot 1/2\,H_2O$ as further products. The wet process is also the main industrial source of hydrogen fluoride which is obtained as a by-product. By digestion of phosphate rock with sulfuric acid in a smaller reaction ratio (reaction 6.2.1b), normal superphosphate NSP fertilizer, which is a mixture of gypsum and calcium bis(dihydrogenphosphate), is produced. Concentration by vacuum evaporation of the crude acid to approx. 50 wt% P_2O_5 is required prior to its use in triple superphosphate TSP fertilizer production as described by reaction 6.2.1c. About 95% of phosphoric acid production is obtained by the wet process despite the high cost for concentrating, the highly corrosive reaction medium and some phosphate loss with the precipitated gypsum. Alternatively, phosphoric acid is produced from elemental phosphorus via combustion and subsequent hydrolysis of P_4O_{10} (thermal process; reactions 6.2.2a to c). Thermal phosphoric acid contains only traces of impurities, in particular arsenic, which can be removed by precipitation as arsenic sulfide [16].

$$Ca_5(PO_4)_3(OH)(s) + 5\,H_2SO_4(l) + (x-1)H_2O(l) \xrightarrow{80^\circ\,C} 5\,CaSO_4 \cdot x\,H_2O(s) + 3\,H_3PO_4(l)\ (x = 0.5 \text{ or } 2) \tag{6.2.1a}$$

$$2\,Ca_5(PO_4)_3(OH)(s) + 7\,H_2SO_4(l) + (x-2)\,H_2O(l) \longrightarrow 7\,CaSO_4 \cdot x\,H_2O(s) + 3\,Ca(H_2PO_4)_2(s)\ (x = 0.5 \text{ or } 2) \tag{6.2.1b}$$

$$Ca_5(PO_4)_3(OH)(s) + 7\,H_3PO_4(l) \longrightarrow 5\,Ca(H_2PO_4)_2(s) + H_2O(l) \tag{6.2.1c}$$

$$2\,Ca_5(PO_4)_3(OH)(s) + 10\,SiO_2(s) + 15\,C(s) \longrightarrow 3/2\,P_4(g) + 10\,CaSiO_3(l) + 15\,CO(g) + H_2O(g) \tag{6.2.2a}$$

$$P_4(g) + 5\,O_2(g) \longrightarrow P_4O_{10}(g)\ (\Delta_r H = -3053\,kJ/mol) \tag{6.2.2b}$$

$$P_4O_{10}(g) + 6\,H_2O(l) \longrightarrow 4\,H_3PO_4(l)\ (\Delta_r H = -377\,kJ/mol) \tag{6.2.2c}$$

Orthophosphoric acid is medium strong, triprotic with the dissociation constants $K_1 = 7.1 \cdot 10^{-3}$ mol/L, $K_2 = 6.3 \cdot 10^{-8}$ mol/L, $K_3 = 4.7 \cdot 10^{-13}$ mol/L. Aqueous solutions of dihydrogenphosphates are weakly acidic, while those of hydrogenphosphates and tertiary phosphates show weak or strong alkaline behavior, respectively [16].

6.2.3 Phosphate anions: From isolated orthophosphate to 3D networks

Phosphorus(V) exhibits exclusively tetrahedral coordination by oxygen in contrast to the octahedral PF_6^- anion and the trigonal-planar NO_3^-. Therefore, the ratio O/P of a phosphate provides immediate structural information (see Table 6.2.1). Phosphates with O/P > 4 generally contain O^{2-} ions besides PO_4^{3-} groups (e.g. $Ti_5P_4O_{20}$ which is better formulated as $Ti_5O_4(PO_4)_4$ [17]) In contrast, O/P < 4 is found for phosphates containing condensed PO_4 tetrahedra (Table 6.2.1). In the Q^n nomenclature introduced for the structural description of silicates by Liebau [18] n denotes the number of tetrahedral groups linked to a central tetrahedron. In phosphates $0 \leq n \leq 3$ is found. Based on the Q^n nomenclature systemizing the wealth of phosphate anions is straightforward (see Table 6.2.1). For metaphosphates ($n = 2$) further distinction between cyclic and chain-like anions is necessary, e.g. *cyclo*(tetrametaphosphate) for the $P_4O_{12}^{4-}$ anion and *catena*(metaphosphate) for PO_3^- chains. Phosphates containing chain-like anions with Q^1 and Q^2 groups are regarded as oligo- and polyphosphates, respectively. This grouping is based on the analytical distinction by ion chromatography [19] and ^{31}P-MAS-NMR spectroscopy with "oligo" being used for chains with less than 50 phosphate tetrahedra and "poly" for even longer chains. The power of the latter method for characterizing phosphates in both, the solid and liquid state, is shown by examples from the more recent literature [20].

Table 6.2.1: Structure and nomenclature of phosphate anions.

O/P ratio	Name	Q^n	Niggli[a)]	Examples
>4	oxide-phosphate	Q^0 and O^{2-}	$(PO_{4/1})^{3-}$ and O^{2-}	$Ca_{10}(PO_4)_6O$ [21], $Ti_5O_4(PO_4)_4$ [17]
4	ortho- or mono-phosphate	Q^0	$(PO_{4/1})^{3-}$	Na_3PO_4 [22]
3.5	pyro- or diphosphate	Q^1	$(PO_{3/1}O_{1/2})^{2-}$	$Na_4P_2O_7$ [23], SiP_2O_7 [24]
3.5 > O/P > 3	oligophosphate or polyphosphate	Q^1 and Q^2	$(PO_{3/1}O_{1/2})^{2-}$ and $(PO_{2/1}O_{2/2})^-$	$Na_5P_3O_{10}$ [25]
3	metaphosphate (cyclic or chain-like)	Q^2	$(PO_{2/1}O_{2/2})^-$	$NaPO_3$ [26][b)], $Na_3(P_3O_9)$ [27]
3 < O/P < 2.5	ultraphosphate	Q^2 and Q^3	$(PO_{2/1}O_{2/2})^-$ and $(PO_{1/1}O_{3/2})^0$	CuP_4O_{11} [28], NdP_5O_{14} [29]
2.5	phosphorus(V) oxide	Q^3	$(PO_{1/1}O_{3/2})^0$	3 polymorphic forms of P_2O_5 [30–32]

a) Structured formula according to Nigglis formalism [33]. A terminal oxygen O_t is counted as 1/1, each O_b as 1/2.
b) Maddrells salt

In condensed phosphates, one can distinguish between O_b, oxygen atoms bridging two phosphate groups, and O_t, terminal oxygen atoms. Coordination of phosphate anions to metals is always through O_t. There are no examples for O_b acting as ligands. The geometric structure of orthophosphate groups is generally close to that of a tetrahedron with $d(P{-}O) = 1.53$ Å. In condensed phosphates $d(P{-}O_t) = 1.49$ Å is typically found and can be as short as 1.45 Å. The distance to bridging oxygen atoms $d(\mathrm{P{-}O_b})$ ranges from 1.58 to 1.63 Å [34, 35]. Angles $\angle(\mathrm{O,P,O})$ in phosphate groups follow the VSEPR model with Q^1 phosphate being comparable to isolated $(PO_3F)^{2-}$ anions, Q^2 to SO_2F_2 or anionic $(PO_2F_2)^-$ and Q^3 to POF_3.

The degree of condensation of phosphate groups expressed by the Q^n formalism is not only providing geometric information. In terms of the Lux-Flood acid-base concept (O^{2-} donating compounds are regarded as bases, O^{2-} acceptors as acids [36]) reactions in particular (but not only) in phosphate melts changing Q^n of the phosphate anion can be understood as acid/base reactions (e.g. reactions 6.2.3a to c; note that 6.2.3c represents a combination of Lux-Flood acid-base and redox reaction). Obviously, P_4O_{10} is a strong Lux-Flood acid, while PO_4^{3-} anions are weak bases. This concept has been applied successfully to understanding quantitatively a wide variety of redox and acid-base reactions in (glass) melts. It is also in line with the classification of oxides in glass chemistry into network builders and modifiers by Zachariasen [37].

$$6\,Na_2O(s) + P_4O_{10}(s) \rightarrow 4\,Na_3PO_4(s) \qquad (6.2.3a)$$

$$4PO_4^{3-}(melt) + P_4O_{12}^{4-}(melt) \rightarrow 4P_2O_7{}^{4-}(melt) \qquad (6.2.3b)$$

$$4\,V^{IV}O(PO_3)_2(s) + P_4O_{10}(g) \rightarrow 4\,V^{III}(PO_3)_3(s) + O_2(g) \qquad (6.2.3c)$$

6.2.4 Monophosphates

Monophosphates of sodium (Na_3PO_4, Na_2HPO_4, NaH_2PO_4) are usually produced from phosphoric acid and soda ash (Na_2CO_3) or caustic soda (NaOH). For production of food-grade phosphates, very pure thermal phosphoric acid is used. The sodium monophosphates are known as anhydrous salts as well as hydrates with varying amounts of crystal water [38]. They show high solubility in water. Dilute aqueous solutions (1%) of NaH_2PO_4 and its commercially available dihydrate are weakly acidic (pH = 4.5). Due to their acidity they are used for phosphating of metal surfaces. Coatings on steel may consist of a variety of mixed-metal phosphates [e.g. $Fe_3(PO_4)_2 \cdot 8\,H_2O$ (vivianite), $(Fe,Mn)_5(HPO_4)_2(PO_4)_2 \cdot 4\,H_2O$ (hureaulite), $Zn_2Fe(PO_4)_2 \cdot 4\,H_2O$ (phosphophyllite) and $Zn_3(PO_4)_2 \cdot 4\,H_2O$ (hopeite)]. Depending on the bath constituents and the formation conditions, chemical and mechanical stability as well as the porosity of the coating can be varied [39]. NaH_2PO_4 is used in acidic detergents and in a mixture with Na_2HPO_4 as buffer in the pH range 5 to 8. Monosodium phosphate is also added for lowering the pH of water purified by ion exchange. Eventually, NaH_2PO_4 serves as pH stabilizer (soups, juices) in nutrition products. Disodium hydrogen phosphate Na_2HPO_4 is used in similar applications as food additive yet it stabilizes the pH at higher values. A dilute aqueous solution (1 wt%) shows pH = 9.5. It is also added in the production of milk powder to prevent coagulation during the evaporation process. Both NaH_2PO_4 and

Na_2HPO_4 are used as starting materials for the production of condensed phosphates (see Section 6.2.5).

Trisodium phosphate Na_3PO_4 exhibits a strong alkaline reaction (pH = 11.5; eq. 6.2.4) in water due to hydrolysis and thus causes saponification of fats. This behavior is the basis for its fat- and dirt-solving properties with application as industrial cleansing agent.

$$PO_4^{3-}(aq) + H_2O \rightarrow HPO_4^{2-}(aq) + OH^-(aq) \quad (6.2.4)$$

Commercially, trisodium phosphate is available as dodecahydrate which crystallizes with small amounts of NaOH, thus forming a solid solution $Na_3PO_4 \cdot 12\,H_2O \cdot x$ NaOH $(1/7 \leq x \leq 1/4)$ [40]. Instead of NaOH, incorporation of NaOCl is also possible, leading to bleaching and bactericidal properties.

Formal substitution of one O^{2-} in Na_3PO_4 by F^- leads to sodium monofluoridophosphate Na_2PO_3F for which various types of synthesis are known (eqs. 6.2.5a to d [16]). Depending on temperature und water vapor pressure, eq. (6.2.5d) can be reversed, leading to hydrolysis of the monofluoridophosphate anion. Na_2PO_3F finds widespread application in toothpaste, carrying fluoride, which is regarded as a caries prophylactic [41]. In special concrete Na_2PO_3F is used as a setting retarder [42].

$$NaPO_3 + NaF \rightarrow Na_2PO_3F \text{ (reaction in a melt)} \quad (6.2.5a)$$

$$Na_4P_2O_7 + 4\,NaF + P_2O_5 \rightarrow 4Na_2PO_3F \quad (6.2.5b)$$

$$Na_4P_2O_7 + 2\,HF \rightarrow 2Na_2PO_3F + H_2O \quad (6.2.5c)$$

$$Na_2HPO_4 + HF \rightarrow Na_2PO_3F + H_2O \quad (6.2.5d)$$

6.2.5 Condensed sodium phosphates: synthesis and properties

Condensed phosphates are formed by thermal dehydration of two or more hydrogenphosphate/dihydrogenphosphate units (see reaction 6.2.6). Figure 6.2.2 shows as an example the phase diagram for the binary system Na_2O-P_2O_5, which includes several condensed phosphates besides the orthophosphate. For a detailed account on the crystal chemistry of condensed phosphates, the reader is referred to the book by Durif [43]. Structurally, the family of condensed phosphates comprises pyrophosphates, oligophosphates, polyphosphates, metaphosphates, and ultraphosphates (see Table 6.2.1). These are all phosphates with $1 \leq n \leq 3$ for Q^n. Most technically relevant condensed phosphates are obtained by thermal treatment of NaH_2PO_4 and of intermediates derived thereof as is summarized in Figure 6.2.3.

$$2\,Na_2HPO_4(s) \rightarrow Na_4P_2O_7(s) + H_2O(g) \quad (6.2.6a)$$

$$NaH_2PO_4(l) \rightarrow NaPO_3(l) + H_2O(g) \quad (6.2.6b)$$

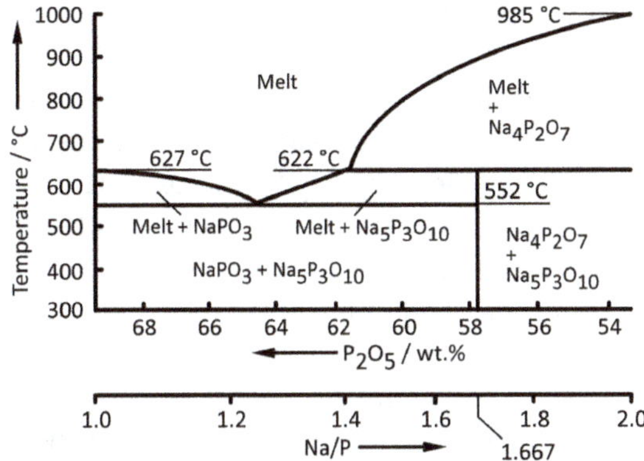

Figure 6.2.2: Phase diagram Na_2O-P_2O_5 between $NaPO_3$ and $Na_4P_2O_7$ [44].

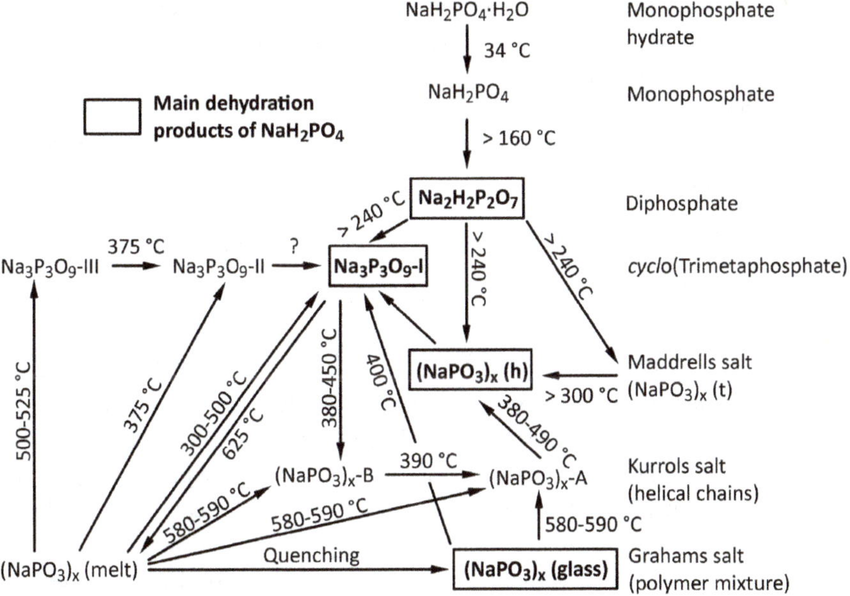

Figure 6.2.3: Condensed phosphates that can be prepared from NaH_2PO_4 by heating and their interrelationship [47].

$$3\,NaH_2PO_4(l) \rightarrow Na_3(P_3O_9)(l) + 3\,H_2O(g) \tag{6.2.6c}$$

Condensed phosphoric acids, which can be obtained from the corresponding sodium phosphates by ion exchange, contain two types of OH groups that differ greatly in their degree of dissociation. One is strongly acidic while the terminal (Q^1) groups bear a second type that is only weakly acidic. Reliable dissociation constants for di- and triphosphoric acid are given in literature (pyro- or diphosphoric acid $H_4P_2O_7$: $K_1 = 1.0 \cdot 10^{-1}$ mol/L, $K_2 = 5.2 \cdot 10^{-3}$ mol/L, $K_3 = 2.0 \cdot 10^{-7}$ mol/L, $K_4 = 4.0 \cdot 10^{-10}$ mol/L, triphosphoric acid $H_5P_3O_{10}$: $K_1 = 1.0 \cdot 10^{-1}$ mol/L, $K_2 = 6.3 \cdot 10^{-3}$ mol/L, $K_3 = 5.0 \cdot 10^{-3}$ mol/L, $K_4 = 2.0 \cdot 10^{-6}$ mol/L, $K_5 = 3.2 \cdot 10^{-9}$ mol/L [45]). Consequently, long-chain ("poly") phosphoric acids and *cyclo*(metaphosphoric) acids behave as strong monoprotic acids. This leads to a decreasing pH for aqueous (1 wt%) solutions along the series of sodium phosphates Na_3PO_4 (pH = 11.5), $Na_4P_2O_7$ (10.2), $Na_5P_3O_{10}$ (9.7), Graham's salt $NaPO_3$ (6.0); see also Figure 6.2.4a.

By reversal of the formation reaction, condensed phosphates undergo hydrolysis in aqueous solution (e.g. 6.2.7a). The rate of hydrolytic decomposition depends strongly on temperature, pH, and the type and concentration of metal ions present in the solution [46, 47]. For sodium triphosphate in aqueous solution (pH = 8, ambient temperature), the half-life is more than a year, which is reduced to a few minutes at pH = 3 and 100 °C. Three types of decomposition reactions (two hydrolytic, one rearrangement) are reported for condensed phosphates: successive cleavage from the chain ends (eq. 6.2.7a), cleavage in the middle of a chain (preferably in strongly acidic solution with formation of tri- and tetraphosphate; eq. 6.2.7b), and rearrangement after intra-chain cyclization (6.2.7c) [47]. According to recent research, intra-chain cyclization beginning from the chain ends with the predominant formation of *cyclo*(trimetaphosphate) is the most important decomposition reaction for water-soluble polyphosphate anions [20a].

$$H_2P_3O_{10}^{3-}(aq) + H_2O \rightarrow H_2P_2O_7^{2-}(aq) + H_2PO_4^-(aq) \qquad (6.2.7a)$$

$$H_2P_6O_{19}^{6-}(aq) + H_2O \rightarrow 2\,H_2P_3O_{10}^{3-}(aq) \qquad (6.2.7b)$$

$$H_2P_{12}O_{37}^{12-}(aq) \rightarrow H_2P_8O_{25}^{8-}(aq) + P_4O_{12}^{4-}(aq) \qquad (6.2.7c)$$

Condensed phosphates of the alkali metals are generally well water-soluble, while phosphates of cations with higher charge are less or even insoluble. These solutions are containing cyclic and chain-like, but no branched (Q^3) phosphate anions [47]. The *catena*(phosphates) behave like soluble ion exchangers for cations with higher ionic charge and prevent these cations from precipitation. Even though the cation binding is reasonably well described as chelating complexation by the oligo- and polyphosphate anions, formation of well-defined complexes is not observed. Quite remarkably, *cyclo*(trimetaphosphate) and *cyclo*(tetrametaphosphate) show no cation exchange behavior and complex formation in aqueous solution. The cation binding behavior of a condensed phosphate is quantified by its lime binding capacity [48], which depends on the chain length of the dissolved phosphate (Figure 6.2.4b). The ion exchange and binding capacity of condensed phosphate solutions is used in numerous applications (detergents, water softener, setting retarder for gypsum and cement, cheese processing, . . .) to mask and inactivate interfering cations. Despite not being surface active by themselves, dissolved condensed phosphates strongly influence the colloidal behavior and reduce the viscosity of aqueous suspensions. Thus, high solid content and low viscosity suspensions of insoluble inorganic solids (e.g. $BaSO_4$ white pigment particles; kaolinite) are obtained by addition of small amounts of Graham's salt (see Figure 6.2.3), which is a water-soluble, true oligophosphate with chain length up to 25 phosphate groups. In the same way, the soil antiredeposition effect of polyphosphates in the laundry process can be understood as suspension stabilization. Polyphosphates are also used to improve homogenization (prevent coagulation) during processing of cheese and meat. Polyphosphates are able to partly dissolve/digest proteins and to convert it to a sol. As shown in Figure 6.2.4c, the dispersive activity of polyphosphates depends on the chain length.

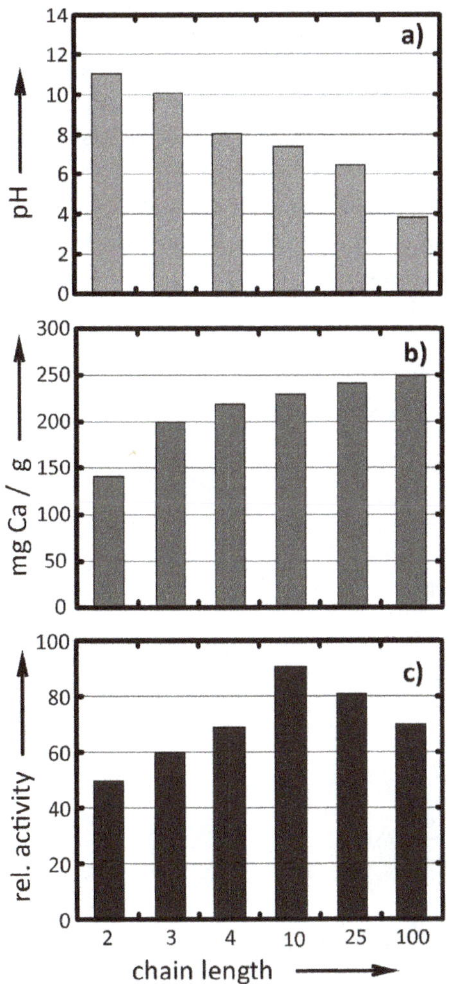

Figure 6.2.4: Properties of condensed phosphates vs. chain length (number of phosphate groups). pH-value (a), lime binding capacity (b), dispersing activity (c).

6.2.6 Condensed phosphates: applications

Setting of pH, cation binding capacity, dispersing activity, and colloidal effects are the most important chemical and physical properties which determine technical applications of condensed phosphates.

$Na_4P_2O_7$, containing the simplest condensed phosphate anion, is used in detergents and cleansers, in metal surface treatment, and in oil well drilling fluids for viscosity adjustment [49]. In food production, sodium pyrophosphate is used in cold puddings and ice cream for controlling their consistency. The stabilization of toothpaste based on calcium hydrogenphosphate is also facilitated by adding $Na_4P_2O_7$. Similar application as detergent, in soaps and cleansers (in particular for automatic dishwashers) are found for sodium triphosphate, which is commercialized as anhydrous salt and as hexahydrate $Na_5P_3O_{10} \cdot 6\,H_2O$. These applications are based on its unique combination of properties, which include water softening, dispersing behavior for various types of dirt, and fat hydrolysis due to moderately alkaline reaction in water.

Solid $Na_2H_2P_2O_7$ is used as baking aid. As acid carrier in dry state it does not react with $NaHCO_3$ and the mixture is stable during transport and storage. Only upon mixing with water, CO_2 is released (eq. 6.2.8). The effect of CO_2 release during baking of profiteroles is shown in Figure 6.2.5.

$$H_2P_2O_7^{2-}(aq) + 2\,HCO_3^-(aq) + \rightarrow P_2O_7^{4-}(aq) + 2\,CO_2(g) + 2\,H_2O \qquad (6.2.8)$$

Figure 6.2.5: Effect of an increased (right to left) amount of $Na_2H_2P_2O_7/NaHCO_3$ mixture as baking aid in the production of profiteroles [50].

In leather production, polyphosphates act in several advantageous ways. They prevent coagulation of blood and reduce the hardness of water during soaking of hides and skins. The liming step becomes shorter and more effective due to stabilization of small solid lime particles, which penetrate quicker the organic structure. Polyphosphates do also react with proteins in a mild tanning process, similar to commonly used chromium salts and classical vegetable tannage. Eventually, polyphosphates prevent staining of the leather during the washing steps by complexation of iron and calcium ions.

In paint production, oligophosphates with chain lengths of six to ten phosphate tetrahedra improve the dispersion of the solid components (color pigment, filler) in water. Thus, better homogeneity during storage and even better scrubbing resistance of the dried paint is achieved.

In construction industry, small amounts of short-chain oligophosphates are used as setting retarders for concrete, cement-based mortar and gypsum plaster. In addition, a positive influence on the fluidity of the mortar is observed. Thus, optimum workability at an increased open time (time prior to setting) is achieved.

6.2.7 Ammonium phosphates

Several phosphates have been identified and crystallographically characterized in the system NH_3-P_2O_5-H_2O [51]. Of these, $(NH_4)H_2PO_4$ and $(NH_4)_2HPO_4$, which are obtained by neutralization of phosphoric acid by ammonia, are of economic importance as fertilizers. In contrast to the corresponding alkali salts, ammonium oligo- and polyphosphates $(NH_4PO_3)_x$ cannot be obtained by thermal condensation of monophosphates due to release of ammonia at the required higher temperatures. Therefore, only slightly above ambient temperature solid sources of ammonia (urea, melamine; eqs. 6.2.9a and b) are reacted with concentrated phosphoric acid.

$$xH_3PO_4(l) + xCO(NH_2)_2(s) \rightarrow (NH_4PO_3)_x(s) + x\,CO_2(g) + x\,NH_3(g) \qquad (6.2.9a)$$

$$xH_3PO_4(l) + x/6\,C_3N_3(NH_2)_3(s) \rightarrow (NH_4PO_3)_x(s) + x/2\,CO_2(g) \qquad (6.2.9b)$$

Ammonium phosphates show good flame retarding properties [52] (see also Chapter 1.6), especially for wood and cellulose. This behavior can be traced back to at least three different types of retarding mechanisms: endothermic degradation (removal of heat from the substrate), thermal shielding (solid phase) by formation of a thermal insulation barrier due to intumescent properties of phosphate together with the organic substrate and, third, dilution of the oxygen containing gas phase. In recent years it was shown that nanocomposites containing zirconium phosphates like α-$Zr(HPO_4)_2 \cdot H_2O$ possess even superior flame retarding properties [53].

6.2.8 Phosphates of the alkaline earth metals, boron and aluminum

Digestion of phosphate rock with sulfuric or phosphoric acid leads to moderately water soluble hydrogenphosphates $Ca(H_2PO_4)_2$ and insoluble $CaHPO_4$. For both, several hydrates are known. Calcium hydrogenphosphates are used in fertilizers. Pure calcium phosphates are predominantly used as feedstuff additives, as anti-caking additives (e.g. for salt) and as polishing compounds in toothpaste due to their abrasive behavior. The production of a certain calcium phosphate from aqueous solution depends strongly on temperature and pH. Complete neutralization of H_3PO_4 by lime leads to precipitation of hydroxyapatite $Ca_5(PO_4)_3OH$. Tricalcium phosphate $Ca_3(PO_4)_2$ can only be obtained via high-temperature processes ($T > 900$ °C), e.g. from $Ca_5(PO_4)_3OH$ and a source of additional P_2O_5.

BPO_4 and $AlPO_4$ in their dense forms require synthesis at temperatures up to 1000 °C with phosphoric acid, boric acid and Al_2O_3, respectively as starting materials. Aluminum orthophosphate $AlPO_4$ (mineral berlinite [54]) adopts a quartz-like crystal structure while BPO_4 crystallizes as homeotype of the cristobalite structure [55, 56]. Open-framework structures ("molecular sieves") based on aluminophosphate ($AlPO_4$), silicoaluminophosphate (SAPO) or metal aluminophosphates (MeAPOs) represent an important group of inorganic materials with large potential as adsorbents and catalysts. By incorporation of transition metals, the Lewis-acidic behavior of nano-porous $AlPO_4$ and of SAPOs can be enhanced. In addition, redox functionality for selective oxidation of short-chain olefins can be achieved. The reader is referred to the literature for details [57] and Chapter 7.3.

For aluminum, as for several other trivalent cations, various acid (e.g. $Al(H_2PO_4)_3$ [58]) and condensed phosphates (e.g. $AlH_2P_3O_{10}$, $Al_2P_6O_{18}$, $Al_4(P_4O_{12})_3$, $Al(PO_3)_3$) are known. These have applications as components in glass production, as hardeners for water glass and as additives in silicate plaster [59].

6.2.9 Transition metal phosphates

For all transition metals, except technetium, at least one anhydrous phosphate has been crystallographically characterized up to date. While for some only one is known (e.g. $AuPO_4$ [60], OsP_2O_7 [61]), others like titanium [62], vanadium [63] or iron [64] form many anhydrous phosphates with compositions ranging from oxide-phosphates (e.g. $Fe_9O_8(PO_4)$ [65]) to P_2O_5-rich ultraphosphates (e.g. FeP_4O_{11} [66]). Structurally, on one side of the compositional range, this reflects in typical "oxide structures" with an occasional phosphate group strewn in, while for phosphates rich in P_2O_5, networks comprising Q^2 and Q^3 groups with well separated metal cations are observed. Vanadium (oxidation states +II, +III, +IV, +V), molybdenum (+III to +VI), and rhenium (+IV to +VII) show an exceptionally wide variety of oxidation states in phosphate environment. This stabilization of high as well as rather low oxidation states reflects the redox stability of phosphate anions, which is the prerequisite for their use as cathode material in lithium ion secondary batteries (see Section 6.2.10). This high electrochemical stability goes along with low polarizability of phosphate groups, which makes phosphates with quite high band gaps suitable for optical materials. Anhydrous transition metal phosphates are insoluble in water and even at elevated temperatures, hydrolysis proceeds only slowly (with the exception of ultraphosphates). Another consequence of the rather high lattice energy of transition metal phosphates is the kinetic stability of their frameworks. This stability leads to application-relevant properties: a) frequently encountered formation of thermodynamically metastable polymorphic forms (e.g. $VOPO_4$ [67], $FePO_4$ [68]; b) requirement of rather high temperatures for recrystallization and equilibration during synthesis (e.g. niobium(V) phosphates $NbP_{1.8}O_7$ [69], $Nb_3(NbO)_2(PO_4)_7$ [70]). Subsequently, examples will be given for the industrial application of transition metal phosphates as color pigments, in heterogeneous catalysis, as cathode material in lithium ion secondary batteries, negative (or zero) thermal expansion materials, and for second harmonic generation in laser materials.

6.2.9.1 Phosphates as color pigments

One of the most striking properties related to transition metal ions is their color due to excitation of d-d electronic configurations. In phosphates, these ions are embedded in a chemically and thermally stable environment. This enables their use as color pigments (see also Chapter 1.4) in conditions detrimental for organic materials. Furthermore, phosphates allow some compositional variability, in contrast to the corresponding binary oxides. While there is not much flexibility in the color of the typical green pigment Cr_2O_3 [71], various ways of synthesis (eq. 6.2.10) are known for the green pigment chromium(III) phosphate (historically named Arnaudon's, Dingler's, Schnitzer's or Plessy's green). Depending on whether the anhydrous compound or one of the two hydrated forms are used, the color ranges from green over bluish-green to greyish-purple [72]. This variation reflects the variability of ligand fields that can be created by a phosphate environment. A striking example for this variability is found with anhydrous titanium(III) phosphates ($TiPO_4$: green, $Ti_4[Si_2O(PO_4)_6]$: purple-red, $Ti(PO_3)_3$: light-blue [73]), even though their molar absorption coefficients and brightness are not sufficient for application.

$$2\,(NH_4)_2HPO_4 + K_2Cr_2O_7 \rightarrow 2\,CrPO_4 + 2\,KOH + 2\,NH_3 + 5\,H_2O + N_2 \qquad (6.2.10a)$$

$$\begin{aligned} 10\,Na_2HPO_4 + 5\,K_2Cr_2O_7 + 3\,C_4H_6O_6 \rightarrow\ & 10\,CrPO_4 + 10\,K^+ + 20\,Na^+ \\ & + 12\,CO_3^{2-} + 6\,OH^- + 11\,H_2O \end{aligned} \qquad (6.2.10b)$$

Over the last decades, efforts were made to find environmentally uncritical color pigments. One example is the replacement of cobalt violet ($Co_3(PO_4)_2$ [74]) by ammonium manganese(III) pyrophosphate $NH_4Mn^{III}P_2O_7$ [75] (Manganese Violet; Figure 6.2.6). It has been shown for the solid solution $In_{1-x}Mn_x(PO_3)_3$ that the distortion of the octahedral chromophore $[Mn^{III}O_6]$ (Mn^{3+}, Jahn-Teller active d^4 electron configuration) can be easily modified by second-sphere ligand field effects leading to a color shift from purple-red ($x = 1$) to violet-blue ($x = 0.6$) [76].

In recent years, a whole new class of copper-containing pigments based on Apatite-type phosphates of the heavy alkaline earth metals was introduced. Formal substitution of X^- by $(Cu^IO)^-$ according to $Ca_5(PO_4)_3Cu_yO_{y+\delta}(OH)_{0.5-y-\delta}X_{0.5}$ (X = OH, F, Cl) yields blue-violet to pink color hues [77] (Figure 6.2.6). The bright color of these solids suggests that intervalence charge transfer (Cu^+/Cu^{2+}) and ligand-to-metal charge transfer ($O^{2-} \rightarrow Cu^{+/2+}$) rather than d-d electronic transitions are responsible for light absorption. Strontium Phosphate Violet with the chemical formula $Sr_5(PO_4)_3Cu_{0.3}O$ is a commercially available violet pigment. Incorporation of lanthanum into the copper-containing apatites $Ca_{10-x}La_x(PO_4)_6O_2H_{1.5-x-y-\delta}Cu_y$ ($x = 0$ to 1.79; $y = 0$ to 0.57) leads to colors varying from pink to pale yellow and blue-grey tints [78].

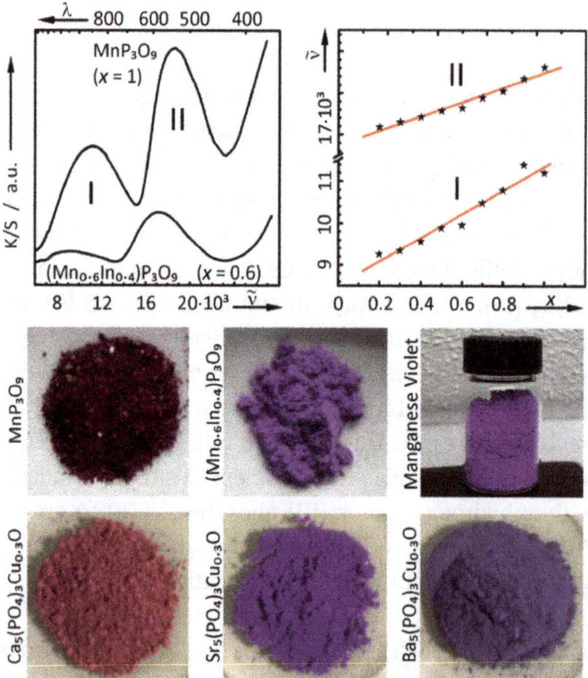

Figure 6.2.6: Color pigments Manganese Violet ($NH_4Mn^{III}P_2O_7$), $Ca_5(PO_4)_3Cu_{0.3}O$, $Sr_5(PO_4)_3Cu_{0.3}O$, and $Ba_5(PO_4)_3Cu_{0.3}O$. The colors of $Mn(PO_3)_3$ and $(Mn_{0.6}In_{0.4})(PO_3)_3$ demonstrate the susceptibility of the Jahn-Teller elongated $[Mn^{III}O_6]$ chromophore to small changes in the crystal chemical environment with shifts of the energies of the electronic transitions I: $^5B_{1g} \rightarrow {}^5A_{1g}$, $d(z^2) \rightarrow d(x^2-y^2)$; II: $^5B_{1g} \rightarrow {}^5B_{2g}$, ^5E_g, $d(xy), d(xz,yz) \rightarrow d(x^2-y^2)$; depending on the compositional parameter x in $(Mn_xIn_{1-x})(PO_3)_3$.

6.2.9.2 Titanium(IV) phosphates

Titanium(IV) pyrophosphate $Ti^{IV}P_2O_7$ [79] shows a cubic crystal structure that can be derived from that of pyrite $Fe^{II}S_2$ by substituting the S_2^{2-} anions by pyrophosphate groups. Upon heating, the bridging oxygen atoms within the $P_2O_7^{4-}$ anions are oscillating with increasing amplitude perpendicular to the P–P vector of the pyrophosphate group. This motion corresponds to a tilting of the two Q^1 tetrahedra towards each other, which in effect reduces the average distance between the terminal oxygen atoms within the anion (Figure 6.2.7). Overall for TiP_2O_7 this leads to lower than typically expected expansion of the unit cell volume with increasing temperature. For bigger tetravalent cations, e.g. Ce^{4+}, the octahedra $[M^{IV}O_6]$ become more flexible and impose less restrictions to the tilting of the phosphate tetrahedra ("quasi rigid unit motion" [80]). Thus, for CeP_2O_7 above 450 °C, shrinking of the unit cell with rising temperature is observed [81]. This negative thermal expansion (NTE) behavior [82], which is also shown by other pyrophosphates $M^{IV}P_2O_7$, is used in glass ceramics, e.g. CERAN (Schott AG), to achieve composite materials with zero thermal expansion over a wide temperature range. In such composites, the normal thermal expansion of the glass matrix is compensated by the negative expansion of the crystalline phase. As a result, materials are obtained that will not change shape over a wide temperature range and that can sustain repeated and quick temperature changes without thermal strain and fracturing.

A second titanium phosphate of technical importance is the non-linear optics (second harmonic generating SHG) material potassium titanyl(IV) phosphate $KTiOPO_4$ (KTP). Its non-centrosymmetric crystal structure (space group $Pna2_1$ [83]) comprises chains of corner sharing, highly distorted octahedra $\left[(Ti^{IV}{\equiv}O)O_4O\right]$. These show a short bond within the titanyl-group and a rather long distance to the second oxide ion (in contrast to four oxygen atoms from coordinating phosphate groups). The anisotropic Ti–O bond distribution seems to be a prerequisite for the SHG effect. $KTiOPO_4$ crystals (hydrothermal or oligophosphate flux growth) are used in diode pumped solid state lasers [84], which found wide-spread application in green laser pointers. In there, near infrared light ($\lambda = 1064$ nm) of a Nd:YAG laser (neodymium-doped yttrium aluminum garnet), is up-converted to its second harmonic, thus providing green light of 532 nm wavelength. KTP has a relatively high threshold to optical damage $\left(\sim 15\,\mathrm{J/cm^2}\right)$, an excellent optical nonlinearity and high thermal stability. However, it is prone to photochromic damage during high-power 1064 nm second-harmonic generation.

6.2.9.3 Vanadyl(IV) pyrophosphate

Vanadyl(IV) pyrophosphate $(V^{IV}O)_2P_2O_7$ [85] is a rare example for a chemically pure, single-phase solid used as catalyst material [86] (see also Chapter 7.2.). In contrast to the strong Lewis-acidity of zeolite-type AlPO-catalysts used for cracking of alkanes, here the peculiar redox behavior of $(V^{IV}O)_2P_2O_7$ provides the basis for its application. The pyrophosphate as well as active materials derived thereof still represent the sole example of a commercialized material for the catalytic oxidation of an alkane. Since its introduction in 1983 [87], when Monsanto started the world's first plant for selective oxidation of *n*-butane to maleic anhydride (MAN), the annual production has climbed to approx. $2{\cdot}10^6$ t/a in 2020. MAN is used as a chemical intermediate in the synthesis of resins, poly-thf (polytetramethylene ether glycol), fumaric and tartaric acid, certain agricultural chemicals, dye intermediates, and pharmaceuticals. While the overall oxidation reaction for *n*-butane to MAN is simple, the individual steps, and even more so, the importance of the crystallographic and chemical properties of $(VO)_2P_2O_7$ for these steps are still disputed. Many reviews and original papers have focused on the process and the catalyst involved [88]. The actual catalyst is obtained *in situ* (within the reactor) either from $(V^{IV}O)HPO_4 \cdot 1/2\ H_2O$ (Figure 6.2.7) or $(V^{V}O)PO_4 \cdot 2\ H_2O$. These precursor materials allow by a process with carefully chosen conditions (T, $p(O_2)$, time) synthesis of the active catalyst with high specific surface area (up to 30 m^2/g), a peculiar, lamellar crystallite shape, as well as the setting of the optimum chemical composition of the surface layers (average oxidation state of vanadium slightly above +4; small excess of phosphorus oxide). SCHLÖGL pointed out [89] that under *operando* conditions this amorphous surface layer shows about 1 nm thickness and high dynamics with respect to chemical composition. The projection of the unique bulk crystal structure of $(V^{IV}O)_2P_2O_7$ [85] as to the structure of the surface is obviously too much of a simplification. For decades, the vanadyl groups, $(V^{IV}O)^{2+}$ and $(V^{V}O)^{3+}$, were regarded as the catalytically active centers [90], being responsible for the rate determining

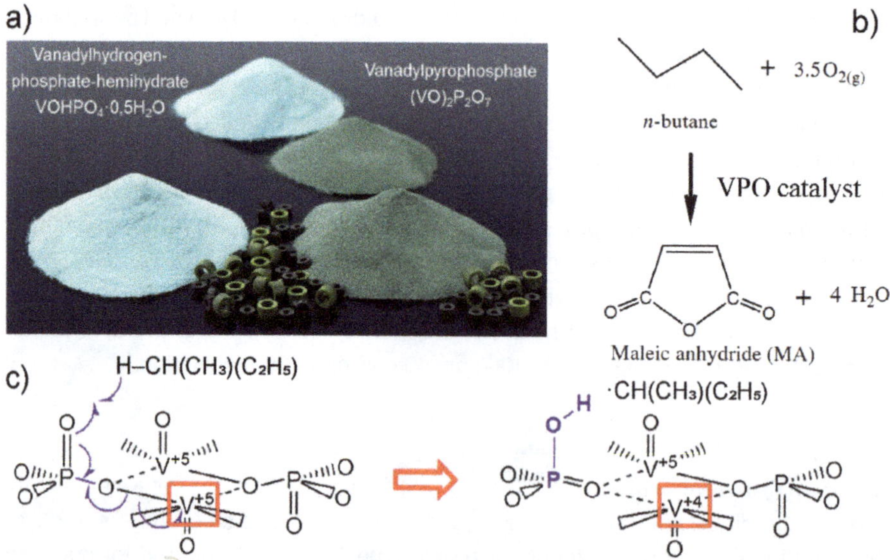

Figure 6.2.7: Precursor and vanadyl(IV) pyrophosphate powders as well as impregnated catalyst carrier (a). Reaction scheme for catalytic oxidation of *n*-butane (b). Visualization of splitting of the first C-H bond of *n*-butane according to the ROA [91] mechanism with formation of a butyl radical (c).

step, the abstraction of the first hydrogen from a *n*-butane molecule. The pyrophosphate groups were merely seen as spacers between these active centers. In recent years, based on extensive quantum mechanical studies a much more active role was attributed to the terminal oxygen atoms of the phosphate group. Though, even this model (reduction-coupled oxo activation, ROA) relies on enhanced activity of the phosphate-oxygen by neighboring V^{5+} ions [91].

6.2.10 Phosphates in electrochemical applications

The rather high electrochemical stability of phosphate ions towards oxidation and reduction is the prerequisite for application of solid phosphates in secondary batteries (see Section 5.1). Since the groundbreaking work of Goodenough [92], the crystal chemical properties and the electrochemical performance of $LiFePO_4$ has been used as a blueprint for further development of cathode materials [93]. Only recently, phosphorus-containing anode materials shifted into the focus, with black phosphorus and transition metal phosphides attracting most attention [94]. Yet, in addition to the well-established ternary, spinel-type oxide $Li_4Ti_5O_{12}$ [95], even some phosphates are considered as anode materials (e.g. PNb_9O_{25}, which can be lithiated electrochemically to an approximate composition $Li_{13.5}PNb_9O_{25}$ [96]). For high-voltage secondary batteries, some multinary phosphates are also considered as solid electrolytes (e.g. $Li_{1+x}La_xTi_{2-x}(PO_4)_3$; see reviews in [97] for details).

Despite their electrochemical stability, oxidation of the various monophosphate anions is even possible in aqueous solution (eq. 6.2.11a to c). Peroxomonophosphoric acid H_3PO_5, peroxodiphosphoric acid $H_4P_2O_8$, and their respective salts can be obtained in good yields (eq. 6.2.11e and f). Various side reactions, which eventually lead to the formation of molecular oxygen, can be restricted by the use of boron-doped diamond electrodes [98]. Even though the reasonably stable alkali metal peroxodiphosphates are used as oxidants in organic synthesis [99], as desinfectants, for bleaching, and in phosphatizing baths for treatment of metal surfaces, they have been crystallographically characterized only recently [100].

$$H_2PO_4^- \rightarrow (H_2PO_4)^\bullet + e^- \tag{6.2.11a}$$

$$HPO_4^{2-} \rightarrow (H_2PO_4^-)^\bullet + e^- \tag{6.2.11b}$$

$$PO_4^{3-} \rightarrow (PO_4^{2-})^\bullet + e^- \tag{6.2.11c}$$

$$H_2O \rightarrow OH^\bullet + H^+ + e^- \tag{6.2.11d}$$

$$2(H_2PO_4)^\bullet \rightarrow H_4P_2O_8 \tag{6.2.11e}$$

$$(H_2PO_4)^\bullet + OH^\bullet \rightarrow H_3PO_5 \tag{6.2.11f}$$

The redox behavior of phosphates is relevant for the synthesis of photochromic and electrochromic glasses [101]. Melts of condensed sodium phosphates can be considered as consisting of large polymeric oxo-anions (cyclic and chain-like) and mobile Na^+ cations. According to Section 6.2.3, the degree of condensation (average n with respect to the Q^n nomenclature) is given by the O/P ratio of the melt. At 700 °C, electrolysis of such melts leads to the cathodic formation of gaseous P_4 and depolymerisation of the condensed phosphate (eq. 6.2.12). Apparently, orthophosphate, pyrophosphate, and triphosphate anions are easier reduced than the pure $NaPO_3$ melt [102].

$$4\,PO_4^{3-} + 20\,e^- \rightarrow P_4(g) + 16\,O^{2-} \qquad (6.2.12a)$$

$$P_4O_{12}^{4-} + 4\,O^{2-} \rightarrow 4\,PO_4^{3-} \qquad (6.2.12b)$$

References

[1] Food and Agriculture Organization of the United Nations. World fertilizer trends and outlook to 2020 – Summary Report, Rome, 2017.

[2] Zapata F, Roy RN. (Eds.) Use of phosphate rocks for sustainable agriculture, Fertilizer and Plant Nutrition Bulletin 13, Food and Agriculture Organization of the United Nations, Rome, 2004.

[3] van Kauwenbergh SJ. World Phosphate Rock Reserves and Resources, published by IFDC, Muscle Shoals, Al. USA, 2010. Download from https://pdf.usaid.gov/pdf_docs/PNADW835.pdf

[4] Hendricks SB, Jefferson ME, Mosley VM. Z Kristallogr, 1932, 81, 352–69.

[5] Náray-Szabó St. Z Kristallogr, 1930, 75, 387–98. https://doi.org/10.1515/zkri-1930-0129

[6] Suetsugu Y, Takahashi Y, Okamura FP, Tanaka J. J Solid State Chem, 2000, 155, 292–97.

[7] Altschuler ZS, Clarke Jr RS, Young EJ Geochemistry of Uranium in Apatite and Phosphorite, Geological Survey Professional Paper 314-D, U. S. Department of the Interior, Washington 1958 (https://pubs.usgs.gov/pp/0314d/report.pdf).

[8] a) Steiner G, Geissler B, Haneklaus N. Environ Sci Technol, 2020, 54, 1287–89. b) Kelley R, Fedchenko V. Phosphate N. Fertilizers as a Proliferation-relevant Source of Uranium, Non-Proliferation Papers 2017, 59, 1–13; c) Intl. Atomic Energy Acency, The Recovery of Uranium from Phosphoric Acid, IAEA-TECDOC-533, Vienna 1989.

[9] a) Blount AM. Am Miner, 1974, 59, 41–47. b) Majzlan J, Speziale S, Duffy TS, Burns PC. Phys Chem Min 2006, 33, 567–73. doi:10.1007/s00269-006-0104-z

[10] a) Araki T, Zoltai T. Z Kristallogr, 1968, 127, 21–33. b) Capitelli F, Della Ventura G, Bellatreccia F, Sodo A, Saviano M, Ghiara M, Rossi M. Miner Mag, 2014,78, 057–1070. doi:10.1180/minmag.2014.078.4.16; c) Kampf AR, Adams PM, Barwood H, Nash BP. Am Miner, 2017, 102, 909–15. doi:10.2138/am-2017-5948.

[11] Jasinski SM (US Geological Survey), Annual Publication 2021 on Phosphate Rock Statistics and Information, U.S. Department of the Interior 2021. Download from: https://www.usgs.gov/centers/nmic/phosphate-rock-statistics-and-information.

[12] Edixhoven JD, Gupta J, Savenije HHG. Recent revisions of phosphate rock reserves and resources: A critique. Earth Syst Dyn, 2014, 5, 491–507. https://doi.org/10.5194/esd-5-491-2014

[13] Schröder JJ, Cordell D, Smit AL, Rosemarin A. Sustainable Use of Phosphorus, EU Tender EN V. B.1/ETU/2009/0025, Plant Research International, Business Unit Agrosystems, Wageningen, The Netherlands, 2010.
[14] Jasinski SM. U.S. Geological Survey, Mineral Commodity Summaries – Phosphate Rock, 2021. Download from: https://pubs.usgs.gov/periodicals/mcs2021/mcs2021-phosphate.pdf.
[15] Rosemarin A. "Global Status of Phosphorus", seminar presented at "Phosphorus a Limited Resource – Closing the Loop" Malmö, Sweden 2016. Download from: https://d1pdf7a38rpjk8.cloudfront.net
[16] Staffel T, Klein T. Phosphoric acid and phosphates. In: Ullmann's Encyclopedia of Industrial Chemistry, 6th edition, Wiley-VCH, 1999.
[17] Reinauer F, Glaum R. Acta Crystallogr, 1998, B54, 722–31. 10.1107/S0108768198003590
[18] Liebau F. Nomenclature and structural formulae of silicate anions and silicates. In: Structural Chemistry of Silicates, Springer, Berlin, Heidelberg, 1985. https://doi.org/10.1007/978-3-642-50076-3_5
[19] a) Matsuura N, Lin T, Kobayashi Y. Bull Chem Soc Jpn, 1970, 43, 2850–58. b) Thermo Scientific: Determination of polyphosphates using ion chromatography with suppressed conductivity detection, DIONEX Application Note 71 and Application Update 172, 2016; c) Greenfield S, Clift M. Analytical Chemistry of the condensed phosphates, 1st edition, Pergamon Press, Oxford, 1975.
[20] a) Avila Salazar DA, Bellstedt P, Miura A, Oi Y, Kasuga T, Brauer DS. Dalton Trans, 2021, 50, 3966–78. b) Prabakar S, Wenslow RM, Mueller KT. J Non-Cryst Solids, 2000, 263&264, 82–93; c) Mangstl M, Celinski VR, Johansson S, Weber J, An F, Schmedt auf der Günne J. Dalton Trans, 2014, 43, 10033–39.
[21] Henning PA, Landa-Canovas AR, Larsson AK, Lidin S. Acta Crystallogr, 1999, B55, 170–76. 10.1107/S0108768198012798
[22] Lissel E, Jansen M, Jansen E, Will G. Z Kristallogr, 1990, 192, 233–43. 10.1524/zkri.1990.192.3-4.233
[23] Leung KY, Calvo C. Can J Chem, 1972, 50, 2519–26. 10.1139/v72-406
[24] Tillmanns E, Gebert W, Baur WH. J Solid State Chem, 1973, 7, 69–84. 10.1016/0022-4596(73
[25] Corbridge DEC. Acta Crystallogr, 1960, 13, 263–69. 10.1107/S0365110X60000583
[26] Jost KH. Acta Crystallogr, 1963, 16, 428. 10.1107/S0365110X63001109
[27] Ondik HM. Acta Crystallogr, 1965, 18, 226–32. 10.1107/S0365110X65000518
[28] Glaum R, Weil M, Özalp D. Z Anorg Allg Chem, 1996, 622, 1839–46. 10.1002/zaac.19966221107
[29] Albrand KR, Attig R, Fenner J, Jeser JP, Mootz D. Mater Res Bull, 1974, 9, 129–40. 10.1016/0025-5408(74
[30] Cruickshank DWJ. Acta Crystallogr, 1964, 17, 677–79. 10.1107/S0365110X64001670
[31] Cruickshank DWJ. Acta Crystallogr, 1964, 17, 679–80. 10.1107/S0365110X64001682
[32] Arbib EH, Elouadi B, Chaminade JP, Darriet J. J Solid State Chem, 1996, 127, 350–53. 10.1006/jssc.1996.0393
[33] Müller U. Inorganic Structural Chemistry, 2nd edition, Wiley & Sons Ltd., New York, 2006.
[34] Durif A. Crystal Chemistry of Condensed Phosphates, Springer US Publishing, 1995.
[35] Glaum R. Neue Untersuchungen an wasserfreien Phosphaten (in German), Thesis of habilitation, University of Gießen, 1999. URL: http://geb.uni-giessen.de/geb/volltexte/1999/124/
[36] a) Lux H. Z Elektrochem, 1939, 45, 303–09. b) Lux H, Rogler E. Z Anorg Allg Chem, 1942, 250, 159–72; c) Flood H, Förland T. Acta Chem Scand, 1947, 1, 592–604; d) Duffy JA, Ingram MD, Sommerville ID. J Chem Soc, Faraday Trans, 1 1978, 74, 1410–19; e) Duffy JA. Bonding, Energy Levels & Bands in Inorganic Solids, Longman Group UK Ltd., 1990.

[37] Zachariasen WH. J Am Chem Soc, 1932, 54, 3841–51.
[38] Staffel T, Klein T. Phosphoric acid and phosphates. In: Ullmann's Encyclopedia of Industrial Chemistry, 6th edition, Wiley-VCH, 1999.
[39] Sankara Narayanan TSN. Rev Adv Mater Sci, 2005, 9, 130–77.
[40] a) Bell RN. Ind Eng Chem, 1949, 41, 2901–05. b) Tillmanns E, Baur WH. Acta Crystallogr 1971, B27, 2124–32.
[41] Hellwig E, Klimek J, Lussi A. Oralprophylaxe & Kinderzahnheilkunde 2012, 34, 2ff. Download on June 22nd 2021 from: https://www.zmk-aktuell.de/fachgebiete/prophylaxe/story/fluoride-wirkungsmechanismen-und-empfehlungen-fuer-deren-gebrauch__977.html
[42] Staffel T, Brix G. Con Chem J, 1996, 3, 93–95.
[43] Durif A. Crystal Chemistry of Condensed Phosphates, Plenum Press, New York, U.S.A, 1995.
[44] a) Turkdogan ET, Maddocks WR. J Iron Steel Inst London, 1952, 172, 1–15. b) Morey GW, Ingerson E. Am J Sci 1944, 242, 4; c) Osterheld RK, Bahr EW. J Inorg Nucl Chem, 1970, 32, 2539–41; d) Phase Diagrams for Ceramists, Levin EM, McMurdie HF, Hall FP. (Eds.). The American Ceramic Society, 1956, Fig. 567 and 568; e) Huang CT, Lin RY. Metal Trans 1989, 20B, 197–204.
[45] Harris DC, Lucy CA. Quantitative Chemical Analysis, 10th edition, Macmillan Learning, New York, U.S.A, 2020.
[46] Watanabe M, Matsura M, Yamada T. Bull Chem Soc Jpn, 1981, 54, 738–41.
[47] Thilo E. Angew Chem, 1965, 77, 1056–66.
[48] van Wazer JR, Callis CF. Chem Rev, 1958, 58, 1011–46.
[49] Sikorski CF, Weintritt DJ. Oil & Gas J, 1983, 81, 71–78.
[50] Brose E, Becker G, Bouchain W. Chemische Backtriebmittel (in german), Hrsg. Chemische Fabrik Budenheim, Budenheim, 1996.
[51] Wendrow B, Kobe KA. Chem Rev, 1954, 54, 891–924.
[52] LeVan SL. Chemistry of fire retardancy. In: Rowell RM. (Ed.) The Chemistry of Solid Wood. American Chemical Society Advances in Chemistry Series 207, Chapter 14, Washington, DC, 1984. 10.1021/ba-1984-0207.ch014
[53] Li K, Lei H, Zeng X, Li H, Lai X, Chai S. RSC Adv, 2017, 7, 49290–98.
[54] Ng HN, Calvo C. Can J Phys, 1976, 54, 638–47. https://doi.org/10.1139/p76-070
[55] Nieuwenkamp W. Z Kristallogr, 1935, 92, 82–88.
[56] Schulze GER. Z Phys Chem, 1934, B24, 215–40.
[57] Hartmann M, Elangovan SP. Adv Nanoporous Mater, 2010, 1, 237–312.
[58] Brodalla D, Kniep R, Mootz D. Z Naturforsch, 1981, 36b, 907–09.
[59] Staffel T, Wahl F, Weber S, Glaum R. Farben & Lacke, 2002, 103–09.
[60] Panagiotidis K. Neue Phosphate der Edelmetalle, Dissertation, Bonn University, 2009. https://nbn-resolving.org/urn:nbn:de:hbz:5N–18379
[61] Wolfshohl A Untersuchungen zur Synthese nanoskaliger Lithium-chrom(III)-phosphate und zur Deinterkalation von Silber-chrom(III)-phosphaten mit einer Ergänzung zur Synthese und Charakterisierung von Osmium(IV)-phosphaten, Dissertation, Bonn University 2019. https://nbn-resolving.org/urn:nbn:de:hbz:5n–54020
[62] Schöneborn M, Glaum R, Reinauer F. J Solid State Chem, 2008, 181, 1367–76.
[63] Benser E, Glaum R, Droß T, Hibst H. Chem Mater, 2007, 19, 4341–48.
[64] a) Modaressi A, Kaell JC, Malaman B, Gerardin R, Gleitzer C. Mater Res Bull, 1983, 18, 101–09. b) Zhang L, Schlesinger ME, Brow RK. J Am Ceram Soc 2011, 94, 1605–10.
[65] Venturini G, Courtois A, Steinmetz J, Gerardin R, Gleitzer C. J Solid State Chem, 1984, 53, 1–12.
[66] Weil M, Glaum R. Eur J Solid State Inorg Chem, 1998, 35, 495–508.

[67] Gautier Ro, Gautier Ré, Hernandez O, Audebrand N, Bataille T, Roiland C, Elkaïm E, Le Pollès L, Furet E, Le Fuer E. Dalton Trans, 2013, 42, 8124–31.
[68] Song Y, Zavalij PY, Suzuki M, Whittingham MS. Inorg Chem, 2002, 41(22), 5778–86.
[69] a) Zah-Letho JJ, Verbaere A, Jouanneaux A, Taulelle F, Piffard Y, Tournoux M. J Solid State Chem, 1995, 116, 335–42. b) Fukuoka H, Imoto H, Saito T. J Solid State Chem, 1995, 119, 98–106.
[70] Zah-Letho JJ, Jouanneaux A, Fitch AN, Verbaere A, Tournoux M. Eur J Solid State Inorg Chem, 1992, 29, 1309–20.
[71] Liang S-T, Zhang H-L, Luo M-T, Luo K-J, Li P, Xu H-B, Zhang Y. Ceram Intl, 2014, 40, 4367–73.
[72] Eastaugh N, Walsh V, Chaplin T, Siddall R. Pigment Compendium: A Dictionary and Optical Microscopy of Historical Pigments, Butterworth-Heinemann/Elsevier, Amsterdam, 2008.
[73] Glaum R, Hitchman MA. Aust J Chem, 1996, 49, 1221–28.
[74] Anderson JB, Kostiner E, Miller MC, Rea JR. J Solid State Chem, 1975, 14, 372–77.
[75] Begum Y, Wright AJ. J Mater Chem, 2012, 22, 21110–16.
[76] Thauern H, Glaum R. Inorg Chem, 2007, 46, 2057–66.
[77] a) Kazin PE, Karpov AS, Jansen M, Nuss J, Tretyakov YuD. Z Anorg Allg Chem, 2003, 629, 344–52. b) Kazin PE, Zykin MA, Romashov AA, Tretyakov YuD, Jansen M. Russ J Inorg Chem, 2010, 55, 145–49; c) Kazin PE, Zykin MA, Tretyakov YuD, Jansen M. Russ J Inorg Chem, 2008, 53, 362–6.
[78] Pogosova MA, Kalacheva IL, Eliseeva AA, Magdysyuk OV, Dinnebier RE, Jansen M, Kazin PE. Dyes Pigm, 2016, 133, 109–13. https://doi.org/10.1016/j.dyepig.2016.05.044
[79] a Levi GR, Peyronel G. Z Kristallogr, 1935, 92, 190–209, b) Norberg ST, Svensson G, Albertsson J. Acta Crystallogr, 2001, C57, 225–27.. https://doi.org/10.1107/S0108270100018709
[80] Khosrovani N, Sleight AW, Vogt T. J Solid State Chem, 1997, 132, 355–60. https://doi.org/10.1006/jssc.1997.7474
[81] White KM, Lee PL, Chupas PJ, Chapman KW, Payzant EA, Jupe AC, Bassett WA, Zha C-S, Wilkinson AP. Chem Mater, 2008, 20, 3728–34. DOI: 10.1021/cm702338h
[82] Lind C. Materials, 2012, 5, 1125–54. DOI: 10.3390/ma5061125
[83] a) Zumsteg FC, Bierlein JD, Gier TE. J Appl Phys, 1976, 47, 4980–85. b) Thomas PA, Glazer AM, Watts BE. Acta Crystallogr, 1990, B46, 333–43.
[84] Ricci L, Weidemüller M, Esslinger T, Zimmermann C, Hansch T. Optics Commun, 1995, 117, 541–49. https://doi.org/10.1016/0030-4018(95)00146-Y
[85] a) Gorbunova YuE, Linde SA. Dokl Akad Nauk SSSR, 1979, 2453, 584–88. b) Geupel S, Pilz K, van Smaalen S, Büllesfeld F, Prokofiev A, Assmus W. Acta Crystallogr, 2002, C58, i9–13.
[86] Schunk SA. Oxy-functionalization of alkanes, chapter 14.11.3. In: Handbook of Heterogeneous Catalysis, 2nd edition, Wiley-VCH, 2008. 10.1002/9783527610044
[87] Bergman RI, Frisch NW. Production of maleic anhydride by oxidation of n-butane. US Patent, 1964, 3(293), 268.
[88] a) Centi G. Catal Today, 1993, 16, 5–26. b) Hutchings GJ. J Mater Chem, 2004, 14, 3385–95; c) Ballarini N, Cavani F, C. Cortelli, Ligi S, Pierelli F, Trifirò F, Fumagalli C, Mazzoni G, Monti T. Topics Catal, 2006, 38, 147–56.
[89] Schlögl R. Concepts in selective oxidation of small alkane molecules. In: Mizuno N. (Ed.) Modern Heterogeneous Oxidation Catalysis – Design, Reactions, and Characterization, Wiley-VCH, Weinheim, Germany, 2009.
[90] Grasselli RK. Top Catal, 2001, 15, 93–101.
[91] Cheng M-J, Goddard III WA, Fu R. Top Catal, 2014, 57, 1171–87. DOI: 10.1007/s11244-014-0284-6
[92] Padhi AK, Nanjundaswamy KS, Goodenough JB. J Electrochem Soc, 1997, 144, 1188–94.

[93] Whittingham MS. Chem Rev, 2014, 114, 11414–43.
[94] Chang G, Zhao Y, Dong L, Wilkinson DP, Zhang L, Shao Q, Yan W, Sun X (Andy), Zhang J. J Mater Chem, 2020, A8, 4996–5048.
[95] a) Panero S, Reale P, Ronci F, Scrosati B, Perfetti P, Albertini VR. Phys Chem Chem Phys, 2001, 3, 845–47. b) Ohzuku T, Ueda A, Yamamoto N. J Electrochem Soc, 1995, 142, 1431–35.
[96] Patoux S, Dolle M, Rousse G, Masquelier C. J Electrochem Soc, 2002, 149, A391–400.
[97] a) Cui Y. Dissertation KIT 2018. urn:nbn:de:swb:90-548531; b) Campanella D, Belanger D, Paolella A. J Power Sources, 2021,482,228949, https://doi.org/10.1016/j.jpowsour.2020.228949; c) Aono H, Sugimoto E, Sadaoka Y, Imanaka N, Adachi G. J Electrochem Soc, 1990, 137, 1023–27; d) Bachman JC, Muy S, Grimaud A, Chang H-H, Pour N, Lux SF, Paschos O, Maglia F, Lupart S, Lamp P, Giordano L, Shao-Horn Y. Chem Rev, 2016, 116, 140–62. DOI: 10.1021/acs.chemrev.5b00563.
[98] a) Fichter F, Rius y Mirò A. Helv Chim Acta, 1919, 2, 3–26. https://doi.org/10.1002/hlca.19190020102; b) Weiss E, Sáez C, Groenen-Serrano K, Caizares P, Savall A, Rodrigo MA. J Appl Electrochem, 2008, 38, 93–100. DOI 10.1007/s10800-007-9405–2
[99] a) Sánchez A, Llanos J, Sáez C, Cañizares P, Rodrigo MA. Chem Eng J, 2013, 233, 8–13. b) Hasbrouck LJ, Carlin CM, Risley JM. Inorg Chim Acta, 1997,258,123–5
[100] Oeckler O, Montbrun L. Z Anorg Allg Chem, 2008, 634, 279–87. 10.1002/zaac.200700390
[101] Ataalla M, Afify AS, Hassan M, Abdallah M, Milanova M, Aboul-Enein HY, Mohamed A. J Non-Cryst Solids, 2018, 491, 43–54.
[102] Franks E. Electrochemical Studies in Molten Phosphates, Ph. D. thesis, The University of London 1969. https://spiral.imperial.ac.uk/handle/10044/1/15807

6.3 Borate applications

Hubert Huppertz, Raimund Ziegler

The following chapter gives a brief overview about the versatility and ubiquity of borates. The chemical background as well as advantages for specific products will be addressed. Borates are essential for biochemistry and are applicated in the broad field of applied inorganic materials from topics like insulation toward batteries.

6.3.1 Adhesives

Due to its empty p-orbital, the trivalent boron atom of boric acid is a good electrophile. Therefore, it reacts with many nucleophiles to form borate esters [1]. This property is useful for various adhesives, like starch-based or dextrin-based ones. Starch molecules exhibit many hydroxy groups, that can be linked by borate compounds as boric acid (H_3BO_3), borax decahydrate ($Na_2[B_4O_5(OH)_4] \cdot 8\ H_2O$) or sodium metaborate ($NaBO_2 \cdot x\ H_2O$) (Figure 6.3.1). The addition of borates improves the tack, increases the viscosity and creates a higher branched chain polymer. Furthermore, it has positive side effects to fire retardancy and wood protection. Borate-modified starch adhesives are often used by the paper industry to produce grocery and multiwall paper bags, paper boxes and many more [2, 3].

6.3.2 Agriculture

Boron plays an essential role in today's agriculture. Since it is a basic nutrient for plants, it is needed for plant growth and crop quality. The optimum concentration of boron in soil solution for growth depends heavily on the species of the plant, as well as local conditions and the availability of other nutrients. For example, the boron concentration in soil solution for best growth should be less than 0.5 ppm for raspberry, while for cotton it should be 10 ppm. The exact functions of boron in plants are still unknown. However, it affects the translocation and the control of the amount of many organic compounds. Furthermore, it enhances the effect of sugars on the hormone action, the amount of photosynthesis, the rate of CO_2 absorption and the growth of roots of plants. Boron fertilizers are preferably applicated before planting, during dormant periods or after cutting, but they can be used as foliar spray too. An example for a boron-containing fertilizer is borax, that can be applied directly, while boron fertilizers in general can also be mixed with other nutrients [4].

https://doi.org/10.1515/9783110798890-011

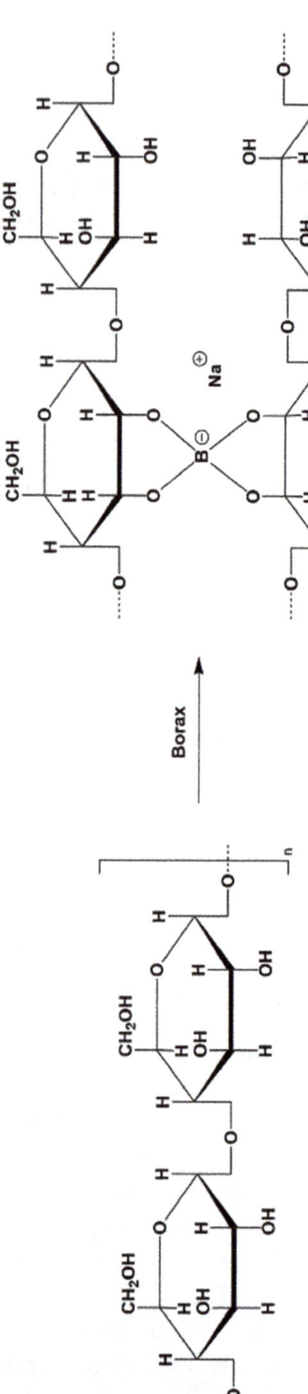

Figure 6.3.1: Linkage of starch molecules with borax.

6.3.3 Energy storage

Borates are very important for different applications in the battery and capacitor industry. One of the most important factors in the production of capacitors is the purity of all used materials, which is crucial for a long durability of the manufactured devices. Therefore, the implemented electrode materials are often cleaned with a boric acid solution to guarantee highest quality standards. Besides that, an aqueous solution of boric acid and ammonium borate is usually taken as an electrolyte in wet-type capacitors. In dry-type capacitors, the electrolyte mixture commonly used includes amine borates, ammonium acetate-borates and glycol-ammonium borates [5]. In the battery industry, lithium batteries play a crucial role. For their production, boron oxide (B_2O_3) is used for many reasons, mostly as reactant or precursor to prepare other battery compounds. The electrolyte of lithium ion batteries often comprises $LiBF_4$ and a solvent mixture of different carbonates. Interestingly, the loss of delivered capacity upon cycling of nonaqueous rechargeable lithium ion batteries can be reduced if the surface of the cathode was treated in the run-up with a small fraction of lithium borates, like $LiBO_2$ [6].

6.3.4 Glass, fibers and insulation

Borate glasses, fibers and other vitreous products make up more than one half of the world-wide boron consumption [7]. Pure borate glass has a low chemical durability and a high water affinity, which limits its use. Therefore, this system is often combined with other oxides like Al_2O_3 or SiO_2 to gain advantages for numerous applications [8]. For example, the combination of boron oxide with silicate glasses leads to the large area of borosilicate glasses. Historically, boron oxide was added to the melts of silicate glasses to improve their homogeneity and therewith to enhance their optical performance [9]. This was followed by the finding that several of these borosilicate glasses exhibited low coefficients of thermal expansion, making applications in the area of thermal shock-resistant glasses possible [10].

From a technical point of view, the production of a borosilicate glass generally starts with silica (SiO_2), which is ground to a very fine powder to speed up the melting process. The additional materials like boric acid, borax, colemanite ($CaB_3O_4(OH)_3 \cdot H_2O$) or limestone are also ground, blended to the desired composition and finally added to the silica melt at temperatures of about 1550–1600 °C. After achieving a homogenous melt, the glass can flow through a submerged throat into the different channels feeding the various production lines of borosilicate products [4].

From a structural point of view, borosilicate glasses possess so-called network formers represented by the silicon and boron atoms. On the atomic scale, these atoms form SiO_4 tetrahedra, trigonal BO_3 units and BO_4 tetrahedra, which are partially connected via common corners (oxygen atoms) to a network. The addition of a modifier in the form of alkali cations (M^+) to the system (e.g. M_2O), leads to an association of the cations either to the SiO_4 network or to the BO_3/BO_4 groups. In the first case, nonbridging oxygen (NBO) atoms are formed, which can be written as SiO^-M^+. In combination with boron, a BO_3 group is converted to a BO_4 group without the formation of an NBO. These insights are originally based on a fundamental NMR (nuclear magnetic resonance) investigation of Dell et al., who predicted the fraction of four-fold coordinated boron atoms (BO_4 tetrahedra) as a function of the concentration of the alkali cations and the ratio $SiO_2 : B_2O_3$ [11]. The possibility to control the degree of polymerization via the amount of tetrahedral borate groups and NBO´s (large impact on the local topology of the network) made it possible to influence several physical properties of borosilicate glasses, namely, the thermal expansion.

Next to others, Corning´s Pyrex and Schott´s Duran are two of the best-known examples for glasses of the family of borosilicates, which are commercially successful. E-Glass, an alumino borosilicate glass, is used for electrical applications (hence the name) and represents the conventional glass type for glass fibers [12]. Table 6.3.1 gives a view on the approximate composition of commercially available borosilicate glasses.

Table 6.3.1: Approximate composition (wt%) of commercially available borosilicate glasses [13].

	Duran, Pyrex	E-Glass
SiO_2	80.6	54.5
B_2O_3	12.6	6.6
Na_2O	4.2	1.0
MgO	0.05	0.6
CaO	–	22.1
Al_2O_3	2.2	14.0
Misc.	0.35	1.2

Glasses with a B_2O_3 content of 1–34% are commonly applied in laboratories and kitchen ware, electron tubes, optical fibers and filters, electronical equipment with seals to metals, vacuum flasks, pharmaceutical applications etc. [4]. Additionally, there are several applications with glasses that have a higher boron oxide amount like sodium vapor lamps (up to 36% B_2O_3) or X-ray absorption glass (up to 83% B_2O_3) [4]. A niche but still important utilization of borosilicate glasses is their use in the vitrification of nuclear wastes to immobilize spent fuel. This is done by several countries with an estimate of more than 90% of the highly radioactive waste [13]. Generally, the advantages of borosilicate glasses are a reduced thermal expansion, good resistance to vibration, high temperatures and thermal shocks, as well as improved toughness, strength and chemical resistance [4]. Another important role of borosilicates is in fiberglass, that is used for insulation, fabrics and reinforcement. Insulations are used in buildings, industrial equipment and automobiles. Fabrics of borosilicate fibers are nonflammable and resistant to sunlight, which make them valuable in higher-temperature industrial applications, as well as curtains and draperies [4]. Fiber-reinforced plastics are widely applied from roofing, shingles, electric appliances or components in cars and aircrafts to sporting equipment. The advantages of fiberglass are light weight, high tensile strength, high modulus of elasticity and high chemical stability [4]. Last but not least, borosilicates are also found as liquid crystal display (LCD) substrate glasses, as touch screens in electronic applications and as glazes and enamels to coat metals or ceramics [7].

6.3.5 Cleaning compounds

Perborates are easily synthesized from borate salts and hydrogen peroxide. The most important perborate is sodium perborate, often abbreviated as SPB. Industrially it is produced by adding hydrogen peroxide to sodium metaborate, which is a product of the reaction of borax with sodium hydroxide. Commercially available are the SPB monohydrate and tetrahydrate, which have the sum formulas $Na_2[B_2(O_2)_2(OH)_4]$ and $Na_2[B_2(O_2)_2(OH)_4] \cdot 6\ H_2O$. The crystal structure of SPB shows a chair conformation of a 1,4-diboratetroxane dianion (Figure 6.3.2).

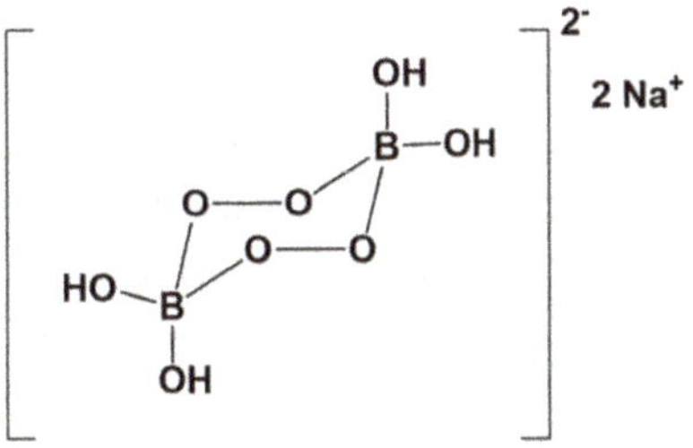

Figure 6.3.2: Chair conformation of the 1,4-diboratetroxane dianion.

In solution, the diboratetroxane dianion hydrolyses to the $[(HO)_3BOOH]^-$ anion, that further on decomposes to $[B(OH)_4]^-$ and H_2O_2, which predominates in solution. SPB acts as an H_2O_2 source, which is the reason for its use as oxidizing agent. In organic chemistry, perborates can be used to oxidize thiols, selenols, amines, oximes and many more organic compounds [14]. In the past, borates in general have been used in many products, like cosmetic creams, hair shampoos, eye drops, bath salts, as well as in cleaning products and laundry detergents for household and industry. The functions of these borates comprise stabilizing enzymes, enhancing stain removal, providing alkaline buffering, softening water, boosting surfactant performance, controlling viscosity and emulsifying waxes or oils [15]. A great benefit of perborates over a liquid hydrogen peroxide solution is their condition of aggregation, since they are solid. Hence, perborates are safe and convenient to handle besides their good stability [14]. However, reproductive toxicity of SPB were proven in recent years. Therefore, it has been banned in the EU for several products. Nowadays sodium percarbonate is often used as an alternative.

6.3.6 Protection of fire, fungi and pests

As already mentioned, boron is an essential micronutrient for plants. However, high concentrations can inhibit the growth of harmful insects, bacteria and fungi. For this reason, borates are added to cellulose insulation, oriented strand board, particle board and other wood products. Another reason for their addition is their beneficial properties as flame retardants. Borates inhibit smoldering combustion, are char formers and drip suppressants. Zinc borates (like $Zn[B_3O_4(OH)_3]$ or $4\ ZnO \cdot B_2O_3 \cdot H_2O$) are often applied in those cases, because they have an unusually high dehydration onset temperature. Furthermore, they are only slowly affected by temperature and do not change the color of plastics, where they are also used [4, 7].

6.3.7 Metallurgy

In metallurgy, elemental boron plays an important role in different alloys and strong permanent magnets, while borates are mainly used during the refining process of different metals (e.g. nickel, aluminium or gold) and alloys. During stainless-steel production, a slag is gained as a side product, that forms a dusty powder of Ca_2SiO_4. By adding boron sesquioxide, this dustiness is prevented, resulting in a stable rock-like material. While refining gold, borates are relevant to facilitate the attack of the ore at a lower temperature, reducing the viscosity of the slag, and therefore easing its handling. Generally, borates are used as cover flux to prevent metal from air oxidation. Furthermore, they reduce wear and tear of melting equipment and minimize the loss of metal through association with metal oxide impurities at adequate low temperatures. Last but not least, borates as flux covers exhibit no combustibility and a minimum fuming tendency [16–19].

6.3.8 Nuclear applications

Boron has two natural isotopes namely ^{10}B and ^{11}B with an isotopic signature of about 20% and 80%, respectively. ^{10}B has a large neutron capture cross-section, which is nearly one million times higher than that of ^{11}B [20]. Therefore, boric acid solutions are used at nuclear power plants for different applications. One of them is in PWRs (pressurized water reactors) as neutron absorber to control the reactor reactivity. Another one is in used fuel pools, where used fuel rods are stored to decay, before they get processed further (e.g. vitrification). Besides that, reserved water for emergency situations also contains boric acid [21]. This storage was and is important for emergencies as it could be seen at the Fukushima Daiichi nuclear disaster, where boric acid solutions were used to retrieve stable reactor conditions.

6.3.9 Medicine

Although boron is an essential micronutrient for plants, in 2004 the European Food Safety Authority (EFSA) classified boron neither as an essential nutrient for higher animals or humans, nor did EFSA state a recommended intake [22]. However, there are publications that disagree with EFSA`s opinion, pointing out the need of further research and discussion of this topic [23, 24]. From a chemical point of view, metabolism of borates is not feasible due to strong boron–oxygen bonds, since they require much energy to be broken. However, low concentrations of inorganic borates react at physiological pH in the aqueous layer of mucosal surfaces to boric acid, which is subsequently absorbed. Animal and human studies show that over 90% of the administered inorganic borate is excreted as boric acid [25]. Because of that, animal studies with boric acid are also relevant for inorganic borates. In 2009, the European Union classified boron compounds as toxic to reproduction [26]. Hence, nowadays boric acid and borates are rarely used as excipients and as active substances in applications. As excipients they are added as antimicrobial preservatives, pH buffers or tonicity-adjusting agents in ophthalmic preparations, ears drops and homeopathic dilutions [27]. As active agent, crisaborole (a benzoxaborole with a boronic acid hemiester) is used in an ointment against eczema (atopic dermatitis), where it blocks the release of certain cytokines. One of them is the tumor necrosis factor alpha, which is involved in the inflammation process. Blocking it eases the inflammation, subsequently relieving the symptoms [28]. Another active compound is bortezomib, that is given during cancer therapy with other medicines. It is used to treat multiple myeloma and cell lymphoma, two types of blood cancer. Bortezomib is a proteasome inhibitor. Proteasome is a system in cells, that breaks down no-longer-needed proteins. Therefore, cells die caused by a build-up of unwanted proteins. This inhibition has a greater impact on cancer cells than on normal ones [29].

6.3.10 Oilfield chemical applications

Borates play an important role in oilfields around the globe. They are widely used as set retardants in oil-well cement. Drilled well bores are stabilized with a steel casing that is reinforced by cement. This cement must have the right properties to be pumped into place without setting previously. Borates adsorb onto the surface, where they react with calcium ions to calcium borates, which delay grain hydration and cement setting. However, they usually are combined for example with sugars, phosphonates or synthetic polymers [30].

Lost circulation is a problem encountered during drilling. To solve the problem, borate cross-linked polymer gels with similar properties to fracturing fluids are used to plug lost circulation zones. Another important use of borates, especially sodium perborates, is as H_2S scavenger. H_2S is a toxic gas that also causes pitting and corrosion of drilling equipment. It originates from pyrite rock or sulphide-producing bacteria that feed from organo-sulphur compounds of the crude oil. When H_2S is present, it can be oxidized by sodium perborates to elemental sulphur and sulphates. Furthermore, the bacteria are also killed by the borate [30].

6.3.11 Industrial fluids

Corrosion of metals is a ubiquitous phenomenon and is classified as an electrochemical reaction. Every electrochemical process can be split into an anodic and a cathodic reaction. To prevent corrosion, one of them needs to be stopped. For cathodic inhibition, a barrier film is usually deposited from solution, while anodic inhibition can be achieved by growing an oxide layer on the surface. The anodic passivation is a controlled corrosion, that, if it persists without a flaw, prevents further unwanted corrosion. Aqueous borate solutions, besides others (like chromate, nitrate, molybdate, silicate or phosphate solutions) are suitable to create those protecting oxide films [31].

Borates are also present in automotive fluids like engine coolants, where they are added to act as buffer to prevent corrosion. As a matter of fact, the most important automotive fluid, the brake fluid, contains borates. Brake fluids require a high boiling point and should tolerate moisture, while keeping the boiling point high. Generally, their main components are triethylene glycol monoalkyl ethers, which are moisture sensitive and therefore have a significantly reduced boiling point in the presence of water. Today, the borate esters of the glycol ether (triethylene glycol monoalkyl ether borate esters) are the main components of brake fluids. That is, because moisture converts the borate ester into boric acid and the glycol ether, which keeps the effect on the boiling point low [31].

To conclude, borates are added to many industrial fluids like lubricants, metalworking fluids, water treatment chemicals or fuel. Their benefits are buffering capacity, reduction of freezing point, increase of boiling point, lubrication, thermal oxidative stability, low sludge formation and low moisture sensitivity [31].

6.3.12 Optical applications

Based on the rich structural chemistry and the characteristic physical properties of borates, it is self-evident that these substances play a decisive role in the area of phosphor materials and modern optoelectronic applications [32]. Especially the wide optical transparency windows of these compounds including high polarizabilities, to realize the simultaneous existence of birefringence and suitable second-order NLO (nonlinear optics) coefficients, make this class of compounds highly interesting for applications in the field of laser materials. In the following, representative examples for applied borates in the fields of phosphors, nonlinear optical materials, laser materials and birefringent materials are given.

6.3.12.1 Phosphors

Borates were applied in various fluorescent lamps and plasma display panels as host for luminescent ions. In this context, an important compound is YBO_3, which can be doped with Eu^{3+}, Ce^{3+}, Tb^{3+} and Bi^{3+} to obtain efficient phosphors. In detail, YBO_3:Eu, was one of the best red phosphors applied in various fluorescent lamps with a high VUV (vacuum ultraviolet) transparency and an exceptional optical damage threshold. Unfortunately, the characteristic emission results from almost equal parts of the $^5D_0$-$^7F_1$ and $^5D_0$-$^7F_2$ transitions, leading to an orange-red emission, where a pure red one was desired [33]. Further phosphors with a relatively high color rendering index are Ce^{3+},Tb^{3+}:$GdMgB_5O_{10}$ and Ce^{3+}, Mn^{2+}:$GdMgB_5O_{10}$. In the first example, the energy transfer ($Gd^{3+}\rightarrow Tb^{3+}$) results in the emission of green light (545 nm) and in the Mn^{2+}-doped compound a broadband red emission centered at 620 nm is observed [34].

6.3.12.2 Nonlinear optical materials

A basic requirement for an implementation of a borate as a nonlinear optical material (see also Chapter 8.6) is the presence of a noncentrosymmetric structure in combination with a preferably high chemical, physical, mechanical and thermal stability. For a high efficiency in frequency doubling, the material has to have large NLO coefficients. They can be determined on sizeable single crystals or on powders via the Kurtz and Perry method [35]. For an application in an optoelectronic device, it is necessary to grow large single-crystals with a compound-specific crystal-growth technique. These large crystals are indexed, cut and polished to receive reasonable plates for the specific device. Under operating conditions, these crystals must stay free of cracks and are not allowed to melt or decompose. From the large number of borates investigated with respect to NLO properties, selected examples (commercially available) are given in the following.

One of the best characterized NLO materials is the alkali metal borate LiB_3O_5 (LBO). The noncentrosymmetric structure ($Pna2_1$) shows ten-membered channels built up from fundamental building blocks $[B_3O_7]^{5-}$ in which the Li^+ ions are located.

Crystals of LiB_3O_5 are grown by the flux method in the system Li_2O-B_2O_3-MoO_3. Long-lasting development led to crystals with dimensions of 160 × 150 × 77 mm^3 and a weight close to 2 kg [36]. The high quality of these as grown crystals enables application for second harmonic generation (SHG) of high-intensity laser radiation, deep-UV sum-frequency generation (SFG), optical parametric oscillation (OPO) applications and ultrafast laser systems. Another important material from the area of alkali metal borates is $CsLiB_6O_{10}$ (CLBO) possessing a three-dimensional network of connected three-membered $[B_3O_7]^{5-}$ rings (space group: $I\bar{4}2d$). The metal cations are located in $[CsO_8]$ and $[LiO_4]$ polyhedra. Due to the fact that $CsLiB_6O_{10}$ melts congruently, large single crystals can be received from the stoichiometric melt (Figure 6.3.3), e.g. by the Kyropoulos method [37]. With this material, it was possible to realize fourth and fifth harmonic generations of a Nd-YAG laser (1064 nm) via type I phase matching, so more possibilities are now given in the generation of UV radiation with short wavelengths. From the field of alkaline-earth metal borates, the commercially available compound β-BaB_2O_4 (BBO) is a well investigated NLO material. The low-temperature phase crystallizes in the noncentrosymmetric space group $R3c$ possessing isolated $[B_3O_6]^{3-}$ rings and Ba^{2+} in an eight-fold coordination [38]. For crystal growth, the top-seeded solution growth method with a NaF flux is most appropriate to receive large sized β-BaB_2O_4 crystals without inclusions [39]. One of the main applications of β-BaB_2O_4 is to double the frequency of the 532 nm laser light, which is produced, for example, by an intracavity frequency-doubled Nd:YVO_4 laser [40]. From the class of rare-earth-metal borates, crystals of the compounds $YCa_4O(BO_3)_3$ and $GdCa_4O(BO_3)_3$ can be grown by Czochralski and Bridgman procedures. By using $YCa_4O(BO_3)_3$ (YCOB), one of the highest conversion efficiencies of about 73% could be achieved from the fundamental source [41]. This compound maybe one of the future materials

Figure 6.3.3: Photograph of large CLBO crystal with dimensions of 14 × 11 × 11 cm^3 for three weeks growth. Reprinted from Mori Y, Kuroda I, Nakajima S, Sasaki T, Nakai S. New nonlinear optical crystal: Cesium lithium borate. Appl Phys Lett 1995,67,1818–20, with the permission of AIP Publishing.

in high-power laser systems. From the field of halide-containing metal borates, the compound $KBe_2BO_3F_2$ (KBBF) is important to be mentioned, which crystallizes in the space group $R32$ possessing optimally aligned BO_3 groups, which is beneficial for a high optical anisotropy and hyper-polarizability, leading to high NLO coefficients and a large birefringence. $KBe_2BO_3F_2$ is currently one of the best NLO materials in deep-UV applications [32].

6.3.12.3 Laser materials

Borates can act as multifunctional materials, e.g. in so-called self-frequency-doubling (SFD) laser crystals (see also Chapter 8.5). In these crystals, the NLO process of frequency doubling is effectively combined with the laser effect. Precondition therefore is that the SFD crystal is in the first instance a good NLO crystal and secondly a good laser crystal, which has to accept luminescent doping with Yb or Nd [32]. Furthermore, the material must have a broadband absorption for the emission of the diodes, a long luminescence lifetime, large emission cross sections, a high radiative quantum efficiency and a quite good thermal conductivity to extract the heat from the host material. In 1981, Dorozhkin et al. discovered with the compound $Nd{:}YAl_3(BO_3)_4$ the first borate, which could be used as a SFD laser crystal [42]. An optimized ratio Nd:Y led to the development of the first green laser (from 1060 to 530 nm) by Lu et al. [43]. With the compound $Nd{:}GdCa_4O(BO_3)_3$, the SFD output power was increased to 17.91 W at a wavelength of 545.5 nm [44]. This compound can be found as the central device in the green laser pointer on the market in China substituting the currently used combination of $Nd{:}YVO_4$ and $KTiOPO_4$ realizing the function of laser and SHG [32].

6.3.12.4 Birefringent materials

Birefringent materials are important components to modulate the polarization of light. So, they find application in optical materials like beam displacers, prism polarizers and phase compensators. From the commercially available materials, α-BaB_2O_4 is one of the few early materials that can be used for the polarization of light in the deep-UV region. The coplanar arrangement of the $[B_3O_6]$ groups in α-BaB_2O_4 leads to an enhanced optical anisotropy, which is the reason for the substantial birefringence. However, this borate has several drawbacks, so there exists an urgent need for new materials. As already mentioned in the NLO-paragraph, the birefringent crystals also need high stabilities (high laser-damage threshold) including mechanical stability and chemical inertness. From a crystallographic point of view, compounds crystallizing in the cubic crystal system are not suitable, because their isotropic nature leads to a birefringence of zero. Recent developments describe Glan-Taylor polarizers with crystals of $Ba_2Mg(B_3O_6)_2$ and $Ba_2Ca(B_3O_6)_2$ being competitive birefringent materials for practical applications [45, 46].

References

[1] Gadhave RV, Kasbe PS, Mahanwar PA, Gadekar PT. To study the effect of boric acid modification on starch–polyvinyl alcohol blend wood adhesive. J Indian Acad Wood Sci, 2018, 15, 190–98.

[2] Anon. Mummies to Modern Industry: Borates in Adhesives, U.S. Borax, 20 Mule Team, Chicago, IL, USA, 2021.

[3] Anon. Borates in Starch and Dextrin Adhesives, U.S. Borax, 20 Mule Team, Chicago, IL, USA, 2021.

[4] Garrett DE. Borates Handbook of Deposits, Processing, Properties, and Use, Academic Press, San Diego, CA, USA, 1998.

[5] Anon. Borates in Electrolytic Capacitors, U.S. Borax, 20 Mule Team, Chicago, IL, USA, 2021.

[6] Zhang M. Use of lithium borate in non-aqueous rechargeable lithium batteries, US2002/0119375 A1, 2002.

[7] Schubert DM. Borates in industrial use. In: Roesky HW, Atwood DA. (Ed.) Group 13 Chemistry III: Industrial Applications, Springer-Verlag Berlin Heidelberg, Berlin/Heidelberg, 2003, 1–40.

[8] Bengisu M. Borate glasses for scientific and industrial applications: A review. J Mater Sci, 2016, 51, 2199–242.

[9] Faraday M, Wood DV, Goetz EJ. Experimental Researches in Chemistry and Physics, R. Taylor and W. Francis, London, 1859.

[10] Hovestadt H, Everett JD, Everett A. Jenaer Glas und seine Verwendung in Wissenschaft und Technik, MacMillan and Co., London, 1902.

[11] Dell WJ, Bray PJ, Xiao SZ. ^{11}B NMR studies and structural modeling of $Na_2O{\cdot}B_2O_3{\cdot}SiO_2$ glasses of high soda content. J Non-Cryst Solids, 1983, 58, 1–16.

[12] Freudenberg C. Textile Faserstoffe. In: Cherif C. (Ed.) Textile Werkstoffe für den Leichtbau: Techniken – Verfahren – Materialien – Eigenschaften, Springer Berlin Heidelberg, Berlin, Heidelberg, 2011, 39–109.

[13] Youngman RE. Borosilicate glasses. In: Richet P, Conradt R, Takada A, Dyon J. (Eds.) Encyclopedia of Glass Science, Technology, History, and Culture, The American Ceramic Society, 2021, 867–78.

[14] McKillop A, Sanderson WR. Sodium perborate and sodium percarbonate: Cheap, safe and versatile oxidizing agents for organic synthesis. Tetrahedron, 1995, 51, 6145–66.

[15] Anon. Borates in Detergents, Cleaners and Personal Care Products, U.S. Borax, 20 Mule Team, Chicago, IL, USA, 2021.

[16] Anon. Borates in Gold Metallurgy, U.S. Borax, 20 Mule Team, Chicago, IL, USA, 2021.

[17] Anon. Borates in Lead Recycling, U.S. Borax, 20 Mule Team, Chicago, IL, USA, 2021.

[18] Anon. Borates in Metallurgical Applications, U.S. Borax, 20 Mule Team, Chicago, IL, USA, 2021.

[19] Anon. Borates in Steel Slag Stabilization, U.S. Borax, 20 Mule Team, Chicago, IL, USA, 2021.

[20] Wiberg N, Wiberg E, Holleman AF. Lehrbuch der Anorganischen Chemie, 102 ed., Walter de Gruyter & Co., Berlin, 2007.

[21] International Atomic Energy Agency. Processing of nuclear power plant waste streams containing boric acid. IAEA, 1996, 28, 1–67.

[22] European Food Safety Authority. Opinion of the scientific panel on dietetic products, nutrition and allergies [NDA] related to the tolerable upper intake level of boron (sodium borate and boric acid). EFSA J, 2004, 2, 80.

[23] Farfán-García ED, Castillo-García EL, Soriano-Ursúa MA. More than boric acid: Increasing relevance of boron in medicine. World J Transl Med, 2018, 7, 1–4.

[24] Pizzorno L. Nothing boring about boron. Integr Med (Encinitas), 2015, 14, 35–48.

[25] World Health Organization. International Programme on Chemical Safety. Boron. World Health Organization, Geneva, 1998.
[26] European Union. Legislation. OJEU, 2009, 52, L235.
[27] European Medicines Agency. Questions and answers on boric acid and borates used as excipients in medicinal products for human use. EMA CHMP, 2013, 619104, 1–8.
[28] European Medicines Agency. Staquis (crisaborole) An overview of Staquis and why it is authorized in the EU. EMA, 2020, 64649, 1–2.
[29] European Medicines Agency. Velcade (bortezomib) An overview of Velcade and why it is authorized in the EU. EMA, 2020, 178154, 1–3.
[30] Anon. Borates in Oilfield Chemical Applications, U.S. Borax, 20 Mule Team, Chicago, IL, USA, 2021.
[31] Anon. Borates in Industrial Fluids, U.S. Borax, 20 Mule Team, Chicago, IL, USA, 2021.
[32] Mutailipu M, Poeppelmeier KR, Pan S. Borates: A rich source for optical materials. Chem Rev, 2021, 121, 1130–202.
[33] Wei Z, Sun L, Liao C, Yin J, Jiang X, Yan C, et al. Size-dependent chromaticity in YBO_3: Eu Nanocrystals: Correlation with microstructure and site symmetry. J Phys Chem B, 2002, 106, 10610–17.
[34] Welker T. Recent developments on phosphors for fluorescent lamps and cathode-ray tubes. J Lumin, 1991, 48–49, 49–56.
[35] Kurtz SK, Perry TT, Powder A. Technique for the evaluation of nonlinear optical materials. J Appl Phys, 1968, 39, 3798–813.
[36] Hu Z, Zhao Y, Yue Y, Yu X. Large LBO crystal growth at 2kg-level. J Cryst Growth, 2011, 335, 133–37.
[37] Mori Y, Kuroda I, Nakajima S, Sasaki T, Nakai S. New nonlinear optical crystal: Cesium lithium borate. Appl Phys Lett, 1995, 67, 1818–20.
[38] Liebertz J, Stähr S. Zur Tieftemperaturphase von BaB_2O_4. Z Kristallogr, 1983, 165, 91–94.
[39] Chen W, Jiang A, Wang G. Growth of high-quality and large-sized β-BaB_2O_4 crystal. J Cryst Growth, 2003, 256, 383–86.
[40] Kondo K, Oka M, Wada H, Fukui T, Umezu N, Tatsuki K, et al.. Demonstration of long-term reliability of a 266-nm, continuous-wave, frequency-quadrupled solid-state laser using β-BaB_2O_4. Opt Lett, 1998, 23, 195–97.
[41] Tu X, Wang S, Xiong K, Zheng Y, Shi E. Growth and properties of large aperture YCOB crystal for NLO application. J Cryst Growth, 2020, 535, 125527.
[42] Dorozhkin LM, Kuratev II, Leonyuk NI, Timchenko TI, Shestakov AV. Optical second-harmonic generation in a new nonlinear active medium: Neodymium-yttrium-aluminum borate crystals. Sov Tech Phys Lett, 1981, 7, 555.
[43] Lu B, Wang J, Pan H, Jiang M, Liu E, Hou X. Exited emission and self-frequency-doubling effect of $Nd_xY_{1-x}Al_3(BO_3)_4$ crystal. Chin Phys Lett, 1986, 3, 413–16.
[44] Du J, Wang J, Yu H, Zhang H. 17.9 W continuous-wave self-frequency-doubled Nd: GdCOB laser. Opt Lett, 2020, 45, 327–30.
[45] Jia Z, Zhang N, Ma Y, Zhao L, Xia M, Li R. Top-seeded solution growth and optical properties of deep-UV birefringent crystal $Ba_2Ca(B_3O_6)_2$. Cryst Growth Des, 2017, 17, 558–62.
[46] Li RK, Ma Y. Chemical engineering of a birefringent crystal transparent in the deep UV range. CrystEngComm, 2012, 14, 5421–24.

6.4 Slags as materials resource

Oliver Janka, Martin Bertau, Rainer Pöttgen

Most large-scale processes in basic inorganic industry for metal production and recycling (pyrometallurgical processes) as well as incineration processes for the treatment of domestic and industrial waste produce considerable amounts of slags. A convenient way for slag disposal is dumping. This was practiced in many branches over long periods; however, dumping has two severe disadvantages: (i) toxic elements from the slag have an impact on water pollution (heavy metal leaching) [1] and (ii) many slags still have valuable metal contents that call for recycling processes [2]. The diverse slags are usually regrouped into the categories of ferrous and nonferrous slags as well as slags originating from incineration processes. The sum of all these slags is not only large in terms of volume but also with respect to the economic value. Industry as well as basic and applied research has learned quickly how to generate an additional value with these materials with respect to their use as secondary resource for recycling. The generation of well-adapted slag-treating processes is still an active research field. Every slag type has its own composition range (depending on the raw materials used in the process) and requires its tailored treatment. In view of the broad field, herein, we can only present some representative examples from the three slag categories. In times of scarcity of resources, the energetic and ecological evaluation of slags for secondary use is becoming an increasingly important research field.

6.4.1 Ferrous slag

The largest group of slags certainly derives from the iron and steel producing industry: (i) slag from the blast furnace, (ii) slag from the basic oxygen furnace (e.g. the Linz-Donauwitz process) and (iii) slags from electric arc-melting processes. The importance of ferrous slag is tied to the association FEhS – Institut für Baustoff-Forschung (www.fehs.de), an institution with a network across Europe, with respect to slag research and construction materials. Other important associations are Euroslag (www.euroslag.com) and National Slag (www.nationalslag.com).

Clearly, the blast furnace (for the basic inorganic blast furnace chemistry, we refer to standard textbooks [3]) slag concerns the by far largest amount. For one ton of raw iron, depending on the composition of the raw materials, approximately 300 kg of slag is produced. Keeping the annual worldwide iron production of around 1280 million tons in 2019 [4] in mind, one can estimate the large amount of ferrous slag produced.

Blast furnace slag contains about 38–41% CaO, 34–36% SiO_2, 10–12% Al_2O_3 and 7–10% MgO besides small amounts (<2%) of S, TiO_2 and FeO (when not otherwise

https://doi.org/10.1515/9783110798890-012

stated, weight-percentages are listed). Since the residual iron content is negligible, blast furnace slag is not important with respect to iron recovery and is mainly used with respect to its mineral content. If liquid blast furnace slag is quenched in water or air, one obtains a glassy, granular product, the so-called ground-granulated blast furnace slag (Figure 6.4.1). It mainly finds application in cement industry (see Chapter 1.1). A niche application of ground-granulated blast furnace slag with a defined particle size concerns blasting abrasives. Slags with high CaO and MgO contents can be re-used as flux materials in sintering processes or as partial substitutes for dolomite and limestone (conserving natural resources) in the blast furnace.

Figure 6.4.1: Ground-granulated blast furnace slag as educt in cement industry. ©HeidelbergCement AG/Steffen Fuchs.

If blast furnace slag is cooled in large quantities, the melt cake can mechanically be broken, and the resulting artificial stones can be sorted according to their grain size, similar to the usual process in natural quarries. Depending on the grain size, these slag fractions find application in road, track and water way construction, as landfill building materials or filling material for gabion baskets. Depending on the raw materials used in the blast furnace, the slag might contain toxic heavy metals. These are usually tightly bound in the slag matrix and not released.

Typical composite materials with added granular slag are asphalt and concrete. Slag addition to bitumen (blast furnace slag has excellent binding behavior to bitumen) leads to very abrasion-resistant road surfaces. For concrete, the slag grains are an equivalent substitute to natural stones. Given the large quantity of these slag grains, the secondary use of slag in the form of artificial stones strongly minimizes the consumption of natural stones. In total, today, around 95% of blast furnace slag is re-used in different fields, and only the remaining 5% are deposited.

Slag that derives from electric arc furnaces (so-called EAF slag) has a higher residual iron content, and iron recovery from the slag might be economically relevant [5]. A typical work-up of such slags is grinding followed by a magnetic separation in order to produce iron concentrates that can be fed back to the blast furnace process. One of the harmful elements in such slags is chromium. Careful analyses of the raw material with respect to the Cr^0, Cr^{III} and Cr^{VI} content are thus necessary. An efficient way to reduce highly toxic Cr^{VI} is reduction with Fe^{II} salts mainly $FeSO_4$ and $FeCl_2$ [6]. Further heavy metal contaminations result from Mn, Cu, Zn, Cd and Pb.

EAF slag might find application as agricultural fertilizer [5]. The high CaO content buffers acidity, and P, K, Mn and Fe sustain plant growth. However, such fertilizers can only be used in the absence of Cd and Pb as heavy metals.

The third point concerns converter steel slag [7] that is produced in the so-called basic oxygen furnace and during the Linz-Donauwitz process. Liquid pig iron or steel is directly treated with oxygen, and the iron accompanying elements are directly oxidized. The slag composition depends on the iron ore used. The following phases occur in different content: Ca_2SiO_4, $Ca_2(Al,Fe)_2O_5$, $(Fe,Mg)O$, Fe_3O_4, Ca_3SiO_5, CaO, MgO, $Ca_3Mg(SiO_4)_2$, Al_2O_3, MnO, P_2O_5 and TiO_2 and eventually Cr_2O_3 and V_2O_5. Carbon and sulfur are released from the furnace as gaseous CO and SO_2. One ton of iron or steel leads to approximately 100 kg of converter slag.

The converter lime that is free of hazardous metals can be used as mineral supplier in agriculture. In former times, phosphate-rich iron ores were used in the blast furnace process, and consequently, the basic oxygen furnace slag was phosphate-rich and sold under the trade name *Thomas phosphate* as fertilizer (see Chapter 6.6).

6.4.2 Aluminum salt slag

Aluminum salt slags are formed during the recycling of aluminum, so-called secondary aluminum (see Chapter 2.3). It mostly contains aluminum oxide, various halide salts, metallic aluminum and impurities such as carbides, nitrides or phosphides. Depending on the feedstock, 200 to 500 kg of salt slag is produced per tonne of secondary aluminum [8].

Two main furnace types are employed for the recycling of scrap aluminum: reverberatory and rotary melting furnaces. While the advantage of the latter one is that also highly contaminated scrap can be used, the reverberatory furnace can handle up to 45 tons of material at once. The furnaces also produce, besides the desired molten metallic aluminum, two side products. The first being off-gas and the second being a semisolid layer of skim on top of the melt. The skim can be further separated into salt and non-salt skim, the latter being referred to as dross [9, 10]. Non-salt dross is produced in processes that do not employ protective salt layers, and it is called white or gray dross in the US and black dross in Europe [11]. Here, the oxidation of the molten aluminum leads to the formation of the skim and in processes handling secondary aluminum, the oxide often originates from the scrap itself. Since the formation of dross is equivalent with a loss of the desired product, this must be reduced by any means possible. One of the key aspects is to avoid any surface disruptions both by handling the melt and by e.g. the reduction of the temperature to reduce convection inside the crucible. The non-salt dross manly consists of Al_2O_3 with metallic aluminum encapsulated in the porous material with an Al content varying between 15 and 80%. Na_3AlF_6 can be found in the dross from the primary Al generation, and MgO and $MgAl_2O_4$ are observed when recycling Mg-containing scraps [11].

Usually, the oxidation of the aluminum is suppressed via a salt flux that floats on top of the molten metal generating the so-called salt dross. Typically, NaCl/KCl mixtures along with CaF_2 are employed here. Besides the oxidation protection, the salt flux enables an enhanced heat transfer, and contaminants such as oxides, nitrides or carbides readily dissolve in the flux, thus cleaning the metal melt [12]. Furthermore, the molten salt disrupts the Al_2O_3 framework in which molten Al is entrapped, leading to the formation of Al droplets that sink to the molten metal phase [13].

One significant problem with these slags is that they are classified as toxic and hazardous waste and are considered to be highly flammable, irritant, harmful and leachable. The Al content within the salt slag reaches up to approx. 24%; however, only roughly 7% of them are metallic Al. Due to its amphoteric nature, metallic aluminum can be dissolved by both acids and bases, always leading to the formation of highly flammable hydrogen. Since about 50% of the slag are soluble salts, the reaction with water can furthermore leach toxic metal ions from the slag and e.g. metal nitrides or phosphides can form NH_3 or PH_3 that get released [2]. Therefore, in contrast to the other types of slags, a disposal is not easily possible, leading to the need for a comprehensive recycling strategy [14].

To properly treat and recover the respective valuable components of the salt slag, a five-stage process, similar to the B.U.S. process (Berzelius Umwelt Services AG, Germany), is employed. In the first step, a dry process, the metallic Al is separated from the oxides and salts by crushing. Afterwards, a leaching step takes place in which the salt is dissolved while the insoluble oxides remain followed by a separation of liquid and solid in the third step. The two final steps are the treatment of the evolved gases like NH_3, CH_4, PH_3 and others followed by a final crystallization step. The gases are neutralized using H_2SO_4 followed by a multi-stage absorber treatment. The oxide-free brine solution finally gets heated in order to grow halide crystal seeds, enabling precipitation from solution. Interestingly, the separated oxidic compounds are not separated further; however, Al_2O_3 and $MgAl_2O_4$ can be obtained via a heat treatment step [2].

As an example, we present the processing scheme (Figure 6.4.2) for salt slag employed by K + S. Figure 6.4.3 shows the raw slag and the recycled product fractions.

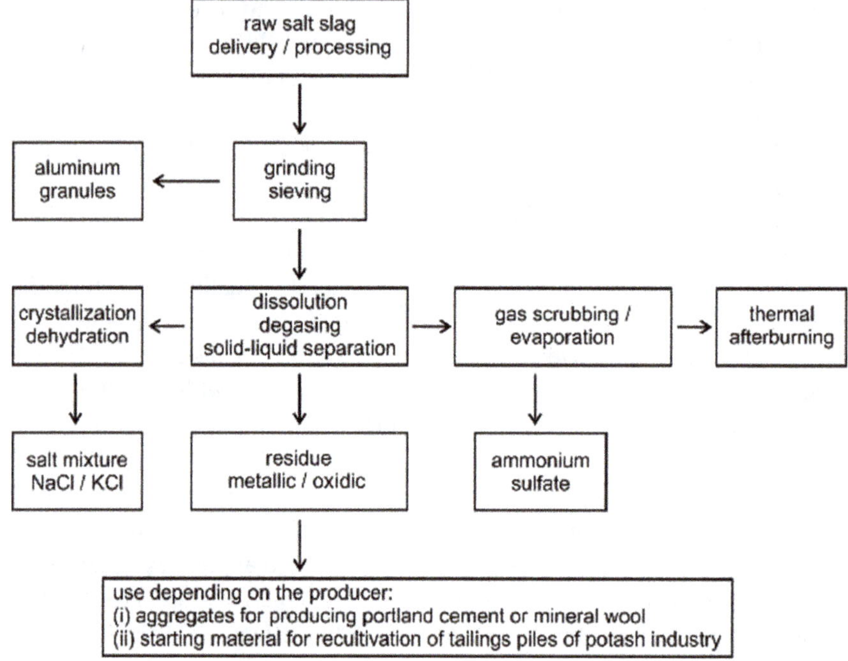

Figure 6.4.2: Process scheme for salt slag treating.

Figure 6.4.3: Fractions of the salt slag processing. The samples were kindly provided by K + S. Fotos by Thomas Fickenscher.

6.4.3 Copper slag

Like for almost all other metals, elemental copper is produced from primary sources (ores) as well as scrap parts, the so-called secondary sources. The ores get separated into oxidic and sulfidic ones, which are subsequently processed differently. While the oxidic ores are treated by a liquid-based process, the sulfides are handled by a thermal process. In this step of the copper production, the copper slags are formed

[15]. It is interesting to note that for one ton of elemental Cu, between 2.2 and 3 tons of slag are produced [2, 16]. The slag consists of oxidic compounds that float on top of the sulfide melt, called copper matte (Cu_2S), which occurs during the pyrometallurgical production of copper. The produced slag contains remaining Cu concentrations between 0.5 and 3.7% [2]. In order to minimize the Cu concentration in the slag, the oxidic copper components can be converted to Cu_2S, enriching the copper concentration in the melt even further. In order to achieve the conversion, FeS is added leading to the reaction shown in eq. (6.4.1), leading to an enrichment of FeO in the slag.

$$FeS + Cu_2O \rightarrow FeO + Cu_2S \tag{6.4.1}$$

In order to get a better separation of slag and melt, SiO_2 is added as slag formation aid while $CaCO_3$ and Al_2O_3 are added as stabilizers for the slag structure [16]. The separation of slag and melt takes place between 1000 and 1300 °C.

The subsequent thermal treatment of the slag determines the respective physical properties. While slow cooling leads to a hard and crystalline, basalt-like material, fast cooling by quenching in water results in an amorphous and granular, glassy obsidian-like product. The melt consists of Fe_2SiO_4 (fayalite), Fe_3O_4 (magnetite), $Ca(Fe,Mg)(SiO_3)_2$ along with Cu, Cu_2S and CuS inclusions [16]. Depending on the feed, sometimes, Co and Ni are also found in the slag. Due to the large amount of slag produced in the copper refinement process, different recycling strategies are employed. The remaining metals can be obtained by different processes, e.g. a carbothermic route, or a leaching after flotation. Alternatively, the slag can directly be processed if a metal recovery is not feasible. In the carbothermic process, the slag is heated with carbon, leading to the formation of a metal alloy that contains Cu and Fe along with other constituents of the slag, e.g. Co or Ni. The produced alloy can be treated with hot concentrated H_2SO_4, leading to a Co and Fe-containing solution from which Co can be precipitated as CoS using H_2S [16]. However, the slag can also directly be treated with sulfuric or hydrochloric acid, ammonia or even cyanide in order to leach the metals. Increased efficiency has been reported by the addition of hydrogen peroxide or pressure leaching with elemental chlorine [2]. Finally, the hydrometallurgical route, based on flotation, has been successfully employed. Here, high recovery rates (>80%) have been reported for a two-stage process [17].

In cases where metal recovery is not profitable, still a direct use of the slag is possible. Due to the high hardness of the material, it can be utilized in ceramic abrasive tools [15]. Also, its addition in the form of fine powders to asphalt pavement mixtures is utilized in California (USA), while granulated copper slag is used in Georgia (USA) [15]. Finally, there are reports on copper slag as construction materials. The ones with higher CaO content have been reported to exhibit cementitious properties similar to the ones of Portland cement [15].

As an example, in Figure 6.4.4, we present slag samples that derive from the different steps of copper processing. The three slag fractions were studied by powder X-ray diffraction and X-ray fluorescence (XRF) analyses. Since such slag fractions differ from batch to batch, we only list the main features (neglecting the side products). An additional problem of the solidified slags concerns amorphous fractions that express themselves only as background bumps in the X-ray powder patterns.

The slag that forms directly in the anode furnace (Figure 6.4.4, top) shows Bragg reflections of a cuprite (Cu_2O)/magnetite (Fe_3O_4) mixture. The reddish color indicates the residual copper content. XRF data show a Cu: Fe ratio of ≈2, calling for copper recycling. In the first step towards the converter slag (Figure 6.4.4, middle), the Cu: Fe ratio is reduced to ≈1.2. The much lower copper content is expressed in the dark anthracite color of the slag piece.

The final residual, granulated iron silicate slag contains almost no residual copper (Cu: Fe ≈ 0.02). The powder X-ray diffraction pattern shows an iron-rich spinel besides iron silicate. The residual iron silicate slag is comparable with materials obtained from mining; however, with a much lower ecological footprint – an excellent recycling of raw material [18]. It finds application for road construction and additive for concrete as well as in cement industry.

Figure 6.4.4: (top) Copper slag of an anode furnace, (middle) converter slag and (bottom) granulated iron silicate slag. The samples were kindly provided by Aurubis AG. Fotos by Thomas Fickenscher.

An example where copper slags have found application as road construction material is the theatre square in Dresden, Germany. These paving stones are called slag stones, and the term copper slag stone is used too. They are produced through pouring molten slag into the desired shape and allowing the melt to cool. Slag stones are produced in the dimensions 160 × 160 × 160 mm and 240 × 160 × 160 mm, of which the cube-shaped ones are far more common. Their surface properties can be adapted to road construction use through roughening, which is done by sprinkling them with grit. Slag stones have a dark gray to almost black coloration and achieve high strength values, which is the reason why they are particularly suited for traffic areas with high wheel loads. In addition, their defined shape predesigned them for use in urban architecture, where they are in use not only as paving but also as wall coatings. Mansfeld copper slag stones have been used widely because, in the former GDR, there was intensive mining for copper shale in the South Harz Mansfeld mining district in today's Saxony-Anhalt. The large amounts of copper slag were made through producing copper slag stones, which were installed in the entire country.

The chemical composition of Mansfeld copper slag varies slightly: 45–50% SiO_2, 18–22% CaO, 16–19% Al_2O_3, 4–9% MgO, 3–5% FeO, 3–5% ($Na_2O + K_2O$), 0.2–0.3% Cu and 0.2–0.4% S. With almost 50% SiO_2, the slag is highly acidic, for which reason its properties come close to those of glass. In fact, upon rapid cooling, the slag solidifies purely glassy to give obsidian-like masses. The outstanding chemical and physical properties of the Mansfeld slag, however, were obtained by varying the cooling rate, which allowed for using it in numerous applications. Upon tempering, pyroxene is produced as the predominant crystalline phase. Where there is a controlled cooling of the slag, a porcelain-like structure with very small crystallite sizes (<1 µm) is obtained. Molded parts made from this copper slags are highly resistant to wear, with a compressive strength of approx. 290 MPa. These properties rendered them a highly demanded wall and road construction material as of 1875 [19–24].

The sometimes bluish color of the copper slag stones is attributed to cobalt; yet with average concentrations <1%, there is no technical use. Nevertheless, Mansfeld copper slag intermediary has been considered as potential cobalt source for lithium ion battery production. Presently, the major application for copper slag is cement industry.

Despite its generally recognized properties, the other side of the medal is the high procedural efforts to be taken to shape the slag stones which demands considerable manual labor. In addition, the high-quality specifications for road construction use allow for a maximum of 50% of the overall slag production to be used for construction purposes, which resulted in the need for stockpiling of the residual slag. The last 50 years of Mansfeld mining suffered from geological deposit changes in the copper shale composition, which resulted in quality problems of the slag. For that reason, and in view of its suitability as a cement additive, slag utilization increasingly changed its use in cement industry, for which copper slag is in use till present [25]. Eventually, Mansfeld copper slag stone production ceded as a consequence of the economic upheavals in the course of the German re-unification in 1990.

As a peculiarity of Mansfeld copper slag stones, there had been newspaper reports on a radioactive pollution originating from this material in the early 1990s. In fact, the mined copper shale contained on average 5 g/t of uranium and other radioactive elements. During metallurgical processing for the extraction of copper, these elements partially accumulated in the slag. However, the sum value of natural radionuclides in the slag is rather in the middle range compared to other building materials such as granite, clay, loam, bricks, clinker and concrete. The radiation emitted is within the legal limit values [26, 27].

6.4.4 Lead slag

Slag formation in the lead production process [28] strongly depends on the educts. Lead ore smelting from galena ores (mainly PbS) produces around one tonne of slag per tonne of pure lead, while lead recycling (mainly from lead-acid batteries) has a significantly lower slag amount of around 300 kg per ton [29]. The primary lead production starts with galena that is roasted to PbO, followed by coke-reduction in a blast furnace. The melting temperature reduction is achieved by an addition of lime, silica and ironstone. In the production of secondary lead from batteries, $PbSO_4$, PbO and PbO_2 are reduced with carbon and iron, and limestone is used as flux. The relevant reactions are:

$$PbSO_4 + 2\,C \rightarrow PbS + 2\,CO_2$$

$$PbS + Fe \rightarrow Pb + FeS$$

$$PbO_2 + C \rightarrow Pb + CO_2$$

$$2\,PbO + C \rightarrow 2\,Pb + CO_2$$

Lead slag from both processes mainly contains iron, zinc and residual lead as valuable metals that can be recovered. The iron mainly occurs as Fe_2SiO_4 (fayalite), Fe_3O_4 (magnetite) and Fe_2AlO_4 (hercynite) in the slag. A possible workup (pyrometallurgical route) is the direct reduction followed by a magnetic separation while lead and zinc are volatized and condensed. This process requires high amounts of energy. An alternative is the hydrometallurgical process with a leaching of the slag with (i) chloride, (ii) acetic acid or (iii) nitric acid. The leaching in concentrated NaCl-$FeCl_3$ solutions proceeds as follows:

$$PbS + 2Fe^{3+} + 2Cl^- \rightarrow PbCl_2 + 2Fe^{2+} + S$$

$$PbCl_2 + 2Cl^- \rightarrow [PbCl_4]^{2-}$$

The tetrachloridoplumbate(II) solution can subsequently be precipitated with sodium oxalate, and PbC_2O_4 is thermally decomposed to PbO. A flow chart for this process is presented in [30]. The acetic acid leaching process leads to $PbSO_4$. Leaching with nitric acid proceeds in combination with Fe^{3+}, leading to $PbSO_4$ and Zn(II)/Ni(II)/Cu(II) mixtures. The slag composition decides on the method used. Many of the hydrometallurgical work-ups are still on the laboratory scale.

After metal recovery, the final lead slag needs to be treated. One possibility is controlled landfilling/deposition. Similar to the ferrous blast furnace slag (vide ultra), lead slag has also been discussed as substitute for natural stone in concrete and cement clinkers; however, the problem of leaching of heavy metals remains. This also applies for the potential use in road construction. A more sophisticated and energy-consuming process is the formation of stable glass ceramics which strongly reduce the leaching process.

6.4.5 Further metal slags

Many other slags from primary metal production contain valuable contents of by-products which make slag work-up an economic process. The first step of the work-up is usually flotation (concentration of the metals). A typical example is pyrite slag which is used as a source for Ag and Au.

The rare metals Nb and Ta can be recovered from tin slag [2]. Leaching of the tin slag leads to separation of soluble metals (e.g. aluminum, calcium, manganese and iron) and a concentrate with niobium and tantalum. Similar to the primary production, this concentrate is chlorinated at high temperature (500–1000 °C) in Cl_2/N_2 or Cl_2/N_2/CO mixtures. The resulting chloride mixture can then be separated by fractional extraction, similar to the primary niobium and tantalum production [3]. Other tin slags are strongly acidic with glassy properties. They contain only trace amounts of tantalum, but significantly more chromium. Major constituents are 55–56% SiO_2, 14–16% ($Na_2O + K_2O$), 9–10% Al_2O_3, 5% CaO, 5% Fe_2O_3, 4% MgO, 1% ZnO, 0.9–1.0% Cr_2O_3 and 300 ppm Ta_2O_5. However, owed to the low chromium content, they are not interesting for exploiting economically, and a utilization as an additive fails through the chromium content being too high [31].

A non-conventional iron slag is obtained from pyrometallurgical recycling of zinc from flue dust originating during the metallurgical work-up of galvanized steel for car bodies. Its chemical composition was found to be: 54% Fe_2O_3, 14% CaO, 10% SiO_2, 5.1% MnO, 3.7% MgO, 3.7% ($Na_2O + K_2O$), 3.6% ZnO%, 3.5% SO_3, 1.1% Al_2O_3 and 1.7% others. The crystalline fraction slag is made up from iron, FeO (wuestite), $CaFeSiO_4$ (kirschsteinite) and Mn_3Si. These slags are interesting for recovering the iron content, yet presently for economic reasons with no activities.

6.4.6 Incineration slag

Municipal solid waste (MSW) is probably the broadest mixture of waste – biomass, metals, plastics, paper and many more. The largest amounts of MSW are incinerated. The incineration process (i) recovers part of the energy bound in the waste (thermal power plant) and reduces the waste volume from one ton of MSW to approx. 300–350 kg of incineration slag. The further use/treatment of the slag strongly depends on its composition which, in most cases, is hardly predictable. The total amount of solid waste from these incineration processes is the sum of the bottom ash, the filter dust and flue-gas residues.

Incineration slags with low metal content and especially with heavy metal contamination can be used for landfilling or might find application as partial replacement for cement in mortar and concrete [32]. The second use concerns the metal contents. These vary from several milligrams to more than 100 g (iron). The most important metal contents in the bottom ash are mainly Fe, Al, Pb, Cu, Sn, Zn and Ti. Iron recovery is the best-developed technique and proceeds through grinding with subsequent magnetic separation and work-up of the iron concentrate. All other elements build a multi-component mixture. Separation strategies are mostly still on the lab scale, but will play an important role in the future since the chemical value of the incineration slags increased in recent years. The total amount of valuable metals is large; however, the dilution is a severe problem. The up-concentration processes considerably consume energy and chemicals [33].

6.4.7 Perspectives: bioleaching

Bioleaching [34] is currently studied under mild laboratory conditions as technique for metal recovery from slags. The strategy of bioleaching is (i) the extraction of hazardous metals from slag in order to release clean slag for deposition or further use and (ii) the recovery of residual, valuable metals from slag.

As compared to chemical leaching, bioleaching is a very complicated process with a high number of parameters: (i) the pH value of the leaching solution and eventual pH changes during the leaching process, (ii) the leaching temperature, (iii) the grain size of the leached suspension and (iv) the use of autotrophic or heterotrophic bacteria. In a final step, the recovered metals from the leached slag are precipitated (hydroxides or sulphides), electrochemically treated (electrolysis), treated by solvent extraction, by ion exchange or biosorbed. Well understood, all these processes are energy consuming since many successive separation steps are required. In addition, bioprocesses are much more time-consuming and require large reactor volumes if not conducted as heap leaching. On the other hand, where the micro-organisms produce the leach-acid themselves, the biotechnical route may be economically superior. At present, there is no general process available, and whether or not bioleaching comes into play is an individual decision.

Tests on the laboratory scale revealed that Zn, Cd and In can be leached in sufficient yield, while there is poor extraction efficiency for Pb, As and Ag. Given the large absolute amount of metal-containing slags, including MSW slags, such alternative strategies might be a solution for future metal recovery.

References

[1] Piatak NM, Parsons MB, Seal II RR. Characteristics and environmental aspects of slag: A review. Appl Geochem, 2015, 57, 236–66. 

[2] Shen H, Forssberg E. An overview of recovery of metals from slags. Waste Manage, 2003, 23, 933–49. 

[3] Holleman AF, Wiberg N. Anorganische Chemie, 103. Auflage, De Gruyter, Berlin, 2016. ISBN 978-3-11-051854-2.

[4] Steel statistical yearbook – 2020 concise version, World Steel Association, Brussels, 2020. worldsteel.org

[5] Teo PT, Zakaria SK, Salleh SZ, Taib MAA, Sharif NM, Seman AA, Mohamed JJ, Yusoff M, Yusoff AH, Mohamad M, Masri MN, Mamat S. Assessment of electric arc furnace (EAF) steel slag waste's recycling options into value added green products: A review. Metals, 2020, 10, 1347.

[6] Reduktion von Cr-VI in Reststoffen wie Aschen, Schlacken und Stäuben. Kronos ecochem®, Technische Information 3.11, Kronos International, Inc., Leverkusen, 2019. www.kronoseco chem.com, accessed december, 27th 2021.

[7] Schollbach K, Ahmed MJ, van der Laan SR. The mineralogy of air granulated converter slag. Int J Ceramic Eng Sci, 2021, 3, 21–36.

[8] Tsakiridis PE. Aluminium salt slag characterization and utilization – A review. J Haz Mater, 2012, 217–218, 1–10.

[9] Satish Reddy M, Neeraja D. Sādhanā, 2018, 43, 124. DOI: org/10.1007/s12046-018-0866-2.

[10] Wibner S, Antrekowitsch H, Meisel TC. Metals, 2021, 11, 1108.

[11] Schlesinger ME. Aluminum Recycling, CRC Press, Boca Raton, 2006. eBook ISBN 9780429122897.

[12] Graczyk DG, Essling AM, Huff EA, Smith FP, Snyder CT. Analytical chemistry of aluminum salt cake. In: Light Metals: Proceedings of Sessions, TMS Annual Meeting. Warrendale, PA, 1997, 1135–40.

[13] Das BR, Dash B, Tripathy BC, Bhattachary IN, Das SC. Production of η-alumina from waste aluminium dross. Miner Eng, 2007, 20, 252–58.

[14] López FA, Sáinz E, Formoso A, Alfaro I. Can Metall Q, 1994, 33, 29–33.

[15] https://www.kupferinstitut.de/kupferwerkstoffe/kupfer/produktionsprozesse/, accessed on November 18th, 2021.

[16] Gorai B, Jana RK. Premchand characteristics and utilisation of copper slag – A review. Resour Conserv Recycl, 2003, 39, 299–313.

[17] Sarfo P, Das A, Wyss G, Young C. Recovery of metal values from copper slag and reuse of residual secondary slag. Waste Manage, 2017, 70, 272–81.

[18] https://www.aurubis.com/verantwortung/umwelt-energie-und-klima/oekobilanz-eisensilikat, accessed on November 18th, 2021.

[19] Stegemann R. Hrsg. Das große Baustoff-Lexikon, Deutsche Verlags-Anstalt, Stuttgart, 1941.

[20] Benedix R. Bauchemie, 5. Aufl., Vieweg Teubner, Wiesbaden, 2011.

[21] Rahmathulla Noufal E, Kasthurba AK, Sudhakumar J. IOP Conf Ser: Mater Sci Eng, 2021, 1114, 012008.

[22] Rajeeth TJ, Shah AH, Makhdomi SM, Wani AN. J Phys: Conf Ser. 2021, 1913. 012066.

[23] Vostal R. Freiberg University of Mining and Technology, personal communication, 2021.

[24] Spitzner J. Die Entwicklung der technischen Verwertung von Kupferhüttenschlacken. www.hel braerleben.de, accessed december 15th, 2021.

[25] Martens H, Goldmann D. Recycling mineralischer Baustoffe und Verwertung von Schlacken und Aschen. In: Martens H, Goldmann D. Recyclingtechnik, Springer, Berlin, 2016, pp. 355–76. DOI: 10.1007/978-3-658-02786-5_10.

[26] Keller G, Hoffmann B, Feigenspan T. Sci Total Env, 2001, 272, 85–89.
[27] Strahlenschutzkommission, Bewertung der Verwendung von Kupferschlacke aus dem Mansfelder Raum – Empfehlung der Strahlenschutzkommission. Bundesanzeiger Nr. 43, 03. März 1992. https://www.bundesanzeiger.de
[28] Pan D, Li L, Tian X, Wu Y, Chen N, Yu H. A review on lead slag genaration, characteristics, and utilization. Resour Conserv Recycl, 2019, 146, 140–55.
[29] Kreusch MA, Ponte MJJS, Ponte HA, Kaminari NMS, Marino CEB, Mymrin V. Technological improvements in automotive battery recycling. Resour Conserv Recycl, 2007, 52, 368–80.
[30] Shu Y, Ma C, Zhu L, Chen H. Leaching of lead slag component by sodium chloride and diluted nitric acid and synthesis of ultrafine lead oxide powders. J Power Sources, 2015, 281, 219–26.
[31] Kraft M, Guhl AC, Bertau M. 2021, unpublished results.
[32] Czop M, Łázniewska-Piekarczyk B. Use of slag from the combustion of solid municipal waste as a partial replacement of cement in mortar and concrete. Materials, 2020, 13, 1593.
[33] Joseph AM, Snellings R, Van den Heede P, Matthys S, De Belie N. Materials, 2018, 11, 141.
[34] Potysz A, van Hullebusch ED, Kierczak J. Perspectives regarding the use of metallurgical slags as secondary metal resources – A review of bioleaching approaches. J Environm Manage, 2018, 219, 138–52.

6.5 Salts in nutrition products

Heiko Hayen

Mineral salts have multi-faceted properties in foodstuff. Human mineral supply depends not only on the total intake of an element but also primarily on its bioavailability. The latter is essentially related to the chemical species, i.e. the specific form of an element defined by the isotopic composition, electronic or oxidation state, and/or complex or molecular structure. Thus, the redox potential and pH value determine the valence state, solubility and, consequently, the absorption. A multitude of food constituents, like proteins, peptides, amino acids, polysaccharides, sugars, lignin, phytin and organic acids, bind minerals, enhance or inhibit their absorption. The importance of minerals as food ingredients depends on their nutritional and physiological roles. They affect the texture of food and contribute to flavor and activate or inhibit enzyme-catalyzed and other reactions. Consequently, inorganic salts can be utilized both as additives to modify food properties and for nutrient fortification. In addition, they are nowadays increasingly used in food supplements. Naturally occurring food ingredients are not discussed, whereas inorganic salts intentionally added to foods are described in detail.

6.5.1 Food additives

A food additive is a substance (or a mixture of substances) which is intentionally added to food. Perhaps, the oldest in history is the addition of salt (NaCl), which was used both to enhance flavor and to preserve food. In general, food additives are involved in food production, processing, packaging and/or storage without being a major ingredient. The main applications are:

- **Nutritive value of food**: Additives such as minerals, vitamins and amino acids are utilized to increase the nutritive value of food.
- **Sensory value of food**: Color, odor, taste and texture may diminish during processing and storage. Such decreases can be addressed by additives, for example, by curing and reddening of meat. Development of off-flavor resulting from lipid oxidation can be suppressed by antioxidants or even unintentionally promoted by the addition of metal ions like iron. Furthermore, food texture can be stabilized by adding minerals or polysaccharides.
- **Shelf life of food**: The extension of shelf life involves protection against microbial spoilage and suppression or retardation of undesired chemical and physical changes in food. The latter is achieved by pH stabilization using buffer additives and the former by antimicrobial agents like nitrites or sulfites.
- **Practical value**: The general trend towards foods, which are easy and quick to prepare, i.e. *convenience foods*, also broadens the use of food additives.

https://doi.org/10.1515/9783110798890-013

In general, food additives at recommended use levels must be non-toxic and have no adverse effects on human health. Furthermore, food additives are only allowed when required for the nutritive or sensory value of food or for its processing or handling. The use of additives is regulated by Food and Drug or Health and Welfare administrations in most countries. In this context, "GRAS" is an important acronym for the phrase "generally recognized as safe" [1].

There are also international institutions that deal with the safety and regulation of food additives. Initially, the Joint FAO/WHO Expert Committee on Food Additives (JECFA) was established as an international scientific expert committee to evaluate the safety of food additives. JECFA serves as an independent scientific expert committee, which performs risk assessments and provides advice to FAO, WHO and the member countries of both organizations as well as to the Codex Alimentarius Commission [2]. As a result, an international numbering system for food additives was implemented by the Codex Alimentarius committee [3]. The Codex Alimentarius or "Food Code" is a collection of standards, guidelines and codes of practice adopted by the Codex Alimentarius Commission. Only part of their food additive compilation is approved for use in the European Union (EU), and others are forbidden [4].

Within the EU, food additives must be approved separately and may only be used if they fulfil the criteria laid down in Regulation (EC) No. 1333/2008. "They must be safe when used and there must be a technological need for their use. Use of food additives must not mislead the consumer and must be of benefit to the consumer. Misleading the consumer includes, but is not limited to, issues related to the nature, freshness, quality of ingredients used, the naturalness of a product or of the production process, or the nutritional quality of the product, including its fruit and vegetable content" [4].

This regulation sets maximum levels for several food additives for the specific foods. Furthermore, the principle "quantum satis" is applied for a large number of additives. "Quantum satis" indicates that no maximum level is specified. However, additives must be used in accordance with good manufacturing practice at a level not higher than necessary to achieve the intended purpose, provided that they do not mislead the consumer. The safety assessment in the EU is the responsibility of the European Food Safety Authority (EFSA) [5].

Regardless of the concentration, food additives must be clearly indicated for the consumer. E-numbers are usually found on food labels throughout the EU, a number code for food additives, where the 'E' prefix stands for Europe.

Please note that naturally occurring food ingredients such as ascorbic acid (vitamin C) are also listed. Ascorbic acid (E 300), sodium ascorbate (E 301) and calcium ascorbate (E 302), for example, are approved as antioxidants by EFSA in the EU.

The most important groups of inorganic food additives are outlined in the following sections.

6.5.1.1 Sodium chloride

Common (table, cooking or kitchen) salt occupies a special position among all spices. It is used in greater amounts to enhance the flavour and taste of food. In addition, some foods are preserved when salted with large amounts of NaCl. Humans require a constant level of intake of sodium and chloride ions to maintain their vital concentrations in plasma and extracellular fluids. The daily requirement is about 5 g of NaCl; an excessive intake is detrimental to health. The salty taste is stimulated together by cation and the anion. In comparison, the sour taste is induced solely by the cation. The pure salt taste is only produced by NaCl. Interestingly, the very next chemical relative, KCl, has a sour/bitter aftertaste. KCl can be used to replace up to about 30% of sodium chloride in many foods, but beyond that concentration, foods may become unsavory [6].

Common cooking salt is nearly entirely NaCl. Impurities are moisture (up to 3%) and other salts, not exceeding 2.5% ($MgCl_2$, $CaCl_2$, $MgSO_4$, $CaSO_4$ and Na_2SO_4). Special salt is also available. Iodized salt (Figure 6.5.1) is established as a preventive measure against goiter, a disease of the thyroid gland. It contains 5 mg kg^{-1} of sodium-, potassium- or calcium iodide (NaI, KI, CaI_2). In addition to iodine, fluorine (e.g. as NaF or KF) can also be added to cooking salt due to its caries-preventive effect [7].

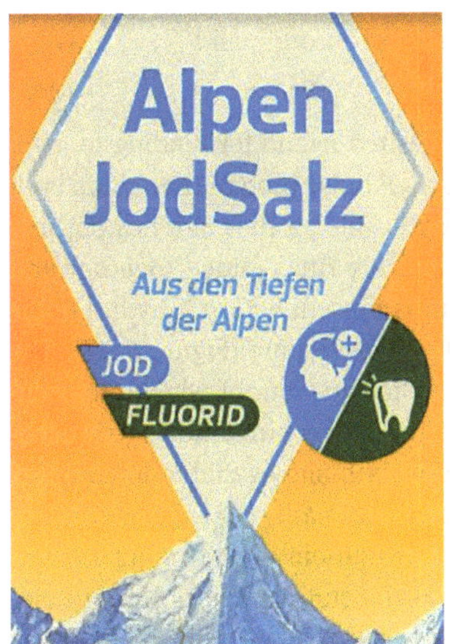

Zutaten: Siedesalz, Kaliumfluorid, Kaliumjodat, Trennmittel Natriumferrocyanid.

Durchschnittliche Nährwerte	pro 100 g	pro Portion (2 g)	% Referenzmenge* pro Portion
Energie	0 kJ/0 kcal	0 kJ/0 kcal	0 %
Fett	0 g	0 g	0 %
davon gesättigte Fettsäuren	0 g	0 g	0 %
Kohlenhydrate	0 g	0 g	0 %
davon Zucker	0 g	0 g	0 %
Eiweiß	0 g	0 g	0 %
Salz	99,9 g	2,00 g	33 %
Mineralstoffe	pro 100 g	pro Portion (2 g)	
Jod	2000 µg (1333 %**)	40 µg (27 %**)	
Fluorid	31 mg (886 %**)	0,62 mg (18 %**)	

*Referenzmenge für einen durchschnittlichen Erwachsenen (8400 kJ/2000 kcal)
**% der Nährstoffbezugswerte (NRV)

Dieses Paket enthält 250 Portionen à 2 g.

Eine abwechslungsreiche und ausgewogene Ernährung sowie eine gesunde Lebensweise sind wichtig.

Figure 6.5.1: Example of cooking salt fortified by iodine and fluorine. Ingredients are evaporated salt (NaCl), potassium fluoride (NaF), potassium iodate (KIO_3) and anticaking agent sodium ferrocyanide (tetrasodium [hexacyanoferrate(II)], $Na_4Fe(CN)_6$).

6.5.1.2 Sodium nitrite and sodium nitrate

The additives, sodium nitrite ($NaNO_2$) and sodium nitrate ($NaNO_3$), are used primarily in a mixture with cooking salt (NaCl) to preserve the red color of meat, to inhibit microorganisms and to develop characteristic flavors. For curing, cooking salt is utilized which is premixed with $NaNO_3$ (1–2%) or $NaNO_2$ (0.5–0.6%) [8]. Nitrite rather than nitrate is apparently the functional constituent. However, some of the nitrate present is reduced to nitrite either by salt-tolerant microorganisms in the brine or by the respiratory enzymes of the muscle tissue.

Nitrites have antimicrobial activity, particularly in a mixture with NaCl. Of importance is their inhibitive action, in non-sterilized meat products, against infections by *Clostridium botulinum* and, consequently, against accumulation of its toxin. Acute toxicity has been found only at high levels of use (formation of methemoglobin). A concern is the possible formation of carcinogenic *N*-nitrosamines, as recently re-evaluated by EFSA [9]. Consequently, the trend is to reduce or eliminate nitrate and nitrite content in food.

No suitable replacement has been found for nitrite in meat processing. However, ascorbic acid or isoascorbic acid can be added to nitrite-containing meats to prevent the formation of *N*-nitrosamines. This is accomplished by forming NO and preventing the formation of undesirable nitrous anhydride (N_2O_3, the primary nitrosating agent) [10]. The proposed reactions are shown here (reaction (6.5.1)):

$$\begin{aligned} &\text{Ascorbic acid} + HNO_2 \rightarrow \text{2-Nitrite ester of ascorbic acid} \\ &\quad \rightarrow \text{Semidehydroascorbate radical} + NO \end{aligned} \tag{6.5.1}$$

The formation of NO is intended because it is the desired ligand for binding to myoglobin to form the cured meat color. Myoglobin is a globular protein that binds both iron and oxygen. The chromophore component responsible for the color is a porphyrin known as heme, which is composed of four pyrrole rings joined together and linked to a central iron atom. The heme porphyrin is present within a hydrophobic pocket of the globin protein and is bound to a histidine residue (Figure 6.5.2). The centrally located iron atom possesses six coordination sites, four of which are occupied by the nitrogen atoms of the tetrapyrrole ring. The fifth coordination site is occupied by the histidine residue of globin, leaving the sixth site available to complex with electronegative atoms donated by various ligands, primarily O_2, NO and CO.

Meat color is determined by the chemistry of myoglobin including its state of oxidation, the type of ligands bounds to heme and the state of the globin protein. The heme iron within the porphyrin ring may exist in two oxidation states: either as the reduced ferrous (Fe^{2+}) or the oxidized ferric (Fe^{3+}) form. This oxidation state of the iron atom in heme must be distinguished from oxygenation of myoglobin. Oxygenation occurs when molecular oxygen binds to myoglobin and oxymyoglobin is formed. When oxidation of myoglobin occurs, the iron atom is converted from the ferrous (Fe^{2+}) to the ferric (Fe^{3+}) state, forming metmyoglobin.

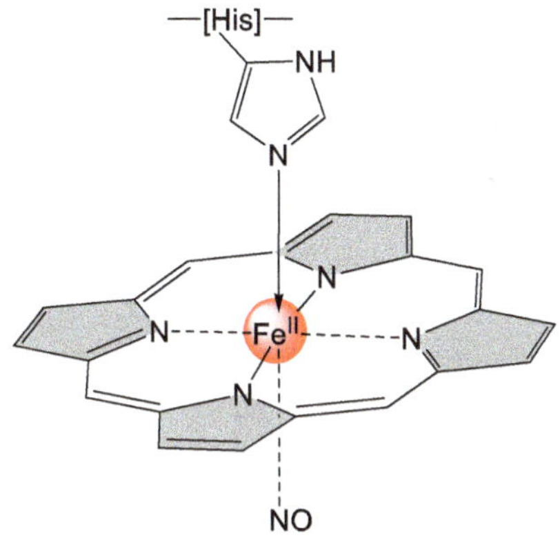

Figure 6.5.2: Octahedral environment of Fe(II)-porphyrin, showing the imidazole ring of globin, histidine residue and nitrogen monoxide. Please note that the side chains of pyrrole rings have been omitted for clarity of presentation.

During the curing process, specific reactions occur that are responsible for the stable pink color of cured meat products. The first reaction occurs between nitric oxide and myoglobin to produce nitric oxide myoglobin, also known as nitrosylmyoglobin. Nitrosylmyoglobin is bright red and unstable. Upon heating, the more stable nitric oxide myohemochromogen (nitrosylhemochrome) forms. This product is responsible for the desirable pink color of cured meats. Heating of this pigment denatures globin, but the pink color persists [11].

6.5.1.3 Ammonium and sodium carbonates

Different carbonates are applied in food production as chemical leavening agents, also known as raising agents. The mode of action is similar to fermentation with Baker's yeast, the release of CO_2. Dough consisting only of flour and water gives a dense flat cake. Baked products with a porous crumb, like bread, are obtained only after the dough is leavened. This is achieved by the addition of yeast, whereas baking powders are used for fine baked products.

The interaction of water, acid, heat and chemical leavening agents (baking powders) releases CO_2. The release of gas may occur in the dough prior to or during oven baking. The agents consist of a CO_2-generating source, often sodium bicarbonate ($NaHCO_3$), and an acid carrier, usually disodium dihydrogendiphosphate ($Na_2H_2P_2O_7$). In baking powder, the two reactive constituents are blended with a filler, which consists of wheat, corn, rice or tapioca starch. The filler content of up to 30% prevents premature release of CO_2 by separating the reaction partners and by creating a dry atmosphere in the baking powder packaging [12].

Mainly ammonium hydrogen carbonate (NH_4HCO_3, ammonium bicarbonate, "ABC") is used, whereas $NaHCO_3$ alone is applied for some flat shelf-stable cookies. Ginger and honey cookies are leavened by NH_4HCO_3, mostly together with potash (K_2CO_3). To a small extent, a 1:1 mixture of NH_4HCO_3 and ammonium carbamate ($H_2NCOONH_4$) is used in some countries. Both decompose above 60 °C to CO_2, NH_3 and H_2O.

6.5.1.4 Ammonium chloride

Salty licorice or salmiac licorice is a variety of licorice flavoured with the ingredient "salmiak salt" (ammonium chloride, NH_4Cl) and is a common confection found in the Nordic countries, Benelux and northern Germany. Salmiak or sal ammoniac is a rare naturally occurring mineral composed of NH_4Cl, which gives salty licorice an astringent, salty taste. Extra salty licorice is additionally coated with salmiak salt or salmiak powder or sometimes cooking salt. Simple licorice products contain 30–45% starch, 30–40% sucrose and at least 5% licorice extract. Higher quality products have an extract content of at least 30%. Licorice roots (generally *Glycyrrhiza glabra* L., Leguminosae) are main ingredients of diverse licorice sweets which are appreciated by consumers all over the world due to their sweet and typical longlasting licorice impression [13].

6.5.1.5 Sulfites and sulfur dioxide

Sulfur dioxide (SO_2) and its derivatives, sodium sulfite (Na_2SO_3) and sodium hydrogen sulfite (sodium bisulfite, $NaHSO_3$), are used as general food preservatives, inhibiting the growth of yeasts, molds and bacteria [14]. The activity increases with decreasing pH and is mostly derived from undissociated sulfurous acid H_2SO_3, which predominates at pH < 3.

In addition, SO_2 and sulfites act as antioxidants and as reducing agents (see also Chapter 11.2). SO_2 gas and the sodium, potassium or calcium salts of sulfite (SO_3^{2-}), bisulfite (HSO_3^-) or metabisulfite ($S_2O_5^{2-}$) are the commonly used forms in foods. Most frequently, the sodium and potassium metabisulfites are used, as they exhibit good stability towards autoxidation in the solid phase.

SO_2 is used in the production of dehydrated fruits and vegetables, fruit juices, syrups, concentrates or purée. During wine fermentation, SO_2 is used at a level of 50–100 ppm, while 50–75 ppm are used for wine storage. Besides inhibiting microbial growth, SO_2 also inhibits discoloration by blocking compounds with a reactive carbonyl group (*Maillard* reaction, non-enzymatic browning) or by inhibiting oxidation of phenols by phenol oxidase enzymes (enzymatic browning) [15].

6.5.1.6 Phosphates and phosphoric acid

Phosphoric acid ($pK_1 = 2.15$; $pK_2 = 7.1$; $pK_3 \sim 12.4$) and its salts account for about one-quarter of all the acids used in the food industry. The main field of use of phosphoric acid is the soft drink industry (cola drinks). It is also used in fruit jellies, processed cheese and as an active buffering agent or pH-adjusting ingredient in fermentation processes. An important application field is baking powder. Acid salts, e.g. $Ca(H_2PO_4)_2 \times H_2O$ (fast activity), $NaH_{14}Al_3(PO_4)_8 \times 4\ H_2O$ (slow activity) and $Na_2H_2P_2O_7$ (slow activity), are used in baking powders as components of the reaction to slowly or rapidly release the CO_2 from $NaHCO_3$ [12]. Please refer to Chapter "6.2.6 Condensed phosphates – applications", and in particular to Figure 6.2.5 and equation (6.2.8), as well as Chapter 6.5.1.3.

6.5.1.7 Silicates and ferrocyanides

Several inorganic salts are used as anticaking agents (Table 6.5.1). Some food products such as common salt, seasoning salt (e.g. a mixture of onion or garlic powder with common salt), dehydrated vegetable and fruit powders, soup and sauce powders and baking powder tend to cake into a hard lump. Lumping can be avoided by using a number of compounds that either absorb water or provide protective hydrophobic films. Anticaking compounds include sodium, potassium and calcium hexacyanoferrate(II), calcium and magnesium silicate, and calcium carbonate [16].

Table 6.5.1: List of permitted anticaking agents in the EU in the order of their E-numbers.

Name	Formulae	E-number
Calcium carbonate	$CaCO_3$	E 170
Natrium ferrocyanide	$Na_4[Fe(CN)_6]$	E 535
Kalium ferrocyanide	$K_4[Fe(CN)_6]$	E 536
Calcium ferrocyanide	$Ca_2[Fe(CN)_6]$	E 538
Silicium dioxide	SiO_2	E 551
Calcium silicate	$CaSiO_3$	E 552
Magnesium silicate, Magnesium trisilicate	$MgSiO_3$	E 553a
Talcum	$Mg_3Si_4O_{10}(OH)_2$	E 553b
Natrium aluminium silicate	$MgAl(SiO_3)_2$	E 554
Kalium aluminium silicate	$K_2Al(SiO_3)_2$	E 555
Calcium aluminium silicate	$CaAl(SiO_3)_2$	E 556
Aluminium silicate (Kaolin)	$Al_2(SiO_3)_3$	E 559

6.5.2 Fortification and enrichment

Fortification of foods with minerals is intended to improve human health. Supplementation of the conventional nutrition has been applied for quite a long time, especially with iron. For example, various governments have introduced state-sponsored programs to reduce iron deficiency anemia in their population long ago. In the United States, for example, fortification of the food supply began in 1924 with the addition of iodine to salt to prevent goiter, a prevalent public health problem in the United States at the time. In the early 1940s, food fortification was expanded further, when it became apparent that many young adults were failing Army physical exams due to poor nutritional status. In 1943, the government issued an order making it mandatory to enrich flour with iron (along with riboflavin, thiamine, and niacin) [17].

Since the introduction of fortification back in the 1920s, the prevalence of many nutrient deficiency diseases in the United States has declined dramatically, including iron, iodine, niacin and vitamin D deficiencies. In the United States today, most foods containing refined cereal grains (e.g. white flour, white rice and corn meal) are enriched with iron, niacin (nicotinic acid), riboflavin, thiamine and folic acid [17].

6.5.2.1 Iron

Despite the abundance of iron in the environment, iron deficiency in humans is widespread due to low concentrations in crops and some farm animals and thus in the human diet. The WHO estimates that iron deficiency is the most prevalent nutritional disorder in the world [18]. Iron deficiency is a major cause of anemia, which is characterized by low red blood cell counts and low blood hemoglobin concentrations. Due to the importance of an adequate iron supply, EFSA proposed dietary reference values of 11 mg/day [19].

The risk of systemic iron overload from dietary sources is mostly negligible with normal intestinal function. Chronic iron overload may occur in individuals affected by hemolytic anemias, hemoglobinopathies or one of the hemochromatosis and results in increasing deposition of iron in ferritin and hemosiderin in all tissues throughout the body [19].

Iron plays many key roles in biological systems, like oxygen transport (hemoglobin and myoglobin), respiration and energy metabolism (e.g. cytochromes) or depletion of hydrogen peroxide (hydrogen peroxidase and catalase). Many of the aforementioned proteins contain heme, a complex of iron with porphyrin (Figure 6.5.2). The involvement of iron in many of these metabolic reactions depends on its ability to readily accept or donate an electron, that is, to easily undergo a redox cycle between the Fe^{2+} and Fe^{3+} forms.

The bioavailability of heme iron is relatively unaffected by composition of the diet and is significantly higher than that of non-heme iron. The bioavailability of non-heme iron varies markedly depending on the composition of the diet [20].

Addition of iron to foods is a difficult balancing act because some forms of iron catalyze oxidation of unsaturated fatty acids and vitamins A, C and E [21]. These oxidation reactions and other interactions of the added iron with food components may produce undesirable changes in color, odor, taste or even degradation of vitamins. In many cases, forms that are highly bioavailable are also the most active catalytically, and forms that are relatively chemically inert tend to have poor bioavailability, making balancing even more difficult. In general, the more water soluble the iron compound, the higher its bioavailability and the greater the tendency to adversely affect sensory properties of foods. To address these issues, various iron species are offered for use in food supplements like

Fe(II)-bisglycinate ($C_4H_8FeN_2O_4$),
Fe(II)-carbonate ($FeCO_3$, protected from oxidation by admixture with sugar),
Fe(II)-citrate ($C_6H_6FeO_7$),
Fe(II)-fumarate ($C_4H_2FeO_4$),
Fe(II)-gluconate (as Fe(II)-gluconate-2-hydrate, $C_{12}H_{22}FeO_{14} \times 2H_2O$),
Fe(II)-lactate (as Fe(II)-lactate-2-hydrate, $C_6H_{10}FeO_6 \times 2H_2O$ or Fe(II)-lactate-3-hydrate $C_6H_{10}FeO_6 \times 3H_2O$),
Fe(II)-phosphate ($Fe_3(PO_4)_2$),
Fe(III)-pyrophosphate (Fe(III)-diphosphate, $Fe_4(P_2O_7)_3 \times H_2O$),
Fe(III)-saccharate ($C_{12}H_{22}O_{11}Fe$),
Fe(II)-sulfate (Fe(II)-sulfate-7-hydrate, $FeSO_4 \times 7H_2O$) or
Fe(III)-sulfate (Monsel salt, $Fe_4(SO_4)_5(OH)_2 \times 10H_2O$) [22].

6.5.2.2 Iodine

Iodine (see also Chapter 13.1.3) is an essential nutrient required for the synthesis of the thyroid hormones. These hormones, thyroxine (3,4,3′,5′-tetraiodothyronine, designated as T4) and 3,5,3′-triiodothyronine (T3), have multiple functions in the body [23]. They influence neuronal cell growth, physical and mental development in children and basal metabolic rate. Goiter is the most widely known iodine deficiency disorder, but many other disorders may result from iodine deficiency including decreased fertility, increased rates of perinatal mortality, growth retardation in children and impaired mental development. Iodine deficiency occurs mainly in regions where soil iodine content is low due to leaching caused by melting glaciers, heavy rainfall and flooding, which is favored by the high solubility of iodine salts. There are only low concentrations of iodine in most foods. Good sources are milk, eggs and, above all, seafood. Drinking water contributes little to the body's iodine supply [24].

Several countries with iodine-deficient areas employ prophylactic measures by iodization of common salt with potassium iodate (KIO_3), potassium iodide (KI), sodium iodate ($NaIO_3$) or sodium iodide (NaI). In the United States, a program for iodization of salt was started in 1924. Despite the relatively simple process for adding iodine to salt and the success of the program in the United States and other developed economies, salt iodization was not common in many emerging economies. Therefore, in 1993, the WHO adopted an intervention strategy called Universal Salt Iodization to tackle the problem [25].

In some regions and for some life stages in Germany and Austria, an insufficient iodine intake is prevalent. Considering influencing factors such as iodine content in food or drinking water, it was decided to retain the previous recommended value of 200 µg/day for adults below 51 years of age and 180 µg/day from 51 years onwards [26].

However, also maximum amounts in food supplements are recommended, as higher amounts of iodine are toxic and, as shown with rats, disturb the normal reproduction and lactation of the animals and can cause diseases of the thyroid gland in humans [27].

6.5.2.3 Zinc

Given the apparently widespread occurrence of marginal zinc deficiency, zinc fortification of foods is advocated as a strategy for addressing this problem. In the United States, five zinc compounds are listed as "GRAS": zinc sulfate ($ZnSO_4$), zinc chloride ($ZnCl_2$), zinc gluconate, zinc oxide (ZnO) and zinc stearate [28]. ZnO is mainly used for food fortification. It is more stable in foods, in part due to its lower solubility. In the same study, a fortification level of 20–50 mg Zn/kg is recommended in Mexiko for corn flour. Following a request from the European Commission, the Panel on Dietetic Products, Nutrition and Allergies (NDA) derived Dietary Reference Values for zinc for the EU, considering the inhibitory effect of dietary phytate on zinc absorption [29]. Most dietary zinc is absorbed in the upper small intestine. The luminal contents of the duodenum and jejunum, notably phytate, can have a major impact on the percentage of zinc that is available for absorption. Therefore, estimated Population Reference Intakes for zinc are provided for phytate intake levels of 300, 600, 900 and 1200 mg/day, which cover the range of mean/median phytate intakes observed in European populations. Dietary Reference Values for zinc range from 7.5 to 12.7 mg/day for women and from 9.4 to 16.3 mg/day for men [29].

6.5.2.4 Fluorine

Fluoride performs no essential function in human growth and development, and no signs of fluoride deficiency have been identified. Although fluoride is not essential for tooth development, its role in the prevention of dental caries has been known for many years. Epidemiological studies have shown an inverse correlation between the presence of fluoride in drinking water and the prevalence of dental caries in children [30]. The positive effect of fluorine on dental caries is well established. The addition to drinking water of 0.5–1.5 ppm fluorine in the form of NaF or $(NH_4)_2SiF_6$ inhibits tooth decay and is applied in some countries. Its beneficial effect appears to be in retarding solubilization of tooth enamel and inhibiting the enzymes involved in the development of caries. The level of fluoride in drinking water must be closely monitored as too much fluoride can be harmful. The World Health Organization recommends a maximum level of 1.5 mg/L [31]. Therefore, the beneficial effect of fluoridating drinking water is a controversial topic of mineral supplementation. In Germany, fluoridating of drinking water is not conducted. As mentioned earlier in Chapter 6.5.1.1, cooking salts with additional fluoride are also available.

6.5.3 Mineral salts in food supplements

Nutritional supplements play an increasingly important role in the modern diet. The EU Directive No. 2002/46/EC defines food supplements as "foodstuffs the purpose of which is to supplement the normal diet and which are concentrated sources of nutrients or other substances with a nutritional or physiological effect, alone or in combination, marketed in dose form, such as pills, tablets, capsules, liquids in measured doses, etc.". A wide range of nutrients and other ingredients might be present in food supplements, including but not limited to minerals, vitamins, amino acids, essential fatty acids, fiber and various plant and herbal extracts.

Food supplements are intended to correct nutritional deficiencies, maintain an adequate intake of certain nutrients or to support specific physiological functions. They are not medicinal products and as such cannot exert any pharmacological, immunological or metabolic action. Therefore, their use is not intended to treat or prevent diseases in humans or to modify physiological functions.

Food supplements are generally available over the counter and are advertised to the consumer accordingly. However, advertising claims regarding effectiveness and health benefits are subject to strict constraints in the EU. The EFSA is responsible for evaluating the scientific proof of such claims. Regulation (EU) No. 1924/2006 sets out the requirements for health claims in detail [32].

In the EU, food supplements are regulated as foods. Harmonized legislation regulates the vitamins and minerals and the substances used as their sources, which can be used in the manufacturing of food supplements. For ingredients other than vitamins and minerals, the European Commission has established harmonized rules to protect consumers against potential health risks and maintains a list of substances which are known or suspected to have adverse effects on health and the use of which is therefore controlled [33].

Several minerals are approved according to the EU Nutrition and Health Claims Regulation [34]. A public EU Register of Nutrition and Health Claims lists all permitted nutrition claims and their conditions of use, as well as authorized health claims, their conditions of use and applicable restrictions [35]. The following selection again shows the importance of inorganic salts in human nutrition and for human health:

- Blood formation, blood clotting, blood function (Fe, Ca)
- Blood pressure (K)
- Bone health (Ca, Mg, Mn, P, Zn, F)
- Glucose metabolism (Cr)
- Healthy hair (Cu, Se, Zn, I)
- Healthy skin (Cu, I, Zn)
- Immune system (Cu, Fe, Se, Zn)
- Nervous system/mental & cognitive health (Mn, Ca, Mg, I, Fe, Zn, Se)
- Maintenance of tooth mineralization (F)

- Muscle function (Ca, Mg, K)
- Sexual health/male hormone balance (Zn)
- Thyroid function (I)

As already stated above at example of Fe (*vide ultra*), the bioavailability of these elements depends on the chemical species and therefore must be considered when choosing the administration form.

References

[1] U.S. Food and Drug Administration, 2022. (Accessed January 4, 2022, at https://www.fda.gov/food/food-ingredients-packaging/generally-recognized-safe-gras)
[2] Joint FAO/WHO Expert Committee on Food Additives (JECFA), (2022. Accessed January 4, 2022, at https://www.who.int/groups/joint-fao-who-expert-committee-on-food-additives-(jecfa)/about)
[3] Codex General Standard for Food Additives (GSFA) Online Database, Updated up to the 42nd Session of the Codex Alimentarius Commission (2019). (Accessed January 4, 2022, at https://www.fao.org/gsfaonline/additives/index.html)
[4] Commission Regulation (EU) No 1129/2011 of 11 November 2011 amending Annex II to Regulation (EC) No 1333/2008 of the European Parliament and of the Council by establishing a Union list of food additives. (Accessed January 4, 2022, at http://data.europa.eu/eli/reg/2011/1129/oj)
[5] European Food Safety Authority, 2022. (Accessed January 5, 2022, at https://www.efsa.europa.eu/en/topics/topic/food-additives)
[6] Israr T, Rakha A, Sohail M, Rashid S, Shehzad A. Salt reduction in baked products: Strategies and constraints. Trends Food Sci Technol, 2016, 51, 98–105.
[7] European food safety authority: Scientific opinion on dietary reference values for fluoride. EFSA J, 2013, 11, 3332.
[8] Frede W. Handbuch für Lebensmittelchemiker, Lebensmittel – Bedarfsgegenstände – Kosmetika – Futtermittel, 3rd revised edition, Springer, Berlin, Heidelberg, New York, 2010.
[9] European food safety authority: Risk assessment of nitrate and nitrite in feed. EFSA J, 2020, 18, 6290.
[10] Liao M-L, Seib P. Selected reactions of L-ascorbic acid related to foods. Food Technol, 1987, 31, 104–07.
[11] Suman SP, Joseph P. Myoglobin chemistry and meat color. Annu Rev Food Sci Technol, 2013, 4, 79–99.
[12] Vetter JL. Leavening Agents, Encyclopedia of Food Sciences and Nutrition, 2nd edition, Elsevier Science, 2003, 3485–90.
[13] Kitagawa I. Licorice root. A natural sweetener and an important ingredient in Chinese medicine. Pure Appl Chem, 2002, 74, 1189–98.
[14] Ramis-Ramos G. Synthetic antioxidants. In: Encyclopedia of Food Sciences and Nutrition, 2nd edition, Elsevier, Amsterdam, 2003, 265–75.
[15] Belitz H-D, Grosch W, Schieberle P. Food Chemistry, 4th revised edition, Springer, Berlin, Heidelberg, New York, 2008, 453–54.
[16] Belitz H-D, Grosch W, Schieberle P. Food Chemistry, 4th revised edition, Springer, Berlin, Heidelberg, New York, 2008, 464.

[17] Carpenter KJ. Episodes in the history of food fortification. Cereal Foods World, 1995, 42, 54–57.
[18] World Health Organization (2022). Micronutrient deficiencies. (Accessed January 4, 2022, at https://www.who.int/health-topics/anaemia)]
[19] European food safety authority: Scientific opinion on dietary reference values for iron. EFSA J, 2015, 13, 4254.
[20] Blanco-Rojo R, Pilar Vaquero M. Iron bioavailability from food fortification to precision nutrition. A review. Innov Food Sci Emerg Technol, 2019, 51, 126–38.
[21] Miller DD. Iron fortification of the food supply: A balancing act between bioavailability and iron-catalyzed oxidation reactions. In: Lyons TP, Jacques KA. (Eds.) Nutritional Biotechnology in the Feed and Food Industries, Nottingham University Press, Nottingham, England, 2002.
[22] Homepage of Dr Paul Lohmann – high value mineral salts. (accessed January 18, 2022, at https://www.lohmann4minerals.com/index.php/fe.html)
[23] Elmadfa I, Leitzmann C. Ernährung des Menschen, 6th edition, Ulmer Verlag, Stuttgart, 2019.
[24] Delange F, de Benoist B, Pretell E, Dunn JT. Iodine deficiency in the world: Where do we stand at the turn of the century? Thyroid, 2001, 1, 437–47.
[25] World Health Organization (2022). Micronutrient deficiencies. (Accessed January 10, 2022, at https://www.who.int/elena/titles/salt_iodization/en/)
[26] Deutsche Gesellschaft Für Ernährung, Österreichische Gesellschaft für Ernährung, Schweizerische Gesellschaft für Ernährungsforschung, Schweizerische Vereinigung für Ernährung), Referenzwerte für Die Nährstoffzufuhr, Neuer Umschau Buchverlag, Frankfurt/Main,Germany, 2013.
[27] European food safety authority: Scientific opinion on dietary reference values for iodine. EFSA J, 2014, 12, 3660.
[28] Rosado JL. Zinc and copper: Proposed fortification levels and recommended zinc compounds. J Nutr, 2003, 133, 2985S–9S.
[29] European food safety authority: Scientific opinion on dietary reference values for zinc. EFSA J, 2014, 12, 3844.
[30] European food safety authority: Scientific opinion on dietary reference values for fluoride. EFSA J, 2013, 11, 3332.
[31] NHS Digital (2015) Child Dental Health Survey 2013, England, Wales and Northern Ireland – Executive Summary. (Accessed January 10, 2022, at https://digital.nhs.uk/data-and-information/publications/statistical/children-s-dental-health-survey/child-dental-health-survey-2013-england-wales-and-northern-ireland
[32] European Commission – Nutrition and Health Claims. (Accessed January 10, 2022, at https://ec.europa.eu/food/safety/labelling-and-nutrition/nutrition-and-health-claims_en
[33] European Food Safety Authority: Food supplements. (Accessed January 10, 2022, at https://www.efsa.europa.eu/en/topics/topic/food-supplements)
[34] Regulation (EC) No 1924/2006 of the European Parliament and of the Council of 20 December 2006 on nutrition and health claims made on foods. (Accessed January 10, 2022, at https://ec.europa.eu/food/safety/labelling-and-nutrition/nutrition-and-health-claims_en)
[35] Public EU Register of Nutrition and Health Claims. (Accessed February 8, 2022, at https://ec.europa.eu/food/safety/labelling_nutrition/claims/register/public/?event=register.home)

6.6 Fertilizers

Robert Glaum

For soil, a typical chemical composition (mass%; without organic matter) of SiO_2 57.6, Al_2O_3 15.3, Fe_2O_3 2.5, FeO 4.3, MgO 3.9, CaO 7.0, Na_2O 2.9, K_2O 2.3, TiO_2 0.8, CO_2 1.4, H_2O 1.4, MnO 0.16, P_2O_5 0.22 can be assumed [1]. Thus, various types of silicates form the mineral basis of soil. Among those, layered (alumino-)silicates (clay) are the most important for plant growth, due to their ion exchange and water retention properties. As a fertilizer, any material, organic or inorganic, natural or synthetic, is regarded, and is placed on or incorporated into the soil to supply plants with one or more of the chemical elements necessary for normal growth. For many centuries, manure was the most common type of fertilizer, and in nineteenth-century literature "manure" was used to designate any fertilizer [2, 3].

6.6.1 History

Lavoisier (1743–1794) realized that malnutrition of livestock is related to limited nutrient supply for plants on infertile soil. He identified a shortage of plant-available nitrogen and suggested as action against this deficit crop rotation including leguminous plants (e.g. beans, soybeans, peas, chickpeas, peanuts, lentils, lupins, clover) [4, 5]. These were left on the fields to decay, thus forming humus and replenishing nitrogen in the soil. Nowadays, it is well established that most of these plants have symbiotic nitrogen-fixing bacteria in structures called root nodules [6]. In the early-nineteenth century, Thaer (1752–1828) formulated his humus theory explicitly [7], arguing "plants can only feed on organic nutrients/decaying plants". This theory stated that plants lived on humus-derived extracts containing simple water-soluble compounds of C, H, O and N, from which they were able to rebuild more complex plant tissue. By Thaer, some fertilizer substances like salts and lime were considered useful for plant growth, but only because they promoted the decomposition of humus and the dissolution of organic matter in the soil solution. Eventually, humus theory

---

Note: In preparation of this Section (6.6) consultation and cross-referencing of three treatises on the subject is particularly acknowledged: a) Nielsson FT. Manual of Fertilizer Processing, Vol 5, In: Fertilizer Science and Technology, Taylor & Francis, Boca Raton, 2018, b) Lee RG, Kopytowski JA. (Eds), Fertilizer Manual by United Nations Industrial Development Organization and International Fertilizer Development Center, 3rd ed, Kluwer Academic Publishers, Dordrecht, 1998, c) Amelung W, Blume H-P, Fleige H, Horn R, Kandeler E, Kögel-Knabner I, Kretzschmar R, Stahr K, Wilke B-M. with contributions by Gaiser T, Gauer J, Stoppe N, Thiele-Bruhn S, Welp G. Scheffer/Schachtschabel: Lehrbuch der Bodenkunde, 17th ed., Springer Spektrum, Berlin, 1994. Several Figures have been taken from "Informationsserie Pflanzenernährung: Ernährung – Wachstum – Ernte", Fonds der Chemischen Industrie im Verband der Chemischen Industrie e. V. (FCI), Frankfurt, 2013.

https://doi.org/10.1515/9783110798890-014

was in some parts refuted and in others further developed by THAERS' student and later assistant SPRENGEL (1787–1859), who realized that the mineral (inorganic) nutrients including nitrogen (as NH_4^+ or NO_3^-), phosphorus ($H_2PO_4^-$, HPO_4^{2-}) and potassium (K^+) are essential for plant growth and that the least available nutrient will limit plant growth (series of papers 1826–1829 [8, 9]). This first formulation by SPRENGEL of the law of the minimum (German: Minimumgesetz; Figure 6.6.1a) was some ten years later adopted and scientifically promoted by LIEBIG (1803–1873). He systematized all available knowledge on fertilizers and extended it by the results of his own experiments and ideas. LIEBIG published his conclusions in the groundbreaking treatise "Die organische Chemie in ihrer Anwendung auf Agricultur und Physiologie/Application of organic chemistry in agriculture and physiology" [10]. The insights by SPRENGEL and LIEBIG can be summarized as follows:

- Plant growth requires, in addition to CO_2 and water, further nutrients, in particular nitrogen, phosphorus and potassium (macronutrients). Calcium, magnesium and sulfur are regarded as secondary macronutrients, since they are essential, however required in lower amounts than N, P and K. These are taken up as water-soluble inorganic ("mineral") substances via the roots.
- Plant growth is limited by the least available nutrient (LIEBIG'S law of the minimum [11], see Figure 6.6.1a).
- Nutrient loss by crop harvest has to be replaced to maintain soil fertility.

Eventually, in the early twentieth century, MITSCHERLICH formulated the law of the decreasing soil fertility as an expansion to the law of the minimum. MITSCHERLICH found that for each plant nutrient in soil an optimum amount exists, beyond which crop production will actually decrease (Figure 6.6.1b). For a concise summary of the historical development in soil science and the use of mineral fertilizers, the reader is referred to an excellent review [12].

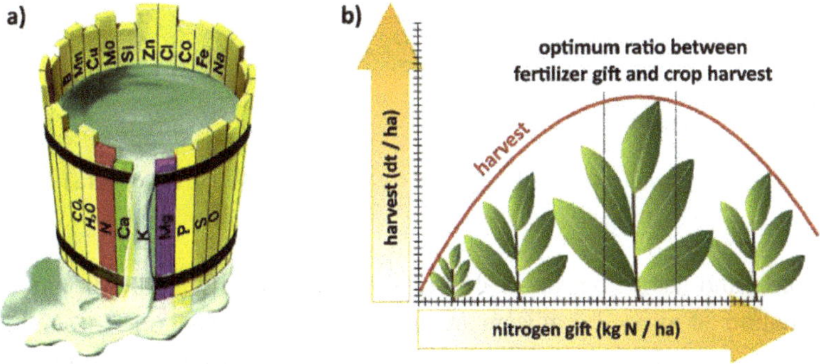

Figure 6.6.1: a) Sprengels/Liebigs law of the minimum visualized by a barrel with staves of different length (source [13]). b) Mitscherlichs law of optimum fertilizer application (Figures adapted from [13]).

Beginning in mid-nineteenth century, the limiting-factor paradigm ("law of the minimum"), which is frequently visualized by a barrel with staves of different length (Figure 6.6.1a) led to an increasing demand for water-soluble inorganic compounds of nitrogen, phosphorus and potassium. While an abundance of potassium containing minerals (see Section 6.6.2.4) was and still is available for fertilizer production, mineral deposits containing phosphorus ("phosphate rock" see Chapter 6.2 on "Phosphates") are in general badly water-soluble and require chemical digestion (see Sections 6.2.2 and 6.6.3.2) prior to their use as fertilizer. Eventually, in the late-nineteenth century, the biggest problem was encountered in providing sufficient amounts of chemically reactive (plant available) compounds of nitrogen. Though the beneficial use of guano and Chile saltpeter (hygroscopic, deliquescent $NaNO_3$) as fertilizer was well established, its supply was rather limited and costly. All this prohibited widespread use of mineral fertilizers at that time, and despite the discovery of some revolutionary industrial processes that solved these problems, it was not until the 1950s that global use of mineral fertilizer became economically reasonable. Figure 6.6.2 shows global mineral fertilizer production from 1900 to 2022. The diagram demonstrates impressively how global population growth during the last 120 years from about 1.6 billion people in 1900 to close to 8 billion in 2022 is directly related to increased fertilizer production. In fact, it is estimated that synthetic nitrogen fertilizer today secures the nutrition basis of approximately half of the global population [14–16].

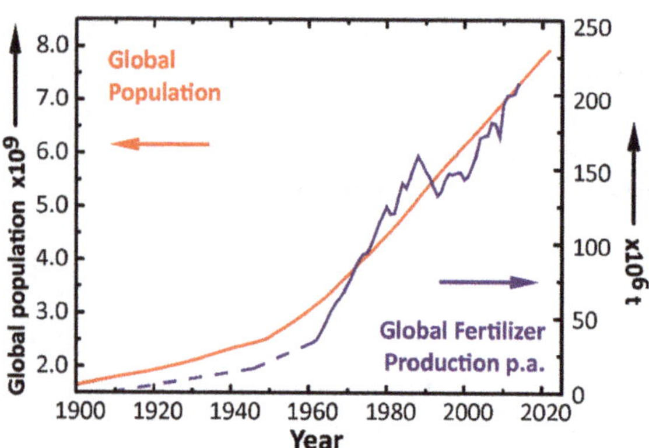

Figure 6.6.2: Global population and total (N + P + K) fertilizer production 1900 to 2020 (data from [17, 18]).

The main chemical and technological breakthroughs that eventually allowed production of huge amounts of mineral fertilizers are summarized in Table 6.6.1. These processes all represent historic landmarks in the development of industrial inorganic chemistry. Yet, none of these had a more dramatic impact than the ability to produce synthetic nitrogen fertilizer [15, 16]. Figure 6.6.3 provides a summary on global fertilizer production and consumption from 1961 to 2020 distinguished by geographic region.

Table 6.6.1: Scientific and technological achievements as the basis for the development of industrial fertilizer production.

Year	Process	Product; fertilizer relevance	Ref.
1861	Solvay	Soda Na_2CO_3; for potassium nitrate	[19]
1875 /1927	Contact	Sulfur trioxide SO_3; sulfuric acid for digestion of phosphate rock	[20]
1877	Linde	Air liquification/distillation; di-nitrogen for NH_3	[21]
1895	Frank-Caro	Nitrolime (Kalkstickstoff) $Ca(CN_2)$	[22]
1902	Ostwald	Nitric oxide NO_2; for nitric acid	[23]
1908 /1913	Haber-Bosch	Ammonia NH_3; for ammonium salts, nitric acid and urea	[24]
1909	Birkeland-Eyde	Nitric oxide NO_2; for nitric acid	[25]
1922	Bosch-Meiser	Urea	[26]
1933	Steam-reforming	Hydrogen; for ammonia synthesis	[27]

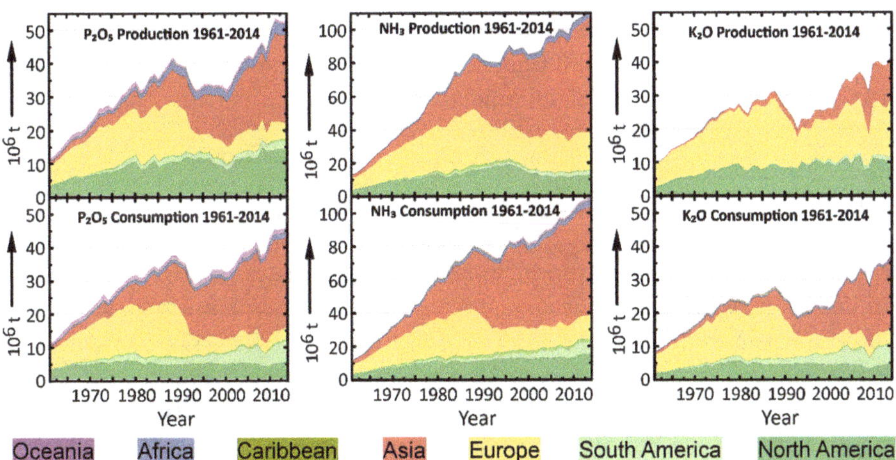

Figure 6.6.3: World fertilizer production and consumption (data taken from [28, 29]).

Several facts on mineral fertilizer production/consumption are visualized by Figure 6.6.3 and should be particularly emphasized:

- Total (N + P + K) global fertilizer production had reached 250 million tons in the year 2020. This is comparable to global sulfuric acid production, but only about 1/20 of crude oil production.
- Economic reforms during the 1990s in Eastern Europe and Eurasia had a tremendous effect on fertilizer use in these regions. During a short period of time, fertilizer use decreased by 30%–70% in many countries.
- Debt crises, foreign exchange shortages and balance-of-payment difficulties led to restricted fertilizer supplies and therefore decreased fertilizer use in many developing countries, especially in Africa and Latin America.
- Policy reforms introduced under structural adjustment programs such as the devaluation of domestic currency, subsidy removal and privatization (sudden withdrawal of governmental organizations) have had a negative impact on fertilizer use in several developing countries.
- Depressed crop prices resulting from grain surpluses and acreage reduction programs in the developed markets have contributed to a decline or stagnation in fertilizer use [3].

Among the world's largest fertilizer producing companies in the year 2021 were Nutrien (North America, merge of Agrium and Potash Corp.), The Mosaic Company (North America), Yara International (former Norsk Hydro; Norway), Israel Chemicals ICL (Israel), Uralkali PJSC (Russia), BASF (Germany), CF Industries (U.S.A), K&S (Germany) and SAFCO (Saudi Arabia) [30].

6.6.2 Plant nutrients (macro)

The nutrients taken up by plants with by far the highest masses are CO_2 and H_2O. These are followed by the primary macronutrients nitrogen, phosphorus and potassium, which also constitute the major components of fertilizers.[1] The secondary macronutrients calcium, magnesium and sulfur are essential for plant growth, too, which means they cannot be substituted by other nutrients. However, the natural supply for these elements in soil, e.g. due to weathering processes, was sufficient to ensure plant growth (Figure 6.6.4). With intensified farming and cultivation of agricultural crops of special nutritional requirements these were to be replenished by mineral fertilizer gifts.

1 Fertilizer grade is used to classify different fertilizer materials on the basis of the content of the three major nutrients. The nutrient content is expressed traditionally for nitrogen in elemental form (N) and for most other nutrients in oxide form (e.g. P_2O_5, K_2O). Thus, a fertilizer grade of NPK 7-28-14 contains 7 wt% N, 28 wt% P_2O_5 and 14 wt% K_2O. Note: Even for potassium chloride containing fertilizers the equivalent content of K_2O is given (e.g. pure KCl with 52.4 wt% K and 47.6 wt% Cl is represented by the grade 0-0-60).

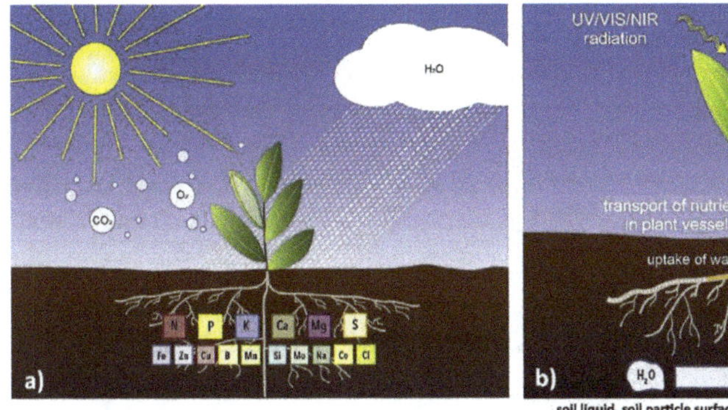

Figure 6.6.4: Plant macro and micronutrients (a) and scheme visualizing nutrient uptake (Figures adapted from [13]).

6.6.2.1 Carbon, hydrogen and oxygen

Carbon, hydrogen and oxygen are provided by the CO_2 from air (41 ppm CO_2 on average in air in the year 2021 [31]) and water. Plants transform CO_2 and water via photosynthesis into oxygen and carbohydrates "$C_n(H_2O)_n$" (e.g. sugars like glucose and cellulose; reaction 6.6.1). The energy for this highly endergonic process, which formally can be compared to splitting of carbon dioxide into the elements, is mainly taken from solar radiation. Figure 6.6.5 visualizes the carbon cycle in nature. Carbohydrates are used by plants for energy storage and tissue formation. Carbon dioxide and water are not regarded as fertilizers, yet an interesting aspect lies in the effect of higher CO_2 supplies, in particular in the context of rising average CO_2 levels. Recent research suggests that this will indeed increase biomass formation. However, this will cause lower nutrient content [32].

$$\mathrm{nCO_2}(g) + \mathrm{nH_2O}(l) \rightarrow \text{“}\mathrm{C_n(H_2O)_n}\text{”}(s) + \mathrm{nO_2}(g) \qquad (6.6.1)$$

6.6.2.2 Nitrogen

Nitrogen is the critical limiting element for crop production and plant growth in general. It is a major component of chlorophyll, the unique pigment needed for photosynthesis. Nitrogen is also required for synthesis of amino acids, the key building blocks of proteins, and other important biomolecules such as ATP and nucleic acids. Even though nitrogen is one of the most abundant elements (predominantly in the form of di-nitrogen gas N_2 in the Earth's atmosphere), plants can only utilize chemically activated forms of this element, mainly as ammonium (NH_4^+) or nitrate (NO_3^-). Plants acquire "active" or "combined" nitrogen by: 1) the addition of ammonia and/or nitrate fertilizer or manure to soil, 2) the release of these compounds during organic matter

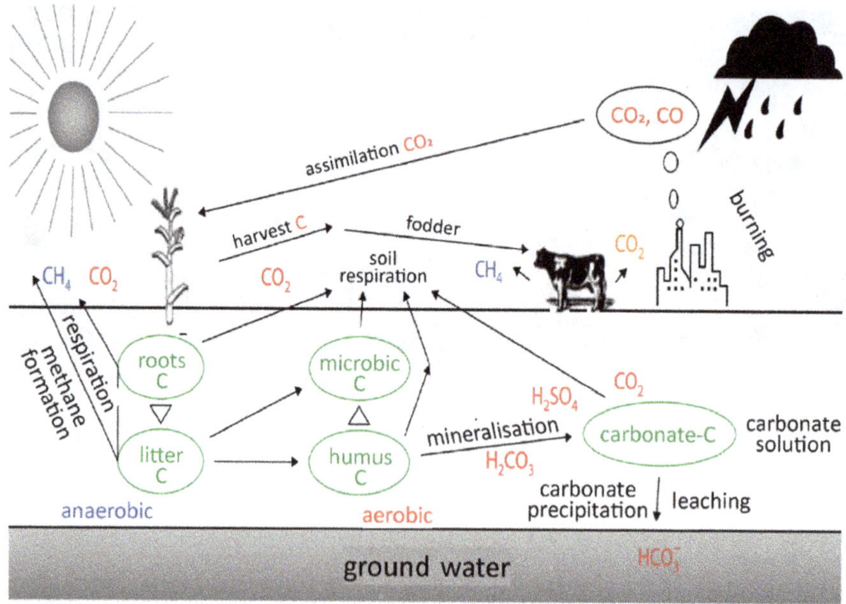

Figure 6.6.5: Visualization of the carbon cycle (graphics adapted from [1]).

decomposition, 3) the conversion of atmospheric nitrogen into the compounds by natural processes, such as lightning, and 4) biological nitrogen fixation by bacteria in the root nodules of leguminous plants [6, 33]. Due to their generally high reactivity, volatility (N_2, NH_3, NO_x) or solubility (NO_3^-), nitrogen compounds in the soil are in rather short supply, in contrast to other nutrients. Therefore, introduction in mid-nineteenth century of mineral ("industrial") nitrogen fertilizers immediately proofed beneficial for crop yield. At that time, Chilean saltpeter ($NaNO_3$) and ammonia liquor (dil. ammonia solution) as well as ammonium sulfate as by-product from coking coal were the only available N-fertilizers and of high cost. The chemical and technological basis for access to cheaper N-fertilizers was laid in the late-nineteenth and early twentieth centuries by a series of inventions (see Table 6.6.1).

Calcium cyanamide ($CaCN_2$, nitrolime, "Kalkstickstoff") was the first industrial N-fertilizer. Its synthesis (see eqs. 6.6.2a and b) by the Frank-Caro process (1895) [22] requires limestone ($CaCO_3$) and nitrogen from the Linde process [21, 34] (see also Section 4.4) and huge amounts of electric energy [35]. This is the reason why historically nitrolime production was restricted to sites with ample supply of (hydro-)energy (e.g. Norway, Tennessee Valley in the U.S.A., Bavaria). Nitrolime shows some peculiar fertilizer properties. It serves as a source for N **and** Ca, rises the pH of soil (in contrast to most ammonium salts), shows delayed nitrogen release due to stepwise hydrolysis and nitrification, and eventually improves soil fertility. The latter follows from bactericide and fungicide behavior of some of the intermediates during hydrolysis (e.g.

cyanamide CN-NH_2, its tautomer carbodiimide HN=C=NH, and its dimer dicyandiamide DCD). Due to these intermediates, which are also known to cause alcohol intolerance [36], careful handling of nitrolime is mandatory to avoid health risk.

$$CaO(s)+3C(s) \xrightleftharpoons{>2200\,°C,\text{ electric arc}} CaC_2(s)+CO(g)\ (\Delta_r H_{298}= +461\,kJ/mol) \quad (6.6.2a)$$

$$CaC_2(s)+N_2(g) \xrightleftharpoons{1100\,°C} Ca(CN_2)(s)+C(s)\ (\Delta_r H_{298}= -286\,kJ/mol) \quad (6.6.2b)$$

The **Birkeland-Eyde process** ("combustion of air" [25, 37]), introduced in 1909, was another energy-intensive way to activate di-nitrogen from air. In the three reaction steps (eqs. 6.6.3a to 6.6.3d) nitric acid is formed. The first step is highly endothermic and carried out at 3300 °C, at best giving a yield of 5%. The nitric acid is reacted with limestone to yield calcium nitrate. The hygroscopic nitrate crystallizes from water as tetra-hydrate $Ca(NO_3)_2 \cdot 4\,H_2O$, which for fertilizer applications is converted into double salts with KNO_3 or NH_4NO_3 (e.g. $5Ca(NO_3)_2{\cdot}NH_4NO_3{\cdot}10H_2O$)

$$N_2(g)+O_2(g) \xrightleftharpoons{>3300\,°C,\text{ electric arc}} 2NO(g)\quad (\Delta_r H_{298} = +178.5\,kJ/mol) \quad (6.6.3a)$$

$$2NO(g) + O_2(g) \xrightarrow{>50\,°C} 2NO_2(g)\quad (\Delta_r H_{298} = -114\ kJ/mol) \quad (6.6.3b)$$

$$2NO_2(g) \longrightarrow N_2O_4(g)\quad (\Delta_r H_{298} = -57\ kJ/mol) \quad (6.6.3c)$$

$$4\,NO_2(g) + 2\,H_2O(l) + O_2(g) \longrightarrow 4\,HNO_3(aq) \quad (6.6.3d)$$

Due to the high-energy consumption, chemical nitrogen fixation as $Ca(CN_2)$ and calcium nitrate was only economically viable where and when electric energy was available at low cost. Thus, even as late as 1950, many agriculturists advocated that principal reliance for nitrogen supplies should be placed on leguminous plants grown in rotation with other crops instead of mineral nitrogen fertilizer [3]. Only with the widespread availability of cheap hydrogen from steam-reforming [27] (see Chapter 4.1), oxygen-free nitrogen (Linde process [21], see Chapter 4.4), the optimization of the Haber-Bosch process (eq. 6.6.4, see Chapter 4.6 for details and [15, 24, 38] for further reading) and improved nitric acid production by combustion of ammonia (Ostwald process [23, 39], eq. (6.6.5) followed by reactions 6.6.3b–d; see Chapter 4.7 "nitrogen oxides" for details) various inorganic nitrogen fertilizers became affordable. The importance of the **Haber-Bosch process** to mankind has already been emphasized and is underscored by the fact that over a period of 90 years three Nobel prizes have been awarded with close relation to the synthesis of ammonia: FRITZ HABER (1918, ammonia synthesis from the elements), CARL BOSCH (1931, mastering high-pressure techniques [40]) and GERHARD ERTL (2007, mechanistic understanding [38]).

$$N_2(g)+3H_2(g) \underset{}{\overset{500\,°C,\,300\,bar,\ Fe\ catalyst}{\rightleftharpoons}} 2NH_3(g)\ \ (\Delta_r H_{298} = -92.28\,kJ/mol) \tag{6.6.4}$$

$$4\,NH_3(g) + 5\,O_2(g) \xrightarrow{800-900\,°C,\ Pt/Pt-Rh\ catalyst} 4\,NO(g) + 6\,H_2O(g)\ \ (\Delta_r H_{298} = -906\,kJ/mol) \tag{6.6.5a}$$

$$2\,NO(g) + O_2(g) \xrightarrow{<50\,°C} 2\,NO_2(g)\ \ (\Delta_r H_{298} = -114\,kJ/mol) \tag{6.6.5b}$$

$$2\,NO_2(g) \longrightarrow N_2O_4(g)\ \ (\Delta_r H_{298} = -57\,kJ/mol) \tag{6.6.5c}$$

$$2\,N_2O_4(g) + 2\,H_2O(l) + O_2(g) \longrightarrow 4\,HNO_3(aq) \tag{6.6.5d}$$

The current importance of hydrocarbon and coal feedstocks for ammonia manufacture should be noted. These perform a dual role in ammonia synthesis, as sources of both hydrogen and energy and there supply creates approx. 70% of the total production costs. Thus, it is not surprising that world ammonia production has shifted toward regions with abundant, low-cost natural gas (Saudi Arabia, China, India). Providing sufficient eco-friendly hydrogen and regenerative energy will be the great challenge for future ammonia production.

Depending on plant nutrition requirements, gaseous ammonia (by direct injection into the soil) and various ammonium salts ($NH_4H_2PO_4$ MAP, $(NH_4)_2HPO_4$ DAP, NH_4NO_3, $(NH_4)_2SO_4$) are applied as N-fertilizers. Ammonium salts are frequently used as components of compound fertilizers ("Volldünger", see Section 6.6.4). Since ammonium salts lead to acidic reaction with water, compensating this effect in the soil liquor by simultaneous application of limestone is frequently encountered. Heterogeneous mixtures of NH_4NO_3 (75 wt%) and $CaCO_3$ (25 wt%) are known as **Nitrochalk** ("Kalkammonsalpeter") and are popular N- and Ca-fertilizers [41].

Apart from liquid, pressurized ammonia (N content 82 wt%) and all other aforementioned N-fertilizers (incl. $CaCN_2$) are low-analysis materials (15–21 wt% N), with only ammonium nitrate (34 wt% N) at a slightly higher grade. This is one of the reasons why urea ($OC(NH_2)_2$, 46 wt% N) has become in recent years world's leading N-fertilizer [3, 41]. The high N-content of urea makes it efficient to transport to farms and apply to fields. In addition, easy handling of solid urea and its pH-increasing effect on the soil liquor are reasons for its extended use. Urea slowly hydrolyzes in the soil to give ammonium, which is taken up by the plant. In some soils, the ammonium is oxidized by bacteria ("nitrification") to give nitrate, which is also a plant nutrient.

Urea ("Harnstoff") synthesis: Historically, WÖHLERS urea synthesis from silver cyanate and ammonium chloride, with ammonium cyanate, $NH_4(NCO)$, as intermediate, was of high importance since it showed unequivocally that for formation of an "organic" molecule no peculiar force ("vis vitalis") was required [42]. The production of urea fertilizer involves controlled reaction of $NH_3(g)$ and $CO_2(g)$ at elevated temperature and pressure. This process, as originally suggested by BASAROFF [43], was first translated into industrial manufacture in 1922 (BOSCH-MEISER urea process [26]). There are two main reactions involved in the synthesis of urea from carbon dioxide and ammonia; the fast and highly exothermic formation of ammonium carbamate (T_m = 155 °C) from carbon dioxide and ammonia, and the slow, endothermic conversion of ammonium carbamate into urea. The molten urea is formed into spheres with specialized granulation equipment or hardened into a solid prill while falling from a tower. The reactions can be represented by the following eqs. (6.6.6a) and (6.6.6b). In the presence of water ammonium bicarbonate will also form [44, 45]. Side reactions, which produce biuret (eq. 6.6.6c) and eventually melamine $C_3N_3(NH_2)_3$ (cyanamide trimer) have to be avoided by thoroughly controlling the process parameters, in particular temperature, since biuret impairs plant growth [46].

$$2NH_3(g) + CO_2(g) \xrightleftharpoons{110\,atm,\,160\,^\circ C} NH_4CO_2(NH_2)(l) \quad (\Delta_r H = -117\,\text{kJ/mol}) \tag{6.6.6a}$$

$$NH_4CO_2(NH_2)(l) \xrightarrow{160-180\,^\circ C} (NH_2)CO(s) + H_2O(l) \quad (\Delta_r H = +15.5\,\text{kJ/mol}) \tag{6.6.6b}$$

$$2\,(NH_2)_2CO(s) \xrightarrow{\Delta} (NH_2)CO(NH)CO(NH_2)(l) + NH_3(g) \tag{6.6.6c}$$

Urea transformation in the soil liquor to ammonium cyanate is very slow, while urease enzyme catalyzed hydrolysis via the unstable intermediate carbamic acid is very fast (eqs. 6.6.7a and b) [47]. NH_4^+ from urea hydrolysis is subsequently transformed to NO_3^- ("nitrification") in most soil conditions by a two-step process via hydroxylamine and nitrite (eqs. 6.6.8a to c) [48]. Nitrification reactions in soil are catalyzed by two different types of bacteria (AOB, NOB). Already urea fertilizer additives are in use to reduce urease and bacteria activities to slow the rate of ammonia production and nitrification. Under certain conditions, this can help reduce ammonia loss to the atmosphere [48]. The N-cycle in soil is visualized in Figure 6.6.6.

$$(NH_2)_2CO(aq) + H_2O(l) \xrightarrow{\substack{\text{urease}\\ \text{enzyme}}} \{HC(NH_2)O_2\} + NH_3(aq) \tag{6.6.7a}$$

$$\{HC(NH_2)O_2\} \longrightarrow CO_2(aq) + NH_3(aq) \tag{6.6.7b}$$

$$NH_4^+(aq) + 1/2\,O_2(g) \xrightarrow{AOB} \{H_2N(OH)\} + H^+(aq) \tag{6.6.8a}$$

$$\{H_2N(OH)\} + O_2(g) \xrightarrow{AOB} NO_2^-(aq) + H_2O(l) + H^+(aq) \tag{6.6.8b}$$

$$NO_2^-(aq) + 1/2\,O_2(g) \xrightarrow{NOB} NO_3^-(aq) \tag{6.6.8c}$$

6.6.2.3 Phosphorus

Phosphorus plays a vital role in photosynthesis, facilitating the capture and transfer of energy into chemical bonds. DNA and RNA are built around a backbone of P atoms, and P plays a major role in the metabolism of sugars and starches, all critical to cell division and growth processes [41]. Due to limited solubility of most

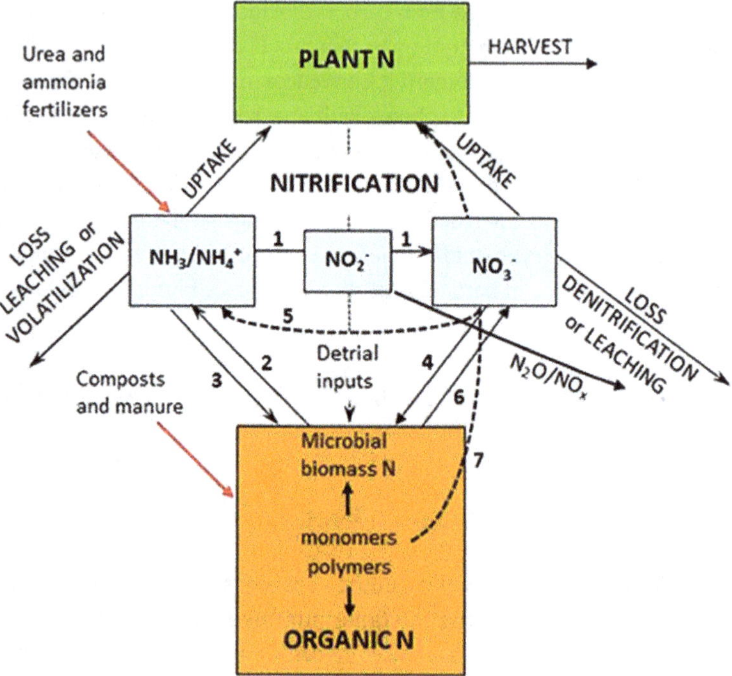

Figure 6.6.6: Visualization of the nitrogen cycle in soil (graphics from [48]). (1) nitrification (including comammox), (2) mineralization (ammonification), (3) ammonium immobilization, (4) nitrate immobilization, (5) dissimilatory NO_3^- reduction to NH_3 (DNRA), (6) heterotrophic nitrification, and (7) plant uptake of monomers.

phosphates, the concentration of inorganic phosphate ions ($H_2PO_4^-$, HPO_4^{2-}) that can be used by plants is usually low in soil liquor. Yet, readily, less-readily and very slowly available P sources are in equilibrium with the soil solution and are keeping the phosphate concentration at a low, however, constant level. Clay minerals can immediately release or adsorb phosphate. Higher phosphate concentration can lead to precipitation. Such deposits as well as decaying organic matter (humus) provide less-readily available P. Phosphate minerals in the soil, e.g. hydroxyapatite are only very slowly available to plants. Phosphate minerals taken from the soil with the crop harvest (approx. 0.3 wt% P_2O_5 in plant ash) have to be compensated by fertilizer gifts. On the other hand, the goal of nutrient management is to avoid loss of P by soil erosion, runoff and leaching and improve P use efficiency.

Phosphate fertilizers: Digestion of phosphate rock (approx. $Ca_5(PO_4)_3(OH)$; see Chapter 6.2 for details) with sulfuric acid produces a mixture of calcium sulfate and calcium bis(dihydrogenphosphate) (molar ratio 7:3; fertilizer grade[2] 0–20-0). This mixture is commercially available as single or normal superphosphate (SSP, NSP) and it contains with Ca and S two further macronutrients. Digestion of phosphate rock with phosphoric acid leads to gypsum-free triple superphosphate (TSP, $Ca(H_2PO_4)_2 \cdot H_2O$, grade 0–45-0 due to by-phases). Phosphoric acid production is described in Chapter 6.2 in detail. Neutralization of phosphoric acid by ammonia yields monoammonium phosphate (MAP, $NH_4H_2PO_4$, typical grade 11–52-0) and diammonium phosphate (DAP, $(NH_4)_2HPO_4$, 18–46-0). MAP is frequently used in liquid fertilizer formulations. DAP is currently the most widely used P-fertilizer. By thermal treatment of MAP condensation of the $H_2PO_4^-$ yields chain-like polyphosphates. Compared to MAP these have a higher P_2O_5 content, which becomes only slowly available to plant by hydrolysis. Furthermore, ammonium polyphosphate supports the uptake of many trace elements. For testing soil for its P_2O_5 content, extraction with water, citric acid and formic acid are used. Thus, P availability under various conditions is determined. Nitrophosphate, a mixture of NH_4NO_3, $CaCO_3$ or $CaSO_4$ and $CaHPO_4$ (di-calcium phosphate, DCP) is obtained via digestion of phosphate rock by nitric acid with subsequent addition of ammonium carbonate or ammoniuim sulfate [49]. Historically, Thomas flour ("Thomasmehl"), a basic slag formed in steel making as by-product, was also used as P-fertilizer. This calcium silicate-phosphate with the approximate formula $Ca_3(PO_4)_2 \cdot Ca_2SiO_4$ contains 15 wt%. P_2O_5 and 45 wt% CaO. It was valued due to its basic properties (rising pH of soil), its contents of various micronutrients and trace elements (Fe, Mn, Mg), and its retarded release of phosphate. The presence of toxic chromium/chromate, however, created some problem. Sintering phosphate rock at around 1200 °C with limestone and alkali silicate yields Rhenania phosphate (named after the company that had developed this process) [50], which is in many respects similar to Thomas flour.

6.6.2.4 Potash/potassium/kalium

Traditionally, extraction from plant ash, which is on average containing approx. 10 wt% K_2O, was the source of K_2CO_3 ("potash"; for the etymology of potassium/Kalium see [51]). This is approx. ten times the amount of Na_2O in plant. Potassium is, in contrast to all other plant nutrients, almost always present as K^+ in cellular liquids only. There, it facilitates transport of sugars through the plant. Potassium acts in protein synthesis and in enzyme activation. Sufficient supply of K^+ ensures growths of healthy and strong plant stalks, resistance against plant diseases, and quality of harvested products. Its concentration in soil liquor is generally rather low; however K^+ is contained in the various inorganic solids in soil in quite high amounts. Still, by intensified farming, soil is rapidly depleted of potassium. This loss has to be remedied by the use of mineral fertilizers.

All major mineral deposits containing potassium are of marine origin, formed by seasonal evaporation of seawater. These deposits are summarized as "evaporite minerals" [52]. Potassium is contained in particular in the minerals sylvite KCl, carnallite $KMgCl_3 \cdot 6\ H_2O$, polyhalite or polysulfate $K_2Ca_2Mg(SO_4)_4 \cdot 2\ H_2O$, langbeinite K_2Mg_2

$(SO_4)_3$, kainite $KMg(SO_4)Cl \cdot 3\ H_2O$. Besides, the heterogeneous evaporites contain large amounts of sodium chloride (halite) and magnesium sulfate ($MgSO_4 \cdot H_2O$, kieserite). Significant evaporite deposits for fertilizer production are found in northern parts of Germany. On a global scale, potash mines are mostly concentrated in North America and Eurasia. These two regions account for over 90% of the existing potash capacity. Canada, Russia and Belarus are dominant producers.

Separation of suitable fertilizer minerals KCl (muriate of potash, MOP), sulfate of potash (SOP; chloride-free potash fertilizer) and potassium magnesium sulfate from halite can be achieved by various processes which are summarized in Figure 6.6.7. Separation of KCl from $NaCl/MgSO_4 \cdot H_2O$ is based on the differences in the temperature dependence of the solubility of these salts in water (Figure 6.6.7a). Subsequently, the mixture $NaCl/MgSO_4 \cdot H_2O$ is split into its components by a flotation process (Figure 6.6.7b).

By applying a strong electrostatic field, the finely milled crude salt NaCl/KCl can also be separated into its components (ESTA process, Figure 6.6.7c [52, 53]). The development of these refining methods gradually increased the grade of commercial K-fertilizers from around 20 wt% K_2O in late-nineteenth century to high-grade potassium chloride (60–62% K_2O; see footnote 2 for fertilizer grading), which is today the main product. Potassium sulfate, potassium magnesium sulfate and potassium nitrate are the principal non-chloride potash fertilizers. Their production from KCl requires additional chemical treatment with sulfuric or nitric acid. Thus, they are more expensive and hence are used primarily on crops or soils for which the chloride is unsuited.

6.6.2.5 Secondary macronutrients (Mg, Ca, S)

The essential and beneficial mineral nutrients for plants are summarized in Table 6.6.2. In contrast to the primary macronutrients N, P, K, which are in soil typically in only limited supply, the secondary macronutrients Ca, Mg, S required for healthy plant growth are typically provided in sufficient amounts by soil minerals [54]. Only with intensified crop production, cultivation of special crop or in very nutrient-poor soil, these elements have to be provided as mineral fertilizers. Nutrient depletion of the topsoil ("Mutterboden") and lowering of soil fertility in general is caused by various effects, namely soil erosion, leaching by excessive rainfall, droughts. Soils in humid tropical and subtropical climates are more often deficient in secondary and micronutrients. Deficiency in Ca, Mg and S can easily be remedied by provision of limestone, dolomite, gypsum and kieserite. Apart from the latter, these are frequently contained as by-products in higher-value fertilizers, e.g. sulfur in single superphosphate and Ca/Mg in nitrochalk. In most industrial countries, millions of tons of sulfur from fuel combustion escape to the atmosphere and subsequently return to Earth in rainfall. In most such areas the deliberate addition of sulfur to fertilizer has not proved necessary, but there are important exceptions. As the removal of sulfur from stack gas becomes widespread

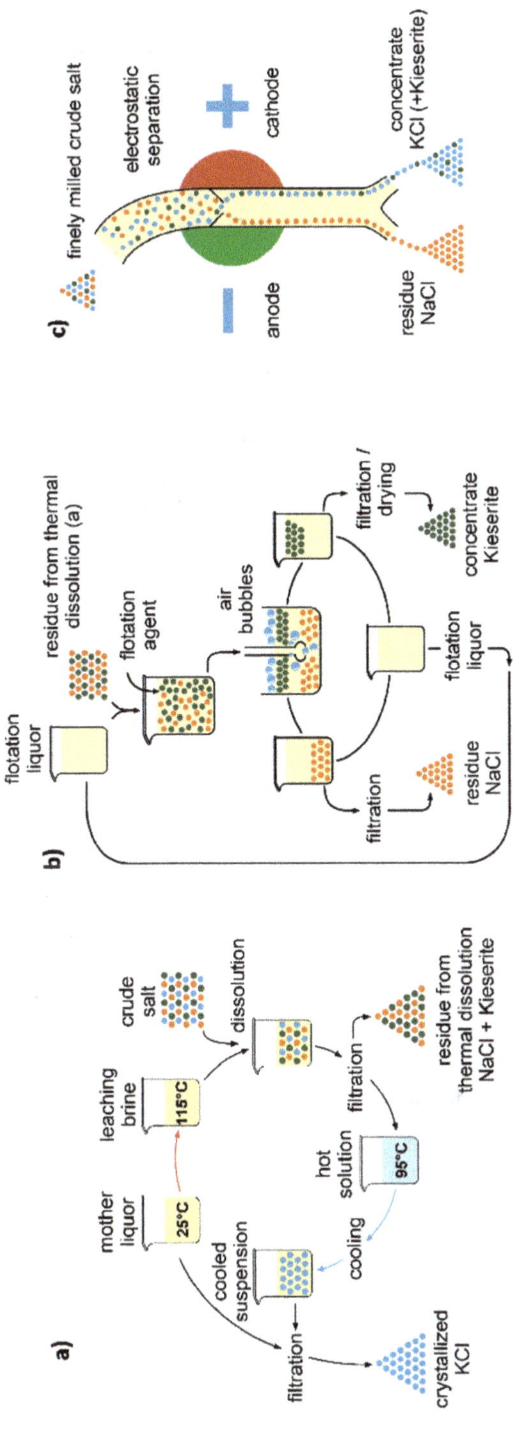

Figure 6.6.7: Separation processes for enrichment of potassium chloride from raw-salt deposits (graphics adapted from [52]).

for the prevention of atmospheric pollution, more sulfur may need to be added to fertilizers. In many less-industrial areas, the need for sulfur is already well known [3].

6.6.3 Plant nutrients (micro)

In addition to the macronutrients, there are currently recognized eight essential nutrients needed by plants in small amounts, called the micronutrients (see Table 6.6.2). Cobalt (Co), silicon (Si) and aluminum (Al) are essential, or at least beneficial, to some plant species, but not required by all [41]. It is a noteworthy consequence of the use of high-grade ("chemically pure") fertilizers that more attention has to be given by

Table 6.6.2: Essential and beneficial mineral nutrients for plants [41, 54].

Category	Nutrient	Primary form of uptake	Main form in soil reserves	Relative # atoms in plants
Macro	Nitrogen	Nitrate NO_3^- Ammonium NH_4^+	Organic matter	1000000
Macro	Phosphorus	Phosphate HPO_4^{2-}, $H_2PO_4^-$	Organic matter, minerals	60000
Macro	Potassium	Potassium ion K^+	Minerals	250000
Macro	Calcium	Calcium ion Ca^{2+}	Minerals	125000
Macro	Magnesium	Magnesium ion Mg^{2+}	Minerals	80000
Macro	Sulfur	Sulfate SO_4^{2-}	Organic matter, minerals	30000
Micro	Chlorine	Chloride ion Cl^-	Minerals, rainfall	3000
Micro	Iron	Ferrous ion Fe^{2+}	Minerals	2000
Micro	Boron	Boric acid H_3BO_3	Organic matter	2000
Micro	Manganese	Manganese ion Mn^{2+}	Minerals	1000
Micro	Zinc	Zinc ion Zn^{2+}	Minerals	300
Micro	Copper	Copper ion Cu^{2+}	Organic matter, minerals	100
Micro	Molybdenum	Molybdate MoO_4^{2-}	Organic matter, minerals	1
Micro	Nickel	Nickel ion Ni^{2+}	Mineral	1

farmers to sufficient supply of micronutrients in soil, since emphasis on high-analysis fertilizers has "squeezed out" other elements. Thus, better diagnosis of plant nutritional needs has identified hitherto unsuspected deficiencies in micronutrients [2]. All these elements are required by plants in rather small amounts only. Yet, their low solubility leads to very low concentrations in the soil liquor and availability to plants. Plants have developed strategies to remedy this limitation by: a) lowering the pH around the roots and b) segregation of chelating agents (e.g. sugars, polycarboxylic acids), which allows increased solubility of these micronutrients and their uptake by the plant.

Iron [56, 57]: Of all micronutrients, iron is the metal required in largest quantities. Iron easily changes between its oxidation states +2 and +3, permitting its main function in the transfer of electrons in cells. Iron is a part of the cytochromes, which perform their function in cellular respiration in the mitochondria and, during photosynthesis, in the chloroplasts. As iron plays a vital role in many enzymatic reactions of chlorophyll synthesis, an iron deficiency causes the chlorophyll production to decrease. There is often a large amount of iron in the soil; however, the major part of it is present only in almost insoluble forms (e.g. oxides, hydroxides, phosphates). Free iron ions (Fe^{2+}, Fe^{3+}) that can be absorbed by plants are only found in trace amounts. Some plants release a chelator (mugineic acid, phytosiderophore "iron carrier") into the soil, which forms complexes with iron(III). These complexes allow iron uptake by the roots and transport inside plant tissues. In fertilizers, iron is bound to stabilizers or chelators like EDTA, DTPA, HEEDTA (see Figure 6.6.8), EDDHA, EDDHSA, citrate, gluconate or ascorbate to keep it dissolved and to make it available to plants [57].

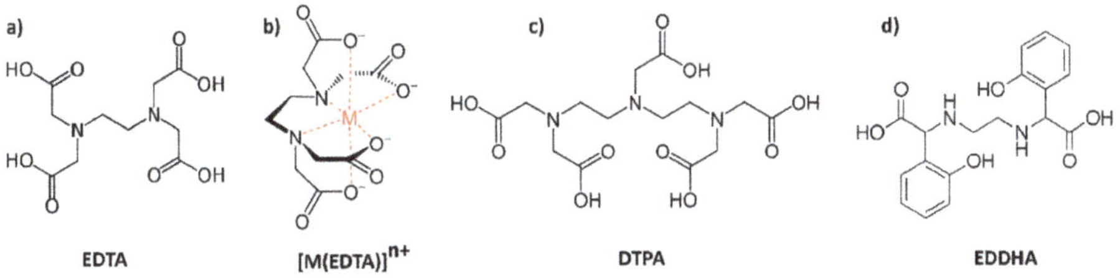

a) EDTA b) $[M(EDTA)]^{n+}$ c) DTPA d) EDDHA

Figure 6.6.8: Various chelating agents used in stabilizing micronutrients in aqueous solution [55, 56].

Manganese [57–59]: Manganese is absorbed by plants as Mn^{2+} cation and can also be bound and carried by chelators, very similar to iron. Its functions within the plant are mostly related to its redox chemistry with quite high standard redox potential for Mn^{3+} and Mn^{4+}. Manganese is a co-factor or forms part of many enzymes that are e.g. needed in the citric acid cycle, the process of water splitting and the transport of electrons during photosynthesis or during nitrogen assimilation (enzyme nitrate reductase). As it also plays a role in the formation of aromatic amino acids phenylalanine and tyrosine and thus of phenolic acids and alcohols, manganese is also important for repelling infectants (e.g. fungi, bacteria). Eventually, manganese is needed for the synthesis of chlorophyll and for protecting chloroplast tissue from free radicals. The uptake of manganese by plants is higher at lower pH of the soil liquor. Liming the soil thus may lead to a manganese deficiency. High levels of other divalent transition metal cations (Fe^{2+}, Cu^{2+}, Zn^{2+}) in the soil might also hinder the uptake of manganese.

Copper [57]: Copper is widely known as a strong cell poison (see also Chapter 13.1.4). However, plants need trace amounts as micronutrient. Copper is absorbed as hydrated Cu^{2+} cation (e.g. from $CuSO_4 \cdot 5\,H_2O$ added to fertilizer mixtures) or bound by a chelator. In the plant cell, free Cu^{2+} ions do not occur, they are always bound to proteins. The slightly oxidizing Cu^{2+} ions would otherwise harm the cell due to oxidative stress and denaturation of proteins. Copper is part of a number of enzymes in redox systems, e.g. it is contained in phenoloxidases, which play a part in the defense against microorganisms by catalyzing the transformation of phenols to brown colored quinones (which is e.g. visible when apple or potato cuttings turn brown). Copper is needed during photosynthesis for transportation of electrons between photosystem II and photosystem I. Ascorbic acid oxidase contains copper, too. It catalyzes the oxidation of ascorbic acid (vitamin C) to dehydroascorbic acid. The copper-based enzyme diamine oxidase is important for the synthesis of lignine and thus for the formation of wood in terrestrial plants. Copper is an iron and manganese antagonist in nutrient absorption; thus, high copper concentrations in the ground may lead to a deficit in these nutrients in plants.

Zinc: Zinc is absorbed by plants as hydrated Zn^{2+} cation, bound to chelators, and, when the pH is sufficiently high, as $ZnOH^+$ ion. This metal forms part of or is a cofactor in many enzymes involved in important metabolic processes. For example, Cu/Zn-superoxide dismutase (Cu/Zn SOD) protects the cellular membranes from damage by free oxygen radicals. In the cell plasma and in the chloroplasts, carboanhydrase (CA), which also contains zinc, regulates the reaction of CO_2 and water to carbonic acid. This enzyme thus acts as a buffer and moreover causes an increase of the CO_2 concentration in the chloroplasts. This raises the CO_2 fixation rate of the enzyme RuBisCO during the dark reaction of photosynthesis. With the help of carboanhydrase, water plants are able to make use of hydrogen carbonate ions (HCO_3^-) as a source of carbon which is fixed during the Calvin cycle [60]. Zinc plays a decisive role in stabilizing the ribosomes and thus for the formation of proteins. As a part of RNA and DNA polymerases it is also crucial for the reproduction of RNA and DNA molecules and consequently for cell partition and protein synthesis. Eventually, the most important energy transmitter, ATP, is formed with the help of the zinc-dependent enzyme hexokinase. High zinc concentrations in the soil can lead to iron deficiencies in terrestrial plants because too much zinc may have a negative influence on the reduction of Fe^{3+} to Fe^{2+} on the roots.

Boron: [57, 61] Boron is mainly absorbed by plants as undissociated boric acid $B(OH)_3$, but also as borate ion $B(OH)_4^-$. In nature, boron is released by withering rock minerals like glimmer or tourmaline, among others. Acidic rocks like granite are poor in boron, whereas the boron concentration in seawater is very high (approx. 5 mg L^{-1}). In freshwater bodies, the boron concentrations are mostly below 0.1 to 0.5 mg L^{-1}. The importance of boron for plants has been known since the 1930s, and many effects of boron deficiencies have been described, many of them species-specific, e.g. deformations, chloroses and necroses of the leaves and dying-off shoot and root tips. However, its specific functions in the plant organism are still partially in the dark. In crop, boron deficiency is often observed, especially during dryness in combination with strong growth in the summer (e.g. heart and dry rot in sugar beets). Plants boron uptake is hindered when soil pH is above 7 and the soil content in iron and aluminum hydroxides is high. In contrast to many other nutritive elements, boron is not a part of any enzymes. The most boron in a plant is bound in compounds stabilizing the cell wall. Indeed, sufficient boron supply to strawberries yields fruits less susceptible to squashing [62]. An important function of this element is the transport of sugar to the growth zones (meristematic tissues) on the shoot and root tips. The transport of sugar probably also explains the role boron plays in various metabolic processes, like cell partition and differentiation, photosynthesis, the metabolism of nitrogen, phosphorus, hormones and fats, the active absorption of salts, the functionality of the cell membranes as well as the fertilization of the female ovules in the flowers (pollen tube growth). It is also important for the production of starch, from which in turn cellulose is formed, an important material for the formation of the cell walls. Moreover, boron plays a role in the enzymatic reduction of iron to a form that can be used by the plants inside their roots. Boron deficiency leads to reduced uptake of iron and other nutrients like magnesium, calcium, potassium and phosphate, which has many secondary effects on plant growth. Plants can absorb boron very quickly when fertilized over their leaves, the same may be assumed for submerged water plants.

Molybdenum [57]: Of all essential micronutrients, molybdenum is used in the least amount. It is mainly absorbed by plants in the form of molybdate ions MoO_4^{2-}. Together with iron, molybdenum forms part of nitrate reductase, which is decisive for nitrogen utilization (see Section 6.6.2.2). Nitrate absorbed by the plant is reduced to nitrite by this enzyme, and further to ammonium by nitrite reductase, which is in turn used in amino acids. The other three enzymes containing molybdenum in plants are aldehyde oxidase, xanthine dehydrogenase (degradation of purine) and sulfite oxidase (detoxification of sulfite through oxidation to sulfate). Symbiotic bacteria in the root nodules of the legumes fixate aerial nitrogen N_2 with the help of molybdenum-based enzyme nitrogenase and make it usable for the plant. The molybdenum availability for plants rises with an increasing pH value of the soil, and a molybdenum deficiency often shows on acidic soils – in contrast to other metal micronutrients.

Chlorine [57]: Chloride is essential for plants, but only in trace amounts, which make it a micronutrient, however, some plants contain it in large quantities. The soils and water in coastal or dry areas often contain a high amount of chloride (caused by salting of soil). Plants adapted to salt (halophytes) tolerate concentrations of over 0.5% by weight of sodium chloride in the ground water. Chloride is easily taken up and very mobile within the plant. It is not used in any organic structures and is only found as free chloride ion inside the plant. Like potassium it is important for osmoregulation (regulation of the internal cellular pressure, opening and closing of the stomata) and for electric charge balance.

Cobalt and nickel [57]: These two micronutrients are treated in one chapter here as they are only required in trace amounts by plants and do not have to be added to fertilizers. The quantities of nickel and cobalt in fertilizing salts present as "impurities" are sufficient for providing the plants with appropriate amounts of those nutritive elements. These metals are absorbed by plants in the form of Ni^{2+} and Co^{2+} ions. They are part of different enzymes, e.g. urease, which contains nickel and catalyzes the degradation of urea to ammonia and carbon dioxide, or of dipeptidases, containing cobalt and splitting dipeptides into two amino acids. Higher nickel concentrations in the soil are toxic for terrestrial plants, as they cause an iron and zinc deficiency.

6.6.4 Soil pH and redox potential [1]

Soil consists of a large amount of badly soluble minerals (e.g. feldspars, quartz, limestone, . . .), small amounts of solid organic matter, an aqueous phase (soil liquor, "Bodenlösung"), and a gas phase filling pores in the solid material. Depending on soil aeration, oxygen pressure in the pores could be lower and CO_2 pressure significantly higher than in surrounding air. Soil liquor contains dissolved and colloidal substances. Weathering, leaching and interaction with plant roots and microorganisms results in continuous (ion) exchange dynamics between the components of soil [1]. These processes determine pH and redox potential of the soil. The pH of arable soil ranges from 4–9 with a crop optimum between 5 and 8. The pH of soil is well defined and easily measurable from soil extracts, e.g. by using a pH electrode. Due to slow kinetics of many soil-related redox processes, the potential measured at a time does not necessarily represent the equilibrium redox potential. In addition, its measurement using a Pt-electrode is rather susceptible to error. Well aerated, acidic soil (pH ≈ 4) shows redox potentials as high as +0.8 V. The lowest values are observed for anaerobic, neutral to basic conditions ($\varepsilon_0 \approx -0.35$ V). Due to its various components soil shows considerable acid/base and redox buffer capacity. Nevertheless, both, pH and redox potential, show significant fluctuation with the vegetation cycle. As was already pointed out, the solubility and hence the plant availability of nutrients in soil depends strongly on pH and ε_0. Figure 6.6.9a gives a schematic overview of the pH dependence in the availability of various nutrients. For some (Fe, Mn,

Zn, Cu), acidic environment allows higher solubility, while for others (K, S, Mo) solubility is improved by higher pH. The Pourbaix diagram in Figure 6.6.9b (highlighted area) shows that the conditions encountered in soil either favor formation of dissolved Fe^{2+}(aq) ions or precipitation of (amorphous) iron(III) hydroxide. A similar situation is encountered for manganese. It should be stressed that pH and redox potential affect also the macronutrients C, N and S, which are rather easily converted from chemical species dissolved in the soil liquor to evaporating gaseous molecules. Reaction (6.6.9) gives a simplified pH-dependent redox reaction between iron(III) oxide-hydroxide and organic matter (which is represented by "H_2CO").

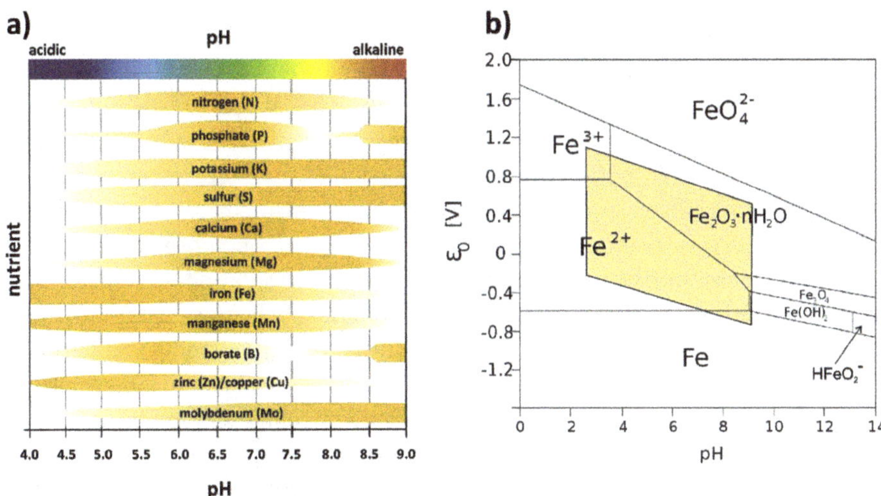

Figure 6.6.9: a) Schematic representation of nutrient availability to plant and pH (graphics modified from [13]). b) Pourbaix diagram showing the influence of redox potential ε_0 and pH on the prevailing iron species. The highlighted area is relevant to the situation encountered in soil (graphics modified from [63]).

$$4\,Fe^{III}O(OH)(s) + H_2CO(l) + 8\,H^+(aq) \longrightarrow 4\,Fe^{2+}(aq) + CO_2(g) + 7\,H_2O(l) \quad (6.6.9)$$

In humid climate, even in natural soil pH shifts slowly to lower values. This tendency is increased by application of fertilizers containing ammonium salts (e.g. $NH_4H_2PO_4$, $(NH_4)_2SO_4$, NH_4NO_3) or some nitrates. The use of ground limestone ("liming") or dolomite, $CaMg(CO_3)_2$, is long established to adjust soil pH in the optimum region. Depending on particle size, application of these low-soluble carbonates will have either an immediate or strongly retarded effect. During the last decades, urea has been introduced as N-fertilizer without a short-term pH lowering effect. Nitrification of ammonia from urea hydrolysis can, however, lead to enhanced nitric acid formation and acidification of soil over a period of several months.

6.6.5 Fertilizer specialties

With the knowledge about nutrition requirements (for plant and crop in particular) being well established, more and more emphasis is put on fertilizer formulations and ways of their application. The driving force behind such efforts is obviously to make optimum fertilization easier for farmers.

Compound fertilizers ("Volldünger"): Already in the 1920s, BASF introduced the first compound fertilizer "Nitrophoska" (typical grade N-P-K 12-8-16) in contrast to straight (single-nutrient) materials [64]. Production of Nitrophoska is based on the nitrophosphate process which relies on digestion of phosphate rock with nitric acid. Thus, gypsum formation is avoided (see eq. 6.6.10). After strong cooling, calcium nitrate precipitates and is partly separated by centrifuging, while the solution, which is rich in phosphoric acid, is neutralized (eq. 6.6.11). Subsequently, ammonium and potassium salts are added to the neutralized solution to adjust desired nutrient contents. After evaporation and granulation, a homogeneous product is obtained that does not show separation of nutrients during storage and application. Figure 6.6.10 summarizes the described processing steps, together with the required resources. In recent years, formulations of compound fertilizers starting from single-nutrient materials have become available, due to improved mixing techniques (Figure 6.6.10, processing ways 3 and 4). Modern compound fertilizers for special applications contain admixtures of secondary macronutrients (Ca, Mg, S) and various micronutrients (Fe, Mn, B, Zn).

$$\mathrm{Ca_5(PO_4)_3F}(s) + 10\,\mathrm{HNO_3}(aq) \longrightarrow 3\,\mathrm{H_3PO_4}(aq) + 5\,\mathrm{Ca(NO_3)_2}(aq) + \mathrm{HF}(g) \tag{6.6.10}$$

$$\begin{aligned} 3\,\mathrm{H_3PO_4}(aq) + 2\,\mathrm{Ca(NO_3)_2}(aq) + 5\,\mathrm{NH_3}(aq) \longrightarrow\ & 2\,\mathrm{CaHPO_4}(aq) \\ & + \mathrm{(NH_4)H_2PO_4}(aq) + 4\,\mathrm{NH_4NO_3}(aq) \end{aligned} \tag{6.6.11}$$

The use of fertilizer solutions (either for application onto the soil or addition to the irrigation, i.e. "fertigation") has become increasingly popular, in particular with production of highly specialized crop (fruit, vegetables). Similarly, foliar spraying ("Blattdüngung") uses fertilizer solutions. These are applied onto the plant leafs, which absorb the nutrients quickly. Foliar spraying in addition to classical fertilizer application via the soil is used to immediately remedy macronutrient deficiencies. The methods also allow accurate dosage of micronutrients to plant without potential losses due to adsorption and precipitation in soil or leaching [66].

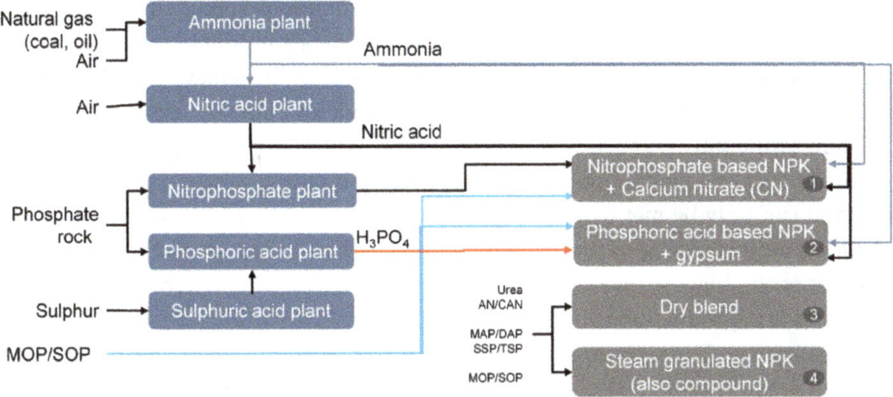

Figure 6.6.10: Flowchart visualizing resources and processing for four different ways (in gray) of modern NPK compound fertilizer production (graphics taken from [65]). MOP muriate of potash (KCl), SOP sulfate of potash (K_2SO_4), CAN calcium ammonium nitrate, MAP mono-ammonium phosphate, DAP di-ammonium phosphate, SSP single superphosphate (approx. $7\ CaSO_4 \cdot 3\ Ca(H_2PO_4)_2 \cdot H_2O$), TSP triple superphosphate ($Ca(H_2PO_4)_2 \cdot H_2O$).

Slow-release N-fertilizer: Since the loss of nitrogenous compounds to the atmosphere and runoff is both wasteful and environmentally damaging, urea is sometimes pretreated or modified to enhance the efficiency of its agricultural use. One such technology is controlled-release fertilizers, which contain urea encapsulated in an inert sealant. Another technology is the conversion of urea into derivatives (e.g. with formaldehyde), which degrade into ammonia at a pace matching plants' nutritional requirements.

6.6.6 Future developments and concluding remarks

When mineral fertilizers were introduced in the late-nineteenth century, they used to remedy a shortage of the primary nutrients N, P and K. In the developed countries, primary nutrients are no longer the most limiting factor and fertilizers are used to supply secondary and micronutrients as well. Actually, in a large number of fields in both developed and developing countries, secondary and micronutrients are now becoming the limiting elements for crop production because farmers have drastically raised primary nutrient concentration in soil by applying mineral fertilizers. Nevertheless, special climate, soil and economic conditions are the reasons, why in several developing countries (in Africa and Asia), N and P are still the limiting elements in crop production. In particular in Europe, economic pressure in crop production and increasing ecologic awareness (paired with law-making to protect the environment) have already made a significant impact on fertilization habits. As a consequence, modern farming gives far more emphasis to MITSCHERLICHS law of the optimum than to SPRENGELS/LIEBIGS law of the minimum (Figure 6.6.1). Fertilization in modern farming might be summarized by the 4R nutrient stewardship concept: right source at the right rate, time and place [54]. To achieve the 4R requires detailed knowledge of biochemical processes in soil and plant, close monitoring of nutrient availability and access to suitable fertilizer formulations.

An increasing global population will demand increasing food production. This demand can only be met by (1) expanding the area of arable land or (2) increasing yields on land currently in production. With respect to (1) one should note that habitat loss is the biggest threat to the world's wildlife, that the potential for putting new land into production is limited, and if new lands are available, these are often less productive. Thus, meeting future food needs with increased crop production through greater yields on existing farm land is the more favorable scenario. Therefore, commercial fertilizer will continue to play a vital role in the future. Nevertheless, increasing agricultural production does not automatically mean a proportionate increase in fertilizer use is needed. Improvements in management and nutrient use efficiency ("4R") will allow productivity to grow relatively faster than the growth rate of inputs, except in regions where fertilizer is underused.

References

[1] Amelung W, Blume H-P, Fleige H, Horn R, Kandeler E, Kögel-Knabner I, Kretzschmar R, Stahr K, Wilke B-M; with contributions by. In: Gaiser T, Gauer J, Stoppe T-BS, Welp G. Scheffer/Schachtschabel: Lehrbuch der Bodenkunde, 17th edition, Springer Spektrum, Berlin, 1994. https://doi.org/10.1007/978-3-662-55871-3

[2] Nielsson FT. Manual of Fertilizer Processing, vol. 5 in Fertilizer Science and Technology, Taylor & Francis, Boca Raton, 2018.

[3] Lee RG, Kopytowski JA. (Eds.) Fertilizer Manual by United Nations Industrial Development Organization and International Fertilizer Development Center, 3rd edition, Kluwer Academic Publishers, Dordrecht, 1998.

[4] Schwenke KD. "Lavoisier und die Anfänge der Agrikulturchemie" Mitteilungen. GDCh Fachgruppe Geschichte der Chemie (Frankfurt/Main), 2012, 22, 20–36.

[5] Feller C. Adv GeoEcol, 1997, 29, 15–46. ISBN-3-923381-40-9.

[6] Ormeño-Orrillo E, Hungria M, Martinez-Romero E. Dinitrogen-Fixing Prokaryotes. In: Rosenberg E ((Ed).) The Prokaryotes, 4th edition, Springer, Berlin, 2013, pp. 427–52. DOI: 10.1007/978-3-642-30141-4

[7] Thaer AD. "Grundsätze der rationellen Landwirthschaft/Principles of rational agriculture". vol. 1–4. Berlin, 1809–12. In: Deutsches Textarchiv https://www.deutschestextarchiv.de/thaer_landwirthschaft01_1809/10, download Feb. 15th 2022.

[8] Sprengel C. Über Pflanzenhumus, Humussäure und humussaure Salze (About plant humus, humic acids and salts of humic acids), Archiv für die Gesammte Naturlehre, 1826, 8, 145–220. ArchPharm, 1827, 21, 261–63. https://doi.org/10.1002/ardp.18270210315

[9] Sprengel C. Von den Substanzen der Ackerkrume und des Untergrundes (About the substances in the plow layer and the subsoil). Journal für Technische und Ökonomische Chemie, 1828, 2, 423–74. and 3: 42–99, 313–352, and 397–421 Online. https://www.digi

tale-sammlungen.de/de/search?filter=volumes%3A%22bsb10707678%2FBV002538144%22 refute of humus theory.

[10] Liebig Jv. "Die organische Chemie in ihrer Anwendung auf Agricultur und Physiologie/ Application of organic chemistry in agriculture and physiology". Vieweg, Braunschweig 1840. Online: Deutsches Textarchiv https://www.deutschestextarchiv.de/liebig_agricultur_1840. download: 16.02.2022.

[11] Mitscherlich EA. Das Gesetz des Minimums und das Gesetz des abnehmenden Bodenertrages/The law of the minimum and the law of decreasing soil fertility. Landwirtschaftliche Jahrbücher, 38, 1909, 537–52. ISSN 0368-8194.

[12] van der Ploeg RR, Böhm W, Kirkham MB. Soil Sci Soc Am J, 1999, 63, 1055–62.

[13] "*Informationsserie Pflanzenernährung Ernährung – Wachstum – Ernte*", Published by Fonds der Chemischen Industrie im Verband der Chemischen Industrie e. V., Frankfurt 2013. Online: https://www.vci.de/fonds/schulpartnerschaft/unterrichtsmaterialien/pflanzenernaehrung-wachstum-ernte.jsp

[14] Smil V. AMBIO, 2002, 31, 126–31.

[15] Smil V. Nature, 1999, 400, 415. https://doi.org/10.1038/22672

[16] Erisman JW, Sutton MA, Galloway J, Klimont Z, Winiwarter W. Nature Geo, 2008, 1, 636–39.

[17] Compiled by IFDC from various issues of FAO Fertilizer Yearbooks and fertilizer data tapes Global fertilizer production.

[18] Gapminder (v6), HYDE (v3.2), United Nations Population Division (2019). download Feb 27th 2022, https://www.gapminder.org/data/documentation/gd003/.

[19] Schallermair C. Die deutsche Sodaindustrie und die Entwicklung des deutschen Sodaaußenhandels 1872-1913. Vierteljahrschrift für Sozial- und Wirtschaftsgeschichte 1997,84,33–67. Retrieved from: https://www.jstor.org/stable/20738772 [Online Resource].

[20] Friedman LJ, Friedman SJ. The History of the Contact Sulfuric Acid Process. Acid Engineering & Consulting, Inc. Boca Raton, Florida 2008. Download from: http://www.aiche-cf.org/Clear water/2008/Paper2/8.2.7.pdf

[21] Linde C. DRP 1250, 1877. Download from: https://worldwide.espacenet.com/patent/search/family/070918076/publication/DE1250C?q=pn%3DDE1250C [Online Resource]

[22] Caro N, Frank A. DRP 88363, 1896. Download from: https://worldwide.espacenet.com/publi cationDetails/originalDocument?CC=DE&NR=88363C&KC=C&FT=D&ND=3&DB=&locale=de_EP# [Online Resource]

[23] Ostwald W. Patent GB190200698: Improvements in the Manufacture of Nitric Acid and Nitrogen Oxides. Applied: January 9th 1902, published: March 20th 1902. Online: https://worldwide.espacenet.com/publicationDetails/biblio?FT=D&date=19020320&DB=&lo cale=de_EP&CC=GB&NR=190200698A&KC=A&ND=3 [Online Resource]

[24] BASF. DRP 235421: Verfahren zur synthetischen Darstellung von Ammoniak aus den Elementen. Applied: October 13th 1908. Download from: https://worldwide.espacenet.com/patent/search/family/005947195/publication/DE235421C?q=pn%3DDE235421C [Online Resource]

[25] Eyde S. J Royal Soc Arts, 1909, 57, 568–76. Download Feb. 28th 2022 from: https://www.jstor.org/stable/i40063334

[26] Bosch C, Meiser W. Patent US1429483: Process of Manufacturing Urea. Applied: July 9th 1920, published: September 19th 1922. Online: https://worldwide.espacenet.com/patent/search/family/023561906/publication/US1429483A?q=pn%3DUS1429483

[27] Schiller G, Wietzel G. Patent US2083795: Production of hydrogen. Applied: April 17th 1933, published: June 15th 1937. Online: https://worldwide.espacenet.com/patent/search/family/024674480/publication/US2083795A?q=pn%3DUS2083795

[28] Ritchie H, Roser M. “Fertilizers”. *Published online at OurWorldInData.org* (2013). Retrieved from: ‘https://ourworldindata.org/fertilizers’ [Online Resource]
[29] Food and Agriculture Organization of the United Nations. World fertilizer trends and outlook to 2020 – Summary Report, Rome, 2017. https://www.fao.org/faostat/en/#data/RA
[30] List given on TrendR website. Retrieved Feb. 28th 2022 from https://www.trendrr.net
[31] Scripps CO_2 Program. The Scripps Institution of Oceanography. Retrieved Feb. 28th 2022 from: https://scrippsco2.ucsd.edu/
[32] Myers SS, Zanobetti A, Kloog I, Huybers P, Leakey ADB, Bloom AJ, Carlisle E, Dietterich LH, Fitzgerald G, Hasegawa T, Holbrook NM, Nelson RL, Ottman MJ, Raboy V, Sakai H, Sartor KA, Schwartz J, Seneweera S, Tausz M, Usui Y. Nat Lett, 2014, 510, 139–43. DOI: 10.1038/nature13179
[33] Wagner SC. Biological Nitrogen Fixation. Nature Education Knowledge 2011,3,15. Retrieved from: https://www.nature.com/scitable/knowledge/library/biological-nitrogen-fixation-23570419/ [Online Resource]
[34] Holleman AF, Wiberg N. Anorganische Chemie, 103, Auflage, De Gruyter, Berlin, 2016. ISBN 978-3-11-051854-2.
[35] Historic American Engineering Record: United States Nitrate Plant Number 2 “Muscle Shoals Alabama”. (https://web.archive.org/web/20140222063133/http://lcweb2.loc.gov/pnp/habshaer/al/al1000/al1051/data/al1051data.pdf) (PDF). Retrieved Feb. 28th 2022.
[36] Directorate-General for Health and Consumer Protection of the European Commission, Scientific Committee on Health and Environmental Risks, Potential risks to human health and the environment from the use of calcium cyanamide as fertilizer, 2016. DOI: 10.2772/66307. Retrieved Feb. 28th 2022.
[37] Birkeland K. Trans Faraday Soc, 1906, 2, 98–116. 10.1039/TF9060200098. Retrieved Feb. 28th 2022.
[38] Ertl G. Angew Chem Int Ed, 2008, 47, 3524–35. 10.1002/anie.200800480
[39] Smith JG. Platinum Metals Rev, 1988, 32, 84–90. Download Feb. 28 2022 from. https://www.technology.matthey.com/article/32/2/83-83-2/
[40] Stoltzenberg D. ChiuZ, 1999, 33, 359–64. DOI: 10.1002/ciuz.19990330607
[41] Reetz HF jr. Fertilizers and Their Efficient Use, 1st edition, International Fertilizer Industry Association, Paris, 2016. Retrieved February 28th 2022 from. https://www.fertilizer.org/public/resources/publication_detail.aspx?SEQN=5221&PUBKEY=811C52B7-F679-472A-B955-3AD6AD18F17D
[42] Wöhler F. Ann Phys Chem, 1828, 88, 253–56. https://doi.org/10.1002/andp.18280880206
[43] Basaroff A. J Chem Soc, 1868, 21, 192–97. DOI: 10.1039/JS8682100192
[44] Brouwer M. *Thermodynamics of the Urea Process*, Urea know How, Process Paper 2009. https://www.academia.edu/8373516. Download February 26th 2022
[45] Clark KG, Gaddy VL, Rist CE. Ind Eng Chem, 1933, 25, 1092–96. https://doi.org/10.1021/ie50286a008
[46] IPNI fact sheets on plant nutrients #14 TSP. Download March 13th 2022 from https://www.ipni.net/specifics-en
[47] Sigurdarson JJ, Svane S, Karring H. Rev Environ Sci Biotechnol, 2018, 17, 241–58. https://doi.org/10.1007/s11157-018-9466-1
[48] Norton J, Ouyang Y. Frontiers MicroBio, 2019, 10, 1931. https://doi.org/10.3389/fmicb.2019.01931
[49] “Nitrophosphate”, fact sheet no. 15, Intl. Plant Nutrition Institute, Norcross, Georgia U.S.A. 2016. Retrieved February 28th 2022 from: www.ipni.net/specifics
[50] Messerschmitt A. Angew Chem, 1922, 35, 537–44. DOI: 10.1002/ange.19220357902

[51] Page „Kalium“. In: Wikipedia – Die freie Enzyklopädie (in German). Version of December 14th 2021, 07:30 UTC. URL: https://de.wikipedia.org/w/index.php?title=Kalium&oldid=218175805 (Download: March 9th 2022, 15:23 UTC)

[52] "Strong Minerals" info brochure by K+S Minerals and Agriculture GmbH, Kassel, Germany. Download March 02nd 2022 from: https://www.kpluss.com/.downloads/agriculture/brochures/en-int-fertiliser-strong-minerals-A4-1611.pdf.

[53] Bock R. Chem-Ing-Tech, 1981, 53, 916–24. 10.1002/cite.330531203

[54] "4R Plant Nutrition: A Manual for Improving the Management of Plant Nutrition", International Plant Nutrition Institute, Peachtree Corners, GA, USA 2016. Download March 10th 2022 from https://plantnutrition.ca/research/4r-plant-nutrition-manual/

[55] Klem-Marciniak E, Huculak-Maczka M, Marecka K, Hoffmann K, Hoffmann J. Molecules, 2021, 26, 1933. 16 pages https://doi.org/10.3390/molecules26071933

[56] Yunta F, García-Marco S, Lucena JJ, Gómez-Gallego M, Alcázar R, Sierra MA. Inorg Chem, 2003, 42, 5412–21. DOI: 10.1021/ic034333j

[57] "Macro- and Micronutrients", Aqua Rebell, Braunschweig, Germany 2021. Download March 10th 2022 from https://www.aqua-rebell.de/nutrient-iron

[58] Millaleo R, Reyes-Diaz M, Ivanov AG, Mora ML, Alberdi M. J Soil Sci Plant Nutr, 2010, 10, 470–81. DOI: 10.4067/S0718-95162010000200008

[59] Alejandro S, Höller S, Meier B, Peiter E. Front Plant Sci, 2020, 11:300. DOI: 10.3389/fpls.2020.00300

[60] Bassham J, Benson A, Calvin M. J Biol Chem, 1950, 185, 781–87. DOI: 10.2172/910351

[61] Pereira GL, Siqueira JA, Batista-Silva W, Cardoso FB, Nunes-Nesi A, Araújo WL. Front Plant Sci, 2021, 11, 610307. DOI: 10.3389/fpls.2020.610307

[62] Lieten P. Acta Hortic, 2002, 567, 451–53. DOI: 10.17660/ActaHortic.2002.567.94

[63] CC BY-SA 2.5, https://commons.wikimedia.org/w/index.php?curid=1219105

[64] Werner W. Z Pflanzenernährung, Düngung, Bodenkunde, 1930, 9, 339–60. DOI: 10.1002/jpln.19300092203

[65] Yara Fertilizer Industry Handbook, 2018. Download March 13th 2022 from: https://www.yara.com/siteassets/investors/057-reports-and-presentations/other/2018/fertilizer-industry-handbook-2018-with-notes.pdf

[66] Fageria NK, Barbosa Filho MP, Moreira A, Guimarães CM. Foliar Fertilization of Crop Plants. J Plant Nutr, 2009, 32, 1044–64. DOI: 10.1080/01904160902872826

6.7 Natural and synthetic gemstones

Lothar Ackermann, Tom Stephan

Gemstones are by definition natural materials and encompass minerals, rocks, natural glasses as well as organic and biogenic materials of great beauty. For facetted gemstones for jewelry purpose transparency, color and hardness are important properties. As there are both quartz and feldspar particles everywhere in our environment it is of some importance that the hardness of gemstones should be at least in the range of feldspar and quartz, which is 6 and 7 on the Mohs hardness scale, respectively. Otherwise, the polish of a cut gemstone may suffer after a short while. The reason for the origin of colors in the majority of gemstones is caused by electronic excitation of elements of the transition metal series, due to the gradual filling of the 3d orbitals. These ions may occupy tetrahedral, octahedral, cubic or even higher coordinated oxygen environments in crystal structures. If one looks into the showcase of the local jeweler, one may identify diamond, ruby, blue sapphire, emerald, tourmaline, garnet or amethyst as well-known gemstones. But there is a vast variety of gemstones from the different mineral groups with all possibilities with regard to the spectrum of colors, but few have become important for technical use. Beneath their application for jewelry purpose, the members of the corundum and the quartz group are of great importance because of their technical applications and will therefore be the main subject in this chapter. For understanding the color variations and transparency of most of the important gemstones we will give a short introduction to electronic absorption theory.

6.7.1 Color of gemstones

The origins of color of gemstones may be interpreted by means of crystal field theory and electronic absorption spectroscopy in the UV/VIS/NIR spectral range. Three types of electronic transitions may cause absorption bands in this spectral range and thus contribute to the color of gemstones [1].

6.7.1.1 Oxygen to metal charge transfer (OMCT)

Electron charge transfer between oxygen ligands and central ions in the polyhedra of the respective crystal structure cause extremely intense bands in the UV range. The absorption band edges may extend into the visible region and thus may influence the color of gemstones. The absorption bands and their maxima can only be visualized in very thin probes (Figure 6.7.1). Contaminations of traces of transition metals such as iron even in the ppb-range are therefore a big issue concerning the transparency of crystals and glasses for UV transparent optics (Figure 6.7.1).

https://doi.org/10.1515/9783110798890-015

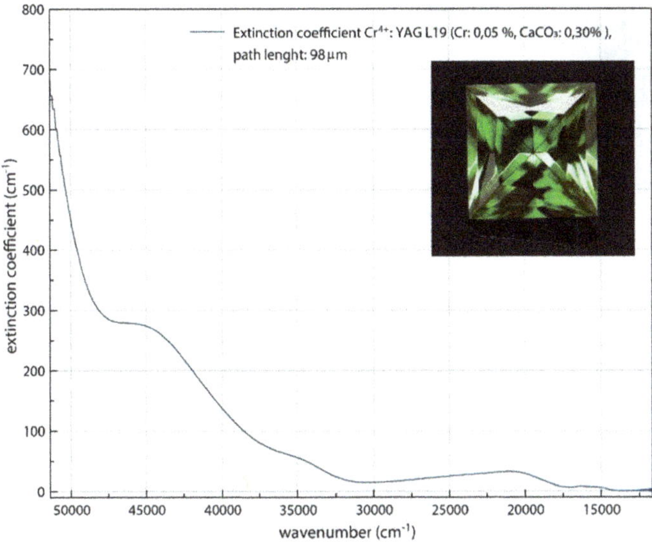

Figure 6.7.1: Absorption spectra of Cr-doped Yttrium-Aluminum-Garnet (YAG, $Y_3Al_5O_{12}$) showing the OMCT absorption band with its absorption maximum at about 45000 cm^{-1}, thickness of probe 89 μm.

6.7.1.2 Metal to metal charge transfer (MMCT)

Electron charge transfer between transition metal ions different in charge, which are located in adjacent polyhedra with common edges, cause absorption bands in the VIS- spectral range. Transitions between ions of the same element, such as Fe^{2+}–Fe^{3+} charge transfer, occur at lower energies than transitions between different transition elements, such as Fe^{2+}–Ti^{4+}, as in blue sapphire (Figure 6.7.2). Characteristic of MMCT-absorption bands is a large band width in the order of 5000–6000 cm^{-1} [2]. The most characteristic property of MMCT-absorption bands is the strict polarization along the MM-vector in the crystal structure [3]. The band intensity and thus also the color intensity of gemstones is related to the concentration product $[M_1] \times [M_2]$ of both metal-metal partners [4].

6.7.1.3 d-d-Transitions

Transitions between electronic configurations involving the incompletely filled d-orbitals split by nonspherical electric fields are responsible for the color as well as the laser properties of some of the most important gemstones such as ruby, Ti-sapphire, alexandrite, emerald or tsavorite (green garnet).

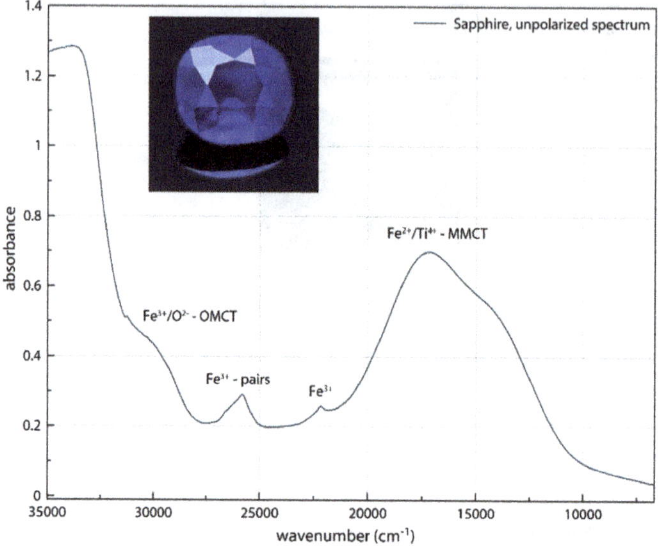

Figure 6.7.2: Absorption spectra showing absorption bands of Fe^{2+} – Ti^{4+} in blue sapphire.

6.7.1.3.1 Crystal field splitting

The five nd orbitals of transition metals are energetically equivalent (degenerated) in a spherical homogeneous field. Incorporated in a crystal field exerted by a non-spherical environment of negatively charged ligands, here in general O^{2-}, the five d orbitals split and thus give the possibilities of transitions between energetically and symmetrically different split terms. In an octahedral field, the lobes of the $d_{x^2-y^2}$ and the d_{z^2} orbitals point toward the negatively charged ligands (Figure 6.7.3). They are repelled and thus lifted to the double degenerate e_g-state. The orbitals d_{xy}, d_{xz} and d_{yz} are energetically lowered to the triple degenerate t_{2g}-state (Figure 6.7.4A). The notations e_g and t_{2g} are from group theory, where e stands for a state with double degeneracy and t for triple degeneracy. The index g indicates a crystal field having a center of symmetry such as in an octahedral field. The energy gap between the split orbitals is denoted as crystal field parameter Δ or Dq. The sum of the energy of the split orbitals is identical with the energy of the degenerate orbitals. As 1 cm^{-1} corresponds to 1.2397×10^{-4} eV, it is favorable to measure absorption spectra in units of cm^{-1}. From the position of the absorption bands, one can easily calculate the crystal field parameters. Following the Laporte selection rule, transitions between two different d-configurations are not allowed, having the same parity, whereas transitions involving p- and d-orbitals are allowed because of their different parity. The Laporte selection rule is overcome by the absence of a center of symmetry in the coordination polyhedron of the transition element in real crystals or by the interaction of d orbitals with odd-parity vibrational modes. By this, spin-allowed transitions where the spin does not change its momentum may become active and give rise to absorption bands much lower in

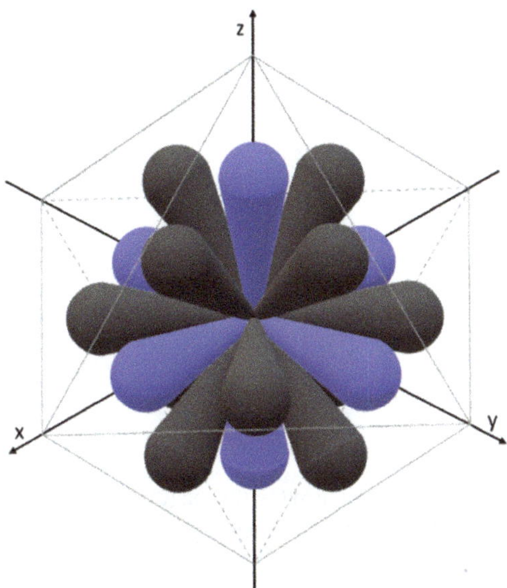

Figure 6.7.3: The d-orbitals of a d^n ion are shown in an octahedral field of negatively charged oxygen ions to explain the energy splitting.

intensity than of OM-charge transfer, MM-charge transfer and dd-transitions of tetrahedral coordinated transition elements, which have no center of symmetry. Spin forbidden transitions, where the of spin momentum is changed, show bands with very weak intensity in electronic spectra and small half width, and therefore play no major role for the color of gemstones. The d^3-ion Cr^{3+} in octahedral field may follow the transition scheme shown in Figure 6.7.4B and 6.7.4C with two spin-allowed transitions resulting in two characteristic absorption bands as in ruby and emerald (Figure 6.7.5). In the crystal structure of ruby, the distances between the oxygen and the central Cr-ion are smaller than in emerald. The respective d-orbitals are therefore more repelled (stronger splitting) and in consequence, the absorption bands occur at higher wavenumbers (energy) than in the emerald spectra. By filling in the d^n-electrons in the split orbitals in accord with the Pauli principle, one can expect two spin allowed transitions in case of N = 2, 3, 7 and 8 and only one transition in case of N = 1, 4, 6 and 9. As most coordination polyhedra in real crystals are distorted, more degenerated states may be lifted resulting in band splitting in the respective absorption spectra [1].

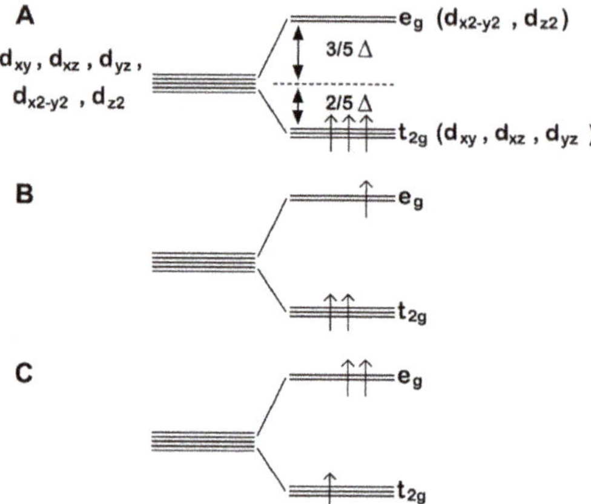

Figure 6.7.4: Crystal field splitting of d-orbitals in an octahedral field (A); (B) and (C) show electronic configurations for the example of n = 3 in the one electron model.

6.7.2 The corundum group (α-Al_2O_3, trigonal)

The name sapphire is widely used for undoped single crystal corundum for technical applications and will thus be used in this article. Because of its outstanding physical properties being the hardest oxide crystal (hardness 9 on the Mohs scale) with outstanding abrasive properties, transparency over a wide spectral range, high-thermal conductivity and its high melting point at 2040 °C, sapphire and its transition metal-doped varieties have a wide range of applications.

6.7.2.1 Ruby

Ruby is one of the well-known and most precious varieties of the corundum group for gemstone application. Most attractive pigeon-blood red color may be found for rubies containing 1–2 wt% Cr_2O_3 where Cr^{3+} occupies octahedral sites in the crystal. The first technical application was the use of gemstone quality ruby as jewel bearings for the manufacturing of clocks and watches. The invention of the Verneuil process also called flame fusion method at the end of the nineteenth century [5] was a breakthrough for the industrial production of jewel bearings and all kind of different colored corundum varieties and spinels for jewel purposes as well as the so-called sapphire glasses for the watch industry. A Verneuil apparatus consists in principle of a vertically arranged oxygen-hydrogen torch that is directed toward a pedestal with a mechanism for lowering the growing boule and that may hold a crystal seed. Above the tube of the torch, there is a container with a mechanism to obtain a controlled smooth feed powder flow, which will be molten while dropping through the vertically downward pointing flame and finally crystallizing by the heat transfer on the molten

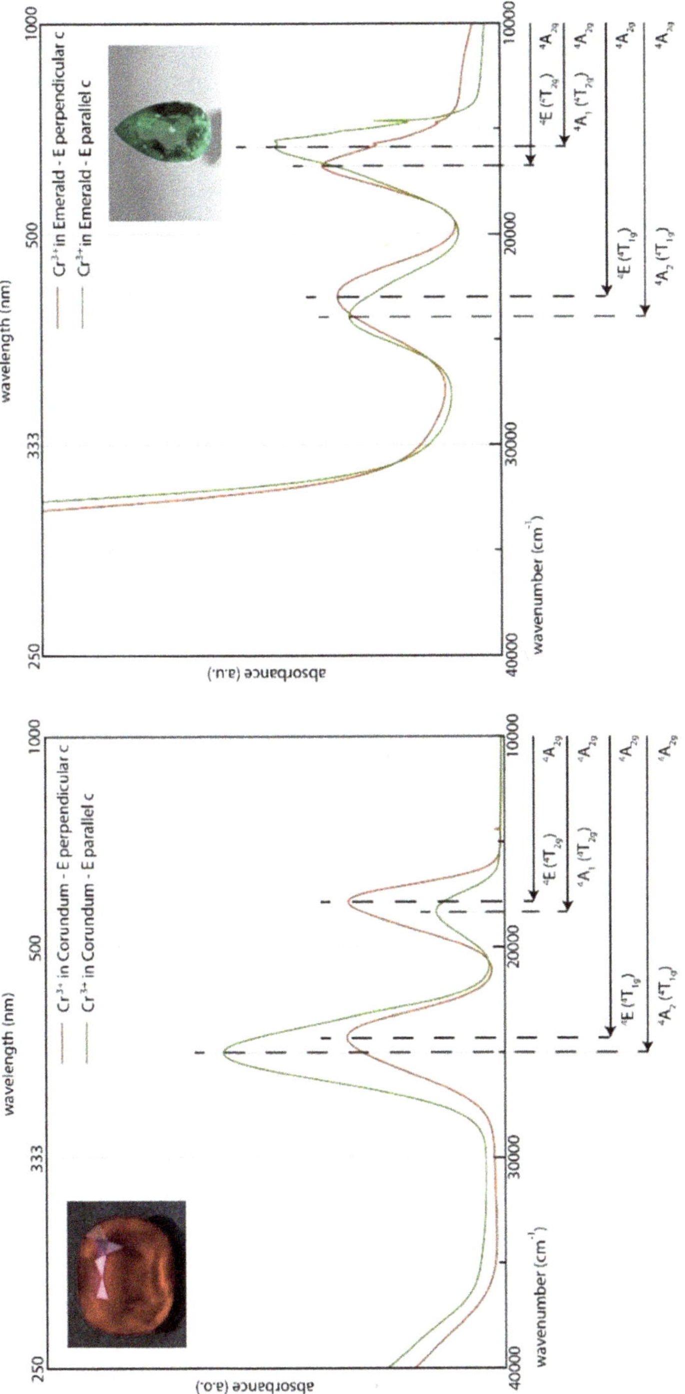

Figure 6.7.5: Absorption spectra showing absorption bands of Cr^{3+} in octahedral field for ruby (left) and emerald (right).

surface of the seed crystal. The crystal growth is achieved by continually lowering the pedestal. One of the advantages of this process is that no container (crucible) is needed for melting the Al_2O_3, its melting point being above 2040 °C. The first solid state laser with a laser wavelength of 694 nm was demonstrated by Maiman in 1960 [6] with a Verneuil ruby. Until 1970, rubies for the manufacturing of lasers (mainly for range finders) were grown by this method. Then, it was finally evident that rubies grown by the Czochralski method gave better optical qualities and thus better laser results. Still today, ruby laser crystals with dopant levels of 0.03 to 0.05 wt% Cr_2O_3 are grown by the Czochralski method (Figure 6.7.6) and commercially used mainly for the manufacturing of lasers for dermatology. The lamp-pumped ruby laser has lost its significance today and is substituted by more efficient lasers.

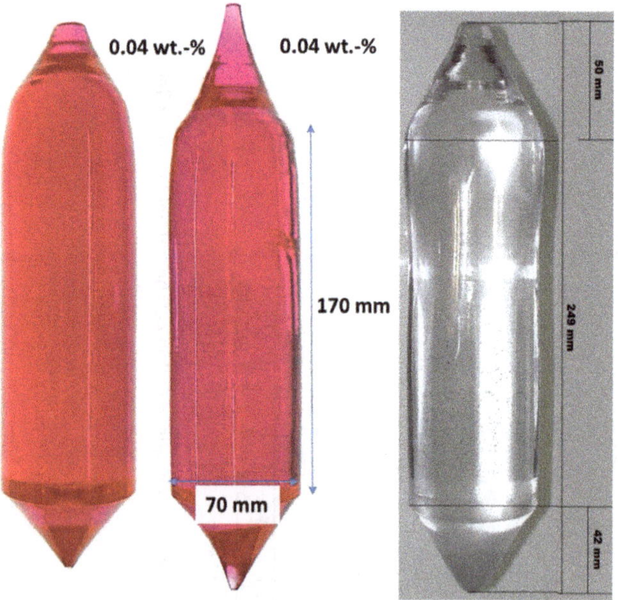

Figure 6.7.6: Ruby and undoped sapphire as grown by the Czochralski method.

6.7.2.2 Ti^{3+} Sapphire

Sapphire doped with 0.03 up to 0.15 wt% Ti_2O_3 is one of the important laser crystals having not only one distinct laser emission wavelength but showing a tunable range of emission wavelengths from 650–1100 nm. Ti-sapphire lasers are very common in nearly all scientific laser labs since they can easily be tuned to the required wavelength to pump test set-ups for new solid-state laser materials or for nonlinear optical experiments. Having a high-gain short-pulsed femtosecond lasers with very high pulse energy can be realized. Therefore, Ti-sapphire lasers play an important role in nuclear fusion experiments. Commercially, Ti-sapphire crystals free of voids and scattering centers are grown by the Czochralski method and heat exchanger method (HEM) and on experimental basis with the Kyropoulos method (Figure 6.7.7).

Figure 6.7.7: Ti-sapphire grown by the Kyropoulos method, diameter 250 mm.

6.7.2.3 Technical sapphire (α-Al_2O_3)

The name sapphire is widely used for single crystalline α-Al_2O_3. Crystallizing in the trigonal crystal system, sapphire is an optical uniaxial negative crystal, which exhibits anisotropy in its physical, optical, thermal and dielectric properties. The hardness ranges from 1525 Knoop perpendicular to the *c* axis to 2000 Knoop parallel to the *c* axis (9 on the Mohs scale). Its dielectric constant ranges from 9.39 perpendicular to the *c* axis to 11.58 parallel to the *c* axis. Sapphire's band gap is with approximately 10 eV one of the largest for oxide crystals and permits a useful optical transmission from about 145 to 5500 nm [7]. Because of its optical and physical properties, sapphire is used in a variety of demanding applications. Due to the specific requirements, such as crystal quality, crystal sizes and economy for the different applications, large efforts have been made to develop and adopt various crystal growing methods. In the context of selecting the correct crystal growth method for its respective application, one has to consider that for thermodynamic reasons crystal growth of sapphire in *c*-axis direction has not yet been successful to produce sapphire crystals with good quality. Therefore, 30 or 60° from the *c* axis have been the preferred growth directions so far. The *a* axis growth of sapphire is the preferred growth direction for crystals grown by the Kyropoulos-method, which will be discussed in the respective chapter.

6.7.2.3.1 Verneuil sapphire

The flame fusion process (Verneuil) is limited to diameters of up to 50 mm. Verneuil grown sapphire crystals are randomly oriented in respect to the crystal axis and show in general small-angle boundaries, as may be observed between crossed polarizers. They may also show an optical biaxial behavior, the angle being up to 5°. Therefore, Verneuil sapphire is mainly used for watch glasses, pistons for HPLC-pumps, balls for special ball bearings or ruby for probe tips in form measuring machines. For the latter applications, it was found that sapphire doped with up to 2 wt% Cr_2O_3 shows a higher hardness and better abrasive resistance as well as a diminishing of the hardness anisotropy [8]. Beside of these applications, some hundred tons of sapphire crackles are produced by the Verneuil process. These crackles are favorable against powder as raw material for the crystal growth processes, which use crucibles as melt containers because one can fill the crucibles in one step.

6.7.2.3.2 Czochralski sapphire

Concerning high-end crystal quality, the Czochralski method is the choice despite having the highest production costs. The price of the iridium crucibles used in this process is one of the driving costs and limits the upscaling for the production. With Czochralski crystal growth, high-quality sapphire crystals with defect densities of $10^3/cm^2$ to $10^4/cm^2$ and high transparency in the UV down to 240 nm have been grown up to 150 mm diameter. Because of the relatively high costs, sapphire crystals grown by this method are meanwhile only used for special applications for special optics, for the manufacturing of wafers for sapphire on wafer (SOS) transistors as well as for solar cell shields in aerospace projects. Transistors and integrated circuits based on these technologies may have high resistance against hard irradiation in space and may be used in satellites. The Czochralski method is of course also the method of choice for ruby- and Ti-sapphire laser crystals because of the high requirements for optical quality (Figure 6.7.6). In a Czochralski process, the material to be grown is melted in an iridium crucible (melting point 2442 °C) by induction heating. The induction coil and the thermal insulation set up being in a water-cooled growth chamber, which enables to adjust a gas atmosphere with adequate oxygen partial pressure to protect the crucible from oxidation while maintaining the right oxidation state for the transition metal ion. A sapphire rod, which serves as seed crystal, is lowered to the surface of the melt by a very precise linear actuator. If the temperature is right and the seed is in equilibrium with the melt, which means it is bathing in the melt without melting or gaining weight, the pulling of the crystal is started. With an electronic weighing device, which holds the seed, the weight gain per time unit during the pulling process is recorded and used in a special algorithm to control the energy input of the generator. Thus, the diameter of the growing crystal is controlled.

6.7.2.3.3 Heat exchanger method (HEM) sapphire

The heat exchanger method was one of the answers to the increasing demand of high-quality large diameter sapphire for various optical applications, such as infrared domes for military missiles and as material for laser windows and transparent armor. This method also proved to be suitable for the growth of large diameter Ti-sapphire laser crystals. Diameters up to 250 mm of sapphire crystals grown by this method have been reported [9]. Because the growth process takes place in high vacuum with resistance heating, comparatively cheap molybdenum or tungsten crucibles can be used, which can be disposed after each growth run. For this method, a seed is placed at the bottom center of the crucible. A tungsten tube, which is centered below the crucible enables to cool the seed by a flow of cold helium. After melting the sapphire crackles in the crucible, the flow of the helium prevents the seed to melt completely. By increasing the helium flow, a steeper thermal gradient is established resulting in a three-dimensional growth from the solid-liquid interface. Thus, the growth speed can be controlled over the complete process. When the content of the crucible is completely crystallized, the sapphire crystal will be annealed at temperatures of about 25 K below the melting point reducing the thermal gradient by decreasing the heat exchanger.

6.7.2.3.4 Kyropoulos sapphire

With this method, the so far largest single crystal sapphires up to 200 kg have been grown. Originally, this crystal growth process was developed in Russia to produce infrared domes for missiles or range-finding devices in jet fighters and transparent armor windows. The development of diodes emitting blue light and the development of phosphor materials to transfer the blue light into white and meanwhile warm white light has led to the replacement of normal light bulbs and neon tubes. Single crystalline GaN is necessary as basic material for the epitaxial growth of the semiconductor design for blue diodes. Unfortunately, GaN-single crystals cannot be grown as bulk material with large diameter as known for silicon. The most successful method for the production of substrates for the design of blue LEDs is the epitaxial growth of GaN layers on *c*-axis-oriented sapphire wafers. Along the *c* axis, the oxygen ions are arranged to have a hexagonal closest packing. Wafers cut perpendicular to the *c* axis (0001 plane) are thus suitable for heteroepitaxial growth of GaN, crystallizing in the hexagonal wurtzite structure [10]. The functional epitaxial layers can be grown on such a buffer GaN-layer, as is shown in Figure 6.7.8. As mentioned before, sapphire with acceptable crystal quality cannot be grown in *c* axis direction. Therefore, it is favorable to grow the sapphire in *a*-axis direction and drill the *c*-axis ingots perpendicular from the side of the *a*-axis-grown crystals (Figure 6.7.9). These ingots are cut into thin wafers and polished to an adequate epitaxial surface quality. Currently, most of the epitaxial growth machines installed worldwide are for 2-inch and 4-inch diameter wafers. A few 6-inch diameter

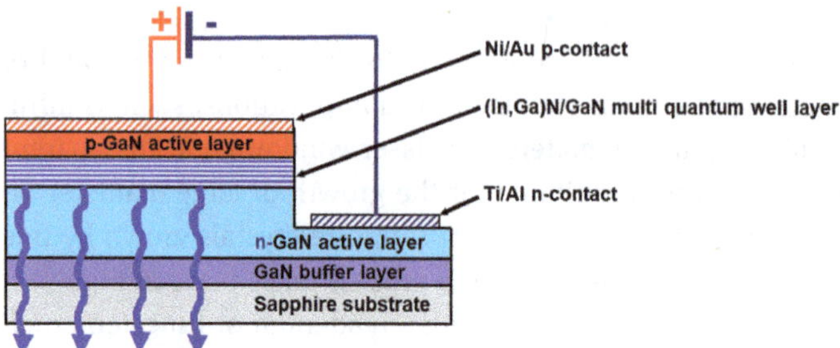

Figure 6.7.8: Schematic structure and electrical contacting of a typical blue emitting (In,Ga)N LED chip.

Figure 6.7.9: Kyropoulos sapphire-boule core drilled and cut into wafers for use as a wafer in blue emitting LED production.

production lines are meanwhile installed but it is not yet clear whether a further upscaling may be feasible in the future because it is most challenging to grow a 200 kg crystal with a minimum of crystal defects than a smaller crystal. The Kyropoulos-growth process of sapphire crystals is performed in a vacuum chamber with a resistance heater. A tungsten crucible filled with sapphire crackles is heated until the sapphire is completely molten. To start the crystal growth process, an *a*-axis oriented seed crystal, which is fixed at a cooled seed holder, is lowered to the center of the melt surface. Sapphire has a high-thermal conductivity and the cooling effect of the seed enables the start of the crystal growth. By gradually lowering the temperature of the melt, the diameter of the growing crystal is increased until it reaches a diameter of a few cm less than the crucible diameter. By pulling the seed holder upward at a very slow pulling rate and lowering the temperature accordingly, the crystal growth in the direction toward the bottom of the crucible until all melt in the crucible is crystallized.

6.7.3 The quartz group

When we talk about quartz, we mean the trigonally crystallizing α-modification of silica (SiO_2), which is stable from room temperature up to 573 °C where it is transformed into β-quartz. The colored varieties purple amethyst, yellow citrine, rose quartz and smoky quartz, occur widely in nature and are used as gemstones for jewelry or gem art sculptures. Natural colorless quartz crystals have been used for a variety of electronic devices since after World War II, when the rising demand for piezo quartz could no longer be satisfied by natural quartz crystals. As a consequence of the intensive studies on hydrothermal crystal growth (during and after the war in several laboratories), commercial production lines could be established worldwide after 1950 [11].

6.7.3.1 Piezo quartz

Because of its piezoelectricity (see also Chapter 9.9), quartz has become one of the most important crystals for various technical applications. One distinguishes between the piezoelectric effect, where electrical charges may appear on nonconductive crystal surfaces by applying mechanical stress such as pressure, shearing or pulling forces, and the inverse piezoelectric effect where mechanical expansion may be caused by applying an electric field.

Today, most electronic instruments use quartz crystal devices such as crystal microphones, piezoelectric speakers, ultrasonic devices and ultraprecise linear actuators in submicron regions. Vibratory quartz platelets are especially used in high-frequency technology for time controlling procedures (as in quartz watches or stabilizing the frequency in electric oscillator circuits). This may be achieved by coupling the mechanical vibrations of quartz platelets through the piezo-effect with an electric oscillator circuit. To keep the changes of the oscillation frequency within defined temperature intervals at a minimum, quartz-platelets of various distinct crystallographic orientations are manufactured. In consequence, quartz is quasi present in nearly all consumer electronic devices.

6.7.3.2 Optical grade quartz

The term optical quartz or optical grade quartz is used for quartz crystals grown at very low growth rates. A growth run may take 6 months up to 12 months, depending on crystal sizes. The most important applications for optical grade quartz crystals are for the production of low-pass filters for digital cameras and quarter wave plates for CD and DVD devices. Low-pass filters for digital cameras consist generally of two quartz plates and one wavelength plate and are positioned directly in front of the sensor of a digital camera, and filter high-frequency waves to prevent pseudo signals. Pseudo signals are generated by image pickup-devices such as CCD chipsets, and cause horizontal lines to look jagged or black and white patterns to look colored. This is caused by high-frequency waves and is also known as Moiré effect.

The function of quarter wave plates in CD and DVD devices take advantage of the effect of circular polarization in quartz crystals. Quarter wave plates used in CD and DVD devices are cut 45° to the *c* axis of the quartz crystal. The thickness of the plates is adjusted to a quarter or a multiple of the wavelength of the light of the used laser diode. The laser diode of a CD or DVD player emits linearly polarized light. A polarizing beam splitter is arranged in a way that the linearly p-polarized laser light passes completely before it is collimated by a lens. The now parallel beams, pass the quarter wave plate and are converted into right circularly polarized beams, which are now focused by an additional lens onto the pits and lands of the CD. The reflected light from the CD will change to be left circularly polarized. When passing through the wave plate, it will be converted again into linearly polarized light but with p-polarization, and will thus be deflected by the beam splitter in the direction of the detector.

6.7.3.3 Hydrothermal growth of synthetic quartz

Since α-quartz exhibits a phase transition to the β-modification at 573 °C, it cannot be grown from the melt and must be grown below the transition temperature. As a consequence, world-wide and dedicated research has led to a hydrothermal commercial process [12, 13], which allows growing large quantities of synthetic quartz in large gas-tight high-pressure autoclaves (Figures 6.7.10 and 6.7.11). The autoclave is filled in the lower part with quartz fragments of natural quartz and water. The filling grade is adjusted according to the pressure-temperature diagram of water to achieve the required pressure. In the upper part of the autoclave, seed plates are fixed on seed holders. The principle is a water solution transport reaction, where the transport of the dissolved SiO_2 is maintained by convection flow, caused by the temperature difference between the feed and the upper seed section of the autoclave. A baffle between the dissolution section and the growing section of the autoclave regulates the upstreaming flow. The growth rate is controlled by the temperature gradient, which can be adjusted. The seed part is heated by a separate heating circuit. A temperature gradient from 10 to 50 K may be used depending on the required growth rate. For quartz crystals used in optical applications, a lower gradient and therefore lower growth rates are chosen for quality reasons. Even at high pressures of about 1000 bar, the solubility of pure quartz between 300 and 400 °C is in the range of 1 to 2 g/1 kg H_2O. The solubility of quartz may be increased by a factor of 10 upon adding NaOH or Na_2CO_3. The use of sodium carbonate pressures up to 1700 bar may be applied, whereas in the sodium hydroxide process, a more moderate pressure of up to 830 bar may be applied during the growth process.

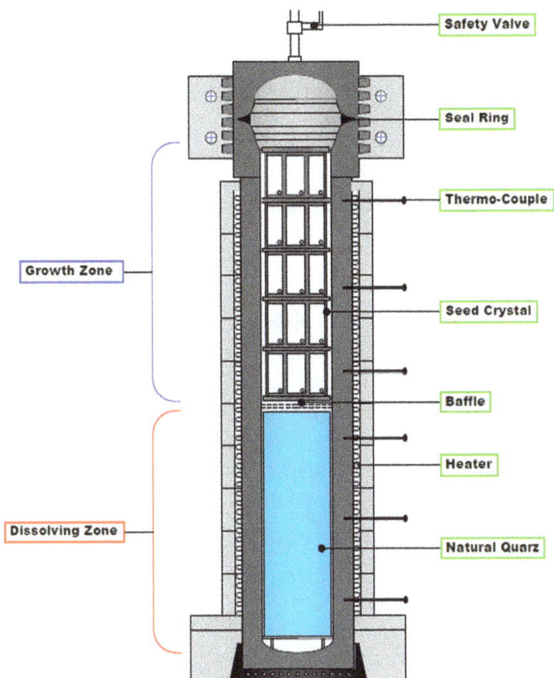

Figure 6.7.10: Schematic drawing of an autoclave for hydrothermal growth of α-quartz.

Figure 6.7.11: Quartz crystals harvested after hydrothermal crystal growth.

References

[1] Burns RG. Mineralogical Applications of Crystal Field Theory, Cambridge University Press, 1983.
[2] Mattson SM, Rossmann GR. Phys Chem Miner, 1987, 14, 94–99.
[3] Platonov AN, Langer K, Chopin C, Andrut M, Taran MN. Eur J Mineral, 2000, 12, 521–28.
[4] Smith G, Strens RGJ. The Physics and Chemistry of Minerals and Rocks, Wiley, New York, 1976, 583–612.
[5] Verneuil A. Ann Chim Phys Ser, 1904, 8, 20–48.
[6] Maiman TH. Nature, 1960, 187, 493–94.
[7] Arsenjew PA, Bagdasarow CHS, Bienert K, Kustow EF, Potjomkin AW. Kristalle in der modernen Lasertechnik, Akademische Verlagsgesellschaft, Geest und Portig, Leipzig, 1980.
[8] Albrecht F. Z Kristallogr, 1954, 106, 183–90.
[9] Schmid F, Khattak CP. September ed, Laser Focus, 1983.
[10] Moustakas TD. ECS Trans, 2011, 41, 3–11.
[11] Nassau K. Gems Made by Man, Gemological Institute of America, Santa Monica, California, 1980, 90404. ISBN 0-87311-016-1.
[12] Rabenau A. Angew Chem Int Ed, 1985, 25, 1026–40.
[13] Kniep R. Kontakte (Merck), 1991, 17–32.

7 Catalytic and active framework materials

7.1 Homogeneous catalysis

Markus Seifert, Jan J. Weigand

7.1.1 Introduction

One of the most attractive applications of various reactive inorganic materials is their use as active catalysts (or support materials) to improve the rate and also the selectivity of chemical transformation reactions in industrial processes. The Nobel awards of 2021 (D.W.C. MacMillan (US) [1] and B. List (GER) [2]) confirm the great importance of materials with inorganic sites as active catalysts for asymmetric organocatalysis in recent developments [3]. From the decomposition of H_2O_2 at low temperature with a platinum sponge (as described by J. J. Berzelius in 1836 [4]) and molecular surface reaction mechanisms in automotive catalysis (as investigated by G. Ertl in 2008 [5]) to selective production with metal complexes and organocatalysts (as discussed by B. List by 2021 [2, 3]), there is an evolution in different fields with certain similarities:

> In a catalyzed process, a substance (catalyst) remains unchanged after the transformation, although it participates in part of the reaction pathways to increase the rate of selected transformation pathways (selectivity). [4]

However, the definition of catalysis and different subgroups still grows with new findings and depends on the perspective (see Chapter 7.2). Two common subdivisions for application focus on the chemical nature of different compounds and the different phases taking part in the process, which both strongly impact the first rough choice of reactor geometries and concepts in lab-scale and industrial-scale production [6]. In general, heterogeneous catalysis (starting with J. J. Berzelius in 1836) describes processes where catalyst and reactants are separated in different phases and the reaction happens at the interphase [5, 7]. In contrast, homogeneous catalysis (as addressed in this chapter) means that the active catalyst and the reactants are in the same phase (gas, liquid, only seldom solid) [4]. According to the chemical nature of the active materials, homogeneous catalysis is divided into four classes:

(i) Soluble acids or bases (acid/base catalysis) [8, 9]
(ii) Organic compounds (organocatalysis) [3]
(iii) Soluble enzymes (enzyme or biocatalysis) [10]
(iv) Soluble transition metal salts or transition metal complexes (transition metal catalysis) [11]

Acknowledgment: The authors are grateful to Mr. Yu-Jian Zhou for polishing figures and schemes.

https://doi.org/10.1515/9783110798890-016

Since the dynamics of homogeneously catalyzed processes mostly do not suffer from slow, rate-determining transport, sorption, or phase transfer steps (see Section 7.1.6), the role of a catalyst *C* can be simplified to a sequence comprising the intermediates described as catalyst-reactant *R-C* and catalyst-product *C-P* complexes, as discussed by Eyring, Evans and Polanyi [12, 13].

$$R + C \rightarrow R - C \rightarrow C - P \rightarrow C + P \tag{7.1.1}$$

The main reaction events in homogeneous catalysis happen at the molecular level, which facilitates developments by control of kinetics, observation of reaction mechanisms and easy reaction engineering for scale-up to industrial level. The typical reactor design in industry is a stirring tank reactor (STR) for batch reactions or a continuously stirring tank reactor (CSTR) for continuous production. Advanced designs such as membrane or filtration reactors are illustrated within Chapters 7.1.4 and 7.1.5.

Since the main feature of a stable and active catalyst is its regeneration at the end of the reaction, the molecular interaction with reactants (as a catalyst-reactant complex) and until the final release of the product could be summarized schematically in a catalytic cycle, as depicted in Figure 7.1.1 [11].

In the case of metal complex catalysis, this is connected to specific reaction steps, such as oxidative addition, reductive elimination, insertion, β-hydride elimination as well as nucleophilic attacks on the ligand [4]. The slowest step(s) within these cycles finally define the overall rate of the catalyzed conversion process.

The performance of a catalyst in general could be judged by its activity for the reaction of interest, its selectivity for particular conversion routes, as well as its stability to endure for upcoming reaction cycles and its regenerability [4]. For the quantification of these performance criteria, there are fixed terms and definitions. The activity, or the conversion rate *r* by a catalyst, compared to the (often neglected) un-catalyzed slow reaction rate, is defined as the amount of reactant converted in relation to the relative amount of catalyst (molar, volume or specific weight). Within a certain period, a catalyst is capable of converting a fixed number of reactant molecules, depending on the reaction conditions (e.g. temperature, pressure, concentration), which is summarized as turn-over frequency (TOF). Ideally, the catalyst recovers after a catalytic cycle, becoming ready for the next reactant molecule. However, after a while and several cycles, real catalysts tend to change their performance due to competing reversible or irreversible deactivation processes (catalyst poisoning) [15, 16]. Consequently, the approximate number of cycles until a deactivation event occurs is quantified as turn-over number (TON). With focus on the desired target product, a definition for the catalyst's productivity (P) follows. In the case of reversible deactivation processes, a regeneration procedure leads to partial or complete recovery of the catalyst activity; however, in most cases, it is necessary to separate the catalyst from the reaction medium, which is challenging in homogeneous processes and often easier in heterogeneous catalysis.

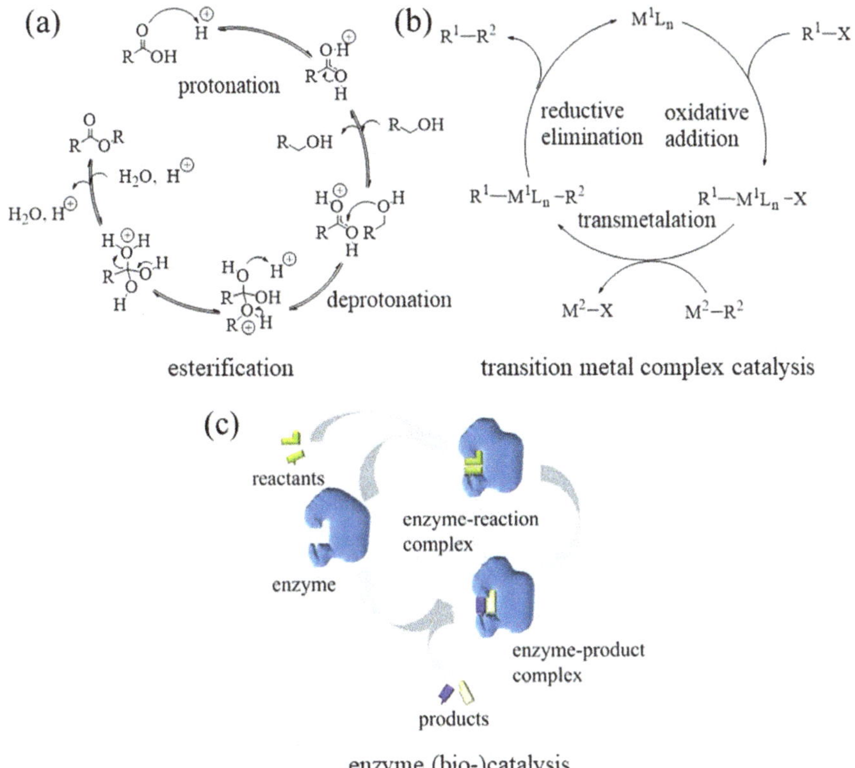

Figure 7.1.1: Mechanistic cycles for homogeneously catalyzed (a) acid-base reactions [14], (b) metal complex reactions [11] and (c) enzyme reactions (Michaelis-Menten) [10].

Especially the activity of a catalyst is ambivalent as highly active materials are less stable and tend to undergo (and also catalyze) competing reactions, which reduces their reusability and catalytic selectivity.

Highly active catalysts are less selective, highly selective catalysts are less active.

From an industrial perspective, the final output, the product yield Y and the productivity P have to be compared economically under the aspects of reactant input (atom efficiency), engineering efforts as well as infrastructural and environmental considerations [17, 18]. The abovementioned parameters (r, TOF, TON, P, Y, conversion X and selectivity S) are defined according to the equations shown below, where v_i corresponds to the stoichiometric coefficient of the reactant i (see eqs. 7.1.2–7.1.9):

$$r=\frac{\frac{1}{v_i}\cdot\frac{d(amount\ reactant)}{d(time)}}{amount\ of\ catalyst} \tag{7.1.2}$$

$$TON=\frac{amount\ product\ [mol]}{amount\ catalyst\ [mol]} \tag{7.1.3}$$

$$TOF=\frac{converted\ reactant\ [mol]}{amount\ catalyst\ [mol]\cdot time\ [h]} \tag{7.1.4}$$

$$P=\frac{target\ product\ [mol]}{amount\ catalyst\ [mol]\cdot time\ [h]} \tag{7.1.5}$$

$$X=\frac{converted\ reactant}{initial\ reactant} \tag{7.1.6}$$

$$S=\frac{actual\ product\ amount}{converted\ reactant} \tag{7.1.7}$$

$$Y=\frac{actual\ product\ amount}{initial\ reactant} \tag{7.1.8}$$

$$Y=X\cdot S \tag{7.1.9}$$

To reveal the origin of the performance of materials for homogeneously catalyzed processes, we have to focus on electrophilicity, nucleophilicity, steric hindrance and coordination chemistry to build a bridge to their reactivity at molecular level (see Figure 7.1.2). The activity of acid-base catalysts is accompanied with protonation/deprotonation reactions as well as the addition or abstraction of hydroxide anions. In some cases, the active species are formed first in the reaction phase, e.g. as in the case of ammonia: $NH_3 + H_2O \rightarrow NH_4^+ + OH^-$. In organometallic catalysis with metal complexes, the design of the interaction between ligand molecules and (transition) metal cations leads to many options by a broad spectrum of different combinations [19]. A free coordination site at the transition metal, often produced in the reactive phase by dissociation or association of loosely bound ligands or additionally reactive ligand molecules themselves, are the actual reactive sites. On the one hand, the electronic configuration at the center and the coordination sphere determine the electrophilic and nucleophilic interaction of reactant molecules with the metal cation. These properties are fine-tuned by electron donating or withdrawing ligand molecules to tailor the catalyst's reactivity, as is also known from coordination and supramolecular chemistry [20]. On the other hand, the geometric environment around the site is also tailored by use of bulky, chelating, bi/multidentate or amphiphilic ligands, which help to tune the residual Tolman cone angle at the site, i.e. the open window ready for a directed reactive attack of the reactants as known from pockets in enzymes [4, 19].

An especially outstanding advantage of ligand/transition metal site combinations in classic homogeneous catalysis is the transfer of stereo-information from the

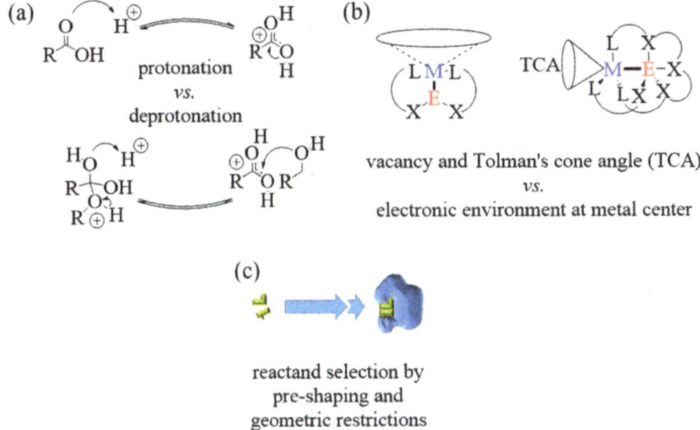

Figure 7.1.2: Concepts about reactivity in homogeneous catalysis: (a) protonation/deprotonation in acid-base catalysis, (b) electronic environment and electro-/nucleophilicity in transition metal catalysis and (c) geometric consideration of the attack of reactants at active catalyst species [4].

catalyst to the product by stereo-selective conversion or production. Especially ligands and their combinations (including intrinsic stereo-information or creating stereo-information by self-assembly and stacking) open the route for highly chemoselective, regioselective, diastereoselective or enantioselective industrial processes, as prominently recognized by the Nobel Prize committee in 2005. Industrially relevant examples are chemoselective metathesis conversions and couplings of olefins, en-in- and hetero-en-metathesis with Grubbs, Schrock and Sonogashira catalysts [4, 21–23]. In general, there is a still up-to-date list of practically relevant challenges in chemical industry, which show the most promising potential for economic and scientific revolutions [19]:

(i) Low-temperature epoxidation of propylene
(ii) Direct synthesis of hydrogen peroxide (H_2O_2) from the elements
(iii) Phenol from benzene and oxygen
(iv) Aromatic amines from ammonia (NH_3)
(v) Selective oxidation of methane to methanol
(vi) Anti-Markovnikov addition of alcohols and amines to alkenes

Most of today's upcoming industrial processes involve catalysts with increasing complexity of designed structures, involved phases and specialized reaction engineering concepts. Therefore, the choice of the right catalyst, phase structure and interphases (and finally of reactor for lab- and industrial-scale conversion reactions) needs a clever pre-choice to avoid costly misleading developments. A short comparison of general advantages and disadvantages of homo- and heterogeneously catalyzed processes is illustrated in Table 7.1.1.

Table 7.1.1: Comparison of advantages and disadvantages of homogeneously and heterogeneously catalyzed processes and catalysts thereof.

Homogeneous catalysis	Heterogeneous catalysis
= reactants, catalyst, products, etc. are present in one single phase	= reactants, catalyst, products, etc. are distributed over different phases with shared interphase surface
Activity and selectivity	
Uniform nature of catalyst after synthesis with direct contact to reactants in homogeneous phase **Tunable** chemo-, regio-, diastereo- or enantioselectivity (e.g. via ligands) [4] **Low** TOF (10^3–10^5 s^{-1}) and TON (~10^3–10^6) because of lower T tolerance (typically room temperature to 250 °C) [4] and due to lower stability of (often) organic materials (see also MOF in Chapter 7.4)	Often **heterogeneous catalyst surface/interphase region,** which leads to product distribution **Harder** to achieve chemo-, regio-, diastereo- and enantioselectivity [4] **High** TOF (~10^6–10^9 s^{-1}) and TON (up to 10^9) because of higher T tolerance (typically 300–900 °C) [7] and higher stability of (often) inorganic catalyst materials (see also Zeolites in Chapter 7.3)
Reaction engineering and control	
Molecular control due to good mass transfer (not rate determining) **Lower rates** and higher amounts of often costly (organometallic) catalysts (Classic) **stirring tank reactor** STR and continuously CSTR [6] (Advanced) **membrane, filtration and microreactors** with reactant and catalyst recycling [6, 24]	Often **transport limitation** by phase-transfer or (pore/film) diffusion and more complex macrokinetics Easy **high-throughput** continuous production with low amount of (often) cheaper catalysts (Classic) **fixed bed reactor** (also monoliths), fluidized bed reactors [25], multiphase reactors, e.g. trickle bed reactor [6] (Advanced) monolith or **impregnated microreactors** [6, 24]
Product purification and recyclization	
Complicated separation of catalyst, because of necessary solubility and low thermal resistance of (partially) organic compounds [19] Often **no options** for catalyst regeneration or reuse [16]	**Easy separation,** e.g. by simple filtration for solid catalysts and fluid reactants/products or by thermal separation (e.g. rectification) in the case of inorganic catalyst Often **available regeneration procedures** of mainly solid catalysts [15]
Main costs for development and production on industrial level	
Easier scale-up by control of molecular processes, facilitates upscaling from lab-scale to industrial scale [26] **Often costly** catalyst ligands and (transition) metals [19] + **costly purification** and catalyst separation	**Multistep process development,** because scale-up leads to complex changes in molecular and heat transport [27] **Often costly** precious metal containing catalysts, **but options for regeneration** and **easier workup** by catalyst separation

Thus, current developments in academia and in chemical industry tend to combine the advantages of both (1) by heterogenization of catalysts from homogenous catalysis [28–30] or (2) by use of (pH, *T* or *p* dependent) separation of reactants, products and catalyst with phase-transfer processes [26, 31], as displayed in Figure 7.1.3. In the context of resource-strategical concerns, the use of alternative non-fossil feedstock sources, the reduction of the amounts of (costly) noble (transition) metals as well as the close-up of production and recycling loops have become politically, but also economically relevant aspects [18].

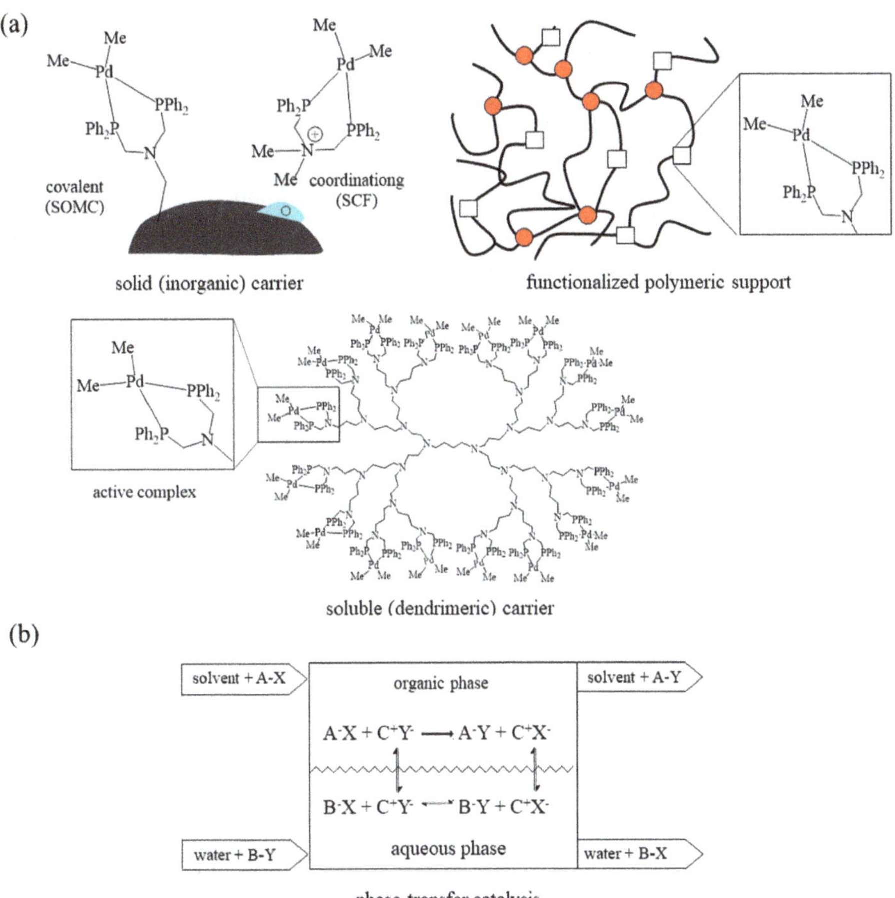

Figure 7.1.3: General developments about catalysis in academia and industry: (a) heterogenization of homogeneous catalysis [29, 30], (b) catalyst separation by phase-transfer processes [26, 31].

The first approach, namely the heterogenization of homogeneous catalysts, has been already reported for zeolite chemistry (see Chapter 7.3), which described the introduction and immobilization of organometallic coordination compounds in porous materials as "ship-in-a-bottle" complexes (SBC) [28]. A covalent immobilization of catalyst complexes is summarized as surface organometallic chemistry (SOMC) [29], which describes the chemical bonding of metal complexes on solid (inorganic) supports of carrier particles or organic and inorganic (ceramic) membranes [19]. Moreover, the enhancement of molecular weight of a catalyst for easier separation is possible by the formation of soluble catalyst dendrimers [26] as liquid supports, or supramolecular catalysts with self-assembled capsules, pockets and cages [30, 32]. A fourth option is the immobilization of active organometallic complexes in porous metal-organic frameworks (MOF) (see Chapter 7.4) as ordered supramolecular catalyst or covalently bonded organic framework (COF) [33]. As already known from enzyme immobilization, the linkage of molecules to a support may change its conformation as well as electronic situation at the active pocket, which reduces, in most cases, the overall productivity P. The heterogenization offers the opportunity to link different catalysts for tandem reactions, which leads to fast one-pot multistep reactions to avoid work-up in-between [4]. Moreover, there is a vast potential for the industrial use of inorganic supports with higher thermal resistance for higher rates and easier product purification [20].

The second strategy involves more complex engineering and switchable catalyst solubility (e.g. by *p*H, *T*, ionic strength) for phase-transfer processes. This enhances yields, e.g. for equilibrium reactions at low feed concentration, and provides a direct and selective transfer of products, educts or catalyst itself to another phase for easy separation. However, more complex engineering strategies (such as filtration reactor designs in liquid-liquid systems) are needed to overcome the additional phase transfer limitations (see Chapters 7.1.4 and 7.1.5) [19, 26, 34]. A direct link to combined electrochemical conversion (reduction/oxidation) and phase transfer steps controlled by relative solubility ratios has led to new combined options for an electrically controlled industrial production. A current example is the use of Ce^{3+}/Ce^{4+} systems as sulfates with different *p*H and temperature-dependent solubility for a redox-dependent salt precipitation [35].

7.1.2 Syntheses of catalysts for homogeneous catalysis

As explained in the previous subchapter, catalysts for classic homogeneous catalysis can be classified according to Behr *et al.* [4] as (1) soluble acids or bases (acid/base catalysis) [9, 10], (2) soluble organic compounds (organocatalysis) [3]), (3) soluble enzymes (enzyme or biocatalysis) [10] and (4) soluble transition metal salts or transition metal complexes (transition metal or organometallic catalysis) [11]. With respect to the aim of this book, this chapter focuses on the (2) organic compounds with heteroatoms and (4) soluble transition metal complexes as precursors or active compounds for their application in inorganic framework materials.

The ***synthesis of organocatalysts*** includes the broad complexity of organic chemistry, which itself is a topic of numerous books and subareas [14]. Small organic molecules predominantly composed of C, H, O, N, S and P accelerate different chemical reactions, which themselves are formed by organic and inorganic bulk reactions. An often-desired example published in 2018 is the anti-Markovnikov hydroamination of alkenes by small organocatalysts (see Figure 7.1.4) [36]. In this special case, a system of catalytic phthalimidyl-radicals (together with an over-stoichiometric amount of $P(OEt)_3$ as oxygen acceptor) leads to the desired regioselectivity. The syntheses of N-hydroxy phthalimide PhthN-OH (from phthalic acid anhydride and hydroxylamine phosphate at 130 °C) as well as the formation of $P(OEt)_3$ (from white phosphorus and chlorine *via* PCl_3 and subsequent conversion with ethanol) are well-known bulk-reactions in chemical industry.

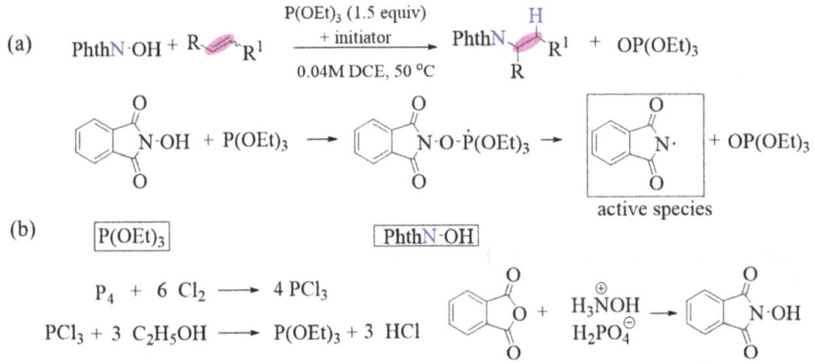

Figure 7.1.4: Application of an organocatalyst for anti-Markovnikov hydroamination of alkenes [36]: (a) chemical equation and stereochemistry, (b) synthesis steps of catalyst precursor PhthN-OH and reducing agent $P(OEt)_3$.

The ***preparation of transition metal complex catalysts*** is subdivided into two steps, i.e. (i) the synthesis of the (organic) ligand and (ii) the synthesis of the complex from a ligand and a transition metal salt.

> While the first step shows a similar variety as the synthesis of organocatalysts, the formation of the active complex in solution is a question of dynamic supramolecular self-assembly by addition of (stoichiometric) amounts of the ligand and the transition metal salt (or oxide), separately or in-situ through the reactant mixture. [4]

The process itself strongly depends on the charge and redox state of the ligands and the transition metal cation, which shows a strong dependency on the temperature as well as on the system's polarity (solvent), *p*H and ionic strength. In some cases of complexes with multiple ligands, the ordering as well as the protonation state of the ligands determines the structure, composition and performance of the catalyst. Often discussed materials are multi***copper*** catalysts for oxidation and hydrocarbonylation of alkanes, ***molybdenum*** complexes for olefins epoxidation, ***manganese*** and ***iron*** complexes for bleaching (cellulose) and oxidation (epoxidation), ***iron-oxo*** species for oxidation reactions in aqueous phase [11] or ***ruthenium***-based metathesis catalysts [37]. The last one, the so-called Grubbs-I-type catalyst, is an industrially relevant example, which is formed in a separate one-pot reaction, leading to air-stable, storable complexes (see Figure 7.1.5).

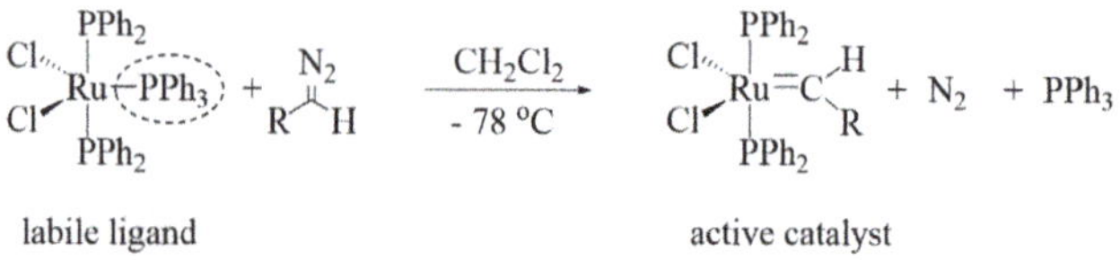

$$Cl_2Ru(PPh_2)_2(PPh_3) + N_2{=}CHR \xrightarrow[-78\,^{\circ}C]{CH_2Cl_2} Cl_2(PPh_2)_2Ru{=}CHR + N_2 + PPh_3$$

labile ligand

active catalyst

Figure 7.1.5: Synthesis of a transition metal Grubbs-I type catalyst for metathesis reactions [37].

The ligand often determines regioselectivity (e.g. for Ziegler-Natta catalysts) as well as chemoselectivity (e.g. for cyclopropanation reactions); see Chapter 7.1.3. Nowadays, there is a trend toward mimicry of biocatalysts (enzymes) with similar stereoselective behavior, but higher molecular weight. The synthesis of the bigger catalyst complexes is generally characterized by (i) supramolecular assembly [30, 38], (ii) formation of inorganic clusters or use of supports [11, 29] or (iii) the molecular weight enlargement and connectivity enhancement of ligands and, therefore, the formation of multinuclear transition metal complexes (see Figure 7.1.6) [20, 26, 29].

The ***encapsulation of the reactants*** and the location of the active metal center in pockets is inspired by the active center shape in enzyme catalysis. In the case of organometallic catalysts, the supramolecular assembly and the chelating effect are driving forces for the efficiency of these approaches [38]. A special application is the use of resorcinarenes and other macrocycles to coordinating cage-like superstructures for shape-selective cyclization reaction of sesquiterpene substrates [30]. In case that the synthesis of the complex by ligand organo-synthesis and combination of transition metal salt is comparable, the interaction between macro-ligand molecules have to be handled even more carefully in the context of reaction parameters (T, pH, solvent ionic strength, etc.) to ensure a controlled assembly of the pockets.

The use of multiple-centered complex nanoclusters in multi***copper*** catalysts for oxidation and hydrocarbonylation of alkanes [11] and the immobilization of classic homogeneous catalysts *via* **covalently bonded or impregnated surface organometallic catalysts** (SOMC) have already reached industrial level [29]. A special example is the MACBETH project by Evonik Performance Materials GmbH, which demonstrates the immobilization of standard catalysts on inorganic membranes for efficient hydroformylation, hydrogen production or propane dehydrogenation reactions [39]. In most cases, the covalent bonding to (mostly silica-based) supports or membranes (in SOMC by condensation reaction) generally changes the catalysts

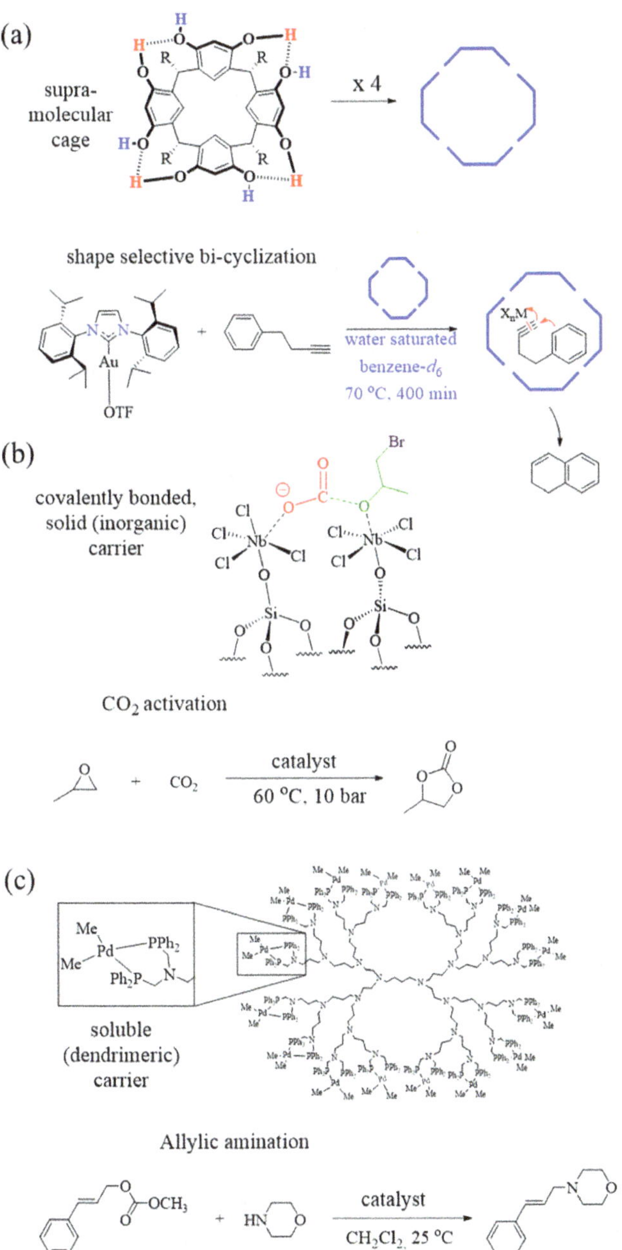

Figure 7.1.6: Strategies for the improvement of selectivity and reactant-specific conversion in homogeneous catalysis with applications (here, mimicry of enzyme pockets): (a) supramolecular encapsulation [30], (b) use of inorganic clusters and supports (SOMC) [29, 39], (c) use of multisite catalyst dendrimers [26, 40].

reactivity, which sometimes leads to a preferential use of surface-coordinating fragments (SCF), which are just loosely bound. The latter one are produced in situ, i.e. the formation of the active complex and the coverage of the support surface constitute a one-pot reaction, whereas a chemical preconditioning of the support surface and a subsequent chemical reaction to the active catalyst are necessary for covalent bonding [29].

The third strategy is the ***enhancement of the molecular weight of ligands and the use of organic or inorganic polymers*** as soluble support materials [20, 26]. The synthesis of the organometallic catalysts is often still a two-step procedure of (i) ligand synthesis and (ii) formation of the complex. However, in this case, the control of the molecular weight (during poly- or oligomerization reactions of bulky ligands or supports) becomes more difficult. As a practical example, a new type of dendrimeric Pd-DAB catalysts for the Pd-catalyzed allylic amination reaction, for instance, has to be built from the inner core to the outer (Pd)-shell in a controlled multiple-step synthesis.

The separate production of the amine backbone is followed by the connection of the terminating PPh_2 moiety as well as by the final coordination of the Pd center, and has to be handled in three, rather than in the usual two, steps to achieve a narrow distribution of molecular weight (see Figure 7.1.6c) [26].

7.1.3 Characterization of catalysts for homogeneous catalysis

For the characterization of homogeneous catalysts, some points must be emphasized: (1) The active catalyst (with focus on metal complexes) has to be activated, e.g. by release of a labile ligand and oxidative addition of a reactant. The composition and conformation under reaction conditions may differ from the as-synthesized or solid crystalline state. Thus, results from *ex situ* and *in situ* methods have to be used with care and discussed together. As a previously mentioned example, the activity of the PhthN-OH species in the hydroamination of alkenes is due to the formation of a radical-type PhthN-O catalyst (see Figure 7.1.4) [36]. (2) Homogeneous catalysis is mainly determined by the rate of the molecular processes. Performance tests have to be handled with particular care to guarantee reliable process parameters if transport processes are fast (e.g. by mixing) and molecular processes are rate-determining. In the special case of modern trends introducing additional phases in phase transfer catalysis (PTC) (see Chapter 7.1.5) and immobilized catalysts, similar transport issues and interface reactions have to be considered, as in the case of heterogeneous catalysis (see Chapter 7.2). (3) The behavior of a catalyst under reactive conditions is additionally affected by external influences, such as solvent molecules adsorbed to and in-between catalyst, reactants and products. An example for the long-range interaction of molecules by solvents is the unusual fast proton transfer mechanism between molecules in protic solvents, as discussed by Grotthus [41].

Therefore, this subchapter describes (i) the quantification of performance parameters and catalytic control, (ii) the characterization of the molecular structure and functionality and (iii) additional methods to characterize the catalytic, the chemical properties and the morphology of the increasing number of heterogenized homogeneous catalysts.

The ***evaluation of performance parameters*** starts with the quantification of the specific conversion rate (activity), the TON and the TOF in lab-scale test reactions.

The most common types are non-steady-state batch reactions with time-dependent sampling and steady-state continuous flow reactions with residence-time-dependent sampling of the product stream. To reduce the gap between lab-scale and pilot-scale tests, and in order to implement new trends such as phase transfer processes (see Chapter 7.1.5), there are specialized test geometries including filtration reactors, micro-reactors, etc. [6, 34]. The first objective after quantification of reaction rates is the analysis of the ranges of temperature and concentrations, where the targeted molecular kinetics constitute the rate determining steps (see Figure 7.1.7). This becomes increasingly important on the route to more complex multiphase and heterogeneous systems (see Chapter 7.2, Figure 7.2.10).

The investigation of trends and reasons for different catalytic performances is coupled to the ***investigation of molecular structural characteristics*** of the catalyst-reactant and catalyst-product complexes. If molecular processes are the slowest steps during the conversion, the transition-state theory by Eyring [12] describes the energy barriers for the formation of the reactive complex, with the conversion and the release of the product as the rate determining quantities. Besides spectroscopic methods such as UV/vis absorption and IR/Raman spectroscopy as well as mass spectroscopy, a special focus is laid on NMR spectroscopy and X-ray diffractometric analysis for crystal structure determination. While ^{1}H, ^{13}C, and heteroatom-specific NMR spectroscopies give hints about the atomic structure and inter-atomic connection of catalyst molecules as well as complexes in solution, the X-ray diffractometric structure determination on single crystals reveals geometric details, such as bond lengths, angles and coordination geometries, which are often similar (or related) if comparing fluid solutions and packed crystals. Together with quantum-chemical calculations, there are limited options to answer the questions regarding what the molecular structure looks like, how a catalytic reaction proceeds at molecular level and why the specific catalytic reaction is preferred over others (see Figure 7.1.8).

With increasing complexity and consideration of additional phases (PTC), additional supports (SOMC), or other methods to form heterogenized catalysts in solution (dendrimers), the morphology of the solids and related transport processes through the surface or solvent-catalyst interphase become increasingly important. As known from heterogeneous catalysis (see Chapter 7.2.3), the size, accessibility and chemical nature of the catalyst surface or reactant-catalyst interphase become rate determining features. A deeper look into sorption methods of different guest molecules, such as nitrogen, bases like ammonia (acid sites), or CO_2 (base sites),

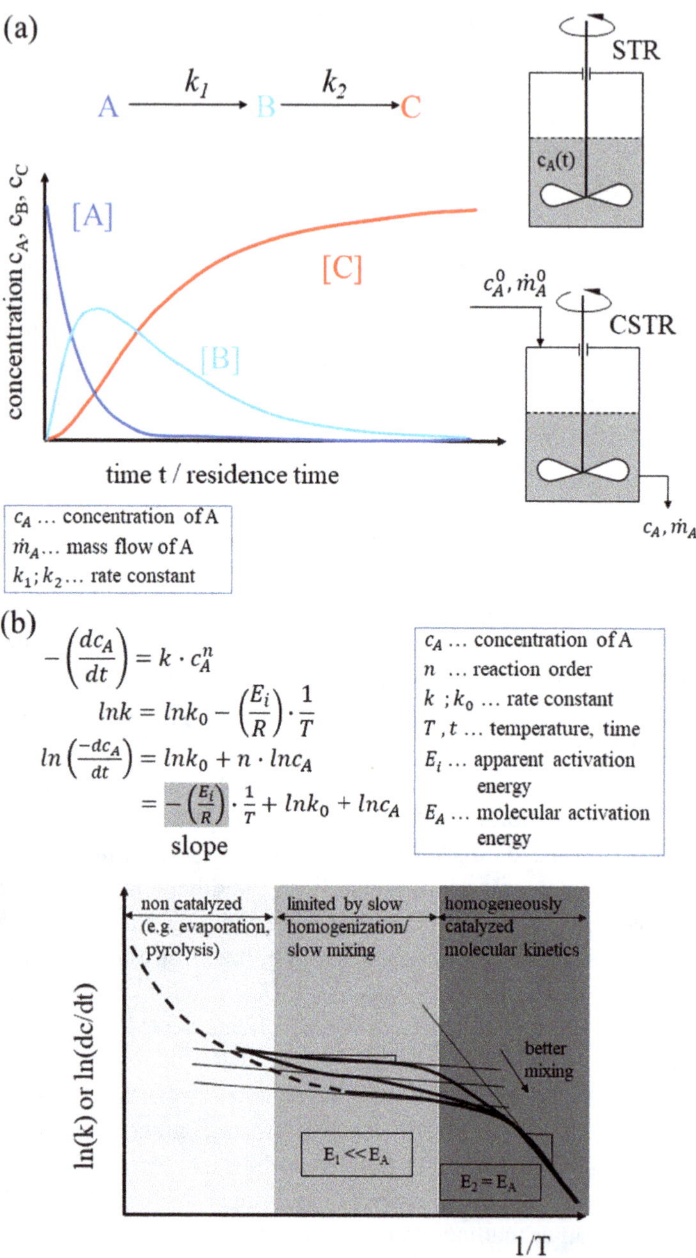

Figure 7.1.7: Performance tests and catalytic evaluation in homogeneous catalysis: (a) concentration-, time- and residence time dependencies in batch (STR) and continuous test reactors (CSTR); (b) Arrhenius plot, T- and mixing dependency to identify ranges of rate-determining molecular kinetics [6].

(a) Solid crystal structure by single-crystal X-ray diffraction

	Bond length [Å]
Ru1-C1	19.451(9)
Ru1-C12	2.0946(9)
Ru1-C22	1.8399(1)
Ru1-O2	2.3320(7)
Ru1-O3	2.2301(6)

(b) Species in solution by liquid-phase NMR spectroscopy

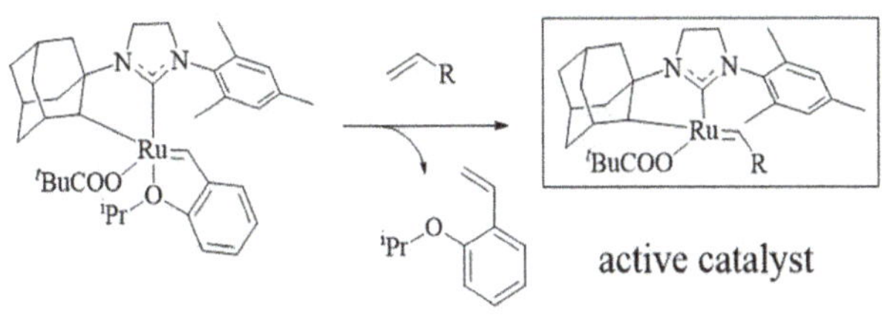

(c)

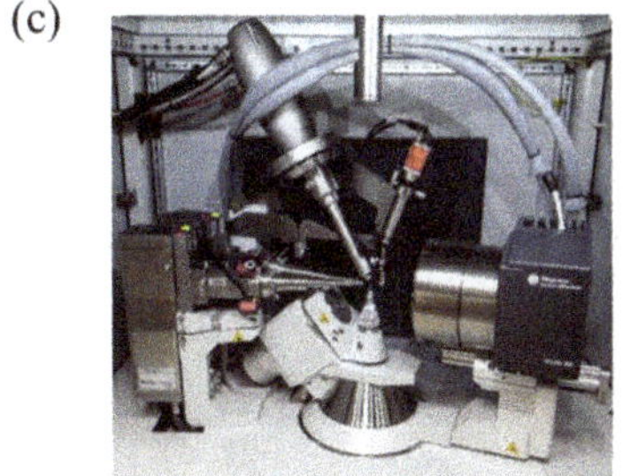

Supernova by Rigaku Oxford Diffraction

Ascend 500 by Bruker

Figure 7.1.8: Structural characterization of a Grubbs-Hoveyda-I catalyst for homogeneous catalyzed alkene metathesis (a) by X-ray diffractometry of single crystalline complexes and (b) by NMR of the active catalyst in liquid phase. Both (c) are used to estimate the atom connectivity, geometric parameters including ligand coordination or Tolman's angle [42].

gives a quantitative impression of the active surface, which has to be correlated with the catalytic performance. A qualitative impression of the local geometric topology can be achieved by the use of different electron microscopy techniques with frozen solutions, dried solid catalysts or resin-embedded materials, including scanning electron microscopies (SEM, TEM, STEM, etc.). Increasing complexity leads to the implementation of methods known from the solid-state chemistry related to porous zeolites (see Chapter 7.3) or MOFs (see Chapter 7.4).

7.1.4 Industrial application of inorganic acids, bases and metal complexes

In homogeneous catalysis, reactants, products and catalyst are typically mixed in one (fluid) phase. Consequently, a tank reactor (either in batch or continuous mode) is the first choice including a mechanical stirring system [6]. However, first advancements to overcome the drawback of costly catalyst and product separation have led to the incorporation of membrane filtration, by reaction and continuous recyclization of unconverted reactants. The use of organic (e.g. Nafion) or inorganic membranes (e.g. ceramics) gives the option to restrain or immobilize the catalyst for continuous operation without the use of costly and degrading thermal or extractive separation processes [26]. Compartmentalization of a reactor or immobilization of a catalyst on a solid (membrane) support involves further phase transfer tasks that are familiar from heterogeneous catalysis and has required the use of more sophisticated mixing and transport engineering strategies, such as in micro-reactor technology [24]. The different types are displayed in Figure 7.1.9.

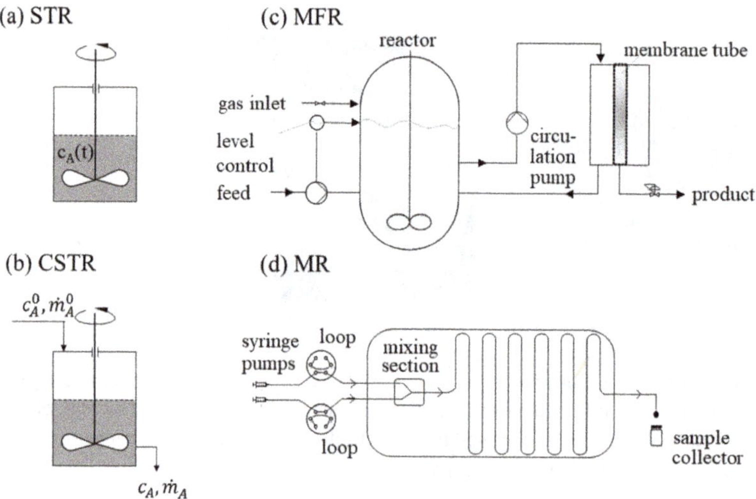

Figure 7.1.9: Pictograms showing the general principle of reactors and related catalysts used in homogeneous and heterogenized homogeneous catalysis: (1) STR – stirred tank reactor, (2) CSTR – continuous STR, (3) MFR – membrane filtration reactor, (4) MR – micro-reactor [6, 24, 26, 34].

For a suitable choice of a reactor in homogeneous or heterogenized homogeneous catalysis, some general aspects for evaluation are summarized in Table 7.1.2 together with industrial examples to give reasons for different applications.

In acid-base catalysis, a typical product (acrylic acid butyl ester as softener in acrylic resins) is produced using sulfuric acid as catalyst in a batch process, e.g. by BASF Schwarzheide GmbH. Security issues and the separation of the ester in the organic phase from the by-product (water) and the catalyst (sulfuric acid) lead to a

Table 7.1.2: Comparison of advantages and challenges of different reactor types in homogeneous and heterogenized homogeneous catalysis and catalysts thereof.

Reactor and catalyst system	Advantages	Challenges	Industrial example
STR (stirring tank reactor) [6]	Easier **reaction control** (mixing, fast release) Multifunctional application (on-demand production) **Lower investment** and easy upgrade (e.g. by addition of sensors) **No start-up times**	**Set-up times** between different batches (cleaning, preparation, etc.) **Limited homogenization** with rising volume **limits throughput** **Demanding** product and catalyst workup (ultrafiltration)	High pricing specialty chemicals On-demand production or enhanced requirements for reaction control (e.g. polymerization of specialty varnishes and adhesives) [43]
CSTR (continuous STR) [6]	Higher **throughput** Easier **process control** (mixing, heat transfer) No **set-up time**	Enhanced **start-up time** Enhanced **investments** **Lower flexibility** **Demanding** product and catalyst workup (ultrafiltration)	Dangerous chemicals (toxic, explosive) in bigger batches (e.g. nitration of toluene in cascades [44])
MFR (membrane filtration reactor) [6, 26]	**Decoupling residence time** of catalyst (fixed on or retained by membrane) and reactants [26] Use of soluble support for catalyst and **on-line separation** from smaller reactants or those with different polarity	**Costly** membranes, hard workup **Difficult replacement** of spent (inactive) catalyst (as soluble support or immobilized on membrane) **Transport issues** by membrane plugging or film formation (parallel flow) **Pressure issues** due to membrane for cross- or dead-end flow [26]	For costly catalysts (e.g. Grubbs-Hoveyda catalysts with Ru salts and costly ligands [45]) For environmentally harmful or toxic catalysts (e.g. Al, Ti or Zn organyles in Ziegler-Natta catalysis [46] or different coupling reactions, e.g. epoxides with CO_2 [47])
MR (micro-reactor) [24]	**Intensification** of mass transport and heat transfer Easy scale enhancement by **numbering up** instead of scale-up Good **reuse of catalyst** (immobilization at reactor wall)	**Complex fine-tuning** of peripheral equipment, numerous feed-pipes **Pressure drop** in small pipes **Complex cleaning** catalyst recovery and workup Risk, **unusual flow** (Taylor diffusion)	For on-demand processes; with high requirements regarding mixing and heat transfer (e.g. also nitration of toluene with higher yields of trinitrotoluene (TNT) [44])

favorable production using a STR but bearing additional costs and setting-up time. A cost factor is the purification of sulfuric acid from organic compounds and the stoichiometric amount of water formed during the process. Alternative developments have emerged, including the use of solid acids, such as heteropolyacids (e.g. polyphosphoric acid) for easier separation and reuse of the catalyst, which in turn raises the technical demands (including phase contact and mixing) [48]. The trend in homogeneous acid-base catalysis goes to the use of small-sized polyanions, such as inorganic polyoxometalates, which precipitate as salts with changing *p*H and temperature, in order to combine the good phase contact of homogeneous, soluble catalysts with the easy separation of solids by filtration [49].

In the case of the "captains of homogeneous catalysis" [4] (namely the transition metal complex catalysts), the development of the industrially relevant ***Grubbs-type metathesis*** catalysts displays some general principles and trends for the future [21–23, 45]. Ru- (and Mo, W)-based alkylidene and aryloxide complexes serve as catalysts for olefins metathesis *via* metal carbene intermediates at closing (CM), ring closing (RCM), ring opening (ROM) as well as enantioselective (EROCM) or asymmetric ring opening and closing metathesis (AROCM) reaction or polymerization (ROMP) [45].

An example for a typical enantioselective olefin metathesis reaction is adopted from Dawood and Nomura [45], which displays the geometric restrictions in Figure 7.1.10.

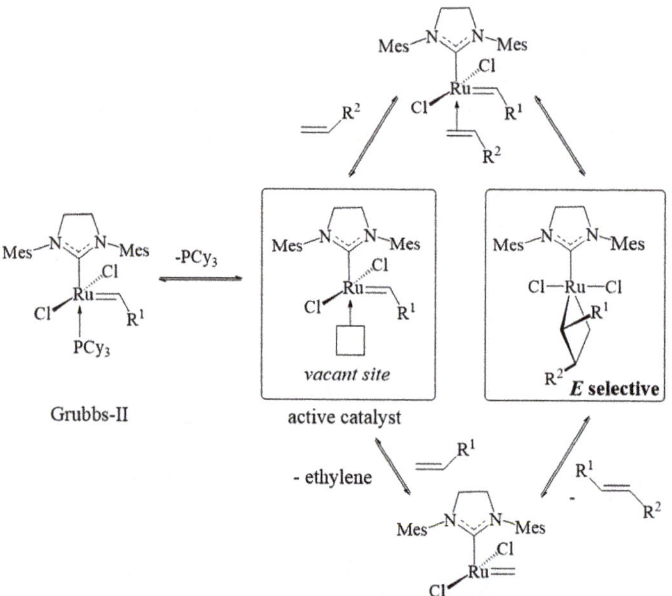

Figure 7.1.10: Scheme of an *E*-selective alkene metathesis by a ruthenium-carbene catalyst (Grubbs-type), adapted from Dawood and Nomura [45].

A labile ligand releases a **free coordination site** for the oxidative addition of an olefin. Rearrangement of the carbene moiety directed by the bulky ligands leads to the formation of an **alkylidene complex by C–C bond formation, subsequent rearrangement** and final regeneration of the active catalyst species.

Since the first reports about metathesis by Calderon in 1967 [50], the development of industrially relevant Grubbs-type metathesis catalysts started in 1996 with the first reported Ru-based catalyst for different RCM and ROM reactions (see Figure 7.1.11) [37]. Introduction of bulky, strong coordinating and sometimes prochiral imidazolidinyl-ligands has led to enantioselective metathesis reactions and higher catalyst stability using Grubbs-type II catalysts [51]. In a third step, the stability and selectivity of the catalyst rise again replacing a phosphane ligand by an aromatic ether attached as

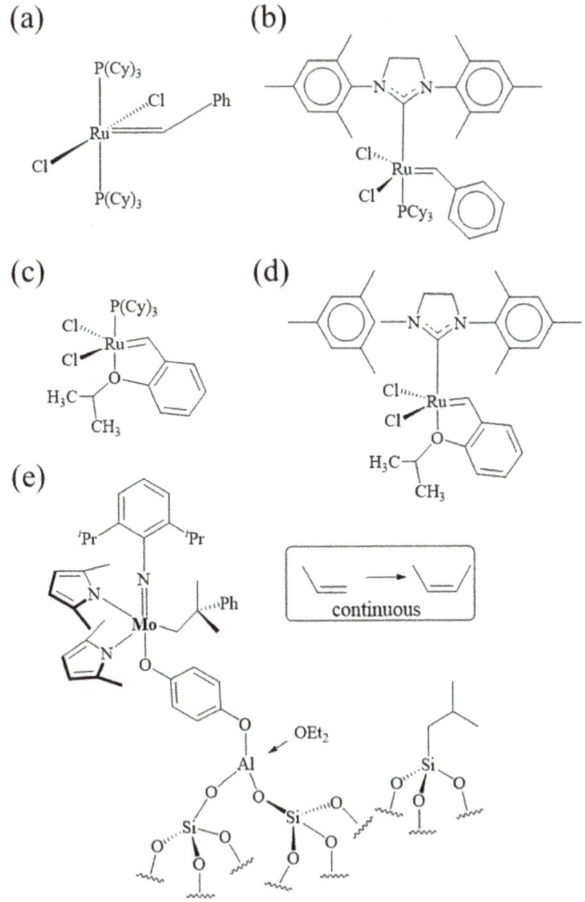

(a) (b) (c) (d) (e)

continuous

Figure 7.1.11: Structure of the industrially relevant olefin metathesis catalysts: (a) Grubbs I, (b) Grubbs II, (c) Grubbs-Hoveyda-I, (d) Grubbs-Hoveyda-II, (e) immobilization and cheaper transition metal (Mo) for continuous production [45].

a carbene and oxo-functionality to force a different coordination geometry and to produce another reactive vacancy at the Ru center. The respective Ru-ligand combinations are known as Grubbs-Hoveyda type I and type II catalysts, respectively [52]. Today, Ru-based Grubbs, Grubbs-Hoveyda and Mo-based Schrock, W and other transition metal metathesis catalysts are often applied in pharmaceutical production, due to their insensitivity toward numerous other functional groups and due to their good applicability in different solvents, which justifies the Nobel Prize for R. H. Grubbs, R. R. Schrock and Y. Chauvin in 2005.

In general, a more rigid and bulky coordination sphere including prochiral ligands improves the enantioselectivity within metathesis and alkylidene transfer reactions and enhances the catalyst's stability. Meanwhile, excellent catalyst solubility has to be achieved during the process to avoid transport limitations, which otherwise complicates subsequent product workup and catalyst reuse. Therefore, even in this case, the trend goes to immobilization of Grubbs-type catalysts for easier separation thereof [53].

As soon as Ru and costly ligands become price drivers, the use of cheaper metals (such as Cu, Mo, Mn, Fe, despite their lower performance) became the focus of transition metal catalysis [11]. A parallel trend combines the advantages of selective and uniform catalysts for homogeneous phase applications with the good separation, thermal resistance and activity of solid catalysts to compensate the process performance (see Chapters 7.2 and 7.3).

A famous strategy is the **mimicry of uniform catalysts** by **single atom or single site surface catalysts**, e.g. for direct epoxidation reactions of ethylene and propylene, formerly mentioned as some main industrial challenges (see Chapter 7.1.1) [19].

Even though well-dispersed, small alumina-supported silver particles of 100–150 μm are suitable species for direct ethylene epoxidation, there is a need for new direct propylene epoxidation technologies as efficient alternatives to the current H_2O_2-consuming HPPO process (hydrogen peroxide to propylene oxide) by Evonik and Uhde [54]. In this trend, the common nanoparticles on solid supports shrink to single atom or few atom ensemble sites at the surface, which show a mixed behavior between classic carbene chemistry and sigma-, delta- and my-adsorption complexes at low reaction temperature (Cu_2O at ~100 °C) [55].

Including an additional phase (during the product formation) leads to a further industrial driving horse, namely the polymer production, especially the case of the ***Ziegler-Natta catalysts*** for olefin polymerization [46]. With increasing molecular weight of the product, the solubility drops until the product can be separated directly as a solid. In the special case of Ziegler-Natta catalysts, the main industrially relevant products are HDPE (high-density polyethylene) and isotactic PP (polypropylene) [56, 57]. It constitutes an industrial example on how the chirality and orientation of the reactive complex determines the stereochemistry of the chain growth (isotactic, syndiotactic) [58].

The intermediary **structure of Ti-, Zn- or Hf-based catalyst and Al-based co-catalysts** in **Ziegler-Natta catalysis** determines the geometric environment, bite angle and orientation of the polymer chain, which promotes isotactic or syndiotactic over atactic chains.

The change in viscosity and solubility with increasing molecular weight has to be handled by special reaction engineering and control of residence (conti) or reaction time (disconti) [59]. To enhance the reusability and separation of the environmentally harmful Al- and Ti-based catalysts, again inorganic-supported materials, such as $MgCl_2$ or $TiCl_3$, are the focus of developments toward heterogenization (see Figure 7.1.12) [56].

7.1.5 Phase-transfer catalysis with inorganic frameworks and polymers

If we consider the trend of catalyst heterogenization as a strategy to combine (by chemical design) the desirable properties of homogeneous and heterogeneous catalysis [26], the introduction of different reactor compartments by additional phase barriers for selective transfer can be a suitable strategy toward reactor and process engineering [19, 34]. Phase transfer catalysis (PTC) describes the transfer of reactants, products or a catalyst-reactant complex through a phase border for simultaneous reaction and separation. The principle mimics nature, where biochemical catalysts (enzymes) are often fixed onto or within biological membranes to use natural gradients (passive transport) or chemical energy (active transport) for selective conversion reactions in low concentrated media, while achieving product and catalyst separation. Most promising examples are the proton pump in cells and the cytochrome complexes for the electron transport chain in photosynthesis [10].

Everything considered, the introduction of natural or artificial liquid-liquid barriers by solubility differences or membranes (and sometimes also including immobilized catalysts) has the significant benefit of heterogeneous catalysis (namely easy and simultaneous separation of reactants, catalyst and products), which enables high-throughput continuous reactor technologies [34] (see Figure 7.1.13). It is increasingly applied if [60]:

(i) educts or intermediates are toxic, explosive or show low solubility (work with low concentrated media);
(ii) equilibrium reactions show low conversion; the removal of the product, therefore, supports significantly higher yields;
(iii) separation and recirculation of reactants, side products or catalysts is needed. In the case of low temperature and highly selective reactions, higher conversion and yields can be achieved by recirculation of the reactants.

Sophisticated designs include dead-end and cross-flow filtration and membrane reactors to ensure good mass transfer through mixing inside the container (see CSTR), to settle the phases afterward and to separate product and catalyst for recirculation (see

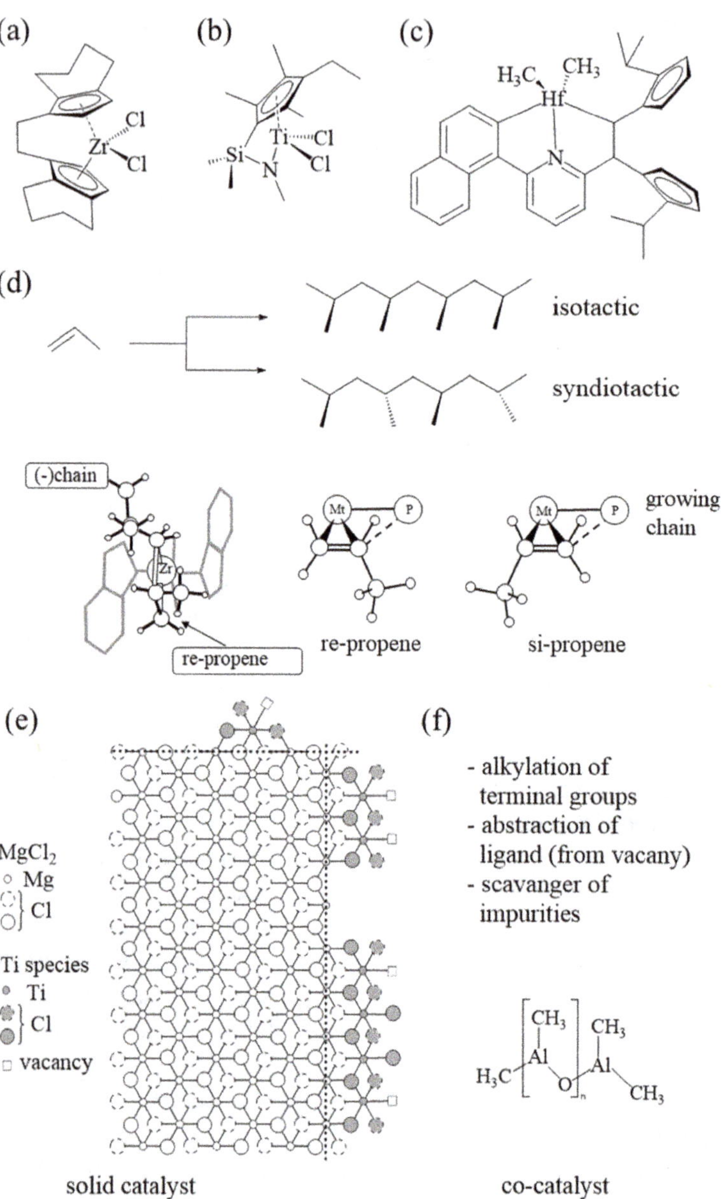

Figure 7.1.12: Structure of the industrially relevant Ziegler-Natta catalysts developed for stereoselective polymerization of olefins: (a-c) available catalysts, (d) stereochemistry of the reactive complex, (e) trend of heterogenized catalysts, (f) methylaluminate as co-catalyst [46, 57, 58].

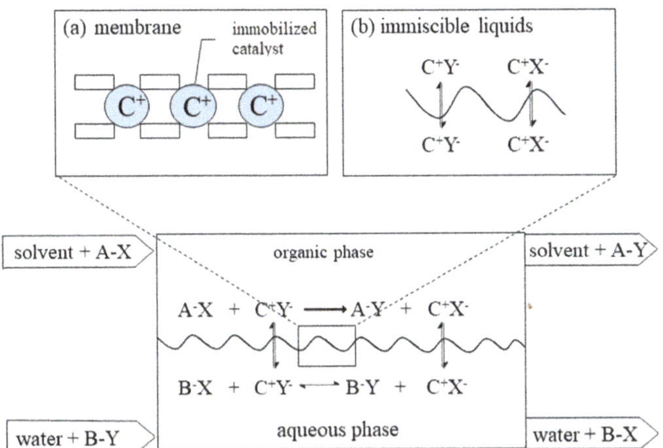

Figure 7.1.13: Scheme of a general phase transfer catalysis process (PTC); focus on the phase transition of catalyst and products through the liquid-liquid or membrane-separated interphase [31].

Chapter 7.1.4) [26]. Most common examples for phase-transfer materials are quaternary ammonium salts, which act as surfactants and form dynamic membranes and micelles for successful compartmentalization [31]. A next step are 0D-phase transfer catalysts, i.e. encapsulated nanoparticles as carriers for immobilized catalysts, which show capsule-dependent solubility and phase-transfer behavior at the interphase between two immiscible liquids [32]. Moreover, there are several smart inorganic polymers, which act as surfactants with flexible pH, T or p-dependent adsorption capacity, solubility and permeability for reactant, product or catalyst molecules [20].

A very successful industrial application of PTC is the ***cyclopropanation of olefins with diazo compounds*** catalyzed by ***Ru-ligand catalysts*** [61] (see Figure 7.1.14). Cyclopropane derivatives are often used for the production of different herbicides and pharmaceuticals such as antibiotics or antipsychotics. A typical example is the herbicide profluralin. The similar structure of the catalysts used for cyclopropanation and Grubbs metathesis accounts often for competing reactions involving those two conversions. Even in the case of cyclopropanation, enantioselective reactions are possible depending on the catalyst (ligand) design, as discussed for metathesis reactions in Chapter 7.1.3. The higher investments of a filtration reactor design in this example are justified by the use of highly toxic and explosive diazo compounds, which are less dangerous at lower concentrations.

The separation of the product and the selective transfer of the desired catalyst-product complex still lead to high yields of the desired cyclopropanation product, despite the competing metathesis side reactions. Particularly ligands with heteroatoms linked to Ru-phosphane complexes (or ruthenacarboranes) are used to achieve enantioselectivity [61].

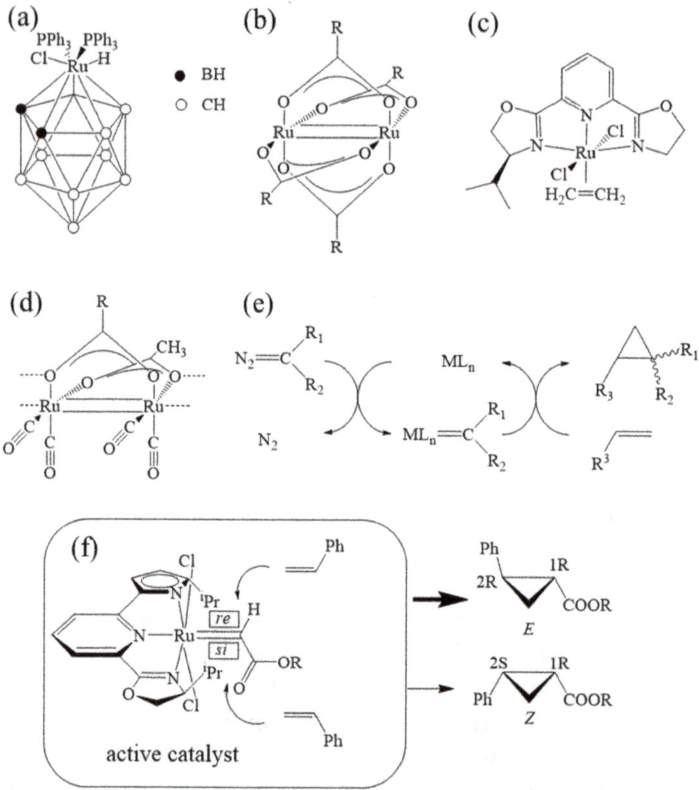

Figure 7.1.14: Application of phase transfer catalysis (PTC) in the cyclopropanation of olefins with diazo compounds [61]: (a–c) catalyst development with excurses (d) to polymeric or heterogenized species, (e) catalytic cycle and (f) example for stereochemistry at the example from (c).

> Again, also in the case of cyclopropanation and PTC, the trend goes to heterogenization by polymerization or resin bonding to enhance the catalysts molecular weight for an easier separation [61, 62].

Nowadays, new trends in sustainable chemistry and resource-strategic engineering make use of electrochemical approaches to reduce the energy consumption (if compared with plain heating), and to enhance atom efficiency (by controlled phase-transfer and redox reaction for product separation [63] or even for catalysis [64]). The phase-transfer ability and sensitivity to other parameters (e.g. *p*H, solubility or morphology, and even the reactivity) of a catalyst are modified by changing the redox state involving an electrode. In the case of electrocatalysis, the electron itself participates on the process.

Taking into account the aspect ratio, shape or size selection of reactants for PTC inaugurates the transition to porous (solid) materials known from classic heterogeneous catalysis (e.g. zeolites in Chapter 7.3) or modern metal organic frameworks (see Chapter 7.4).

For a preliminary conclusion, PTC, MFR and MR technologies on the one hand, and catalyst heterogenization (e.g. SOMC, SCF, soluble carriers) on the other hand, represent modern trends for the development of processes including inorganic and organic framework clusters, as a transition between homogeneous and heterogeneous catalysis (see Figure 7.1.15).

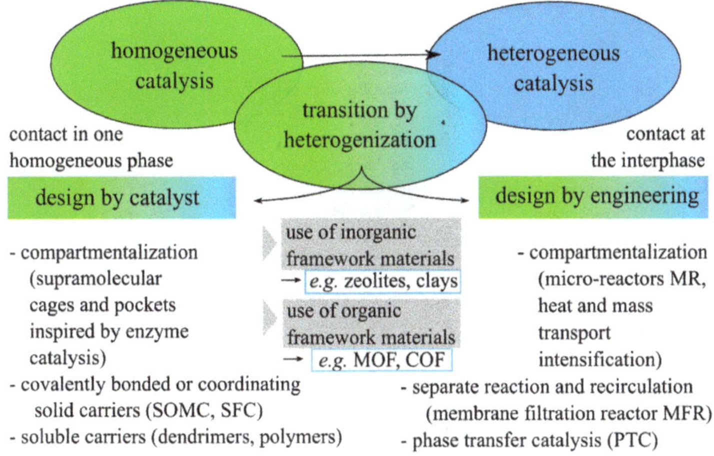

Figure 7.1.15: Development of the transition from homogeneous to heterogeneous catalysis by heterogenized homogeneous catalysts (design by catalyst) and phase transfer catalysis (PTC) (design by engineering).

7.1.6 Thermodynamics and kinetics in homogeneous catalysis

By definition, homogeneous catalysis involves the direct contact of reactants, catalyst and products in one phase, and therefore markedly differs from heterogeneous catalysis. As described in the former subchapters, there is a trend to profit from heterogeneous systems increasing the complexity of different phases while collecting benefits, such as easier catalyst and product (or costly catalyst) separation as well as ensured conversion (even at low concentrations) of harmful reactants (see Chapter 7.1.5).

Due to the use of liquid solvents and the tendency to use (partially) organic compounds, there is a limited temperature range (between approx. 20 and 250 °C) along with an absolute pressure span (from 2 to 30 bar) to achieve an economic production. In most cases, transport properties are less controllable due to their relatively low temperature or pressure dependence. Consequently, homogenization by mixing or conversion at lower temperature leads to rate-determining molecular kinetics and process control. Considering the balance equations by Damköhler for mass, heat and momentum, they simplify in the case of classic homogeneously catalyzed processes to simple molecular kinetics (see Figure 7.1.16). Since the general kinetic considerations by Eyring, Evans and Polanyi, and respective rules, the speed of the molecular steps are commonly described by the activation energy barrier of the slowest (rate determining) step [12, 21].

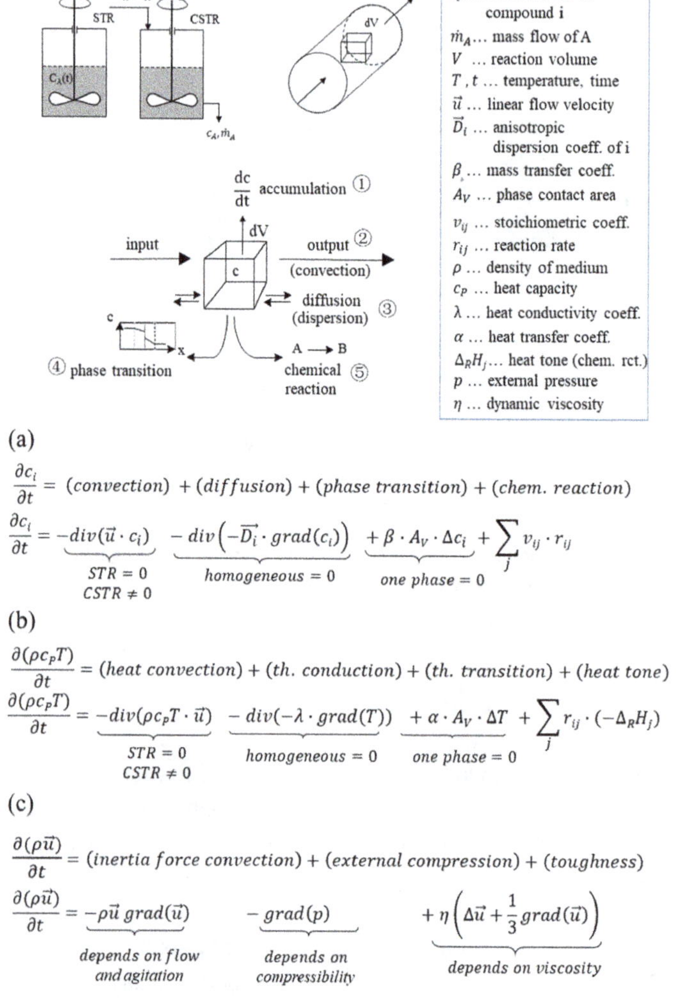

Figure 7.1.16: Balance equations by Damköhler for transport and chemical reaction in chemical reactors (STR, CSTR) for homogeneous catalysis including (a) mass balance, (b) heat balance and (c) momentum balance [47].

From Arrhenius kinetics, there is an option to access the rate-determining step in a reactor system by change of the reaction temperature (see Figure 7.1.17). From this, it is possible to find a suitable parameter set for micro-kinetic reaction control [4].

In a next step, product evolution and reactant consumption along with (residence) time paired with different (assumed) reaction networks can be used altogether to determine the reaction dynamics at a molecular level.

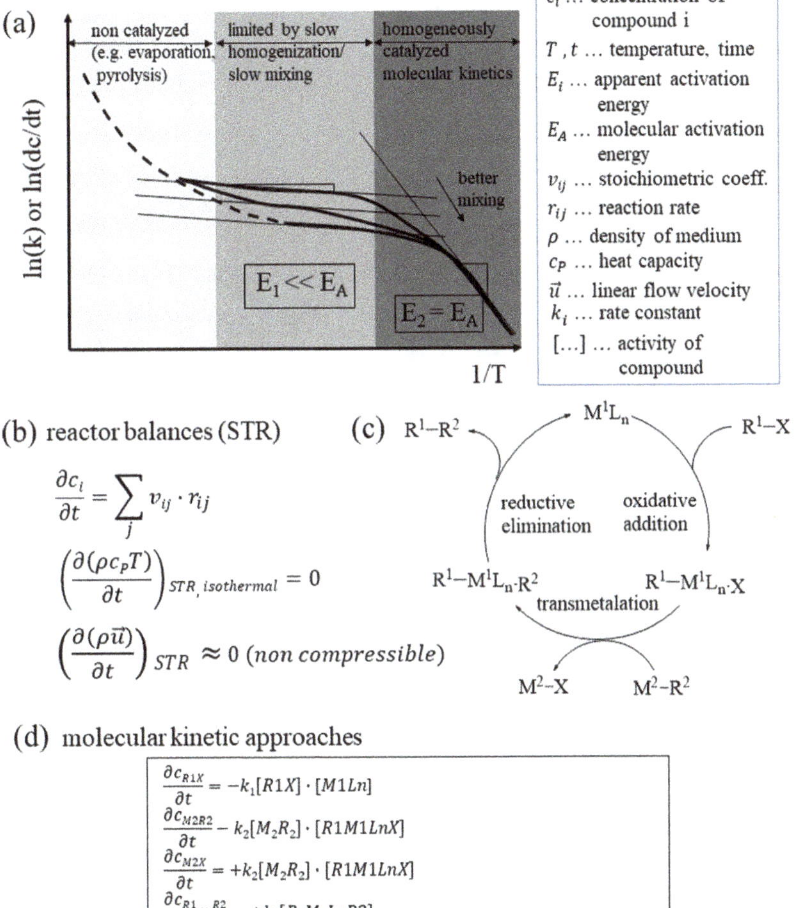

Figure 7.1.17: General example of the steps for a kinetic consideration of a homogeneously catalyzed batch reaction (STR): (a) analysis of mixing and T-dependent initial rates to reach molecular controlled parameter regime, (b) approaches for reactor balances (simplified for well stirred isothermal STR), (c) kinetic approach for catalytic cycle and (d) solution to molecular approach for mass balance [6].

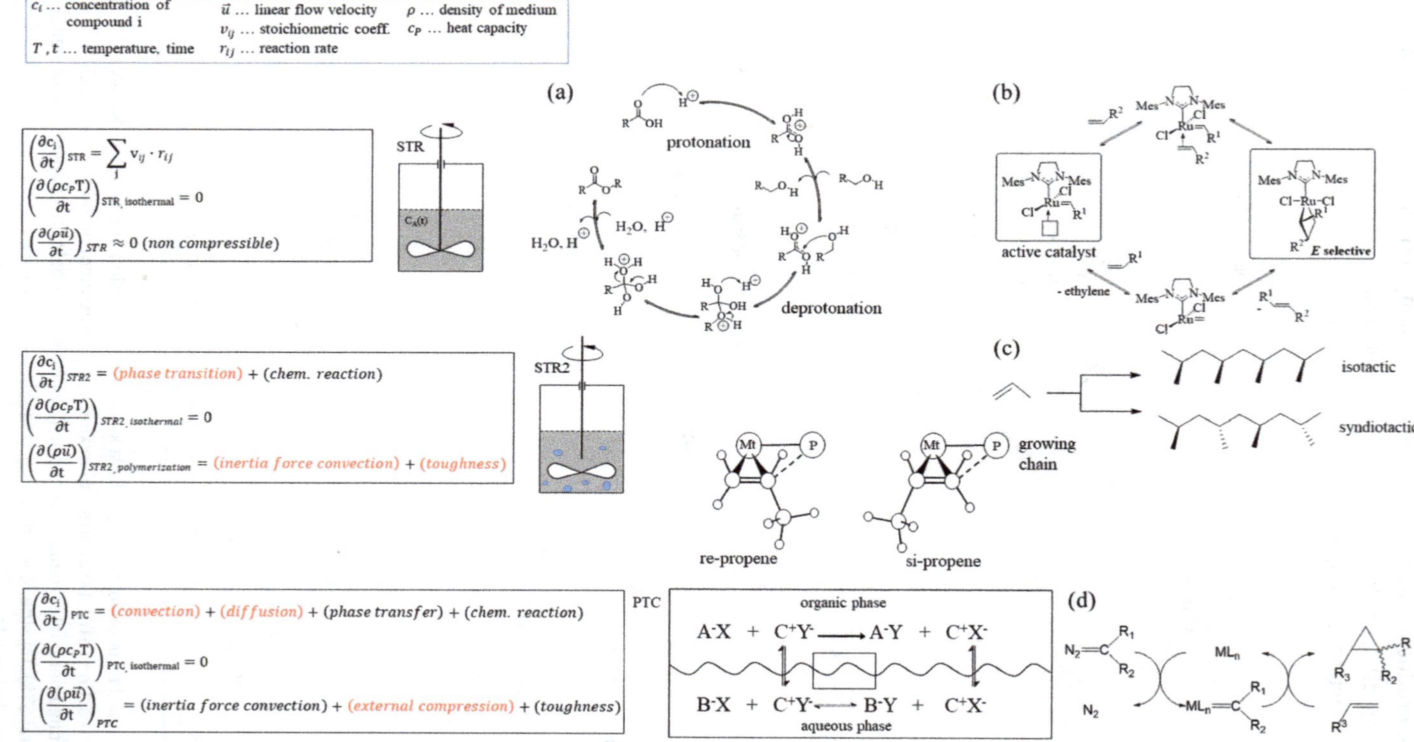

Figure 7.1.18: Application strategy of molecular approaches for macroscopic consideration of kinetics and balance equations for (a) acid catalyzed ester formation from acrylic acid and butanol, (b) stereo-selective metathesis alkene reaction with a Grubbs-II catalyst [22], (c) isotactic Ziegler-Natta polymerization of propylene [56] and (d) phase-transfer catalyzed cyclopropanation of diazo-compounds with Ru-based catalysts [31, 61].

Using our four examples from the last subchapters, there is an increasing level of complexity (see Figure 7.1.18): Starting from simple ester formation (of acrylic acid with butanol and sulfuric acid, A → B), going through the suppression of cyclopropanation and parallel stereo-selective metathesis reaction (involving Grubbs catalysts, A → B; A → C, A → D) [22, 61], as well as through the precipitation of a new phase (Ziegler-Natta polymerization, n A → B) [56], it culminates in more complex, almost heterogeneous phase-transfer kinetics including phase-transfer limitation (as for cyclopropanation of diazo-compounds) [38].

Increasing complexity and number of compartments in a reaction system lead to a mimicry of biocatalysts including their dynamic behavior in chemical conversion. Therefore, in some cases, Michaelis-Menten-like kinetics are observed, if the adsorption of species on bulky or heterogenized catalysts becomes rate determining or an equilibrium reaction state [6]. This leads to the description of some inorganic (e.g. zeolites) or mixed organic and inorganic framework materials (e.g. MOF) as "chemozymes", which could be translated as artificial enzymes (see Chapters 7.3 and 7.4).

References

[1] MacMillan DWC, David WC. MacMillan – Facts. https://www.nobelprize.org/prizes/chemistry/2021/macmillan/facts/ (accessed 2021/11/25).

[2] List B. Benjamin List – Facts. https://www.nobelprize.org/prizes/chemistry/2021/list/facts/ (accessed 2021/11/25).

[3] List B. Asymmetric Organocatalysis, Edition 1, Springer-Verlag, Berlin Heidelberg, 2009.

[4] Behr A, Neubert P. Applied Homogeneous Catalysis, Vol. 1, Wiley-VCH, Weinheim, 2012. Verlag GmbH & Co. KGaA: Weinheim, 2012.

[5] Ertl G. Angew Chem Int Ed, 2008, 47, 3524–35.

[6] Reschetilowski W. Handbuch Chemische Reaktoren, Edition 1, Springer-Spektrum, Berlin Heidelberg, 2020.

[7] Ertl G, Knözinger H, Schüth F. Handbook of Heterogeneous Catalysis, Edition 2, Vol. 1, Wiley-VCH, Weinheim, 1997.

[8] Rizwanul Fattah IM, Ong HC, Mahlia TMI, Mofijur M, Silitonga AS, Rahman SMA, Ahmad A. Front Energy Res, 2020, 8, 1–17.

[9] Rothenberg G. Catalysis: Concepts and Green Applications, Edition 2, Wiley-VCH, Weinheim, 2017.

[10] Bisswanger H. Enzyme Kinetics: Principles and Methods, Edition 2, Wiley-VCH, Weinheim, 2008.

[11] Eldik R, Hubbard CD. Advances in Inorganic Chemistry: Homogeneous Catalysis, Vol. 65, Elsevier, Oxford, 2013.

[12] Eyring H. J Chem Phys, 1935, 3, 107–15.

[13] Evans M, Polanyi M. Trans Faraday Soc, 1937, 33, 448–52.

[14] Brückner R. Reaktionsmechanismen, Edition 3, Springer Spektrum, Berlin Heidelberg, 2004.

[15] Argyle M, Bartholomew C. Catalysts, 2015, 5, 145–269.

[16] Crabtree RH. Chem Rev, 2015, 115, 127–50.

[17] Sheldon RA, Arends I, Hanefeld U. Green Chemistry and Catalysis, Edition 1, Wiley-VCH, Weinheim, 2007.
[18] Anastas P, Eghbali N. Chem Soc Rev, 2010, 39, 301–12.
[19] Cornils B, Herrmann WA. J Catal, 2003, 216, 23–31.
[20] Hey-Hawkins E, Hissler M. Smart, Inorganic Polymers: Synthesis, Properties, and Emerging Applications in Materials and Life Sciences, Edition 1, Wiley-VCH, Weinheim, 2019.
[21] Vougioukalakis GC, Grubbs RH. Chem Rev, 2010, 110, 1746–87.
[22] Thangavel M, Chin SY. IOP Conf Ser: Mate Sci Eng, 2020, 991(012073), 1–11.
[23] Huang S, Lei Z, Jin Y, Zhang W. Chem Sci, 2021, 12, 9591–606.
[24] Reschetilowski W. Microreactors in Preparative Chemistry: Practical Aspects in Bioprocessing, Nanotechnology, Catalysis and More, Wiley-VCH, Weinheim, 2013.
[25] Sie ST. Appl Catal A, 2001, 212, 129–51.
[26] Müller C, Nijkamp MG, Vogt D. Eur J Inorg Chem, 2005, 4011–21.
[27] Mitchell S, Michels NL. Perez-Ramirez J. Chem Soc Rev, 2013, 42, 6094–112.
[28] Bania KK, Karunakar GV, Goutham K, Deka RC. Inorg Chem, 2013, 52, 8017–29.
[29] Samantaray MK, Pump E, Bendjeriou-Sedjerari A, D'Elia V, Pelletier JDA, Guidotti M, Psaro R, Basset JM. Chem Soc Rev, 2018, 47, 8403–37.
[30] Gaeta C, La Manna P, De Rosa M, Soriente A, Talotta C, Neri P. ChemCatChem, 2020, 13, 1638–58.
[31] Starks CM, Liotta CL, Halpern ME. Phase-Transfer Catalysis: Fundamental, Applications, and Industrial Perspectives, Edition 1, Springer-Science, Dordrecht, 1994.
[32] Sadjadi S. Encapsulated Catalysts, Elsevier, Oxford, 2017.
[33] Rogge SMJ, Bavykina A, Hajek J, Garcia H, Olivos-Suarez AI, Sepúlveda-Escribano A, Vimont A, Clet G, Bazin P, Kapteijn F, Daturi M, Ramos-Fernandez EV, Llabrés I Xamena FX, van Speybroeck V, Gascon J. Chem Soc Rev, 2017, 46, 3134–84.
[34] Cornils B, Herrmann WA, Horvath IT, Leitner W, Mecking S, Olivier-Bourbigou H, Vogt D. Multiphase Homogeneous Catalysis, Wiley-VCH, Weinheim, 2005.
[35] Paulenova A, Creager SE, Navratil JD, Wei YJ. Power Sources, 2002, 109, 431–38.
[36] Lardy SW, Schmidt VA. J Am Chem Soc, 2018, 140, 12318–22.
[37] Schwab P, Grubbs RH, Ziller JW. J Am Chem Soc, 1996, 118, 100–10.
[38] Shirakawa S. Designed Molecular Space in Material Science and Catalysis, Edition 1, Springer Nature, Singapore, 2018.
[39] Richter M. Jetzt wird es groß – neuartiger Membranreaktor auf dem Weg zum Industriemaßstab. https://corporate.evonik.de/de/presse/pressemitteilungen/performance-materials/jetzt-wird-es-gross-neuartiger-membranreaktor-auf-dem-weg-zum-industriemasstab-128880.html (accessed 2021/11/25)
[40] Brinkmann N, Giebel D, Lohmer G, Reetz MT, Kragl U. J Catal, 1999, 183, 163–68.
[41] Agmon N. Chem Phys Lett, 1995, 244, 456–62.
[42] Endo K, Grubbs RH. J Am Chem Soc, 2011, 133, 8525–27.
[43] Ilare J, Sponchioni M, Storti G, Moscatelli D. React Chem Eng, 2020, 5, 2081–90.
[44] Halder R, Lawal A, Damavarapu R. Catal Today, 2007, 125, 74–80.
[45] Dawood KM, Nomura K. Adv Synth Catal, 2021, 363, 1970–97.
[46] Piovano A, Signorile M, Braglia L, Torelli P, Martini A, Wada T, Takasao G, Taniike T, Groppo E. ACS Catal, 2021, 11, 9949–61.
[47] Wulf C, Doering U, Werner T. RSC Adv, 2018, 8, 3673–79.
[48] Mebah JMN, Mieloszynski JL, Paquer D. ChemInform, 1994, 25(23), 1–1.
[49] Kamata K, Sugahara K. Catalysts, 2017, 7(345), 1–24.
[50] Calderon N, Chen HY, Scott KW. Tetrahedron Lett, 1967, 8, 3327–29.

[51] Grubbs RH. Handbook of Metathesis, Edition 1, Wiley-VCH, Weinheim, 2003.
[52] Grubbs RH, Wenzel AG, O'Leary DJ, Khosravi E. Handbook of Metathesis, Edition 2, Wiley-VCH, Weinheim, 2015.
[53] Dewaele A, van Verlo B, Dijkmans J, Jacobs PA, Sels BF. Catal Sci Technol, 2015, 6, 2580–97.
[54] Schmidt F, Bernhard M, Morell H, Pascaly M. Chim Oggi, 2014, 32, 31–35.
[55] Xiong W, Gu XK, Zhang Z, Chai P, Zang Y, Yu Z, Li D, Zhang H, Liu Z, Huang W. Nat Commun, 2021, 12(5921), 1–8.
[56] Kissin YV, Rishina LA. Polym Sci Ser A, 2008, 50, 1101–21.
[57] Claverie JP, Schaper F. MRS Bull, 2013, 38, 213–18.
[58] Corradini P, Guerra G, Cavallo L. Acc Chem Res, 2004, 37, 231–41.
[59] Hungenberg KD, Wulkow M. Modeling and Simulation in Polymer Reaction Engineering, Edition 1, Wiley-VCH, Weinheim, 2018.
[60] Fita P. J Phys Chem C, 2014, 118, 23147–53.
[61] Maas G. Chem Soc Rev, 2004, 33, 183–90.
[62] Leadbeater NE, Scott KA, Scott LJ. J Org Chem, 2000, 65, 3231–32.
[63] Zhou Y, Schulz S, Lindoy LF, Du H, Zheng S, Wenzel M, Weigand JJ. Hydrometallurgy, 2020, 194(105300), 1–11.
[64] Nawaz Khan F, Jayakumar R, Pillai CN. J Mol Catal A, 2003, 195, 139–45.

7.2 Heterogeneous catalysts

Klaus Stöwe

7.2.1 Introduction

The use of catalysis in technical processes goes back a long way in human history: for example, alcohol fermentation from sugar was used by the Sumerians in Mesopotamia as early as 6000 BC, as was acetic acid production from alcohol with the aid of catalytically acting enzymes [1]. In more recent times, Priestley in 1783 was the first to demonstrate the conversion of alcohol to ethene and water on heated clay (catalytic dehydration of alcohol) [2, 3]. A description of the catalytic effect of noble metals such as platinum on the combustion of methane and other compounds such as alcohols was provided first by Davy in 1817 [2]. In his experimental setup, a heated platinum wire was able to oxidize a methane-containing gas mixture flamelessly even below the incandescent temperature and cause the platinum wire to glow. This observation initiated Döbereiner to his development of the so-called Döbereiner lighter formulated in a supplement to his publication, which enjoyed great popularity after its invention [4]. The term "catalysis" (Greek: "καταλισισ" dissolution, annulment, disengagement) was introduced by Berzelius in 1836 in a paper on plant chemistry, recognizing that catalysts can also produce different products starting from a mixture of reactants [5]. The definition of catalyst that is widely used today goes back to Ostwald in 1894, who formulated: "Catalysis is the acceleration of a slowly proceeding chemical process by the presence of a foreign substance" [6]. Ostwald was awarded the Nobel Prize for Chemistry in 1909 for his research work, which forms the basis for many of today's industrial catalytic processes, including the ammonia synthesis process developed in 1909 by Haber, Bosch and Mittasch at BASF in Ludwigshaven (Germany). In the Roadmap for Catalysis Research in Germany the German Catalysis Society GECATS wrote, "Catalysis is the single most important interdisciplinary technology in the chemical industry. More than 85% of all today's chemical products are produced using catalytic processes" including related economic sectors as "processing of raw materials in refineries, during the production of energy e.g. in fuel cells and batteries, as well as in terms of climate and environmental protection" [7]. The journal *CHEManager* has estimated that "the international catalyst market has now reached a volume of over USD 18 billion, with an upward trend. Without catalysts, an efficient, environmentally friendly chemical industry is not possible" [8].

The field of catalysis can be further divided into homogeneous and heterogeneous catalysis. In homogeneous catalysis (see Chapter 7.1), all reaction components, catalyst as well as reactants and products, are in a single common phase, whereas in heterogeneous catalysis the catalyst is in solid phase and the reaction system is thus at least two-phase. In heterogeneous catalysis, gases, liquids or gas-liquid mixtures flow around the solid catalyst, typically in a fixed bed reactor with chemical reactions

https://doi.org/10.1515/9783110798890-017

of the reactants taking place on its surface (outer surface, pore surface, inner surface, see Figure 7.2.1). Homogeneous catalysts, on the other hand, are either liquid acids or bases or metal complexes, in many cases with sterically demanding ligands, which are soluble in the reaction component phase; the typical reactor type for homogeneous catalysis is the stirring tank reactor (STR). Both types of catalysts have their specific advantages and disadvantages: in the case of homogeneous catalysts, their separation from the solution in order to recycle them for further reactions is particularly disadvantageous. This is offset by specific advantages such as designability or selectivity (see Chapter 7.1). Advantages of both subareas can be brought together by heterogenized homogeneous catalysts or surface organo-metallic chemistry (SOMC). Finally, biocatalysis with enzymes should not remain unmentioned, which will play an increasingly important role in the future.

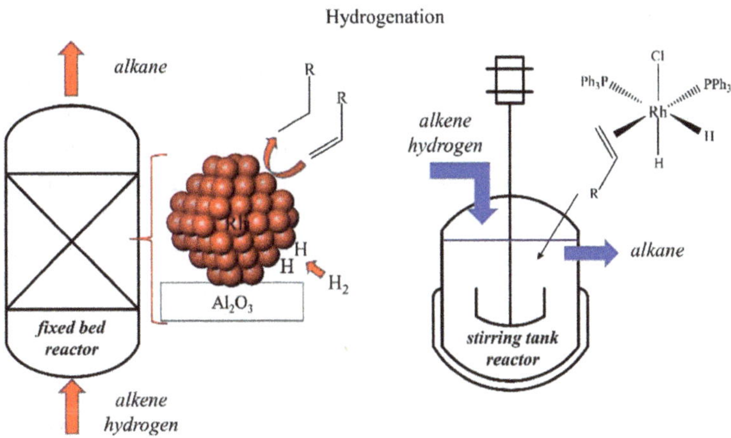

Figure 7.2.1: Heterogeneous and homogeneous catalysis in typical reactor types.

Some basic facts have to be summarized for catalysis:

1. The equilibrium conversion is not changed by catalysis. A catalyst that accelerates the forward reaction also does so for the backward reaction.
2. A catalyst can radically change the selectivity of a reaction by preferentially lowering the activation energy of a particular reaction pathway over all other possible reaction pathways.
3. During the reaction, active centers on the catalyst surface stabilize the transition stages and intermediates of the reaction. Comparatively few catalytically active centers are needed to generate large amounts of product.

The first point describes the fact that equilibrium conversions of catalytic reactions can be calculated from the thermodynamic data of the reactants alone. In practice, this means that exothermic reactions should be carried out at the lowest possible temperatures and endothermic reactions at the highest possible temperatures in order to achieve high equilibrium conversions. However, very low reaction temperatures are often opposed by the kinetics of the reactions, since the reaction rate is too low for technical realization. Thus, in these cases, a compromise must be found between thermodynamics and kinetics. A way out is often the reaction control with e.g. intermediate adsorption of the reaction products and recirculation of nonreacted reactants.

The second point relates in particular to reaction networks and the fact that, as a rule, not only one product is possible, but numerous alternative product molecules can be formed via parallel and consecutive reactions. Typical examples here are partial oxidations such as that of propene, which can lead to acrolein, but also to acrylic acid, acetone, propene oxide, acetic acid, 1,5-hexadiene, benzene, etc., where the end product of the reaction sequence, carbon dioxide, is undesirable as a product of the total oxidation. Here a catalyst lowers as selectively as possible the activation energy of a single reaction path.

The last point shows the course of a heterogeneous catalytic reaction: the catalytic reaction comprises the stages of exothermic adsorption to the active center, the reaction at the active center in the adsorbed state with possible intermediate stages, and finally the desorption of the products. This internal reaction, called microkinetics, is supplemented by external processes such as mass and heat transport through the laminar boundary layer and in the pore system of the catalyst grain, and thus representing macrokinetics. All in all, the catalytic process consists of seven steps, which are shown in Figure 7.2.2.

Basically, heterogeneous catalysts can be roughly divided into full catalysts and supported catalysts. Full catalysts, in turn, can be porous or nonporous. Among the supported catalysts, there are numerous subvariants such as metal oxide versus metal catalysts, precious metal-containing versus precious metal-free catalysts, redox catalysts, exhaust gas catalysts, zeolites, mesoporous catalysts and MOF, just to name a few examples. Among the catalyst supports, too, there are numerous shape variants: monoliths, honeycombs, molded bodies of other geometries, extrudates or foams. Powdered catalysts are not used in large quantities on an industrial scale because they lead to high flow resistance. They are only used for instance in granular form in fluidized bed reactors or in immobilized form as wash coats on monoliths. Single crystals as “supports” with vapor-deposited catalytically active layers in the form of a few monolayers or even sub-monolayers are used in studies under ultrahigh vacuum or certain partial pressures of reactive gases as models of real catalysts in mechanistic studies on reaction kinetics resulting in a pressure gap between study or even in situ and operando reaction conditions. Catalyst supports are made of ceramic materials and can also be divided into dense catalyst supports for shell-type catalysts and porous catalyst supports for impregnated catalysts. Support materials are for instance alumina, silica, zirconia, titania, ceria and other oxides, nitrides, carbides as well as zeolites, mesoporous materials and MOF. Last but not least, carbon

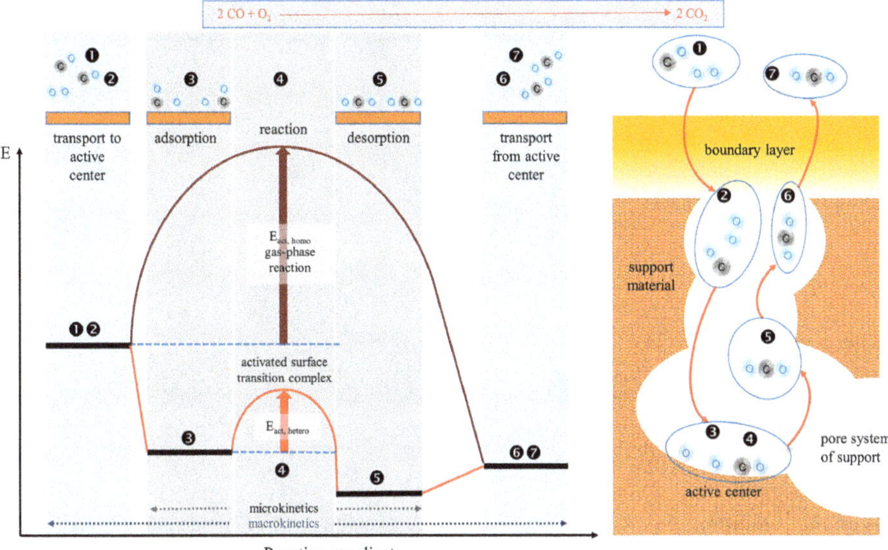

Figure 7.2.2: Seven steps of heterogeneous catalysis.

in the form of activated carbons or carbon blacks must also be listed here, although this support material can only be used for reactions without the participation of oxygen, e.g. in hydrogenations. In cases where the support materials are present in different polymorphic forms, there are large differences between the polymorphs, which shows that they are not only simple supports for active components, but in many cases are also involved in the catalytic reaction in the reaction mechanism (e.g. SMSI effect: strong metal-support interaction). As catalytically active polymorphs of alumina the θ, χ, κ and γ form have been reported [9]. The synthesis routes to the different polymorphs and their transformations based on the fundamental research of Stumpf et al. [10] and others are summarized in Figure 7.2.3.

There are a number of parameters to consider when developing new heterogeneous catalysts, which are summarized in Figure 7.2.4. The discovery and optimization of heterogeneous catalysts is ideally carried out using high-throughput methods (see e.g. [12, 13]).

7.2.2 Syntheses of heterogeneous catalysts

The synthesis of heterogeneous catalysts comprises numerous methods, among them electrostatic adsorption, impregnation, sol-gel processing, hydrothermal synthesis, deposition-precipitation, coprecipitation or cluster immobilization. Not all of them can be discussed in more detail here.

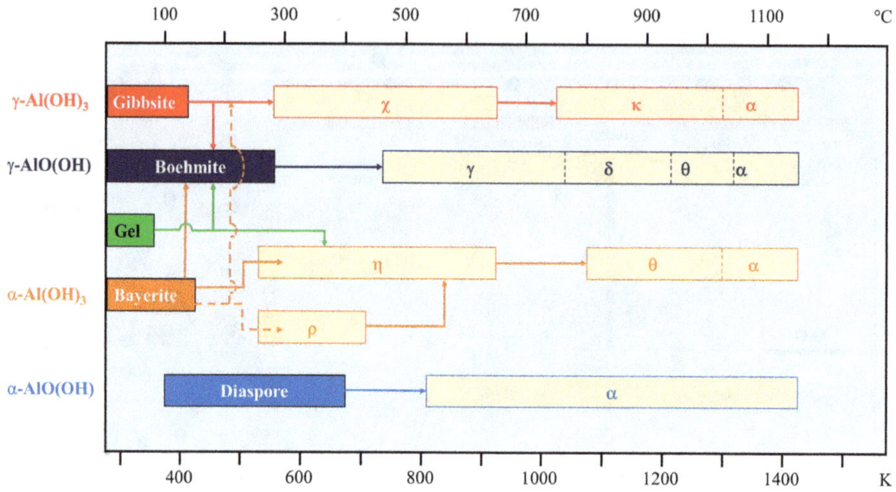

Figure 7.2.3: Transformation sequence from $Al(OH)_3$ and AlOOH to different polymorphic alumina forms (adapted from reference [11]).

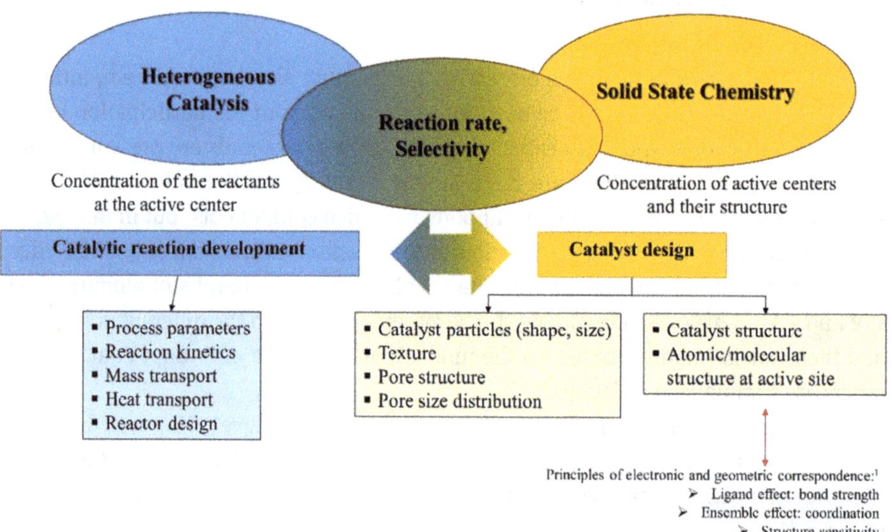

Figure 7.2.4: Development parameters for heterogeneous catalysts.

In electrostatic adsorption, a catalyst support material is dispersed as a powder in a metal salt solution, brought to equilibrium with stirring, and then the support loaded with a metal precursor is filtered off. Special attention must be paid to the pH value of the solution in relation to the surface potential of the support material, which is characterized by the point of zero charge (PZC): at the neutral point of the carrier material, the achievable metal uptake of the carrier by electrostatic interaction is lowest. At lower pH values, the surface of the carrier is positively charged and the electrostatic interaction with metal complex anions leads to the highest loadings. Similarly, at high pH values above the PZC metal (complex) cations interact effectively resulting in a high metal uptake. This is illustrated schematically in Figure 7.2.5 and the PZC values of important support oxides are summarized in Table 7.2.1.

Table 7.2.1: PZCs of common support materials and suitable metal complexes [14].

Support	PZC	Charge
MoO_3	<1	+
Nb_2O_5	2–2.5	+
SiO_2	4	+
Oxidized carbon black	2–4	+
Oxidized activated carbon	2–4	+
Graphitic carbon	4–5	+
TiO_2	4–6	+(/–)
CeO_2	7	+/–
ZrO_2	8	+/–
Co_3O_4	7–9	+/–
Al_2O_3	8.5	(+/)–
Activated carbon	8–10	–
Carbon black	8–10	–

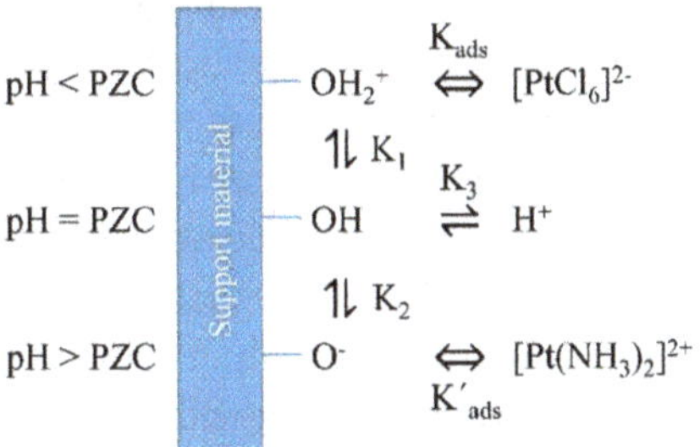

Figure 7.2.5: Electrostatic adsorption principle (point of zero charge PZC).

If higher loadings are to be realized, this can be achieved by impregnation. Impregnation is the soaking of a suitable, i.e. porous, carrier material with a (usually aqueous) solution of the active component(s). To achieve very high loadings or the impregnation of more than one metal component, this can be done either simultaneously or consecutively. The amount of loading per step is limited by the solubility of the metal salt in the solvent. The subsequent post-treatment (calcination, oxidative or reductive) converts the precursor into oxide or metal. Two subvariants can be distinguished in impregnation: firstly, wet impregnation with soaking from supernatant solution with subsequent drying. Diffusion and electrostatic adsorption are the mechanisms used here. Secondly, impregnation with a quantity of liquid corresponding to the pore volume of the substrate, which is called dry or pore volume impregnation ("incipient wetness"). For this purpose, the pore volume of the substrate must be determined in advance with the pure solvent. To do this, impregnating solution is added drop by drop until lumps are formed. The solvent is added until the lumps no longer dissolve. When impregnating with the metal salt solution, a reduced solution volume is then added so that only the pores, but not the particle surface, are coated with active phase. The impregnated substrates are then dried and calcined. During impregnation and drying, effects such as solvent evaporation, convection, adsorption and back diffusion play a very important role. The impregnation of pellets or other molded bodies can therefore result in either egg shell, egg yolk or uniformly impregnated catalysts, depending on the drying rate. In an industrial context at large scale, impregnation is performed in fluidized bed reactors by spraying the impregnation solution into the fluidized bed of the support material. Subsequently, the dried support materials are shaped into the corresponding catalysts form.

Another synthesis method for heterogeneous catalysts is the sol-gel process. Here, either the active component and the carrier material can both be prepared from a single solution, or an active component can be applied on top of a carrier material added to the precursor solution. The first variant has the advantage that only pipetting steps have to be carried out, but no weighing steps, which are much more difficult to automate than weighing. In the sol-gel process, the steps of hydrolysis of the precursor and condensation of the intermediates are carried out in succession, with the pH value determining whether rather less branched, chain-like oligomers and thus condensation networks or particles with a high degree of branching (see also Stöber particles) are formed. For the synthesis of mixed oxides as catalysts, the acidic range is recommended, since here the proton-catalyzed hydrolysis leads to an electrophilic attack on a negatively polarized ligand atom instead of base catalysis with a nucleophilic attack of the hydroxide ions on the metal center. A large number of subvariants of the sol-gel process have been described in the literature, including the classical alkoxide route or so-called polymer-complex routes, in which oligomers to polymers form the gel by addition of polymerizable monomers together with the complexed metal ions. As an example, the Pechini route with citric acid and polyhydroxy alcohols such as ethylene glycol may be mentioned [15]. Figure 7.2.6 gives a summary of the process.

The last process for heterogeneous catalyst synthesis to be mentioned here is the precipitation method. Precipitation, which corresponds to the basic operation of crystallization, proceeds through the stages of nucleation and nucleus growth. An important criterion is the supersaturation of the solution, which influences the particle sizes and the particle formation rate. Numerous other parameters also determine the result of precipitation: the solution composition, including purity of the components used, influences the precipitate composition; the pH and precipitation temperature direct the phase formed; anions, solvents and additives can influence the crystallinity, morphology of the precipitate as well as its texture properties; last but not least, the precipitant also influences the phase formed and its homogeneity. In presence of a reduction agent directly metal or alloy containing catalysts can be obtained (so-called reduction-precipitation). Additional variants as deposition-precipitation and co-precipitation are summarized in Figure 7.2.7.

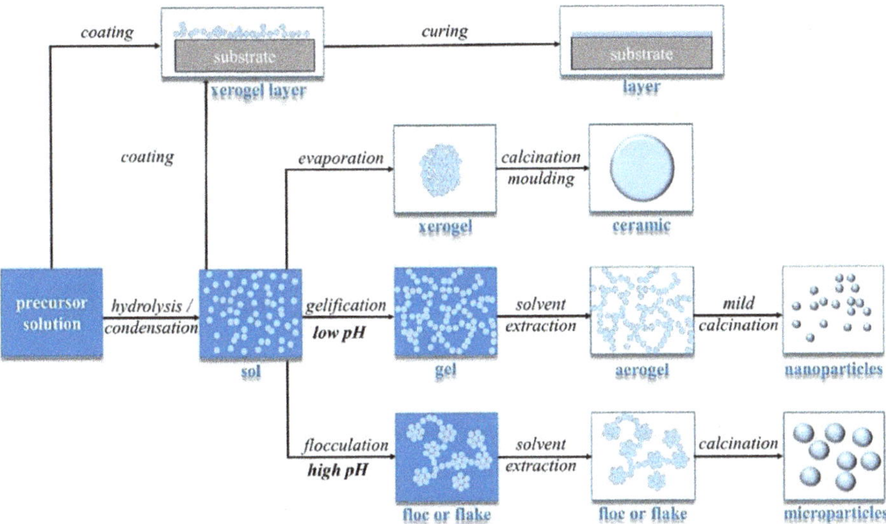

Figure 7.2.6: Sol-gel process for the synthesis of heterogeneous catalysts as powders, molded bodies or thin films on substrates.

Especially impregnation and precipitation are of high relevance for the industrial synthesis of heterogeneous catalysts. Note that Raney-Nickel is a full catalyst material and belongs to the group of skeletal catalysts. They are synthesized by leaching of an aluminum alloy with a base. As they have a high surface area, they are pyrophoric and have to be stored under water.

7.2.3 Characterization of heterogeneous catalysts

In the characterization of heterogeneous catalysts, several points must be particularly emphasized: (1) In many cases, a catalyst must first be activated and is present under reaction conditions in a changed composition and/or morphology; therefore, ex-situ investigations should always be supplemented by in-situ or operando investigations.

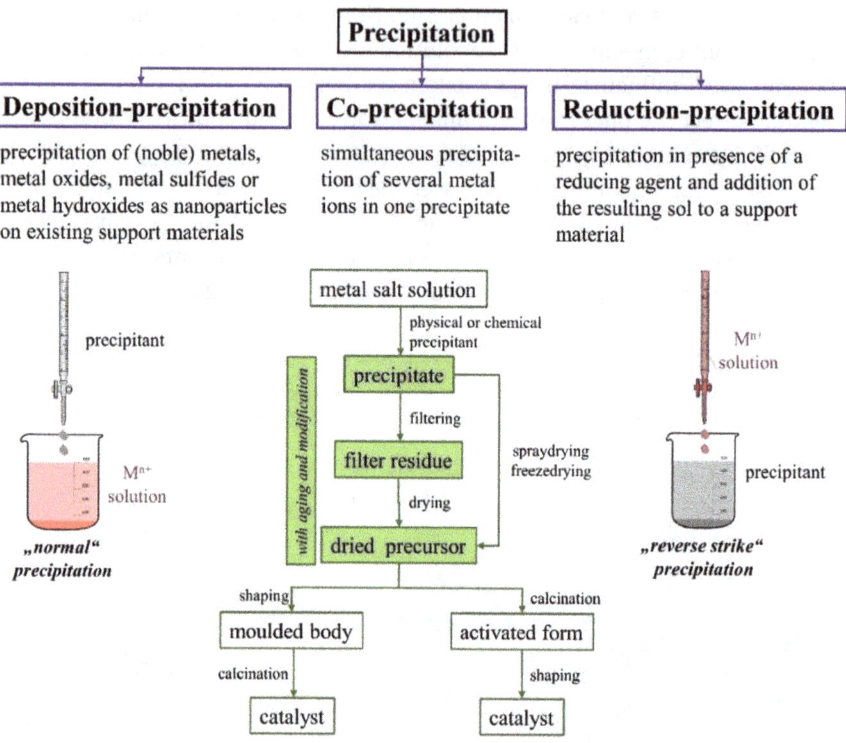

Figure 7.2.7: Precipitation variations for heterogeneous catalyst synthesis.

A very well-known example is the catalyst for the Haber-Bosch process: it is prepared as iron oxide with magnetite crystal structure and reduced to α-Fe under the conditions of the process. (2) Heterogeneous catalytic processes occur at the surface of the catalyst at the fluid/catalyst interface. Therefore, all methods such as powder X-ray diffraction PXRD, X-ray fluorescence XRF analysis, X-ray absorption near edge structure XANES or extended X-ray absorption fine structure EXAFS as typical volume methods reveal only limited information about the reactions taking place at the surface. Nevertheless, they are often used as complementary methods in characterization. Typical surface sensitive methods are on the one hand imaging methods like SEM, here especially ESEM, TEM and HRTEM or the scanning probe methods like STM and AFM, on the other hand spectroscopic methods like photo and Auger electron spectroscopy, XPS as well as UPS and AES, as well as IR spectroscopy for adsorbates and Raman spectroscopy for near surface catalyst states (surface enhanced Raman spectroscopy SERS and related methods), just to name a few. (3) The catalyst system as a whole, consisting of support material and active component, requires special characterization in terms of total physical surface area, pore size, pore size distribution and surface area of the catalytically active component to determine the number of active sites. Only the last point will be addressed here in more detail.

The determination of the total internal surface area of a heterogeneous catalyst with its pore system is performed by physisorption measurements with inert gases such as nitrogen, helium, argon or krypton at cryogenic temperatures. Among these gases, nitrogen at a temperature of 77 K is most commonly used for the measurement. However, because of its quadrupole moment, measurement with nitrogen in liquid nitrogen leads to an excess finding of up to 25% in surface area from the adsorption measurements, so argon at a boiling temperature of 87 K is better used here. Helium is utilized as a probe gas for extremely small micropores [16]. According to the IUPAC Technical Report [16], the manometric method is generally considered the most suitable technique for undertaking physisorption measurements. The basis for the calculation of the internal catalyst surface area was provided by Brunauer, Emmett and Teller in a paper from 1938. For this purpose, the authors assumed that in adsorption measurements, the heat of adsorption for second and higher layers is constant and equal to enthalpy of condensation, and that the top layer is in dynamic equilibrium with gas phase. Also, the number of molecules per layer was assumed constant and combined with a multilayer theory:

$$\frac{1}{N\left[\frac{p}{p_0}-1\right]}=\frac{1}{N_m C}+\frac{C-1}{N_m C}\left(\frac{p}{p_0}\right)=I+S\left(\frac{p}{p_0}\right) \tag{7.2.1}$$

In this linearized form of the BET equation, from the intersection point I and the slope S, the number of molecules/atoms for a monomolecular layer $N_m = 1/(I + S)$ and thus the BET surface area S_{BET} can be calculated from the space requirement of a gas molecule or atom [17]. From the form of the adsorption isotherms including the hysteresis behavior between adsorption and desorption branch, conclusions can be drawn about the pore size, whether the catalyst is microporous, mesoporous or macroporous (see Figure 7.2.8).

From the adsorption isotherms, specifically the desorption branch in this case, the pore size distribution can also be determined via downstream theories. In particular, the theories of Barrett, Joyner and Halenda for mesoporous systems starting from 1.5 nm diameter [18] and Horwâth and Kawazoe for microporous systems [19] should be mentioned. Pore-size analysis of adsorbents over the complete micro- and mesopore range requires physisorption experiments, which span a broad spectrum of pressures (up to seven orders of magnitude) starting at ultralow pressures below 1 Pa [16].

Chemisorption instead of physisorption measurements are necessary if the turnover frequency *TOF* is to be calculated. This requires the knowledge of the total number of active centers *Z*:

$$TOF_i=\frac{|R_i|L_A V_{fluid}}{Z}=\frac{|R_i|L_A \varepsilon_{bed}}{(Z/m_{cat})\rho_{bulk}}\,[\text{molecules/s}] \tag{7.2.2}$$

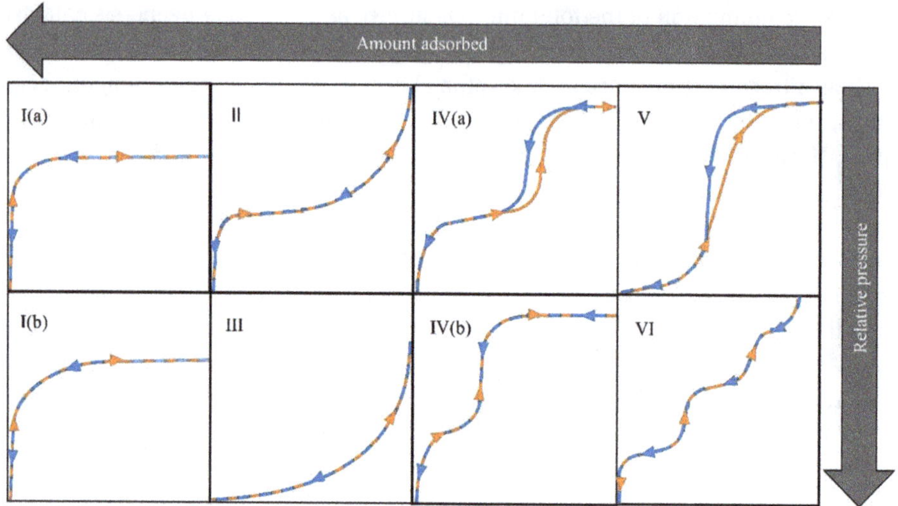

Figure 7.2.8: IUPAC classification of adsorption isotherms: I microporous; II nonporous or macroporous; III no identifiable monolayer formation; weak adsorbent-adsorbate interactions, rare; IV: adsorption hysteresis as in many mesoporous technical adsorbents; V similar to III, e.g. for water adsorption on hydrophobic microporous and mesoporous adsorbents; VI: reversible stepwise isotherm representative of layer-by-layer adsorption on a highly uniform nonporous surface (adapted from reference [16]).

with R_i the mass transfer rate for the species i, L_A the Avogadro number and V_{fluid} the fluid volume. Dynamic chemisorption is typically performed using temperature-programmed flow methods as TPD (temperature programmed desorption) with probe molecules such as CO, CO_2, H_2 or NH_3, the latter especially to determine the number and strength of Brønstedt and Lewis acid sites, TPO (temperature programmed oxidation) or TPR (temperature programmed reduction). Alternatively, the method can be performed by titration using pulses and compared with the results of static volumetric chemisorption. As an example for thermal desorption TPD spectra of a clean and potassium covered platinum single crystal surface Pt(111) at various carbon monoxide CO coverages is shown in Figure 7.2.9.

7.2.4 Kinetics of heterogeneously catalyzed reactions

As shown in Figure 7.2.2 in the seven steps of heterogeneous catalysis, the stages 3–5 (adsorption of the reactive component(s), surface reaction at the active site(s) and desorption of the product(s), respectively) represent the microkinetic part of the overall reaction. The reaction rate of a heterogeneously catalyzed reaction is to be normalized to either the catalyst mass m_{cat} or the surface area A:

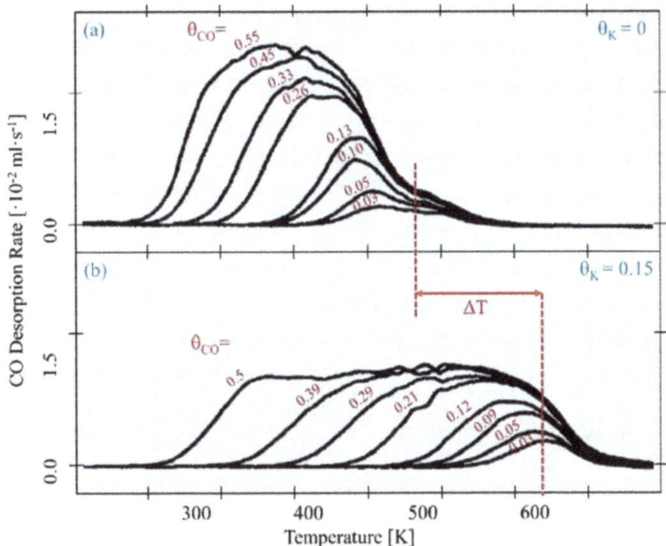

Figure 7.2.9: Thermal desorption TPD spectra of a clean and a potassium covered ($\theta_K = 0.15$ ML) Pt (111) single crystal surface at various coverages of CO θ_{CO}. (Figure reproduced from reference [20] with permission from AIP Publishing and adapted).

$$r = \frac{1}{\nu_i} \cdot \frac{dn_i}{m_{cat} \cdot dt} \qquad r' = \frac{1}{\nu_i} \cdot \frac{dn_i}{A \cdot dt} \tag{7.2.3}$$

If a reaction network is considered, the mass transfer rates R_i must be introduced – i.e. sum of all formation rates minus sum of all consumption rates.

$$R_i = \sum_{j=1}^{j=k} \nu_{i,j} \cdot r_j \tag{7.2.4}$$

Assuming a quasi-steady state (Bodenstein principle), a temporal change in the coverages θ of the surface species of the reactants and products equal to zero, and a reaction rate of reactants equal to the rate of formation of the products, the reaction rate can be calculated for the respective reaction rate-determining step according to Hougen and Watson by an equation of the type [21]:

$$r = \frac{(\text{kinetic term}) \cdot (\text{potential term})}{(\text{adsorption term})^{\mathbf{n}}} \tag{7.2.5}$$

The kinetic term contains rate constant(s) and possibly adsorption equilibrium constants; the potential term describes the driving force of the reaction, i.e. the distance from chemical equilibrium; the adsorption term (chemisorption term) describes the equilibrium adsorption of the components on the catalyst and thus the inhibition of the reaction by covering the catalytically active surface sites with reactants. The exponent n is equal to the number of surface sites involved in the rate-determining elementary reaction. Well-known examples for representatives of the Hougen-Watson equations are the Langmuir-Hinshelwood (eq. (7.2.6), irreversible reaction rate limiting, desorption fast [22]) as well as the Eley-Rideal reaction mechanisms and rate equations, respectively (eq. (7.2.7)) [23]:

$$r = \frac{k' \cdot K_1 \cdot K_2 \cdot p_1 \cdot p_2}{(1 + K_1 \cdot p_1 + K_2 \cdot p_2)^2} \tag{7.2.6}$$

$$r = \frac{k' \cdot K_1 \cdot p_1 \cdot p_2}{1 + K_1 \cdot p_1} \tag{7.2.7}$$

with k' the reaction rate constant of the rate determining step and K_i the equilibrium constants of the upstream adsorption equilibria and p_i the partial pressures of the reactant component(s).

Mass as well as heat transport limitations can lead to apparent erroneous reaction orders in kinetic studies, as shown in Figure 7.2.10.

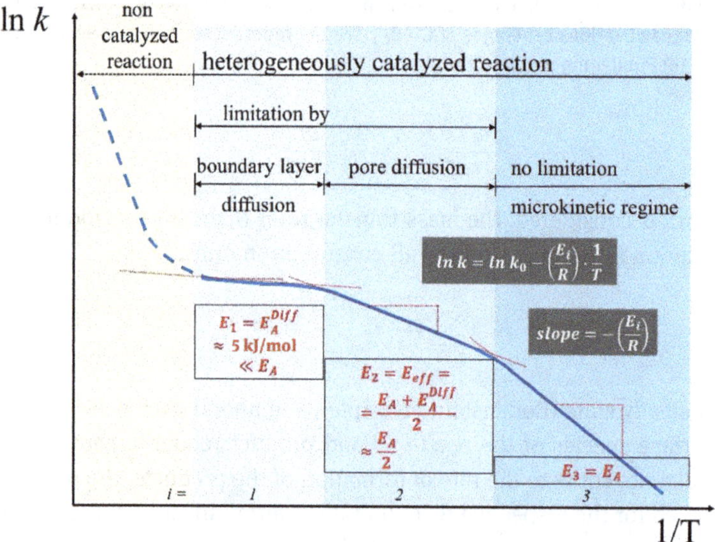

Figure 7.2.10: Temperature dependence of the rate constant of a heterogeneously catalyzed reaction with different regimes (adapted from reference [24]).

In the transport limitation regimes, the slopes in the ln k versus $1/T$ representation of the linearized Arrhenius equation are lower than in the kinetically controlled regime, either roughly one half of the true activation energy E_A (pore diffusion regime) or almost zero (boundary layer diffusion controlled regime). Operating a heterogeneous catalyst in the transport-limited region also means that the catalyst (particle) effectiveness factor η_C is less than one:

$$\eta_C = \frac{k_{eff}}{k_3} = \frac{3}{\phi_c}\left(\frac{1}{tanh\phi_c} - \frac{1}{\phi_c}\right) \tag{7.2.8}$$

with k_{eff} the effective (apparent) rate constant in the limited regime and k_3 the rate constant in the kinetically controlled regime. The Thiele modulus ϕ_c can be expressed for a spherical catalyst particle as:

$$\phi_C = r_P\sqrt{\frac{k_3 S_V c_{1,s}^{m-1}}{D_1^e}} \tag{7.2.9}$$

with r_P the radius of the particle, S_V the specific inner surface area of the catalyst pellet [m^2/m^3], D_1^e the effective diffusion coefficient related to component 1 and m the reaction order. An effectiveness factor η_C significantly below one means that a valuable catalyst is not completely used in the catalytic reaction and is therefore wasted. Especially in the case of catalysts containing precious metals, this must not be accepted and the transport limitations must be eliminated. This can be done, for example, by smaller particle diameters (pore limitation elimination) or increased reactant flow rates with the same modified residence time (boundary layer thickness reduction).

References

[1] Jastram A, Langschwager F, Kragl U. Reaktoren für spezielle technisch-chemische Prozesse: Biochemische Reaktoren. In: Reschetilowski W. (Ed.) Handbuch Chemische Reaktoren: Chemische Reaktionstechnik: Theoretische und praktische Grundlagen, Chemische Reaktionsapparate in Theorie und Praxis, Springer Berlin Heidelberg, Berlin, Heidelberg, 2020, 961–99.
[2] Mittasch A. Naturwissenschaften, 1936, 24, 770–77.
[3] Davy H. Ann Phys Leizpig, 1817, 56, 242–55.
[4] Döbereiner JW. Ann Phys Leipzig, 1823, 74, 269–73.
[5] Berzelius JJ. Jahresbericht über die Fortschritte der physischen Wissenschaften, 1836, 15, 237–440.
[6] Ostwald W. Z Phys Chem, 1894, 15U, 699–708.
[7] Roadmap for the Catalysis Research in Germany http://gecats.org/gecats_media/Urbanczyk/Katalyse_Roadmap_2010_engl_final.pdf (accessed 2021/10/06).
[8] Zukunftstechnologie Katalyse. https://www.chemanager-online.com/news/zukunftstechnologie-katalyse (accessed 2021/10/06).
[9] Rubinshtein AM, Slovetskaya KI, Akimov VM, Pribytkova NA, Kretalova LD. Izv Akad Nauk SSSR Ser Khim, 1960, 30–38.
[10] Stumpf HC, Russell AS, Newsome JW, Tucker CM. Ind Eng Chem, 1950, 42, 1398–403.
[11] Wefers K, Misra C. Oxides and Hydroxides of Aluminum, Alcoa Research Laboratories, 1987.

[12] Maier WF, Stoewe K, Sieg S. Angew Chem Int Ed, 2007, 46, 6016–67, S6016/1.
[13] Potyrailo R, Rajan K, Stoewe K, Takeuchi I, Chisholm B, Lam H. ACS Comb Sci, 2011, 13, 579–633.
[14] Regalbuto JR. Electrostatic adsorption. Synth Solid Catal, 2009, 33–58.
[15] Pechini MP. Method of Preparing Lead and Alkaline Earth Titanates and Niobates and Coating Method Using the Same to Form a Capacitor. US Patent No. 3330697, 1967.
[16] Thommes M, Kaneko K, Neimark AV, Olivier JP, Rodriguez-Reinoso F, Rouquerol J, Sing KSW. Pure Appl Chem, 2015, 87, 1051–69.
[17] Brunauer S, Emmett PH, Teller E. J Am Chem Soc, 1938, 60, 309–19.
[18] Barrett EP, Joyner LG, Halenda PP. J Am Chem Soc, 1951, 73, 373–80.
[19] Horváth G, Kawazoe K. J Chem Eng Jpn, 1983, 16, 470–75.
[20] Whitman LJ, Ho W. J Chem Phys, 1989, 90, 6018–25.
[21] Hougen OA, Watson KM. Ind Eng Chem, 1943, 35, 529–41.
[22] Ertl G. Chem Rec, 2001, 1, 33–45.
[23] Eley DD, Rideal EK. Nature, 1940, 146, 401–02.
[24] Baerns M, Hofmann H, Renken A, Falbe J, Fetting F, Keim W, Onken U. Chemische Reaktionstechnik: Lehrbuch der Technischen Chemie, Band 1, Wiley, Weinheim, 1999.

7.3 Zeolites for ion exchange, adsorption and catalysis

Hubert Koller

This chapter provides a selected view on some of the most important applications of synthetic zeolites. The global zeolite market was estimated at 27 billion US$ in 2020, and analysts project a growth to over 32 billion US$ by 2027 [1]. The largest market segments of synthetic zeolites are detergents, catalysts and adsorbents, while natural zeolites are also applied in public health, livestock, fisheries, agriculture, concrete additive and waste-water remediation. Synthetic zeolites were introduced as catalysts in petrochemical industry in the second half of the last century. The ongoing demand for new zeolites is due, to a large extent, to their high potential in large-scale industrial and environmental catalysis. Therefore, the research and development in zeolite technologies have been growing in industry and academia, sometimes driven by environmental legislation policies, but often motivated by the demand for technological innovations. What makes them so useful is such a variety of applications.

Zeolites are typically open-framework aluminosilicates with pore openings of molecular dimensions. The name "zeolite" was introduced in 1756 by the Swedish mineralogist Axel F. Cronstedt (1722–1765), when he discovered the mineral stilbite [2, 3], whose crystals exhibited intumescence (induced by evaporation of water from the pores) when heated in a blowpipe flame. The name was derived from the Greek words, "zeo" and "lithos", which means "to boil" and "a stone". The synthesis of zeolites is typically a hydrothermal reaction, which was pioneered by Richard M. Barrer in the mid-twentieth century [4, 5].

Zeolites can be described more formally as microporous materials, which means pore diameters of <2 nm [6], that are built from fully connected tetrahedral units, $TO_{4/2}^{m+/-}$, with a variety of possible elements in the T-atom position. Depending upon the nature of the T atom, a neutral or negative charge results for the three-dimensional framework structure. For example, pure silica frameworks are neutral. Positively charged open-framework materials do not exist, and positive local charges (e.g. in $PO_{4/2}^{+}$) are always neutralized by a negative charge, such as for aluminophosphates. The most common variants are negatively charged aluminosilicates, due to $AlO_{4/2}^{-}$ tetrahedra, with the general formula $|x/a\mathrm{M}\,(\mathrm{H_2O})_b|[\mathrm{Al}_x\mathrm{Si}_y\mathrm{O}_{2(x+y)}]$-**TLC** [7]. The species between the vertical bars, |, are those that are not part of the tetrahedral framework ("extra-framework"), and the brackets, **[]**, embrace the framework composition. **TLC** is a **T**hree-**L**etter-**C**ode assigned by the Structure Commission of the International Zeolite Association [8], under authorization of the IUPAC. It describes the unique three-dimensional connectivities of the T centers, leading to the various

Acknowledgment: The author is grateful to Lynne B. McCusker, Christian Baerlocher, Alexander Katz and Cong-Yan Chen for carefully reading the manuscript and helpful advice.

https://doi.org/10.1515/9783110798890-018

framework types. M is an extra-framework cation with valence *a*, and it can be replaced in an ion-exchange process. As an example, the crystal chemical formula for the commercially important zeolite A is $|Na_{12}(H_2O)_{27}|_8$ $[Al_{12}Si_{12}O_{48}]_8$-**LTA**. This can be simplified to $|Na\,(H_2O)|$ $[$Al-Si-O$]$-**LTA** for unknown compositions. To date, there are over 250 known zeolite framework types that have been assigned a TLC [8], and the number of hypothetical zeolites far exceeds 2 million [9]. Although the crystal chemical formula is recommended by the IUPAC, the traditional mineral or synthetic names are also used. One of the most important natural zeolites is clinoptilolite (framework type **HEU**), which occurs in large amounts in the sea floor. In addition to zeolite A (framework-type **LTA**) [10], important synthetic materials include zeolites X or Y (both framework type **FAU**) [11, 12], ZSM-5 (**MFI**) [13, 14], mordenite (**MOR**) [5, 15], beta (***BEA**) [16, 17] and chabazite (**CHA**) [18]. The zeolite beta framework structure is disordered. Ordered layers are stacked in a random fashion (therefore, the "*" in its TLC), leading to intergrowths of various polymorphs [8, 19].

The atomic Si/Al ratios, *y*/*x*, are variable and the compositional range that can be obtained by direct synthesis depends on the framework type. In some cases, post-synthetic modifications can extend this range. Si/Al ratios less than one are not permitted by the Loewenstein rule [20], which rules out Al-O-Al linkages because two charge centers would be too close together. Other examples of negative tetrahedral frameworks are borosilicates, gallosilicates or zincosilicates ($ZnO_{4/2}^{2-}$ introduces two local charges). A combination of pentavalent and trivalent T atoms can form a neutral framework structure, as in the case of aluminophosphates, where a 1:1 ratio of $PO_{4/2}^{+}$ and $AlO_{4/2}^{-}$-building units equalize the charge. Titanosilicates and germanosilicates are also neutral, and some are relevant in industrial applications.

This chapter aims first and foremost to elucidate the link between the inorganic structure and its function in the three main categories of zeolite application: ion exchange in detergents, adsorption and catalysis. A general overview based on a selection of cases that are typical for the impact of zeolites in various fields will be given. The zeolites and examples in this chapter are intended to demonstrate some established applications rather than to report on recent research work. Therefore, the specialized readers will miss some of the latest academic and industrial developments. For more comprehensive information regarding zeolite properties and applications, the readers are referred to a few excellent publications [21–31].

7.3.1 Structures of selected zeolite frameworks: setting the scene

Zeolite A (**LTA**) and zeolites X and Y (**FAU**) are composed of sodalite cages (*sod*), which can be regarded as truncated octahedra (Figure 7.3.1). The natural mineral sodalite (**SOD**) is a body-centered arrangement of face-sharing *sod* cages, and it also occurs as the natural gemstone, lapis lazuli, also known as ultramarine, where sulfur radicals, $\cdot S_3^-$ and $\cdot S_2^-$, in the *sod* cages are responsible for the blue color [32].

The *sod* cages are separated by double 4-rings (*d4r*) in **LTA** and by double 6-rings (*d6r*) in the "faujasite" topology (**FAU**) of zeolites X and Y (Figure 7.3.1), thus creating larger 8-rings and 12-rings, respectively. A zeolite is considered porous, when its biggest pore is larger than 6-rings. Therefore, sodalite is not porous, and water (but no larger molecules) can penetrate the 6-rings. Zeolites with 8-rings as the biggest pores are small-pore zeolites, 10-ring zeolites are medium-pore and 12-rings are found in large-pore zeolites. Odd-numbered pore sizes (>6-rings) are rare. Pores larger than 12-rings are called extra large, and synthetic efforts made them accessible in the last two to three decades [33]. The author of this chapter is not aware of any published applications of extra-large-pore zeolites to date.

Zeolite A (**LTA**) has an 8-ring as its largest pore opening and is classified as a small-pore zeolite, which are wide enough to allow linear, aliphatic alcohols or hydrocarbons to enter, but not branched molecules. Small-pore zeolites have ring apertures of 0.35–0.45 nm [34]. They have gained growing relevance (e.g. chabazite-type zeolites) in the more recent history of zeolite catalysis because of their superior catalytic properties in the commercialization of NO_x removal from exhaust gas and the conversion of methanol to light olefins [30].

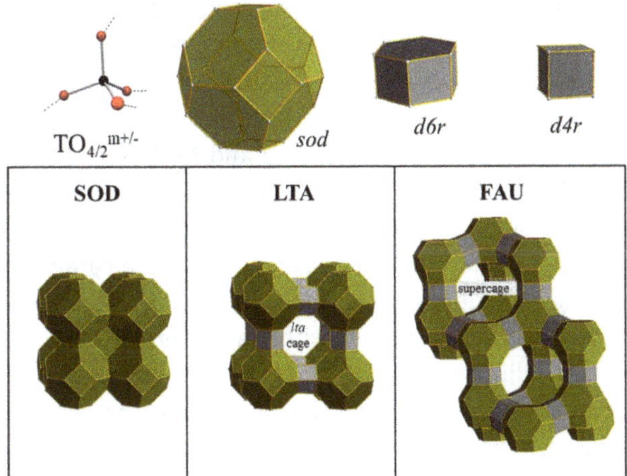

Figure 7.3.1: Cubic zeolite structures composed of the so-called sodalite cage (*sod*). Typically, the oxygen atoms are omitted in structural illustrations of zeolite framework structures, and only the T atoms are shown, connected by a line, thus yielding a representation of cages and pores. The number of $TO_{4/2}^{m+/-}$ building units define the ring size, so a 6-ring is composed of six such units (but has 12 atoms in total, that is 6 T and 6 O). The different connectivities of *sod* cages in **LTA** and **FAU** via *d4r* or *d6r* yield larger cages, which are called *lta* cage (often also named α-cage) and supercage, respectively.

Zeolites X and Y have the same **FAU** framework structure, but they differ in composition: zeolite X is Al-rich (Si/Al ratio of 1–1.8), while zeolite Y is Si-rich (Si/Al > 1.8). The framework structure has 12-ring pores in three dimensions, and it is classified as a large-pore zeolite. Large-pore zeolites have 0.6–0.8 nm pore apertures [34]. Zeolites A, X and Y will be discussed below with selected examples of important, large-scale applications.

Another key zeolite for applications, mainly in catalysis, is ZSM-5 (**MFI**). It belongs to the so-called pentasil family of zeolites, and it is constructed from pentasil (*mfi*) units (Figure 7.3.2a), which form chains along the *c*-direction. ZSM-5 was patented by the Mobil Oil company [13], and its original use was mainly in refinery applications. Today, ZSM-5 is used in a wide variety of industrial manufacturing processes, including organic fine-chemicals and large-scale intermediates, for example, in the plastics industry. ZSM-5 can be synthesized with a variety of Si/Al ratios in the high-silica range, but it is also known as pure silica material. Pure SiO_2 zeolites have a hydrophobic surface, making them interesting for adsorption of nonpolar molecules in adsorption/separation processes. Another important variant of ZSM-5 is the titanosilicate, TS-1. Titanium atoms in the zeolite framework are catalytically active, for example, in selective oxidation reactions or in the ammoximation of ketones [35, 36].

ZSM-5 belongs to the zeolites that are classified as medium-pore zeolites, whose 10-ring pore openings have 0.45–0.60 nm ring apertures [34]. There are straight 10-ring channels in the *b*-direction, intersecting with 10-ring zigzag channels in the *c*-direction (Figure 7.3.2). Because of the channel intersections, diffusion is possible in all three dimensions, so it has a three-dimensional channel system.

All three zeolite topologies discussed above (**LTA**, **FAU** and **MFI**) have three-dimensional pore systems, which is an advantage for molecular traffic when the interior of the crystals hosts the active sites in a catalytic reaction. Nevertheless, pore diffusion is a limitation in the macrokinetics of heterogeneous zeolite catalysis. Newer developments of the last decade are aiming at the synthesis of hierarchical zeolite systems [37]. These materials have mesopores with diamenters of 2–50 nm and macropores (>50 nm) to reduce diffusion limitations, and microporous zeolite channels extend only for a short distance from those larger pores to the reactive centers. This exciting approach is still striving towards higher maturity.

7.3.2 Zeolites as ion exchangers in detergents

Ca^{2+} and Mg^{2+} ions in tap water can cause limescale buildup in washing machines and can damage the device. Therefore, washing powder was augmented with sodium tripolyphosphate as a water softener until its environmental impact became a real problem in the 1970s. High levels of phosphates are fertilizers in natural waters, causing algae to grow and deplete oxygen levels in water (eutrophication), thus killing the flora and fauna. Since phosphate additives in washing powders were replaced by zeolite A in the mid-1980s, this is no longer a problem. Zeolite A is a low-silica zeolite,

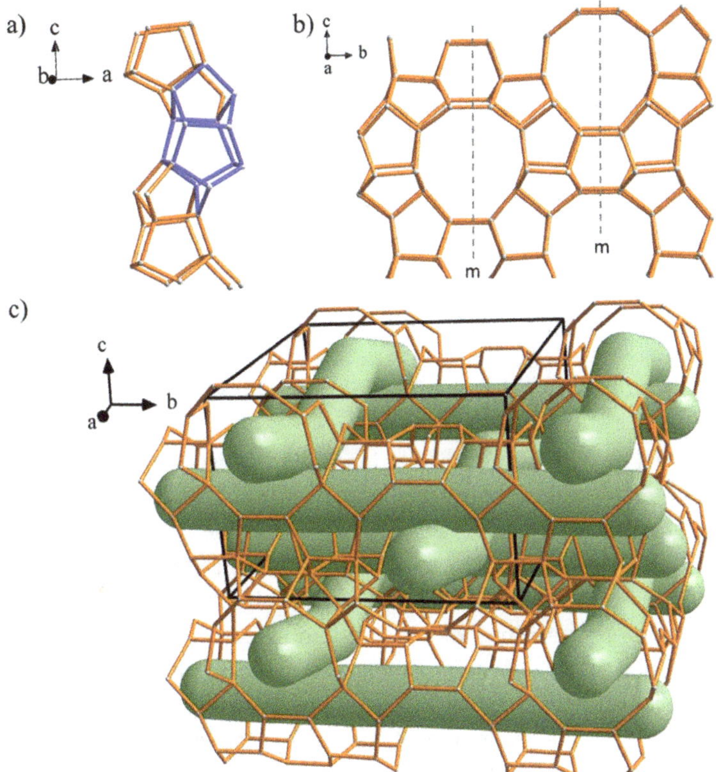

Figure 7.3.2: Structure of zeolite ZSM-5($|Na_x(H_2O)_b|[Al_xSi_{96-x}O_{192}]$-**MFI**). Extra-framework cations other than Na^+ are possible, and only the framework structure is shown; (a) shown in dark blue is the so-called pentasil unit (*mfi*), which is the composite building unit that forms the pentasil chains along the *c*-direction; (b) the pentasil chains can be connected along the *b*-direction via mirror planes, thus forming a pentasil layer; (c) the pentasil layers are fused by inversion centers, from which the three-dimensional structure of ZSM-5 is constructed. It has 10-ring pores, and the green tubes indicate the path through linear channels in the *b*-direction and zigzag channels in the *a*-direction. The structure is monoclinic or orthorhombic, depending on the framework composition and the pore filling.

typically made with a Si/Al ratio of 1, which requires an equivalent number of extra-framework counter ions for charge balance. Zeolite A is manufactured on a large scale with sodium cations, which interact with the oxygen atoms of the framework and water molecules in the pores [38]. These extra-framework cations can be easily ion-exchanged by other cations such as Ca^{2+} and Mg^{2+}, and this is what happens in the washing cycle. This is most likely the number one application of synthetic zeolites in terms of mass production. The concept of ion-exchange isotherms is illustrated in Figure 7.3.3.

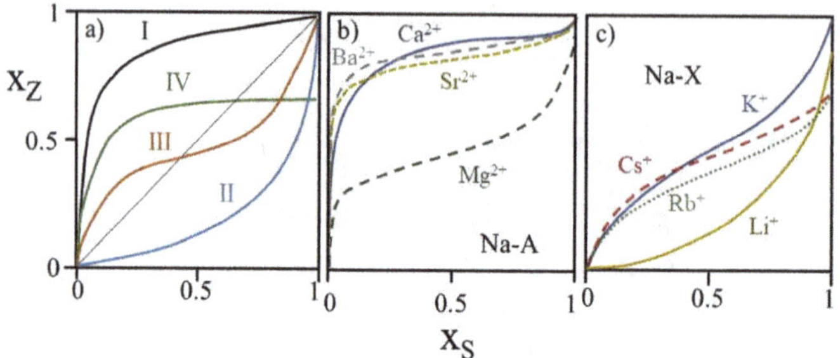

Figure 7.3.3: Ion exchange isotherms: (a) sketch of the four different types of isotherms (see text), (b) exchange isotherms for $|Na_{12}(H_2O)_b|[Al_{12}Si_{12}O_{48}]$-**LTA** (zeolite Na-A) with alkaline earth cations and (c) exchange of zeolite X ($|Na_x(H_2O)_b|[Al_xSi_{192-x}O_{384}]$-**FAU**) with alkali cations [39].

Figure 7.3.3a shows four different types of ion exchange isotherms. The plots show the equilibrium mole fractions of the incoming ions in solution, x_S, and in the zeolite, x_Z. The isotherm of type I shows a fast growth of x_Z for small x_S, and it is typical for a cation, which is preferred by the zeolite over the cation that leaves in the ion-exchange process. The reverse situation is shown by the type II isotherm, where the zeolite prefers the residing ions to stay, and a full ion-exchange only takes place at high concentration of the incoming ion. The sigmoid isotherm III is typical for a cation that is preferred by the zeolite at specific locations until saturation occurs. At higher exchange levels, the zeolite prefers the leaving cation. The type IV isotherm is observed, when the zeolite has inaccessible locations for the ion-exchange, leading to an incomplete exchange even at high concentrations of the incoming cation. Figure 7.3.3b sketches such ion-exchange isotherms for an aqueous alkaline earth cation exchange with zeolite Na-A (the sodium form of zeolite A). Clearly, the Ca^{2+} and Mg^{2+} cations exhibit type I and type III ion-exchange isotherms, respectively, both yielding a full exchange at high solution concentrations. These properties make zeolite Na-A an excellent water softener additive in washing powders. Another zeolite that is used as water softener is high-aluminum zeolite P (**GIS**) [26, 40, 41]. A somewhat different example is shown in Figure 7.3.3c for ion-exchange of zeolite Na-X with various alkali cations. The Na^+ cations within the *sod* and *d6r* cages are inaccessible for an aqueous ion exchange with Rb^+ or Cs^+, thus yielding type IV isotherms. The Li^+-exchanged material will be discussed below for its application in air separation.

7.3.3 Adsorption and separation processes fostered by strong electrostatic interactions

A structural feature beyond their porosity that makes zeolites unique to some extent is the coordination geometry of the extra-framework cations. Alkali cations interact

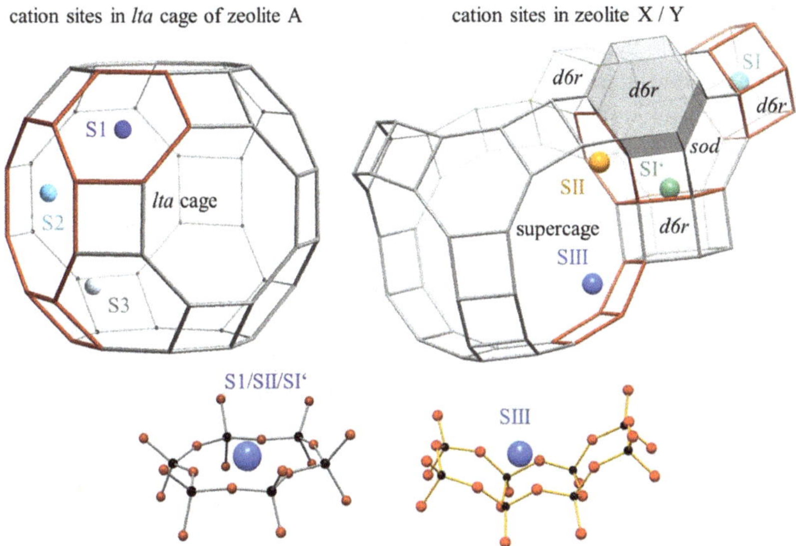

Figure 7.3.4: Cation sites in dehydrated zeolite A (**LTA**) [42] and zeolite X/Y (**FAU**) [43, 44]. For clarity, only one of the symmetry-equivalent positions is shown for each cation site. The rings (or *d6r* cage for position SI) in which cations are located are marked with red lines. Close-ups of some cation geometries are shown at the bottom (oxygen atoms are in red).

with the internal surface of the zeolite pores and cages, and this often makes their coordination numbers and geometries unusual, especially in dehydrated zeolites, compared to their classical coordination patterns in inorganic salts. Figure 7.3.4 shows the locations of Na^+ cations in the *lta* cavity of zeolite A [42]. As a consequence of the high abundance of Al atoms in the framework (Si/Al ratio of 1), a high number of Na^+ cations is required for charge balance. In the hydrated zeolite, these cations interact with the oxygen atoms of the framework and water molecules [38]. Various distinct Na^+ locations were identified for the dehydrated structure (Figure 7.3.4). The space between these Na^+ cations in the center of the *lta* cage is devoid of any matter in a vacuum-dehydrated zeolite A, and it is obvious that charge repulsion between these cations is a dominating factor for the electrostatic energetics and the properties within such cages with a high ion loading. A similar situation exists in the supercage of the **FAU**-type structure (Figure 7.3.4), where the cation positions SII over 6-rings and SIII over 4-rings induce a strong electrostatic field within the supercage. Therefore, the cations are squeezed towards the framework walls in dehydrated zeolites. The strong electrostatic fields within the cages are exploited in various applications. The cation positions SI and SI' in the *d6r* or the sodalite cage of a **FAU**-type structure, respectively, may also contribute to an aqueous ion exchange, but these locations are not all accessible for certain cations (e.g. Rb^+ or Cs^+ in Figure 7.3.3c). Thus, these sites can lead to type IV ion exchange isotherms. Nevertheless, an exchange of Na^+ at SI or SI' positions with large Cs^+ cations is still possible by heat treatment of the dehydrated material, followed by another aqueous ion-exchange [45].

One of the applications of the electrostatic field within the zeolite pores is air separation, which utilizes the different electrostatic properties of N_2 and O_2. Neither molecule has a dipole moment, but N_2 has a quadrupole moment ($(-5.01 \pm 0.08) \cdot 10^{-40}$ Cm^2) almost five times larger than that of O_2 ($(-1.033 \pm 0.027) \cdot 10^{-40}$ Cm^2) [46, 47]. Thus, the interaction of N_2 with extra-framework cations is stronger than that of O_2, and the SIII positions in the FAU structure are superior for a strong interaction with N_2 [44, 48, 49]. In particular, Li^+ cations at SIII positions were identified as those having strong interactions with the N_2 molecules, making Li-X (|Li|[Al-Si-O]-**FAU**) a material of choice for air separation. The separation can be achieved by the so-called pressure-swing adsorption (PSA) process. The first step of this process begins at high adsorption pressure, where N_2 is preferentially adsorbed at the extra-framework cations, and the outcoming gas stream is enriched in O_2. As the adsorption progresses, N_2 will gradually saturate the zeolite bed and then break through in the outcoming gas stream. Then, in step two, the O_2 collector is closed by a valve, and the outcoming gas is let in a separate line, which is accompanied by a pressure drop (pressure-swing) for N_2 desorption. This gas stream is enriched in N_2 due to the lower level of its adsorption isotherm properties at lower pressure compared to the higher pressure in the first step. The PSA process exploits the different equilibrium adsorption properties of N_2 and O_2. These pressure steps can be repeatedly cycled to reach the desired purity of O_2. For purified nitrogen, porous carbons are typically used instead of zeolite adsorbents, but zeolites, for which the adsorption kinetics rather than an equilibrium process is utilized, have also been developed. When the separation of naturally occurring argon from air is not required, then air separation by selective adsorption is more cost-effective than cryogenic distillation for producing up to 50 t of nitrogen or 200 t of oxygen per day [50]. Medical oxygen concentrators are an example of smaller applications. These devices utilize a PSA on a zeolite (typically Li–X) to enrich the oxygen level in breathable air for patients with lung diseases [51].

Zeolites are generally used as versatile adsorbents for separation processes. Table 7.3.1 lists a survey of the main concepts that can be exploited for such separation processes. As mentioned before, all-silica zeolites (pure SiO_2 phases) have a very hydrophobic internal pore surface, which allows the selective adsorption of nonpolar guest molecules to separate them from more polar molecules in mixtures. For more details, the reader should consult more specialized literature sources, such as reference [52]. An impressive example of the potential of kinetic vapor phase adsorption that depends on subtle differences in zeolite adsorbents was reported by Chen and Zones [53]. These authors show how sensitive the adsorption kinetics of *n*-hexane is to effective pore size, which can also be modified by extra-framework counterions in small pore zeolites.

Table 7.3.1: List of various separation concepts by zeolite adsorbents.

Zeolite function	Separation based on
Pore size similar to molecular dimensions	Molecular size of adsorbates
Cations in pores, nature and density thereof can be varied	Selective interaction between cation and adsorbates
Acid sites, nature, density and strength can be varied	Basicity of adsorbates
Hydrophilic or hydrophobic surface	Polarities of adsorbates

An emerging application is heat storage and the use of zeolites for refrigeration [1, 54]. These applications utilize the strong interaction of water molecules with extra-framework cations. For example, the Na^+ cations located at 8-, 6- or 4-rings in zeolites (Figure 7.3.4) are keen to convert their almost planar coordination with framework oxygen atoms in dehydrated zeolites to a more nonplanar (e.g. octahedral) coordination by virtue of a supplemental interaction with adsorbed H_2O molecules. Strong electrostatic repulsion of the cations in dehydrated zeolites is giving rise to an energetically high state, and relaxation is possible by adsorption of such polar molecules. This is obviously the reason why zeolites exhibit a high heat of adsorption for water [55]. Heat storage as well as the use in refrigeration is based on the adsorption of water in zeolites. The cooling effect is generated, when a wet reservoir is brought in contact with a dry zeolite, thus cooling the wet reservoir by the heat of evaporation. Vice versa, heat can be used to dehydrate a zeolite, and this dried material may be stored for an infinite time before it is rehydrated. Rehydration is then accompanied by a rise in temperature, thus retrieving some of the heat that was invested for dehydration.

7.3.4 Catalysis

Catalysis is a dynamic field, and heterogeneous catalysts contribute to about 90% of the industrial chemical processes [56]. In this context, zeolite research contributes to a growing area of catalysis applications in industry. The history of zeolites in catalysis includes their applications in petroleum refineries – the innovation of using zeolite catalysts led to 30% increased gasoline yields. Other high-value-added applications are emerging rapidly involving zeolite catalysts. One recent example is the advent of an entirely new class of zeolite-containing industrial catalysts for pollution abatement in diesel vehicles. It is such emerging applications of zeolites that include, in addition to the production of fuels, chemicals for polymers (and their upcycling and reuse) and pharmaceuticals. Zeolite catalysis has such an enormous impact in commercial applications that it is impossible to provide a comprehensive survey here. Therefore, this chapter is limited to some important conceptual basics along with a few highlighted examples, focusing on the specific function of the zeolite structure.

As zeolites have pore diameters of molecular dimensions, chemical reactions inside the pores often follow the concept of shape selectivity [28, 57–61], which is explained by some examples in Figure 7.3.5. There are three different variants of shape selective catalysis.

Reactant shape selectivity allows molecules that are small enough to enter the pore system of a zeolite to react inside, whereas larger molecules are hindered. Figure 7.3.5a shows the dehydration of an alcohol in zeolite A as an example. Another example would be the hydroisomerization of linear hydrocarbons in the silicoaluminophosphate SAPO-11 (**AEL**) with a one-dimensional, 10-ring channel system, where the bulkier, branched hydrocarbons in the feed are rejected. This industrial "Isodewaxing" process (commercialized in 1993 by Chevron) is exploited to improve the low-temperature properties (e.g. pour point) of lubricant base oils because waxy linear alkanes with high melting points are selectively converted to their branched isomers, which have the desired lower melting points [62].

For product shape selectivity, the reactants enter the pore system and react inside. Then, only those products that are small enough to diffuse through the narrow pore system can leave the zeolite crystals. An example is the acid catalyzed alkylation of toluene with methanol in ZSM-5 (Figure 7.3.5b). The channel intersections (Figure 7.3.2c) provide large reaction spaces. All three xylene isomers form within these channel intersections, but only the para isomer can more readily leave through the narrower channels. Ultimately, the remaining bulkier molecules (*ortho*- and *meta*-xylene) will be converted, again by acid catalysis, to the slim *para*-xylene in xylene isomerization, that is a commercialized process and utilizes product shape selectivity. This is useful because the demand for *p*-xylene is high in the plastics industry (to produce terephthalic acid and then polyethylene terephthalate, PET), whereas that for *o*- and *m*-xylene is not.

Reactant and product shape selectivity can be verified in part by diffusion measurements of the molecules involved. More difficult to prove is the transition state selectivity (Figure 7.3.5c). In this case, a transition state can form in pores large enough to accommodate it and lead to the formation of the associated product, while other potential, larger transition states cannot. As a result, no molecules are produced via the route associated with the latter.

Many chemical reactions taking place inside zeolite pores are influenced by some sort of shape selectivity, and this is a peculiarity of zeolites compared to heterogeneous catalysts without a well-defined pore system. In most cases, the shape-selective effect has either not been investigated in detail or is difficult to prove. The examples shown in Figure 7.3.5 are established examples of the concept of shape selectivity [63]. Different pore sizes present in one material can also be used to tune the catalyst by post-synthesis modification in such a way that only discrete portions of the zeolite are catalytically active [64].

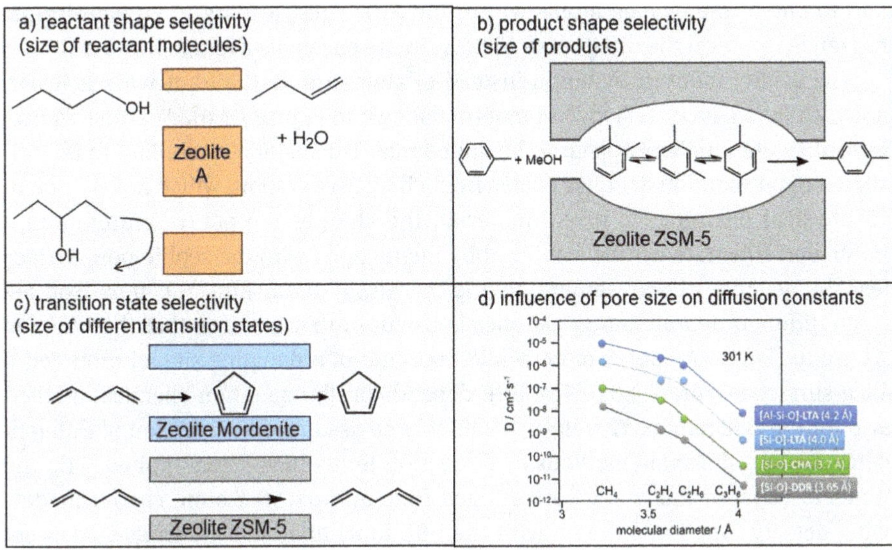

Figure 7.3.5: (a–c) Principles of shape selectivity in zeolite catalysis (examples reviewed in [63]) and (d) the influence of pore size on diffusion [65].

Although it is not considered to be a shape-selective effect in the strictest sense, the dependence of diffusion constants on molecular size and pore size (Figure 7.3.5d) can become significant, and it can contribute to macrokinetics far from the thermodynamic chemical equilibrium of the various products. [Al-Si-O]-**LTA**, [Si-O]-**LTA**, [Si-O]-**CHA** and [Si-O]-**DDR** all have 8-ring pores, but have different diameters of 4.2, 4.0, 3.7 and 3.65 Å, respectively. The diffusion constants for the various molecules shown in Figure 7.3.5d differ by up to three orders of magnitude. Thus, diffusion control is another important property of zeolites. The microkinetics characterizes the local reaction mechanism at the active site, but kinetic experiments will always deliver information on the macrokinetics, which is a combination of microkinetics and diffusion. It is recommended that the interested readers consult the excellent publication by Kärger, Ruthven and Theodorou for further details on diffusion in zeolites [65].

A strong Brønsted acid site is formed when a proton is bound to a bridging oxygen atom between Si and Al (Figure 7.3.6) [66]. This can be for example achieved by replacing Na^+ by an ion exchange with NH_4^+ cations after which NH_3 is removed by thermal treatment. In the protonated state, the O atom adopts larger distances to its bonding partners, Si and Al, and the Al coordination is severely trigonally distorted [67]. The bridging OH group in Figure 7.3.6b can be regarded as a silanol group, which is activated by the interaction with a Lewis acidic AlO_3 site. Therefore, the (calculated) deprotonation energy (DPE) is on the order of 1180 kJ/mol, which is significantly lower than that of a free SiOH group, 1415 kJ/mol, where oxygen only binds to two atoms, Si and H. It should be noted that the proton has four choices to bind to one of the oxygen atoms within the AlO_4 sites, leading to an equilibrium that tends to cancel the differences in DPEs in the ensemble averaged values [68].

The acidity induced by boron instead of aluminum in the framework is lower, indicated by a DPE of 1215 kJ/mol mol (not shown in Figure 7.3.6). At a first glance, these absolute numbers appear to be quite large. The major contribution to the calculated deprotonation energies comes from charge separation, which has no practical physical meaning. In a real material, the charges are not separated, and a protonated substrate will be stabilized by interactions with the zeolite pore surface near the negative charge centers, that is the $AlO_{4/2}^-$ tetrahedra. A comparison between different deprotonation energies is useful in that it shows that the bridging OH group (Figure 7.3.6b) is more acidic than that of a dangling silanol group on a silica surface (Figure 7.3.6a). The DPE depends on the quantum-chemical method used for the calculation. Therefore, it can only be used for a comparison of different values, where the same methods or correction terms have been applied [71]. The numbers given in this work were derived from approx. 50 T atom cluster models using density functional theory with the PBE functional and the triple-ζ basis set def2-TZVP, along with D3 dispersion correction and a three-body term. Further information on the physical significance of the DPE can be found in reference [71].

Acid catalysis has a long history in zeolite applications, starting from the first use of zeolite Y in refineries for hydrocarbon cracking. Ultrastabilized zeolite Y (USY) is a partially dealuminated material with extra-framework Al moieties (creating Lewis acid sites) and other species, such as La^{3+} or other cations, which make the catalyst hydrothermally stable for catalytic runs and the regeneration process, where undesired carbonaceous deposits are first removed by hot steam and then by combustion in air. The regeneration conditions are challenging for the stability of the catalyst, and this is a major problem for the development of new catalysts. USY has proximal Brønsted-Lewis acid pairs [72], and it is still one of the main components of the catalyst in the so-called fluid catalytic cracking (FCC) process, in which a heavy hydrocarbon feedstock is converted to useful products with lower boiling points (e.g. gasoline) by acid-catalyzed cracking [73].

Because of the high catalytic performance and reaction properties, zeolites were regarded as superacids for some time. Superacids are typically liquids and stronger than sulfuric acid (an example is $HSO_3F/HSbF_6$) and George A. Olah received the Nobel Prize in chemistry in 1994 for his pioneering contributions to carbocation chemistry with superacids. The acid strength in a solid is more difficult to characterize because the pK_a value cannot be easily defined and measured [74]. Although zeolites are suitable for catalyzing chemical reactions of hydrocarbons as are superacids, we know today that their catalytic performance is not only due solely to their intrinsic acidity but also to a large extent to the stabilization of

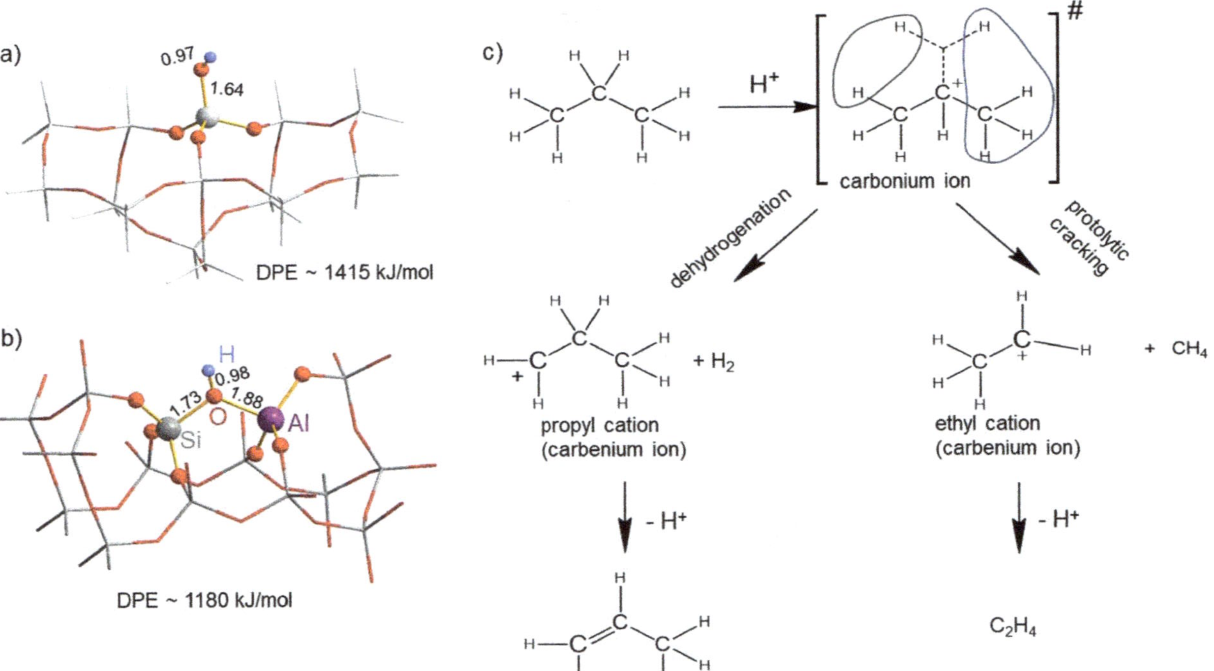

Figure 7.3.6: (a) A SiOH group on a silica surface compared to a (b) bridging OH group of an acidic zeolite catalyst. Interatomic distances are given in Å units. Models were derived from DFT cluster calculations (DFT method: PBE/def2-TZVP with D3 dispersion correction and three-body term). Typical Si-O and Al-O bond lengths in unprotonated zeolites are 1.62 and 1.75 Å, respectively. (c) Principle of dehydrogenation and catalytic cracking (Haag-Dessau mechanism): the zeolite proton from a bridging OH group attacks the C-H σ-bond to result in a pentacoordinated carbon atom in the transition state [28, 69, 70].

transition states and intermediates by internal pore interactions or the additional influence of extra-framework Al [75–80]. Therefore, zeolites are no longer considered to be superacidic [81]. Nonetheless, they still have a tremendous impact on catalytic conversions by providing an ideal reaction environment where the tailoring of transitions states and intermediates is important to achieving high reactivity and selectivity. The most common methods to characterize the zeolite acid strength are the heats of adsorption or temperature-programmed desorption of probe molecules, calculated deprotonation energies and IR frequency or NMR chemical shift changes upon adsorption of probe molecules [71, 74, 82, 83].

Such hydrocarbon reactions with acidic zeolite catalysts run through an ionic mechanism at elevated temperatures (typically 350 to over 500 °C), in which carbon atoms adopt a pentacoordinated transition state via a carbonium ion, which is formed by a proton from the zeolite surface attacking a σ-bond. This could be a C–C bond or as shown in Figure 7.3.6c, a C–H bond. Either hydrogen or methane is formed from the carbonium ion shown in Figure 7.3.6c along with two distinct alkyl cations: a propyl cation or an ethyl cation, respectively [28, 69, 70]. These alkyl cations are also called carbenium ions. They do not occur as free species and are instead stabilized by the interaction with the pore walls of the zeolite. A proton is then separated from the carbenium ions and transferred to the zeolite surface (restoring the bridging OH group) to yield the corresponding alkenes. Alkyl cation intermediates, on the other hand, are also alkylating agents that can attack alkene intermediates in zeolite-catalyzed reactions. This leads to a molecular growth, and ultimately coke deposits that fill the zeolite pore system, resulting in catalyst deactivation. This is for example highly relevant in the FCC process (see above), where the coke is removed by combustion in air, and the regenerated catalyst can then be used again. The variety of reactions taking place in zeolite-catalyzed hydrocarbon chemistry is more complex, and further information can be obtained for example in reference [28].

Acid catalysis over zeolites can be supplemented by an additional function, that is hydrogenation/dehydrogenation over metal particles. Noble metals are used, such as Pd and Pt (or the sulphur-resistent metals, Ni or Mo, when the feed contains organic sulfur compounds). In this bifunctional catalysis, an *n*-paraffin is dehydrogenated to an *n*-olefin first, and via a sequence of subsequent steps (Figure 7.3.7), an *iso*-paraffin or cracking products are formed, depending on the reactivity and shape selectivity of the carbenium ion. The isomerization step takes place via a protonated cyclopropane transition state. The carbenium ions can also break a C–C bond in the β-position (β-scission), which leads to cracking to an olefin and a smaller carbenium ion. More information, for example on the various types of β-scission can be found in reference [28]. For the isomerization step of the aforementioned "Isodewaxing" in the manufacturing of lube oils, where Pt particles are used in the silicoaluminophosphate SAPO-11 [62], this catalyst uses shape selectivity in its narrow pore system to reduce the formation of those bulky (branched, secondary and tertiary) carbenium ions that more readily lead to cracking because they can break apart by β-scission to a (relatively stable) smaller tertiary carbenium ion fragment and an alkene [28]. On the other hand, the carbenium ions formed from waxy *n*-paraffins cannot be converted to tertiary carbenium ions by β-scission, and they are preferably isomerized to their less waxy mono-branched isomers. When Chevron has commercialized this "hydroprocess" with this shape selective catalyst, the yields could be considerably improved over the prior bifunctionally catalyzed cracking technology using ZSM-5, where β-scissions are more dominant due to the larger space in its pore system. Therefore, another area where zeolite catalysts are used industrially includes the synthesis of lubricants, which have become a major business for oil companies, as well as surfactants [84].

Although boron sites in zeolite frameworks are generally considered to be very weak acidic centers, these sites can contribute to some important chemistries such as catalytic naphtha reforming at elevated temperatures when metal functions are present (e.g. Pt) that are for example capable of generating olefin intermediates [85, 86].

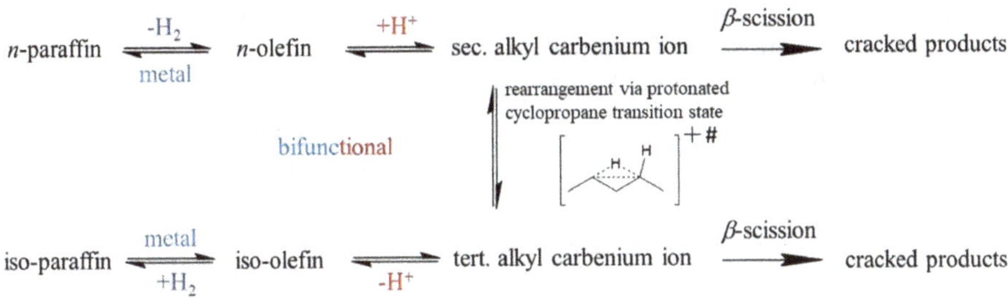

Figure 7.3.7: Principle of bifunctional catalysis with an acid function and a hydrogenation/dehydrogenation function [28, 87–89].

Tuning a zeolite to direct a catalytic reaction towards a specific product or product distribution, while maintaining the catalytic activity for an extended period of time, is a challenging task, requiring a wealth of know-how based on fundamental research [90], sometimes accompanied with intuition. The differences in pore diffusion and shape selectivity depend on pore size and dimensionality, and they require a versatile library of zeolite structures to serve specific applications. With this in mind, it is not surprising that some petrochemical companies are still quite active in the synthesis of new zeolite structures, which contributes to the growth of the number of known zeolite framework types. Larger chemical companies also have specialized sections involved in zeolite research. Zeolites are used as catalysts in processes other than FCC and hydrocracking-hydroisomerization, and more information may be obtained for example in reference [91].

In the following, selected examples of important technical applications beyond refining and petrochemistry are outlined. The methanol-to-olefin process is used to convert an important platform chemical, namely methanol, to low-molecular-weight hydrocarbons [92–95]. The largest MTO unit was announced in 2018, operated by Jiangsu Sailboat Petrochemical Company, Ltd. with an annual capacity of 833000 metric tons [96]. The products are mainly used for the plastics industry. The catalysts for the MTO process are the silicoaluminophosphate SAPO-34, which has the chabazite (**CHA**) framework type or ZSM-5. SAPO-34 is a small-pore molecular sieve, which has the advantage that the formation of products with larger molecular weight is suppressed and coke formation is prevented. The elucidation of the MTO reaction mechanism has required high efforts, and it was not obvious how an acid catalyst can lead to the formation of C–C bonds starting from methanol. It is now widely recognized that cyclic organic species (with 5 or 6 C atoms in the ring), such as hexamethylbenzene (HMB), within the zeolite pores are the reaction centers for the formation of C–C bonds [97]. As an example, HMB is methylated by methanol to yield Me-HMB$^+$. Deprotonation of the methyl group in the para-position of the C6-ring yields an alkene, which can grow in side-chain alkylation reactions until dealkylation leads to the olefin products, and HMB is restored for another catalytic cycle [92]. This mechanism is called the "hydrocarbon pool" mechanism because the C–C bond formation takes place in a co-catalytic pool of cyclic organic species in combination with the zeolite acid sites. The hydrocarbon pool must be assembled within an induction period of the catalytic process. The first C–C bond is formed when methanol is decomposed to formaldehyde and CO, which is then the carbonylation agent for methanol or dimethylether [98].

An important improvement in environmental catalysis in recent history is the selective catalytic reduction of NOx by urea using the zeolite chabazite, which was patented by a team at the BASF Florham Park, N.J. (USA) R&D headquarter [99]. The active site in this catalytic reaction is a Cu^{2+} ion, which is located in the six-rings of the zeolite [100–102]. The catalytic cycle involves the formation of bidentate nitrate/nitrite complexes on the copper by reaction of NO with NH_3 (formed from urea). The nitrite complex then reacts with another NH_3 to form N_2 and H_2O.

Propylene oxide is an important intermediate in polymer chemistry (e.g. for polyurethanes), and it was initially made using the chlorohydrine process, in which large amounts of $CaCl_2$ waste were produced. This process is still in use because the chemical industry needs pathways to consume Cl_2 gas from the chloralkali electrolysis. However, newer technology has been developed to avoid the $CaCl_2$ waste, and an environmentally benign alternative is the HPPO process (hydrogen peroxide propylene oxide), which uses the titanosilicate variant of ZSM-5 [97], TS-1. The process was independently developed by Thyssen-Krupp Uhde/Evonik Industries and BASF/Dow Chemical [105]. The first plants with the HPPO process went in operation in 2008 in Ulsan/South Korea (SKC) with a capacity of 130000 t/a and in Antwerp/Belgium (BASF/Dow) with 300000 t/a. This zeolite-catalyzed conversion is almost 100% selective, and it can be performed near room temperature because the strong Lewis acidity of Ti activates H_2O_2 [24, 104].

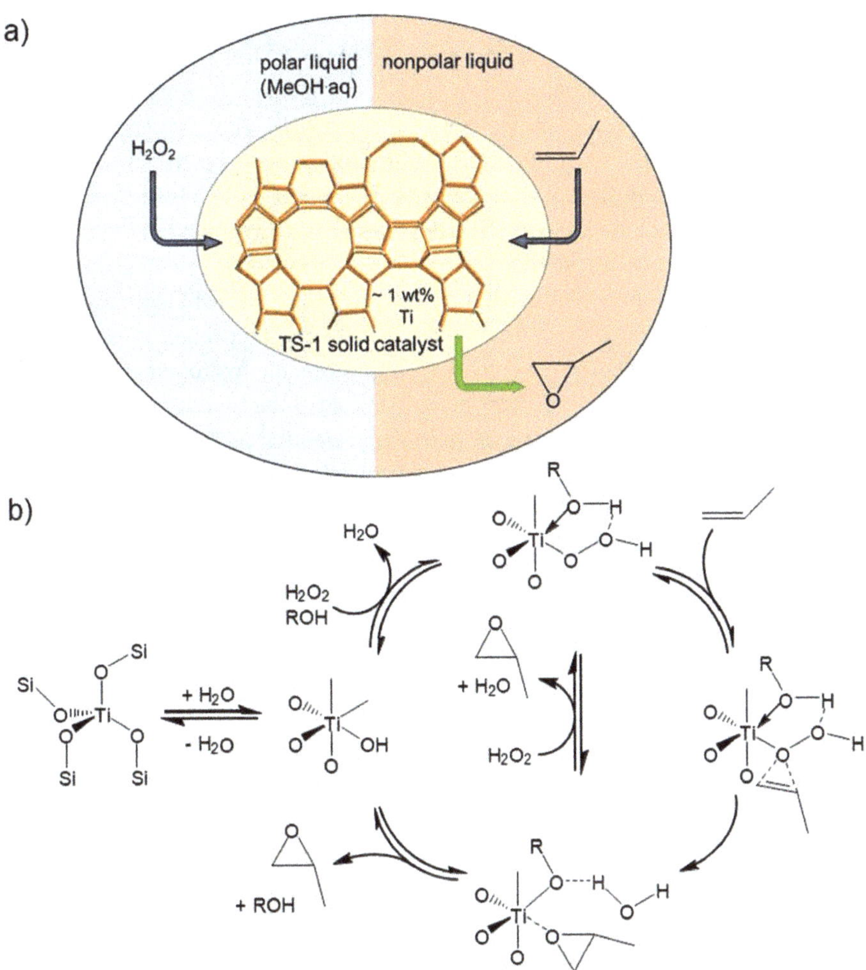

Figure 7.3.8: (a) Principle of propylene oxide batch production in a three-phase process; the propylene reactant is added via a nonpolar liquid phase into the reaction vessel, and the oxidizing agent is supplied via an aqueous phase with some methanol to tune the solvent polarity [24, 103]. (b) Scheme of the catalytic cycle [24, 104].

Propylene is directly oxidized to propylene oxide and H_2O is the only side product. The peculiarity of this process is that a nonpolar reactant, propylene and a polar reactant, H_2O_2, have to be assembled efficiently. The internal pore surface of TS-1 needs to be hydrophobic to reduce water adsorption to a low level and allows the hydrophobic reactant, propylene, to access the active site [106]. Hydrogen peroxide is less polar than water and is miscible with the moderately polar carrier methanol. Therefore, H_2O_2 can also enter the pore system. The batch reaction is a three-phase process (Figure 7.3.8a) with a nonpolar liquid, a polar liquid and the solid catalyst. The conversion is then carried out over a Ti center in the zeolite framework, and the catalytic cycle is shown in Figure 7.3.8b. The Ti coordination number is initially increased from 4 to 6 by water, which subsequently allows H_2O_2 to open a framework Ti-OSi bond. This yields Ti-OOH and SiOH. The oxygen atom closest to the Ti center of the Ti – OOH moiety is electrophilic, and it inserts in the double bond of propylene. The electrophilic property can be promoted by a hydrogen bond to the other oxygen atom of the peroxide. The donor for this hydrogen bond can be either methanol, H_2O_2, TiOH or the silanol group that has formed by opening the Ti-OSi bond [107, 108]. This is left open in Figure 7.3.8b as protic "R-O-H" moiety. Detailed studies have shown that a defect silanol group near the reaction center is essential for this catalytic reaction [106, 109]. Instead of H_2O_2, an organic hydroperoxide can also be used as the oxidizing agent, and here again, the role of the silanol in activating the Ti-bound organic peroxo ligand is crucial [109, 110]. A more recent paper suggests that dinuclear Ti sites are effective as reaction centers for selective alkene epoxidation [111]. This would represent an alternative model to the previously established single-site Ti center mechanism in other studies, which was recently again proven by the high activity of a calix [4] arene-Ti-IV single-site complexes grafted on delaminated zeotype supports, where the effect of external-surface pockets and silanol groups on catalysis was rigorously demonstrated in a comparative study [109].

Polyamides are a family of plastics that are manufactured on a large scale. In the following, several processes are described for which now some manufacturers use zeolite catalysts. An early precursor in the production of certain polyamides is cumene (i.e. *i*-propyl-benzene). It was produced previously using Friedel-Crafts catalysis ($AlCl_3$or H_3PO_4/SiO_2) in the reaction of benzene with propylene. Several manufacturers have replaced these catalysts, which are difficult to handle, with zeolites. Depending on the company, zeolites MCM-22 (**MWW**), zeolite Y, mordenite or beta are used [36]. Dow Chemical has developed a deeply dealuminated mordenite with mesopores, thus utilizing a hierarchical catalyst for improving molecular transport [37, 61].

One of the important monomers that leads to polyamide 6 is ε-caprolactam, which is made from cyclohexanone oxime. Cyclohexanone oxime can now also be made using the EniChem process, where TS-1 ([Ti-Si-O]-**MFI**) is the catalyst [36]. The conversion of cyclohexanone to the oxime is usually carried out with hydroxylamine. The EniChem process avoids the awkward hydroxylamine as a starting material, and the reactants are just NH_3 and H_2O_2. Two alternative reaction pathways were proposed. Either the cyclohexanone reacts with NH_3 first to yield the imine, which is subsequently oxidized by H_2O_2 to the oxime [112], or NH_3 forms hydroxylamine *in situ* by the reaction with H_2O_2 to subsequently convert cyclohexanone to the oxime [35]. The imine route is widely accepted for the gas phase reaction, whereas the hydroxylamine route is considered most likely in the liquid phase reaction.

Traditionally, the conversion of cyclohexanone oxime to ε-caprolactam is carried out with oleum, a highly corrosive inorganic acid. However, this Beckmann rearrangement does not require such a strong acid, and newer technology employs weakly acidic ZSM-5 catalysts either [Si-O]-**MFI** with weakly acidic defect silanol groups or [B-Si-O]-**MFI** with acid sites associated with boron [113]. The conversion of ε-caprolactam by a ring-opening polymerization produces polyamide 6 (trade name Perlon).

Some of these last examples substantiate how zeolite catalysts can contribute to process steps that are safer in handling and/or environmentally more benign than traditional manufacturing routes. This is one of the properties that has stimulated the track record of zeolites in the past, and as the author of this chapter assumes, it will do so in the future.

7.3.5 Summary and outlook

The three main categories of applications of synthetic zeolites are ion exchange, adsorption/separation and catalysis. Ion exchange is an application that has made its way into consumer households worldwide as a water softener in laundry detergents to replace polyphosphates, which had caused problems with eutrophication of natural waters. Adsorption and separation processes cover a wide field of special applications for which zeolite properties such as hydrophobic/hydrophilic surfaces, electrostatic fields caused by extra-framework cations, surface acidity and pore size provide the toolkit to tune the desired function. Catalysis was originally implemented in refining and petrochemical industries for cracking and other applications, among which FCC is still one of the largest units for zeolite catalysis. It is the "grand old lady", as Vogt and Weckhuysen named it in their 2015 review on recent developments of FCC [73]. Today, zeolites are much more than refinery and petrochemical catalysts. They play an important role in many manufacturing processes, for example in plastics industry, and in environmental catalysis, such as automotive and other exhaust gas cleaning applications. This short review does not claim to be exhaustive. It is rather an attempt to highlight the grand potential of zeolites for a wealth of applications in many different areas by restricting this text to some of the basic principles of the functional properties of zeolites.

The author assumes that zeolite catalysts will continue their success story in the future when fossil resources become less available or when they are replaced by other raw materials for environmental reasons. Current research addresses for example biomass utilization for novel synthetic routes to commodity chemicals and consumer products [114–119]. The synthesis of chiral zeolites was developed just recently, and it will be exciting to follow their properties in the future [120].

References

[1] https://www.reportlinker.com/p05960558/Global-Zeolite-Industry.html
[2] Cronstedt AF. Ron och beskriting om en obekant bärg ant, som kallas zeolites. Akad Handl Stockholm, 1756, 18, 120–30.
[3] Flanigen EM, Broach RW, Wilson ST. Introduction. In: Kulprathipanja S. (Ed.) Zeolites in Industrial Separation and Catalysis, Wiley-VCH, Weinheim, 2010.
[4] Barrer RM. Synthesis of a zeolitic mineral with chabazite-like sorptive properties. J Chem Soc, 1948, 33, 127–32.
[5] Barrer RM, White EAD. The hydrothermal chemistry of silicates. Part II. Synthetic crystalline sodium aluminosilicates. J Chem Soc, 1952, 1561–71.
[6] Rouquerol J, Avnir D, Fairbridge CW, Everett DH, Haynes JM, Pernicone N, Ramsay JDF, Sing KSW, Unger KK. Recommendations for the characterization of porous solids (Technical Report). Pure Appl Chem, 1994, 66, 1739–58.
[7] McCusker LB, Liebau F, Engelhardt G. Nomenclature of structural and compositional characteristics of ordered microporous and mesoporous materials with inorganic hosts – (IUPAC recommendations 2001). Pure Appl Chem, 2001, 73, 381–94.
[8] Baerlocher C, McCusker LB. Database of Zeolite Structures. http://www.iza-structure.org/databases/
[9] Pophale R, Cheeseman PA, Deem MW. A database of new zeolite-like materials. Phys Chem Chem Phys, 2011, 13, 12407–12.
[10] Reed TB, Breck DW. Crystalline zeolites .2. Crystal structure of synthetic zeolite, Type-A . J Am Chem Soc, 1956, 78, 5972–77.
[11] Milton RM. Molecular sieve adsorbents. U.S. Patent 2,882,244, 1959.
[12] Breck DW. Crystalline zeolite y. U.S. Patent 3,130,007, 1964.
[13] Argauer RJ, Landolt GR. Crystalline zeolite zsm-5 and method of preparing the same. U. S. Patent 3,702,886, 1972.
[14] Kokotailo GT, Lawton SL, Olson DH, Meier WM. Structure of synthetic zeolite Zsm-5. Nature, 1978, 272, 437–38.
[15] Meier WM. The crystal structure of mordenite (ptilolite). Z Kristallogr, 1961, 115, 439–50.
[16] Wadlinger RL, Kerr GT, Rosinski EJ. Catalytic composition of a crystalline zeolite. U.S. Patent 3,308,069A, 1967.
[17] Newsam JM, Treacy MMJ, Koetsier WT, Degruyter CB. Structural characterization of zeolite-beta. Proc R Soc London, Ser A, 1988, 420, 375–76.
[18] Dent LS, Smith JV. Crystal structure of chabazite, a molecular sieve. Nature, 1958, 181, 1794–96.
[19] Camblor MA, Barrett PA, Diaz-Cabanas MJ, Villaescusa LA, Puche M, Boix T, Perez E, Koller H. High silica zeolites with three-dimensional systems of large pore channels. Microporous Mesoporous Mater, 2001, 48, 11–22.
[20] Loewenstein W. The distribution of aluminum in the tetrahedra of silicates and aluminates. Am Mineral, 1954, 39, 92–96.
[21] Schüth F, Sing KSW, Weitkamp J. Handbook of Porous Solids, Wiley-VCH, Weinheim, 2002.
[22] Čejka J, Corma A, Zones S. Zeolites and Catalysis, VCH-Wiley, Weinheim, 2010.
[23] van Bekkum H, Flanigen EM, Jacobs PA, Jansen JC. Introduction to Zeolite Science and Practice, Elsevier Science B.V., Amsterdam, 2001.
[24] van Santen RA. Modern Heterogeneous Catalysis, Wiley-VCH, Weinheim, 2017.
[25] Moliner M, Martinez C, Corma A. Multipore zeolites: Synthesis and catalytic applications. Angew Chem Int Ed, 2015, 54, 3560–79.

[26] Sherman JD. Synthetic zeolites and other microporous oxide molecular sieves. PNAS, 1999, 96, 3471–78.
[27] Yilmaz B, Trukhan N, Müller U. Industrial outlook on zeolites and metal organic frameworks. Chin J Catal, 2012, 33, 3–10.
[28] Stöcker M. Gas phase catalysis by zeolites. Micropor Mesopor Mater, 2005, 82, 257–92.
[29] Kulprathipanja S. Zeolites in Industrial Separation and Catalysis, Wiley-VCH, Weinheim, 2010.
[30] Dusselier M, Davis ME. Small-pore zeolites: Synthesis and catalysis. Chem Rev, 2018, 118, 5265–329.
[31] Corma A. State of the art and future challenges of zeolites as catalysts. J Catal, 2003, 216, 298–312.
[32] Schwarz KH, Hofmann U. Lazurite and ultramarine. II Z Anorg Allg Chem, 1970, 378, 152–56.
[33] Jiang J, Yu J, Corma A. Extra-large-pore zeolites: bridging the gap between micro and mesoporous structures. Angew Chem Int Ed, 2010, 49, 3120–45.
[34] Flanigen EM. Zeolites and molecular sieves: An historical perspective. In: van Bekkum H, Flanigen EM, Jacobs PA, Jansen JC. (Eds.) . Introduction to Zeolite Science and Practice, Elsevier Science B.V., Amsterdam, 2001, 11–35.
[35] Zecchina A, Spoto G, Bordiga S, Geobaldo F, Petrini G, Leofanti G, Padovan M, Mantegazza M, Roffia P, Kaliaguine S, Krauss HL, Sivasanker S. Ammoximation of cyclohexanone on titanium silicalite – investigation of the reaction-mechanism. Stud Surf Sci Catal, 1993, 75, 719–29.
[36] Bellussi G, Perego C. Industrial catalytic aspects of the synthesis of monomers for nylon production. CATTECH, 2000, 4, 4–16.
[37] Hartmann M, Thommes M, Schwieger W. Hierarchically-ordered zeolites: A critical assessment. Adv Mater Interfaces, 2021, 8, 2001841.
[38] Gramlich V, Meier WM. Crystal structure of hydrated NaA – detailed refinement of a pseudosymmetric zeolite structure. Z Kristallogr, 1971, 133, 134–36.
[39] Schmidt W. Application of microporous materials as ion-exchangers. In: Schüth F, Sing KSW, Weitkamp J. (Eds.) Handbook of Porous Solids, vol. 2, Wiley-VCH, Weinheim, 2002, 1058–97.
[40] Carr SW, Gore B, Anderson MW. (SiAl)-Si-29-Al-27 and H-1 solid-state NMR study of the surface of zeolite MAP. Chem Mater, 1997, 9, 1927–32.
[41] Adams CJ, Araya A, Carr SW, Chapple AP, Graham P, Minihan AR, Osinga TJ. Zeolite map: A new detergent builder. In: Karge HG, Weitkamp J. (Eds.) Stud Surf Sci Catal, vol. 98, Elsevier, 1995, 206–07.
[42] Pluth JJ, Smith JV. Accurate redetermination of crystal-structure of dehydrated zeolite A – absence of near zero coordination of sodium – refinement of Si,Al-ordered superstructure. J Am Chem Soc, 1980, 102, 4704–08.
[43] Olson DH. The crystal-structure of dehydrated NaX. Zeolites, 1995, 15, 439–43.
[44] Feuerstein M, Lobo RF. Characterization of Li cations in zeolite LiX by solid-state NMR spectroscopy and neutron diffraction. Chem Mater, 1998, 10, 2197–204.
[45] Koller H, Burger B, Schneider AM, Engelhardt G, Weitkamp J. Location of Na^+ and Cs^+ cations in CsNaY zeolites studied by Na-23 and Cs-133 magic-angle spinning nuclear magnetic resonance spectroscopy combined with X-ray structure analysis by rietveld refinement. Microporous Mater, 1995, 5, 219–32.
[46] Halkier A, Coriani S, Jorgensen P. The molecular electric quadrupole moment of N_2. Chem Phys Lett, 1998, 294, 292–96.
[47] Couling VW, Ntombela SS. The electric quadrupole moment of O_2. Chem Phys Lett, 2014, 614, 41–44.

[48] Mikosch H, Uzunova EL, St Nikolov G. Interaction of molecular nitrogen and oxygen with extraframework cations in zeolites with double six-membered rings of oxygen-bridged silicon and aluminum atoms: A DFT study. J Phys Chem B, 2005, 109, 11119–25.
[49] Feuerstein M, Lobo RF. Influence of oxygen and nitrogen on Li-7 MAS NMR spectra of zeolite LiX-1.0. Chem Commun, 1998, 1647–48.
[50] Gaffney TR. Porous solids for air separation. Curr Opin Solid State Mater Sci, 1996, 1, 69–75.
[51] Wikipedia. Oxygen concentrator. https://en.wikipedia.org/wiki/Oxygen_concentrator
[52] Valencia S, Rey F. New Developments in Adsorption/Separation of Small Molecules by Zeolites, Springer International Publishing, Heidelberg, 2020.
[53] Chen CY, Zones SI. Investigation of small-pore zeolites via vapor-phase adsorption of n-hexane. Chem Ing Tech, 2021, 93, 959–66.
[54] Vasta S, Brancato V, La Rosa D, Palomba V, Restuccia G, Sapienza A, Frazzica A. Adsorption heat storage: State-of-the-art and future perspectives. Nanomaterials, 2018, 8, 522.
[55] van Alebeek R, Scapino L, Beving MAJM, Gaeini M, Rindt CCM, Zondag HA. Investigation of a household-scale open sorption energy storage system based on the zeolite 13X/water reacting pair. Appl Therm Eng, 2018, 139, 325–33.
[56] Yilmaz B, Müller U. Catalytic applications of zeolites in chemical industry. Top Catal, 2009, 52, 888–95.
[57] Weisz PB. Molecular shape selective catalysis. Pure Appl Chem, 1980, 52, 2091–103.
[58] Csicsery SM. Shape-selective catalysis in zeolites. Zeolites, 1984, 4, 202–13.
[59] Degnan TF. The implications of the fundamentals of shape selectivity for the development of catalysts for the petroleum and petrochemical industries. J Catal, 2003, 216, 32–46.
[60] Gounder R, Davis ME. Beyond shape selective catalysis with zeolites: Hydrophobic void spaces in zeolites enable catalysis in liquid water. AlChE J, 2013, 59, 3349–58.
[61] Garces JM, Olken MM, Lee GJ, Meima GR, Jacobs PA, Martens JA. Shape selective chemistries with modified mordenite zeolites. Top Catal, 2009, 52, 1175–81.
[62] Miller SJ. New molecular sieve process for lube dewaxing by wax isomerization. Microporous Mater, 1994, 2, 439–49.
[63] Davis ME. Catalytic materials via molecular imprinting. CATTECH, 1997, 1, 19–26.
[64] Zones SI, Chen CY, Benin A, Hwang S-J. Opportunities for selective catalysis within discrete portions of zeolites: The case for SSZ-57LP. J Catal, 2013, 308, 213–25.
[65] Kärger J, Ruthven DM, Theodorou DN. Diffusion in Nanoporous Materials, Wiley-VCH, Weinheim, 2012.
[66] Haag WO, Lago RM, Weisz PB. The active-site of acidic aluminosilicate catalysts. Nature, 1984, 309, 589–91.
[67] Koller H, Meijer EL, van Santen RA. ^{27}Al quadrupole interaction in zeolites loaded with probe molecules – A quantum-chemical study of trends in electric field gradients and chemical bonds in clusters. Solid State Nucl Magn Reson, 1997, 9, 165–75.
[68] Jones AJ, Iglesia E. The strength of bronsted acid sites in microporous aluminosilicates. ACS Catal, 2015, 5, 5741–55.
[69] Haag WO, Dessau RM. In 8th Int Congress on Catalysis, Berlin, 1984, Berlin, 1984, 305.
[70] Kotrel S, Knözinger H, Gates BC. The Haag-Dessau mechanism of protolytic cracking of alkanes. Micropor Mesopor Mater, 2000, 35-6, 11–20.
[71] Sauer J. Bronsted activity of two-dimensional zeolites compared to bulk materials. Faraday Discuss, 2016, 188, 227–34.
[72] Schroeder C, Hansen MR, Koller H. Ultrastabilization of Zeolite Y Transforms Brønsted–Brønsted acid pairs into Brønsted–Lewis acid pairs. Angew Chem Int Ed, 2018, 57, 14281–85.
[73] Vogt ETC, Weckhuysen BM. Fluid catalytic cracking: Recent developments on the grand old lady of zeolite catalysis. Chem Soc Rev, 2015, 44, 7342–70.

[74] Haw JF. Zeolite acid strength and reaction mechanisms in catalysis. PCCP, 2002, 4, 5431–41.
[75] Schallmoser S, Ikuno T, Wagenhofer MF, Kolvenbach R, Haller GL, Sanchez-Sanchez M, Lercher JA. Impact of the local environment of Bronsted acid sites in ZSM-5 on the catalytic activity in n-pentane cracking. J Catal, 2014, 16, 93–102.
[76] Eder F, Lercher JA. On the role of the pore size and tortuosity for sorption of alkanes in molecular sieves. J Phys Chem B, 1997, 101, 1273–78.
[77] Svelle S, Tuma C, Rozanska X, Kerber T, Sauer J. Quantum chemical modeling of zeolite-catalyzed methylation reactions: Toward chemical accuracy for barriers. J Am Chem Soc, 2009, 131, 816–25.
[78] Gounder R, Iglesia E. The roles of entropy and enthalpy in stabilizing ion-pairs at transition states in zeolite acid catalysis. Acc Chem Res, 2012, 45, 229–38.
[79] Li SH, Zheng AM, Su YC, Zhang HL, Chen L, Yang J, Ye CH, Deng F. Bronsted/Lewis acid synergy in dealuminated HY zeolite: A combined solid-state NMR and theoretical calculation study. J Am Chem Soc, 2007, 129, 11161–71.
[80] Mezari B, Magusin PCMM, Almutairi SMT, Pidko EA, Hensen EJM. Nature of enhanced Brønsted acidity induced by extraframework aluminum in an ultrastabilized faujasite zeolite: An in situ NMR study. J Phys Chem C, 2021, 125, 9050–59.
[81] Haw JF, Nicholas JB, Xu T, Beck LW, Ferguson DB. Physical organic chemistry of solid acids: Lessons from in situ NMR and theoretical chemistry. Acc Chem Res, 1996, 29, 259–67.
[82] Jiang YJ, Huang J, Dai WL, Hunger M. Solid-state nuclear magnetic resonance investigations of the nature, property, and activity of acid sites on solid catalysts. Solid State Nucl Magn Reson, 2011, 39, 116–41.
[83] Derouane EG, Vedrine JC, Pinto RR, Borges PM, Costa L, Lemos MANDA, Lemos F, Ribeiro FR. The acidity of zeolites: Concepts, measurements and relation to catalysis: A review on experimental and theoretical methods for the study of zeolite acidity. Catal Rev Sci Eng, 2013, 55, 454–515.
[84] Aitani A, Wang JB, Wang I, Al-Khattaf S, Tsai T-C. Environmental benign catalysis for linear alkylbenzene synthesis: A review. Catal Surv Asia, 2014, 18, 1–12.
[85] Chen CY, Rainis A, Zones SI. Reforming with novel borosilicate molecular sieve catalysts. In: Lednor PW, Ledoux M-J, Nagaki DA, Thompson LT. (Eds.) Mater Res Soc Symp – Advanced Catalytic Materials, Material Research Society, 1997, 205–15.
[86] Koller H, Chen CY, Zones SI. Selectivities in post-synthetic modification of borosilicate zeolites. Top Catal, 2015, 58, 451–79.
[87] Corma A. Inorganic solid acids and their use in acid-catalyzed hydrocarbon reactions. Chem Rev, 1995, 95, 559–614.
[88] Weitkamp J, Hunger M. In: Cejka J, van Bekkum H, Corma A, Schüth F. (Eds.) Introduction to Zeolite Molecular Sieves, Elsevier, 2007, 787.
[89] Weitkamp J, Jacobs PA, Martens JA. Isomerization and hydrocracking of C9 through C16 n-alkanes on Pt/HZSM-5 zeolite. Appl Catal, 1983, 8, 123–41.
[90] Haag WO. Catalysis by zeolites – science and technology. In: Weitkamp J, Karge HG, Pfeifer H, Hölderich W. (Eds.) Stud Surf Sci Catal (Zeolites and Related Microporous Materials: State of the Art 1994), vol. 84, Elsevier Science B.V., Amsterdam, 1994, 1375–94.
[91] Corma A. From microporous to mesoporous molecular sieve materials and their use in catalysis. Chem Rev, 1997, 97, 2373–419.
[92] Tian P, Wei YX, Ye M, Liu ZM. Methanol to olefins (MTO): From fundamentals to commercialization. ACS Catal, 2015, 5, 1922–38.
[93] Bertau M, Räuchle K, Offermanns H. Methanol – the basic chemical product. Chem unserer Zeit, 2015, 49, 312–29.

[94] Goeppert A, Czaun M, Jones JP, Prakash GKS, Olah GA. Recycling of carbon dioxide to methanol and derived products – closing the loop. Chem Soc Rev, 2014, 43, 7995–8048.
[95] Gogate MR. Methanol-to-olefins process technology: Current status and future prospects. Pet Sci Technol, 2019, 37, 559–65.
[96] Jenkins S. World's largest single-train methanol-to-olefins plant now operating. https://www.chemengonline.com/worlds-largest-single-train-methanol-to-olefins-plant-now-operating/
[97] Prieto G, Schüth F. The Yin and Yang in the development of catalytic processes: Catalysis research and reaction engineering. Angew Chem Int Ed, 2015, 54, 3222–39.
[98] Liu Y, Müller S, Berger D, Jelic J, Reuter K, Tonigold M, Sanchez-Sanchez M, Lercher JA. Formation mechanism of the first carbon-carbon bond and the first olefin in the methanol conversion into hydrocarbons. Angew Chem Int Ed, 2016, 55, 5723–26.
[99] Bull I, Xue WM, Burk P, Boorse S, Jaglowski WM, Koermer GS, Moini A, Patchett JA, Dettling JC, Caudle MT. Processes for reducing nitrogen oxides using copper CHA zeolite catalysts. U. S Patent, 8,404,203 B2, 2009.
[100] Janssens TVW, Falsig H, Lundegaard LF, Vennestrom PNR, Rasmussen SB, Moses PG, Giordanino F, Borfecchia E, Lomachenko KA, Lamberti C, Bordiga S, Godiksen A, Mossin S, Beato P. A consistent reaction scheme for the selective catalytic reduction of nitrogen oxides with ammonia. ACS Catal, 2015, 5, 2832–45.
[101] Beale AM, Gao F, Lezcano-Gonzalez I, Peden CHF, Szanyi J. Recent advances in automotive catalysis for NOx emission control by small-pore microporous materials. Chem Soc Rev, 2015, 44, 7371–405.
[102] Wang Y, Nishitoba T, Wang YN, Meng XJ, Xiao FS, Zhang WP, Marler B, Gies H, De Vos D, Kolb U, Feyen M, McGuire R, Parvulescu AN, Muller U, Yokoi T. Cu-exchanged CHA-type zeolite from organic template-free synthesis: An effective catalyst for NH3-SCR. Ind Eng Chem Res, 2020, 59, 7375–82.
[103] Behr A, Agar DW, Jörissen J, Vorholt AJ. Einführung in die Technische Chemie, Springer, Berlin, 2016.
[104] Clerici MG, Ingallina P. Epoxidation of lower olefins with hydrogen-peroxide and titanium silicalite. J Catal, 1993, 140, 71–83.
[105] Schmidt F, Bernhard M, Morell H, Pascaly M. HPPO process technology a novel route to propylene oxide without coproducts. Chim Oggi, 2014, 32, 31–35.
[106] Khouw CB, Dartt CB, Labinger JA, Davis ME. Studies on the catalytic-oxidation of alkanes and alkenes by titanium silicates. J Catal, 1994, 149, 195–205.
[107] Neurock M, Manzer LE. Theoretical insights on the mechanism of alkene epoxidation by H_2O_2 with titanium silicalite. Chem Commun, 1996, 1133–34.
[108] Bellussi G, Carati A, Clerici MG, Maddinelli G, Millini R. Reactions of titanium silicalite with protic molecules and hydrogen-peroxide. J Catal, 1992, 133, 220–30.
[109] Grosso-Giordano NA, Schroeder C, Okrut A, Solovyoy A, Schöttle C, Chassé W, Marinkoyic N, Koller H, Zones SI, Katz A. Outer-sphere control of catalysis on surfaces: A comparative study of Ti(IV) single-sites grafted on amorphous versus crystalline silicates for alkene epoxidation. J Am Chem Soc, 2018, 140, 4956–60.
[110] Grosso-Giordano NA, Hoffman AS, Boubnov A, Small DW, Bare SR, Zones SI, Katz A. Dynamic reorganization and confinement of TiIV active sites controls olefin epoxidation catalysis on two-dimensional zeotypes. J Am Chem Soc, 2019, 141, 7090–106.
[111] Gordon CP, Engler H, Tragl AS, Plodinec M, Lunkenbein T, Berkessel A, Teles JH, Parvulescu AN, Copéret C. Efficient epoxidation over dinuclear sites in titanium silicalite-1. Nature, 2020, 586, 708–13.

[112] Reddy JS, Sivasanker S, Ratnasamy P. Ammoximation of cyclohexanone over a titanium silicate molecular sieve, TS-2. J Mol Catal, 1991, 69, 383–92.
[113] Heitmann GP, Dahlhoff G, Hölderich WF. Catalytically active sites for the Beckmann rearrangement of cyclohexanone oxime to epsilon-caprolactam. J Catal, 1999, 186, 12–19.
[114] Climent MJ, Corma A, Iborra S. Converting carbohydrates to bulk chemicals and fine chemicals over heterogeneous catalysts. Green Chem, 2011, 13, 520–40.
[115] Li H, Riisager A, Saravanamurugan S, Pandey A, Sangwan RS, Yang S, Luque R. Carbon-increasing catalytic strategies for upgrading biomass into energy-intensive fuels and chemicals. ACS Catal, 2018, 8, 148–87.
[116] Hu L, Xu JX, Zhou SY, He AY, Tang X, Lin L, Xu JM, Zhao YJ. Catalytic advances in the production and application of biomass derived 2,5-dihydroxymethylfuran. ACS Catal, 2018, 8, 2959–80.
[117] Bermejo-Deval R, Assary RS, Nikolla E, Moliner M, Roman-Leshkov Y, Hwang SJ, Palsdottir A, Silverman D, Lobo RF, Curtiss LA, Davis ME. Metalloenzyme-like catalyzed isomerizations of sugars by Lewis acid zeolites. PNAS, 2012, 109, 9727–32.
[118] Roman-Leshkov Y, Barrett CJ, Liu ZY, Dumesic JA. Production of dimethylfuran for liquid fuels from biomass-derived carbohydrates. Nature, 2007, 447, 5.
[119] Davis ME. Heterogeneous catalysis for the conversion of sugars into polymers. Top Catal, 2015, 58, 405–09.
[120] Brand SK, Schmidt JE, Deem MW, Daeyaert F, Ma YH, Terasaki O, Orazov M, Davis ME. Enantiomerically enriched, polycrystalline molecular sieves. PNAS, 2017, 114, 5101–06.

7.4 Metal-organic frameworks

Christoph Janiak

Metal-organic frameworks, abbreviated as MOFs, are coordination networks with organic ligands containing potential voids. Coordination networks are cross-linked metal coordination compounds extending through repeating coordination entities in two or three dimensions (2D, 3D) [1]. Porosity is usually determined after removal of the templating solvent molecules by measuring a nitrogen (or argon) sorption isotherm at 77 K (78 K) and then calculating an internal surface area and pore size distribution based on the Brunauer-Emmett-Teller (BET) theory and related to 1.0 g material (specific surface area). The four prototypical MOFs MOF-5, Cu-btc, MIL-53 (Al) and ZIF-8 will be briefly presented.

7.4.1 Selected MOF structures

MOF-5: Based on hydrothermally obtained tetranuclear and tetrahedral $\{Zn_4(\mu_4\text{-}O)\}$ units and aromatic para-dicarboxylate ligands such as benzene-1,4-dicarboxylate (bdc^{2-}, terephthalate), a series of similarly constructed (iso-reticular) structures, including $3D\text{-}[Zn_4O(bdc)_3]$, MOF-5, can be obtained [2]. The carboxylate groups span the six edges of the $\{Zn_4(\mu_4\text{-}O)\}$ tetrahedron, and the dicarboxylate bridges are oriented at right angles to each other (Figure 7.4.1).

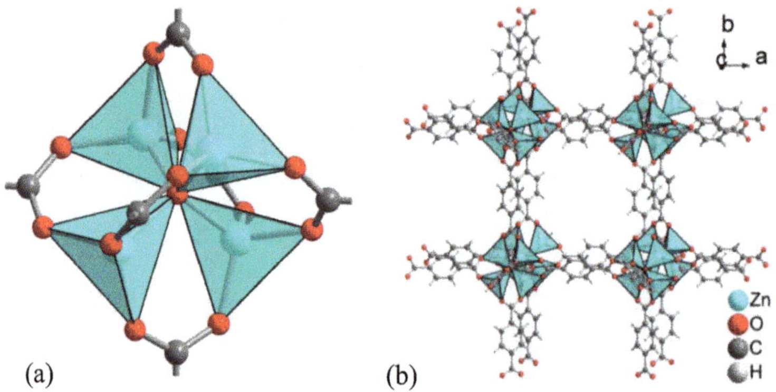

Figure 7.4.1: (a) Tetranuclear $\{Zn_4(\mu_4\text{-}O)\}$ unit of MOF-5 with bridging carboxylate groups. The C atoms of the carboxylate bridge are located at the vertices of an octahedron. Each Zn atom is tetrahedrally coordinated by four O atoms. (b) Packing diagram of $3D\text{-}[Zn_4O(bdc)_3]$, MOF-5. The network is traversed by channels along the *a*, *b* and *c* axis. The specific internal surface area is ~2900 $m^2\ g^{-1}$.

https://doi.org/10.1515/9783110798890-019

Cu-btc: The compound 3D-[$Cu_3(btc)_2(H_2O)_3$] (also called HKUST-1, btc^{3-} = benzene-1,3,5-tricarboxylate, trimesate) contains two copper atoms as the metal atom assembly, which are bridged by four carboxylate groups as in copper acetate ([$Cu_2(CH_3COO)_4$]) to form a "paddle-wheel" unit (Figure 7.4.2). Cu-btc is stable up to 240 °C. The aqua ligands can be removed [3].

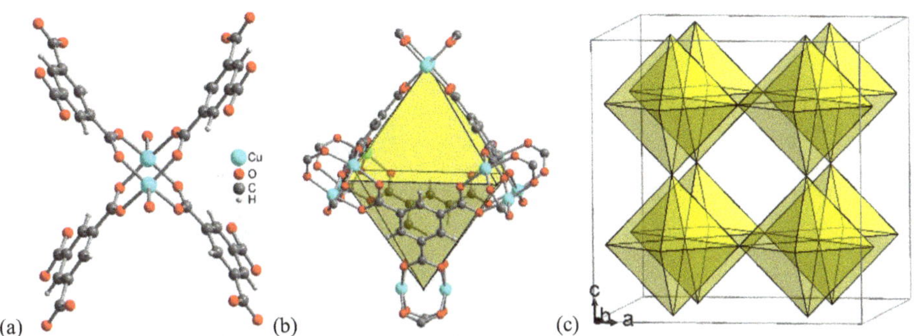

Figure 7.4.2: (a) {$Cu_2(btc)_4$} "paddle-wheel" unit in Cu-btc (HKUST-1). (b) The $\{Cu\}_2$ dumbbells sit at the corners of an octahedron. Four btc ligands each span opposite four of the eight faces of the octahedron. (c) These octahedra form a porous 3D network via vertex linkage. The network is traversed by channels along the *a*, *b* and *c* axis. The specific surface area is ~1300 $m^2\ g^{-1}$.

MIL-53(Al): Many aluminum MOFs with the formula unit 3D-[Al(µ-OH)(dicarboxylate ligand)] such as MIL-53(Al), are built up from slightly angled strands of vertex-sharing $\{AlO_6\}$ octahedra. In the strands, the Al atoms are connected by the oxygen atoms of µ-hydroxido and carboxylate bridges (Figure 7.4.3) [4]. The dicarboxylate ligands connect each strand to four neighboring strands, forming microporous rhombic to square channels in the 3D framework. The dicarboxylate ligands are terephthalate in MIL-53(Al).

ZIF-8: Three-dimensional zeolitic imidazolate frameworks (ZIFs) with the formula 3D-[Zn(imidazolate)$_2$] are obtained from Zn^{2+} and imidazolate ligands. ZIFs are one of the few porous MOFs in which individual metal atoms are bridged by ligands. In ZIFs, each Zn atom is surrounded by four N atoms from four imidazolate ligands (Figure 7.4.4). The linkage leads to zeolite-like structures in which the Zn atom occupies the position of Al/Si and the imidazolate bridges occupy the positions of the oxygen bridges in the zeolite. The best-known ZIF is ZIF-8 with 2-methylimidazolate as ligand [5, 6].

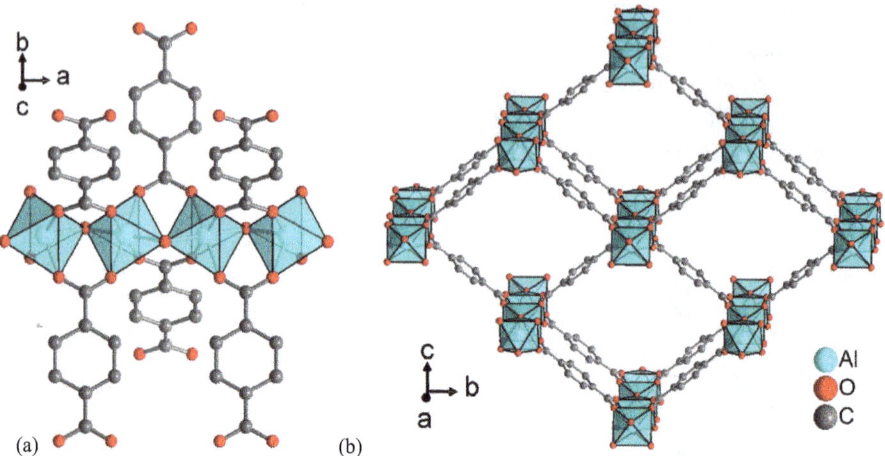

Figure 7.4.3: (a) Vertex-linked $\{Al(\mu\text{-}OH)(O_2C\text{-}C_6H_4\text{-}CO_2)_2\}$ strand in MIL-53(Al) and (b) resulting 3D framework with parallel channels in only one direction (here along a) (hydrogen atoms are not shown). The specific surface area is 1600 $m^2\,g^{-1}$ for (the large-pore form of) MIL-53(Al).

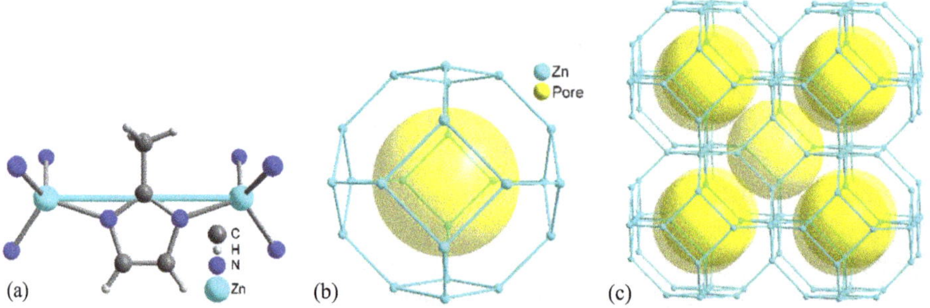

Figure 7.4.4: (a) Zn and 2-metylimidazolate building groups in ZIF-8 and (b) the assembly to a β-cage and (c) sodalite structure showing only the Zn atoms and their topological connections. The pore represented by the yellow sphere inside the β-cage has a diameter of 12 Å; the hexagonal rings as pore openings have a diameter of 3.4 Å (taking the van-der-Waals surface into account). The BET surface area is about 1600 $m^2\,g^{-1}$.

7.4.2 General MOF properties

MOFs are crystalline microporous to mesoporous materials (microporous < 2 nm, mesoporous from 2 to 50 nm pore diameter). In contrast to the amorphous mesoporous materials silica gel and activated carbons, MOFs have defined identical pore systems due to their crystallinity. In contrast to the likewise crystalline and thus uniform microporous zeolites with only limited structural units, the pore size and shape, the hydrophilicity/hydrophobicity, the inner surface functionality up to the chirality can be tailored in many ways in MOFs by the organic bridging ligands. The ligands can be modified by organic-chemical reactions (substitutions, additions) even after the formation of the MOF network. For this purpose, the term "post-synthetic modification" has been coined for MOFs [7].

There are now about 200 zeolite structures compared to over 20000 MOF structures. The BET surface area for zeolites, silica gels and activated carbons is at most about 1000 $m^2\ g^{-1}$, while values of 2000–4000 $m^2\ g^{-1}$ are reproducible and applicable for MOFs. The pore openings or channel diameters in MOFs range from 0.3 to 3.4 nm, with specific pore volumes up to 1.5–2 $cm^3\ g^{-1}$.

From the synthesis, the pores of the MOF scaffolds are filled with solvent molecules, which have a templating effect. Before using the porosity, the solvent guest molecules have to be removed, which is done by evacuation, if necessary after an exchange of less volatile for more volatile solvents. In this way, the MOF is first activated.

Some MOFs are already being produced industrially on a pilot scale in anticipation of applications. Without going into details of MOF synthesis, there is the problem of "greener", cheaper and easily scalable synthesis procedures to provide access to larger amounts if these materials are to be used. The synthetic procedures for fabricating MOFs can require harsh conditions (high temperature/high pressure in autoclaves) and/or expensive organic linkers together with not environmentally benign solvents. There is the need to develop more syntheses at ambient pressures, possibly as continuous processes and without the often-used organic solvents like dimethylformamide [8, 9]. A positive example is the MOF aluminum fumarate ($[Al(OH)(O_2C\text{-}C_2H_2\text{-}CO_2)]$, BASOLITE™ A520) [10, 11], where only aqueous solutions of aluminum sulfate and sodium fumarate are combined at ambient pressure and at a temperature of 60 °C, even in a continuous flow process with a space-time-yield of 97159 kg m^{-3} day^{-1} and a rate of 5.6 kg h^{-1} [12].

From the initial synthesis, MOFs are prepared as microcrystalline powders. Yet, powdered MOFs have limited practical interest owing to poor processability, safety problems from dusting and poor recyclability. The use of MOFs in potential applications most often requires shaping as an important aspect of their prospective use at an industrial scale. Thus, MOFs have to be formulated to obtain handable objects as deposited MOF layers on a substrate, coatings, films or 3D-structured composites such as grains, pellets, beads, areogels, hydrogels and monoliths (Figure 7.4.5) [13–15]. These shaped MOF particles have to be mechanically stable and retain the initial MOF porosity [16]. The simple use of a polymeric binder to assemble and shape MOF particles to a larger object may block the MOF pores and reduce the accessible surface area.

A possible obstacle for applications of MOFs is their often-insufficient hydrothermal stability [17] even if by now, some MOFs are known to be stable against prolonged contact with water and humid air.

Figure 7.4.5: MOF@polymer-monolith composites (with a scanning electron microscopy image) from the MOFs Al-fumarate (left) and MIL-101(Cr) (right) with the polymer poly(vinyl alcohol) (PVA), prepared using a phase separation technique. In the MOF@PVA monoliths, the mass-weighted BET surface area and the water vapor uptake capacity reproducibly reached 60–100% of the neat MOF values with MOF loadings of 50–80 wt%. The monoliths exhibit slightly plastic properties and a high resistance against deformation.

7.4.3 Potential applications of MOFs

In terms of industrial scale applications, MOFs can still be seen as an emerging class of materials. So far, most of the noted "applications" have to be regarded as potential ones and have to do with the porosity. MOFs are intensively investigated for a task- and compound-specific selective adsorption, storage and separation, sensing, catalysis, etc. For many of the applications, it is mandatory to develop robust and inexpensive MOF materials with good hydrolytic stability, excellent adsorption properties and recyclability.

The following paragraphs summarize the most important "application-oriented" developments with MOFs [18], based on review articles, but they do not present a full account of all "possible applications". It should also be noted that there are occasional applications with "MOFs" which have no proven porosity and are rather nonporous coordination polymer/network. The noted possible applications do not only refer to pristine MOFs but also refer to MOF composites, for example, mixed-matrix membranes for gas separation with MOFs as fillers [19, 20].

Early on, MOFs came into the focus for H_2 storage for mobile applications. At 77 K, volume-specific storage curves of hydrogen on different MOF-materials in comparison to the compression curve of hydrogen into empty gas containers showed an up to three times higher H_2 uptake in g(H_2)/L(container). Still, a sufficiently high amount of storage at room temperature could not be attained [21, 22]. For methane, the storage capacity of a tank filled with shaped bodies of the MOF Basolite A520 could reach a 40% increase vs. a conventional container at 35 bar and room temperature [10].

Numerous studies have demonstrated the capture and separation of carbon dioxide, CO_2 from binary mixtures with N_2, which were simplistically viewed as flue gas [23, 24] as well as of SO_2, H_2S, NH_3 and NO_x from gas mixtures [25, 26] and even up to chemical warfare agents [27]. MOFs have addressed the adsorption-based separations of fluorocompounds [28], acetylene/ethylene [29], olefin/paraffin, linear/branched alkanes, xenon/krypton [30], acetylene/CO_2 [31], hydrogen isotopes [32] and adsorption-based purifications, of e.g. ethylene [33].

MOFs have been investigated for the ionic conduction of various charge carriers, such as a proton (>10^{-2} S cm^{-1}), hydroxide ion (>10^{-2} S cm^{-1}), lithium ion, sodium ion and magnesium ion (all about 10^{-4} S cm^{-1}) in their pores [34, 35] as well as for energy applications such as electrocatalysts [36], energy storage and conversion [37], supercapacitors [38], batteries and fuel cells [39].

MOFs have been developed for their possible use as sensors for the detection of many different gases and volatile organic compounds [40, 41], as signal amplification elements for electrochemical sensing [42], for the sensing and imaging of biomolecules [43, 44], environmental pollutants [45], explosives [46], ionic species [47, 48], among many others. Often, the luminescence of MOFs from the linkers, metals, incorporated luminophores or a combination thereof is utilized for the sensing application [49–51].

MOFs are used as size- and enantioselective heterogeneous catalysts [52, 53], including enzyme-MOF composites with MOFs as an enzyme immobilization platform [54, 55], as solid catalysts for liquid-phase continuous flow reactions [56], for the photocatalytic degradation of dyes in water depollution [57], etc.

MOFs are studied for the adsorptive removal of synthetic dyes in textile effluents [58] as well as of pharmaceuticals, personal care products [59] and of fluoride and arsenic [60] from (waste)water.

The porosity of MOFs enables them to be used for (self-sacrificial) drug delivery [61–63], including the sustained release of antibiotic and antimicrobial drugs [64] or bisphosphonate anti-osteoporotic drugs [65].

MOFs are investigated as materials for adsorption and desiccant cooling technologies through cyclic sorption of water or methanol in adsorption-driven heat pumps or adsorption chillers [66–70]. The aspect of cyclic water sorption extends to water harvesting from air for fresh water production [71–73].

In 2016, the first commercial application of a MOF may have materialized. The MOF (by the company MOF Technologies) was used to store and release 1-methylcyclopropene (1-MCP) to slow the fruit ripening process. 1-MCP binds and blocks ethylene receptors in the fruit, with ethylene being a ripening hormone [74, 75].

The MOF start-up NuMat is supplying MOF-filled gas storage cylinders, named ION-X, to store toxic gases (e.g. arsine, phosphine and boron trifluoride), which are used as dopants in the semiconductor industry, and doing this more safely than conventional pressurized cylinders [73, 76].

References

[1] Batten SR, Champness NR, Chen XM, Garcia-Martinez J, Kitagawa S, Öhrström L, O'Keeffe M, Paik Suh M, Reedijk J. Pure Appl Chem, 2013, 85, 1715–24.
[2] Eddaoudi M, Kim J, Rosi N, Vodak D, Wachter J, O`Keeffe M, Yaghi OM. Science, 2002, 295, 469–72.
[3] Chui SSY, Lo SMF, Charmant JPH, Orpen AG, Williams ID. Science, 1999, 283, 1148–50.
[4] Serre C, Millange F, Thouvenot C, Nogues M, Marsolier G, Louer D, Férey G. J Am Chem Soc, 2002, 124, 13519–26.
[5] Huang XC, Lin YY, Zhang JP, Chen XM. Angew Chem Int Ed, 2006, 45, 1557–59.
[6] Park KS, Ni Z, Cote AP, Choi JY, Huang R, Uribe-Romo FJ, Chae HK, O'Keeffe M, Yaghi OM. Proc Natl Acad Sci USA, 2006, 103, 10186–91.
[7] Cohen S. Chem Rev, 2012, 112, 970–1000.
[8] El-Sayed EM, Yuan D. Green Chem, 2020, 22, 4082–104.
[9] Reinsch H, Stock N. Dalton Trans, 2017, 46, 8339–49.
[10] (a) Yilmaz B, Trukhan N, Müller U. Chin J Catal, 2012, 33, 3–10. https://doi.org/10.1016/S1872-2067(10)60302-6. (b) Gaab M, Trukhan N, Maurer S, Gummaraju R, Müller U. Microporous Mesoporous Mater, 2012, 157, 131–6.
[11] Jeremias F, Fröhlich D, Janiak C, Henninger SK. RSC Adv, 2014, 4, 24073–82.
[12] Rubio-Martinez M, Hadley TD, Batten MP, Constanti-Carey K, Barton T, Marley D, Mçnch A, Lim KS, Hill MR. Chem Sus Chem, 2016, 9, 938–41.
[13] Wang LY, Xu H, Gao JK, Yao JM, Zhang QC. Coord Chem Rev, 2019, 398, 213016.
[14] Cheng P, Wang C, Kaneti YV, Eguchi M, Lin J, Yamauchi Y, Na J. Langmuir, 2020, 36, 4231–49.
[15] Meng J, Liu X, Niu C, Pang Q, Li J, Liu F, Liuu Z, Mai L. Chem Soc Rev, 2020, 49, 3142–86.
[16] Bazer-Bachi D, Assié L, Lecoq V, Harbuzaru B, Falk V. Powder Technol, 2014, 255, 52–59.
[17] Leus K, Bogaerts T, De Decker J, Depauw H, Hendrickx K, Vrielinck H, Van Speybroeck V, Van Der Voort P. Microporous Mesoporous Mater, 2016, 226, 110–16.
[18] Silva P, Vilela SMF, Tomé JCP, Paz FAA. Chem Soc Rev, 2015, 44, 6774–803.
[19] Dechnik J, Sumby CJ, Janiak C. Cryst Growth Des, 2017, 17, 4467–88.
[20] Dechnik J, Gascon J, Doonan CJ, Janiak C, Sumby CJ. Angew Chem Int Ed, 2017, 56, 9292–310.
[21] Mueller U, Schubert M, Teich F, Puetter H, Schierle-Arndt K, Pastré J. J Mater Chem, 2006, 16, 626–36.
[22] Paik Suh M, Park HJ, Prasad TK, Lim DW. Chem Rev, 2012, 112, 782–835.
[23] Zhang Z, Zhao Y, Gong Q, Li Z, Li J. Chem Commun, 2013, 49, 653–61.
[24] Sumida K, Rogow DL, Mason JA, McDonald TM, Bloch ED, Herm ZR, Bae TH, Long JR. Chem Rev, 2012, 112, 724–81.

[25] Martínez-Ahumada E, López-Olvera A, Jancik V, Sánchez-Bautista JE, González-Zamora E, Martis V, Williams DR, Ibarra IA. Organometallics, 2020, 39, 883–915.
[26] Martínez-Ahumada E, Díaz-Ramírez ML, de Velásquez-Hernández MJ, Jancik V, Ibarra IA. Chem Sci, 2021, 12, 6772–99.
[27] Islamoglu T, Chen Z, Wasson MC, Buru CT, Kirlikovali KO, Afrin U, Rasel Mian M, Farha OK. Chem Rev, 2020, 120, 8130–60.
[28] Wanigarathna DKJA, Gao J, Liu B. Mater Adv, 2020, 1, 310–20.
[29] Hua GF, Xie XJ, Lu W, Li D. Dalton Trans, 2020, 49, 15548–59.
[30] Adil K, Belmabkhout Y, Pillai RS, Cadiau A, Bhatt PM, Assen AH, Maurin G, Eddaoudi M. Chem Soc Rev, 2017, 46, 3402–30.
[31] Fu XP, Wang YL, Liu QY. Dalton Trans, 2020, 49, 16598–607.
[32] Ju Z, El-Sayed ESM, Yuan D. Dalton Trans, 2020, 49, 16617–22.
[33] Liu S, Dong Q, Zhou Y, Wang S, Duan J. Dalton Trans, 2020, 49, 17093–105.
[34] Sadakiyo M, Kitagawa H. Dalton Trans, 2021, 50, 5385–97.
[35] Bhardwaj SK, Bhardwaj N, Kaur R, Mehta J, Sharma AL, Kim KH, Deep A. J Mater Chem A, 2018, 6, 14992–5009.
[36] Li L, He J, Wang Y, Lv X, Gu X, Dai P, Liu D, Zhao X. J Mater Chem A, 2019, 7, 1964–88.
[37] Wang K, Nam Hui K, San Hui K, Peng S, Xu Y. Chem Sci, 2021, 12, 5737–66.
[38] Pei C, Sung Choi M, Yu X, Xue H, Xia BY, Seok Park H. J Mater Chem A, 2021, 9, 8832–69.
[39] Kuyuldar S, Genna DT, Burda C. J Mater Chem A, 2019, 7, 21545–76.
[40] Yao MS, Li WH, Xu G. Coord Chem Rev, 2021, 426, 213479.
[41] Li HY, Zhao SH, Zang SQ, Li J. Chem Soc Rev, 2020, 49, 6364–401.
[42] Liu S, Lai C, Liu X, Li B, Zhang C, Qin L, Huang D, Yi H, Zhang M, Li L, Wang W, Zhou X, Chen L. Coord Chem Rev, 2020, 424, 213520.
[43] Dong J, Zhao D, Lu Y, Sun WY. J Mater Chem A, 2019, 7, 22744–67.
[44] Duman FD, Forgan RS. J Mater Chem B, 2021, 9, 3423–49.
[45] Rosales-Vázquez LD, Dorazco-González A, Sánchez-Mendieta V. Dalton Trans, 2021, 50, 4470–85.
[46] Dutta A, Singh A, Wang X, Kumar A, Liu J. CrystEngComm, 2020, 22, 7736–81.
[47] Kaur H, Sinha S, Krishnan V, Rani Koner R. Dalton Trans, 2021, 50, 8273–91.
[48] Panda SK, Mishra S, Singh AK. Dalton Trans, 2021, 50, 7139–55.
[49] Lustig WP, Mukherjee S, Rudd ND, Desai AV, Li J, Ghosh SK. Chem Soc Rev, 2017, 46, 3242–85.
[50] Hazra A, Mondal U, Mandal S, Banerjee P. Dalton Trans, 2021, 50, 8657–70.
[51] Zhao Y, Li D. J Mater Chem C, 2020, 8, 12739–54.
[52] Qin JS, Yuan S, Lollar C, Pang J, Alsalme A, Zhou HC. Chem Commun, 2018, 54, 4231–49.
[53] Rogge SMJ, Bavykina A, Hajek J, Garcia H, Olivos-Suarez AI, Sepúlveda-Escribano A, Vimont A, Clet G, Bazin P, Kapteijn F, Daturi M, Ramos-Fernandez EV, Llabrés I Xamena FX, Van Speybroeck V, Gascon J. Chem Soc Rev, 2017, 46, 3134–84.
[54] Lian X, Fang Y, Joseph E, Wang Q, Li J, Banerjee S, Lollar C, Wang X, Zhou HC. Chem Soc Rev, 2017, 46, 3386–401.
[55] Majewski MB, Howarth AJ, Li P, Wasielewski MR, Hupp JT, Farha OK. CrystEngComm, 2017, 19, 4082–91.
[56] Dhakshinamoorthy A, Navalon S, Asiri AM, Garcia H. Chem Commun, 2020, 56, 26–45.
[57] Sharma VK, Feng M. J Hazard Mater, 2019, 372, 3–16.
[58] Parmar B, Bisht KK, Rajput G, Suresh D. Dalton Trans, 2021, 50, 3083–108.
[59] Jin E, Lee S, Kang E, Kim Y, Choe W. Coord Chem Rev, 2020, 425, 213526.
[60] Biswal L, Goodwill JE, Janiak C, Chatterjee S. Sep Purificat Rev, In Press. https://doi.org/10.1080/15422119.2021.1956539.

[61] Wang L, Zheng M, Xie Z. J Mater Chem B, 2018, 6, 707–17.
[62] Chen W, Wu C. Dalton Trans, 2018, 47, 2114–33.
[63] Zhu WJ, Zhao JY, Chen Q, Liu Z. Coord Chem Rev, 2019, 398, 113009.
[64] Kaur N, Tiwari P, Kapoor KS, Kumar Saini A, Sharma V, Mobin SM. Cryst Eng Comm, 2020, 22, 7513–27.
[65] Vassaki M, Papathanasiou KE, Hadjicharalambous C, Chandrinou D, Turhanen P, Choquesillo-Lazarte D, Demadis KD. Chem Commun, 2020, 56, 5166–69.
[66] Steinert DM, Ernst SJ, Henninger SK, Janiak C. Eur J Inorg Chem, 2020, 4502–15.
[67] Liu X, Wang X, Kapteijn F. Chem Rev, 2020, 120, 8303–77.
[68] Rafique MM. Appl Syst Innov, 2020, 3, 26. https://doi.org/10.3390/asi3020026.
[69] Hastürk E, Ernst SJ, Janiak C. Curr Opin Chem Eng, 2019, 24, 26–36.
[70] AL-Dadah RK, Mahmoud S, Elsayed E, Youssef P, Al-Mousawi F. Energy, 2020, 190, 116356.
[71] Kim SI, Yoon TU, Kim MB, Lee SJ, Hwang YK, Chang JS, Kim HJ, Lee HN, Lee UH, Bae YS. Chem Eng J, 2016, 286, 467–75.
[72] Trapani F, Polyzoidis A, Loebbecke S, Piscopo CG. Microporous Mesoporous Mater, 2016, 230, 20–24.
[73] Kim H, Yang S, Rao SR, Narayanan S, Kapustin EA, Furukawa H, Umans AS, Yaghi OM, Wang EN. Science, 2017, 356, 430–34.
[74] Urquhart J. Chemistry World 2016, September 27. https://www.chemistryworld.com/news/worlds-first-commercial-mof-keeps-fruit-fresh/1017469.article (accessed July 27, 2021).
[75] Notman N. Chemistry World 2017, May, Vol. 14 (5), p. 44–7. https://www.chemistryworld.com/features/mofs-find-a-use/2500508.article
[76] Trager R. Chemistry World 2016, October 28. https://www.chemistryworld.com/news/mofs-offer-safer-toxic-gas-storage-/1017610.article?adredir=1 (accessed July 27, 2021).

8 Photofunctional materials

8.1 Solid-state lighting materials

Thomas Jüstel, Florian Baur

The recent transition from traditional lighting toward solid-state lighting (SSL) is regarded as the fourth revolution in lighting technology, while the first three ones were the use of fire from torches to gas lamps, the electrification of lighting by the invention of the incandescent bulb and then the development of gas-discharge lamps at the beginning of the twentieth century. Today, solid-state light sources, namely the phosphor converted LEDs, are the most important "lamps" in general and special lighting and it continues to increase its market share in other fields of application.

8.1.1 Historical development

The first scientific description of electroluminescence was published in 1907 by Henry Joseph Round [1]. Research progressed in the 1920s and 1930s with Oleg Losev and Georges Destriau, who worked independently on the electroluminescent properties of SiC and ZnS. In 1939, Zoltán Bay and György Szigeti filed a patent in Hungary and the US (assigned to General Electric) describing an electroluminescence device comprising SiC that emitted white light [2]. However, these devices were costly to produce and never used commercially. In the 1950s and 1960s, the first III–V semiconductor LEDs were developed. This led to the first commercially available LED, produced by Texas Instruments, which used a GaAs chip to generate 890 nm infrared radiation [3]. The first LEDs that emitted visible light were developed by Allen and Cherry (GaP) [4] and by Holonyak and Bevacqua (Ga(As,P)) [5], respectively, in 1962. The initially high price for infrared LEDs (US$ 130, Texas Instruments) and for red light LEDs (US$ 260, General Electric), decreased steadily and in the 1970s Fairchild Optoelectronics offered LEDs for less than five cents [6]. These LEDs were used as indicator lights and not for general lighting due to their low luminous flux and the lack of efficient blue-emitting LEDs.

This situation changed in 1993 when Shuji Nakamura strongly optimized the growth of blue-light-emitting GaN chips while working at Nichia Corporation in Japan, a discovery for which he was awarded the Nobel Prize in Physics in 2014 together with Amano and Akasaki [7]. To produce white light, the LED chip was coated with the yellow-emitting phosphor $Y_3Al_5O_{12}:Ce^{3+}$, which partially converts the primary blue chip emission to yellow light. The combination of blue and yellow light yields cool-white light. This approach, called phosphor-converted LED (pcLED), was patented by Nichia in 1996 and allowed the LED to be used as a light source in general lighting [8]. However, these LEDs have a low color-rendering index (CRI) and

https://doi.org/10.1515/9783110798890-020

cannot produce warm-white light, as they solely show weak emission in the red spectral range. Therefore, in the following years, research focused on the development of blue-excitable red-emitting phosphors. In 1999 the so-called 258-nitrides $(Ca,Sr,Ba)_2Si_5N_8{:}Eu^{2+}$ were introduced [9], followed by the alumino-silico-nitrides $(Ca,Sr)AlSiN_3{:}Eu^{2+}$ in 2004 [10]. The majority of LEDs for general lighting use one of these Eu^{2+} activated nitrides as a red-emitting component, which gives access to lower color temperatures, i.e. warm-white light, while this measure also increases the CRI. The development of these LEDs led to a strong increase of the usage of LEDs for general lighting. Current development focuses on further increasing the CRI to closely resemble that of an incandescent bulb, which is important for areas such as museum lighting. To achieve this, the blue-emitting LED chip is substituted by an (Al,Ga,In)N chip emitting around 400–420 nm, whose emission is completely converted by phosphors. The advantage of this so-called full-conversion LED (fcLED) is that the primary LED emission is solely responsible for exciting the phosphor layer, whereas in conventional pcLEDs the primary emission is used both as the blue spectral component and as excitation for the phosphors. This makes the fine tuning of the emission spectrum difficult.

8.1.2 Technology of LEDs

The first LEDs were usually 3 or 5 mm wide LEDs. The bonded semiconductor chip is placed on top of a reflector and the chip and reflector are encapsulated in a polymer dome, mostly made from an epoxy resin. The drawback of this technology is the low heat-transfer rate it exhibits. Heat can be transferred by convection, thermal radiation or conduction. Convection cannot occur in an LED due to the encapsulation. Moreover, the low target temperature of the semiconductor of around 100 °C limits the heat transfer by thermal radiation to about 1000 W m^{-2}. For a typical surface area of a chip of 1 mm^2, or 10^{-6} m^2, this corresponds to thermal radiation of about 1 mW from a single chip, which is way too low to cool higher power LEDs. Therefore, thermal conduction is the main mechanism of heat transfer in an LED. Even though LEDs have a relatively high energy efficiency, i.e. modern LEDs convert about 60% of the electrical input power to optical output power, the remainder of 40% is still converted into heat. That means a 3 W LED generates about 1.2 W of heat, which has to be dissipated from the chip. In traditionally wired LEDs this is not possible as the chip is fully encapsulated and – other than through the wires – has no contact to a potential heat sink. The power density of these LEDs is therefore severely limited. To enable LEDs with high power density, as required for general lighting, a different design is used.

Figure 8.1.1 shows an (In,Ga)N flip-chip LED, which will exhibit emission in the blue spectral range. Figure 8.1.7 shows a photograph of an LED with gold wires and a clear epoxy dome. It consists of a sapphire substrate on which layers of III–V semiconductor materials are grown epitaxially. The first contact layer is an n-type GaN, which is usually achieved by doping GaN with Si^{4+}. It is followed by the multiquantum well layer, a region in which the band gap is modulated in order to trap the electrons and holes to increase the probability of radiative recombination. This mechanism is explained in more detail below. Next is the p-GaN contact layer, i.e. GaN doped with Mg^{2+}. The semiconductor chip is contacted to the submount with solder and efficient heat transfer via conduction is possible. The development of this type of LED allowed a much higher power density and enabled LEDs with a luminous flux suitable for general lighting.

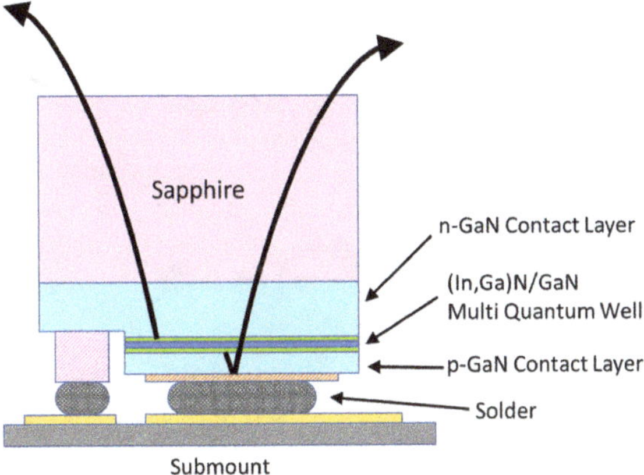

Figure 8.1.1: Schematic design of a high power (In,Ga)N flip-chip LED.

The electroluminescence of the semiconductor chip is caused by radiative recombination of electrons and holes. If a voltage is applied, electrons flow from the n-doped cathode (negatively polarized electrode, i.e. the "–" contact) to the p-doped anode (positively polarized electrode, i.e. the "+" contact) and holes move from the p-doped anode (+) to the n-doped cathode (–). The probability of radiative recombination is low, however, because the electrons and holes have a high mobility. For that reason, a quantum-well layer is used to trap the electrons and holes, which strongly increases the probability of radiative recombination. A simple quantum-well structure is depicted in Figure 8.1.2. It consists of four layers of semiconductors with slightly different compositions and, therefore, slightly different band gaps. The first layer on the left with band gap $E_{G,1}$ is followed by a layer with a slightly smaller band gap $E_{G,2}$. The decrease in band gap energy can be achieved by e.g. increasing the amount of In^{3+} in (In,Ga)N. Similarly, an increase in band gap energy can be achieved by increasing the amount of Ga^{3+}, resulting in the layer with larger band gap $E_{G,3}$. Electrons that move from the left to the right will be trapped in the low band gap layer because they would need to overcome the energy difference U_1 to proceed further. Holes that flow from right to left in the figure will be trapped in the same layer, as movement of a hole requires an electron to occupy its former position. Since the hole is located at a higher energy than the electrons in the valence band of the surrounding layers, it cannot move further anymore without additional energy. Electrons and holes are thus trapped in a thin layer of the semiconductor and will recombine radiatively with a high probability. At low temperature and low current, radiative recombination efficiencies of close to 100% can be achieved by using optimized multiquantum well (MQW) structures [11]. However, at elevated operation temperature, the radiative recombination rate decreases due to various loss processes and reaches approximately 70–80% [12].

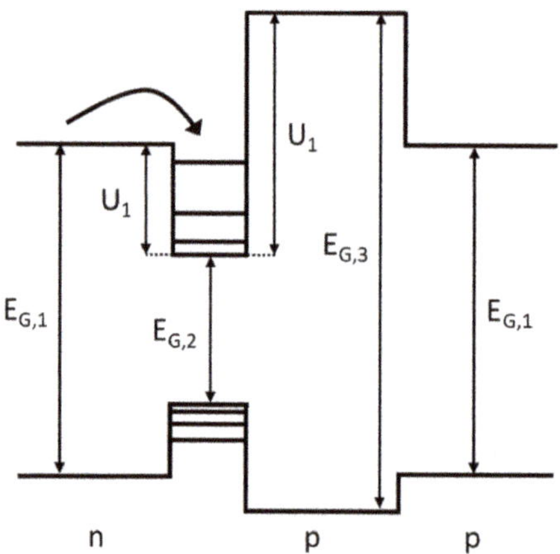

Figure 8.1.2: Electron blocking layer in an inorganic semiconductor LED.

The emission wavelength of the LED is determined by the band gap of the semiconductor material present in the recombination zone. LEDs emit in a narrow spectral range with a full-width at half-maximum (FWHM) of typically 20–30 nm. To generate white light, two approaches can be considered: either several LEDs with different emission wavelengths can be combined or a single type of LED is coated with one or more photoluminescent compounds that convert the primary emission of the LED chip into a spectrum in the visible range. The emission spectra obtained from the combination of continuous different types of LEDs are shown in Figure 8.1.3. For each spectrum the corresponding luminous efficacy (LE) and the color-rendering index (CRI) is given. The LE quantifies the luminous flux per optical watt, in other words it is a measure for the perceived brightness per optical output power of the LED. It should be noted that LE can generally also be expressed in lumen per electrical watt, i.e. the perceived brightness per electrical input power.

If more LEDs are used to compose the spectrum, the LE decreases from about 422 to 324 lm W^{-1} [13]. This is a result of the human eye sensitivity that is highest at 555 nm and decreases strongly in the blue and red range. However, while LE decreases, the additional emission ranges strongly increase the color-rendering index (CRI). The CRI is a measure for the quality of light reflected by illuminated items. It expresses how closely the color of a surface illuminated by the specific light source

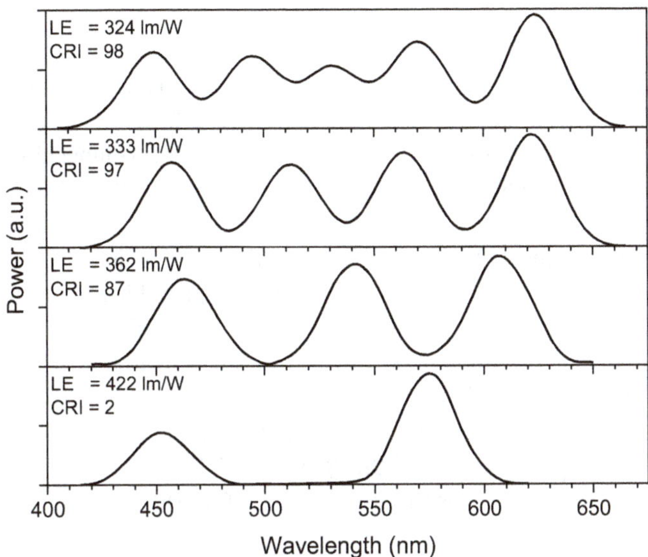

Figure 8.1.3: Spectra of the combined emission of 2 to 5 LEDs (bottom to top) with the corresponding luminous efficacy (LE) and color-rendering index (CRI). Modified after Zukauskas et al. [13].

resembles the color that is perceived if the surface is illuminated by a black body radiator, such as the sun or an incandescent bulb. The 2-LED setup lack vast sections of cyan, green, and deep-red light and cannot faithfully reproduce such colors. Hence, the extremely low CRI of 2. For general lighting, a CRI of 80 to 90 is considered sufficient, whereas for specific application areas such as museum or store lighting, a higher CRI is desired.

While the combination of several LEDs can produce very good LE and CRI values, it is not generally in use. The problem with this approach is the large difference in lifetime, temperature and voltage/current dependence of the different LED types. For instance, red-emitting LEDs show stronger quenching and color point shift with increasing temperature than blue-emitting LEDs. Red-emitting LEDs also usually have a shorter lifetime and thus their emission weakens over time. As a result, the combined emission spectrum will change depending on the temperature and operation time. In theory, this can be managed by continuously monitoring the emission spectrum and regulating the driving current and voltage of each LED type. However, this is complicated and expensive and, therefore, multi-LED lamps have never been widely employed. The sole exception is lamps where the consumer can freely change the color point and temperature of the emitted light. Generally, though, phosphor-converted LEDs are used to generate white light with high LE and CRI.

8.1.3 Semiconductor materials

Current LEDs are equipped with III–V semiconductors, i.e. semiconductors consisting of elements from the third and fifth main groups, respectively [14]. These III–V semiconductors have direct band gaps whose size depends on the chemical composition. Smaller ions will result in larger band gaps, as depicted in Figure 8.1.4. With 6.0 eV (approximately 210 nm), AlN has the largest band gap of the III–V semiconductors, whereas InSb has a band gap of just 0.2 eV (approximately 6200 nm). Solid solutions can be used to modify the band gap to the required energy. Typical blue-emitting semiconductors are (In,Ga)N solid solutions where the band gap of GaN (3.4 eV, 365 nm) is decreased by partial substitution of Ga^{3+} by In^{3+} to values of around 2.7 eV. Efficient green emission is difficult to obtain, as the compositions in question have stability issues and to date cannot be grown with the required high purity. Such lack of efficient LEDs in that spectral region is called the "green gap". For emission in the red to NIR, phosphides or arsenides, such as (In,Ga,Al)As or (In,Ga,Al)P, are used [15].

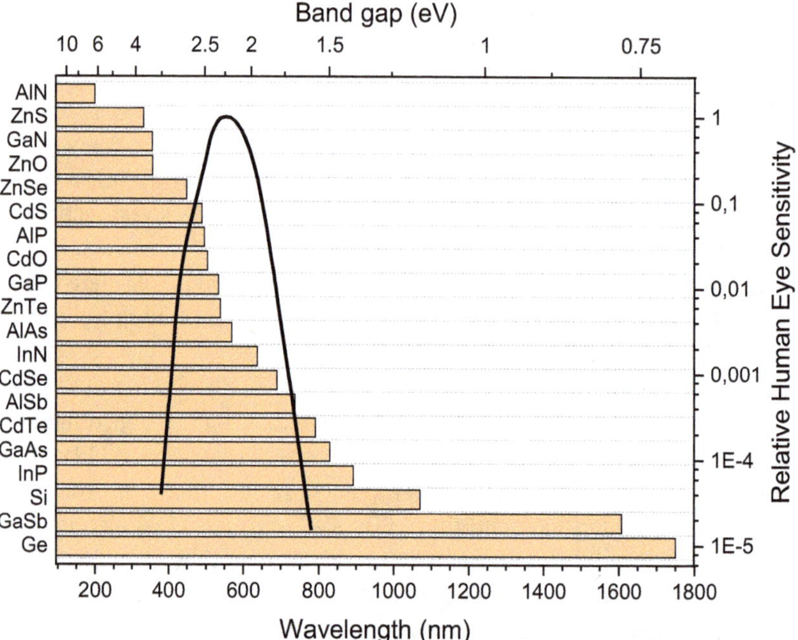

Figure 8.1.4: Band gap in eV and corresponding emission wavelength in nm for various semiconductors. The black line shows the human eye sensitivity curve for a phototopic lighting situation, i.e. the eye is adapted to a bright surrounding. Modified after Zukauskas et al. [13].

Whereas in theory a wide selection of band gaps and, therefore, emission wavelengths can be obtained by simply selecting the proper composition, the difficulty lies in growing high-purity single crystals of the material. State-of-the-art is the "Metal-Organic Chemical Vapor Deposition" (MOCVD), where metal-organic compounds are used as precursors [16]. Blue-emitting LEDs only became widely available as Nakamura developed a technique to efficiently grow (In,Ga)N semiconductor chips on a sapphire substrate [7]. Currently, research activity focuses on the development of efficient UV-emitting LEDs based on (Al,Ga)N semiconductors.

8.1.4 Luminescent materials

As shown above, LEDs show narrow-band emission, which is not suitable for applications where a continuous spectrum is desired, such as general lighting. As discussed in Section 8.1.2, the combination of several LEDs has severe disadvantages. Thus, phosphor-converted LEDs (pcLEDs) are used to generate a continuous spectrum. Here, the semiconductor chip is coated with one or more phosphors dispersed in an epoxy or silicone resin in concentrations between 10 and 15 wt%. Typically, the particle size is between 5 and 15 µm.

The phosphors for pcLEDs need to exhibit sufficiently strong absorption of the primary emission of the semiconductor chip. The phosphors that were used for decades in fluorescent lamps could, therefore, not be utilized as they were developed for efficient excitability at 253.7 nm, the most intense emission line of a mercury gas discharge lamp. The first phosphor to be used by Nichia as a converter in pcLEDs was the yellow-emitting garnet $Y_3Al_5O_{12}:Ce^{3+}$ (YAG:Ce). In such artificial garnets, the excitation band of Ce^{3+} is located in the blue spectral region, overlapping very well with the emission of an (In,Ga)N chip. YAG can be modified by substitution of Y^{3+} with other trivalent cations such as Lu^{3+} or Gd^{3+}, resulting in $Lu_3Al_5O_{12}:Ce^{3+}$ (LuAG:Ce), where the emission is shifted to shorter wavelengths, or $Gd_3Al_5O_{12}:Ce^{3+}$ (GAG:Ce), where the emission is shifted to longer wavelengths. Furthermore, substitution of Al^{3+} with Ga^{3+} also shifts the emission toward the green spectral region. However, with increasing Ga^{3+} concentration, the thermal quenching temperature of the phosphor, i.e. the temperature at which the phosphor's emission intensity begins to decrease significantly, decreases [17]. Table 8.1.1 lists a variety of LED phosphors with their respective emission wavelengths.

Alkaline earth orthosilicates activated with Eu^{2+} show emission from the cyan to red spectral region depending on their composition. The solid solution $(Ba,Sr)_2SiO_4:Eu^{2+}$ delivers excellent green to yellow-emitting phosphors. However, for commercial application, a protective coating is required, as these orthosilicates are slightly sensitive toward moisture. The nitrides $SrSi_2O_2N_2:Eu^{2+}$ and β-SiAlON:Eu^{2+} are also well-known green-emitting phosphors [26, 46].

To achieve warm-white light (CCT < 4000 K) with a high CRI, a red-emitting phosphor is required. Early development resulted in Eu^{2+} activated sulfides such as $SrS:Eu^{2+}$, which shows strong absorption in the blue spectral region and emission

Table 8.1.1: Characteristics of selected phosphors for phosphor converted LEDs.

Phosphor	Emission color	Em_{max} (nm)	CIE 1931 (x, y)	Luminous efficacy [$lm/W_{optical}$]	Ref.
$BaSi_2O_2N_2:Eu^{2+}$	cyan	495	0.076; 0.440	233	[18]
$BaGa_2S_4:Eu^{2+}$	green	504	0.143; 0.506		[19]
$Ba_2SiO_4:Eu^{2+}$	green	505	0.240; 0.626		[20]
$Ca_3Sc_2Si_3O_{12}:Ce^{3+}$	green	507	0.310; 0.590		[21]
$CaAl_2S_4:Eu^{2+}$	green	516	–		[22]
$CaLu_2Al_4SiO_{12}:Ce^{3+}$	green	524	0.329; 0.558		[23]
$Ba_3Si_6O_{12}N_2:Eu^{2+}$	green	527	0.310; 0.625		[24]
$SrGa_2S_4:Eu^{2+}$	green	535	0.270; 0.690	576	[18]
$BaYSi_4N_7:Eu^{2+}$	green	537	–		[25]
$SrSi_2O_2N_2:Eu^{2+}$	green	538	0.337; 0.619	522	[18]
$Lu_3Al_5O_{12}:Ce^{3+}$	green	540	0.350; 0.575	530	[23]
$ZnGa_2S_4:Eu^{2+}$	green	540	0.299; 0.673		[19]
Ca-α-SiAlON:Yb^{2+}	green	545	0.323; 0.601		[26]
$SrYSi_4N_7:Eu^{2+}$	green	550	–		[27]
$CaLaGa_3S_7:Eu^{2+}$	yellow	554	0.390; 0.580		[28]
$Lu_3Al_3MgSiO_{12}:Ce^{3+}$	yellow	555	0.429; 0.538		[29]
$CaSi_2O_2N_2:Eu^{2+}$	yellow	555	0.419; 0.556		[18]
$Y_3Al_5O_{12}:Ce^{3+}$	yellow	556	0.458; 0.530	433	[30]
$CaGa_2S_4:Eu^{2+}$	green	558	0.410; 0.580		[19]
$SrLi_2SiO_4:Eu^{2+}$	yellow	562	0.368; 0.515	388	[31]
$Tb_3Al_5O_{12}:Ce^{3+}$	yellow	562	0.482; 0.511	409	[30]
$Gd_3Al_5O_{12}:Ce^{3+}$	yellow	565	0.496; 0.497		[30]
$Ca_2BO_3Cl:Eu^{2+}$	yellow	570	0.493; 0.496		[32]
$Sr_3SiO_5:Eu^{2+}$	yellow	570	–		[33]
$Ba_2Si_5N_8:Eu^{2+}$	yellow	572	0.521; 0.472	280	[34]
$GdSr_2AlO_5:Ce^{3+}$	yellow	574	0.475; 0.502		[35]
$Y_3Al_3MgSiO_{12}:Ce^{3+}$	yellow	575	0.488; 0.499		[29]
$CaAlSiN_3:Ce^{3+}$	orange	580	–	108	[36]

Table 8.1.1 (continued)

Phosphor	Emission color	Em_{max} (nm)	CIE 1931 (x, y)	Luminous efficacy [$lm/W_{optical}$]	Ref.
$Ba_2AlSi_5N_9:Eu^{2+}$	orange	584	–		[37]
$Y_3Mg_2AlSi_2O_{12}:Ce^{3+}$	orange	600	0.542; 0.452		[38]
Ca-α-SiAlON:Eu^{2+}	orange	603	0.560; 0.436		[39]
$Ca_2Si_5N_8:Eu^{2+}$	orange	605	0.620; 0.380		[40]
$SrS:Eu^{2+}$	orange	610	0., 0.	260	[41]
$SrAlSiN_3:Eu^{2+}$	orange	610	–		[42]
$Sr_2Si_5N_8:Eu^{2+}$	red	625	0.621; 0.368	199	[43]
$SrAlSi_4N_7:Eu^{2+}$	red	635	–		[44]
$CaS:Eu^{2+}$	red	648	0.703; 0.297	84	[41]
$CaAlSiN_3:Eu^{2+}$	red	649	0.647; 0.347	108	[45]

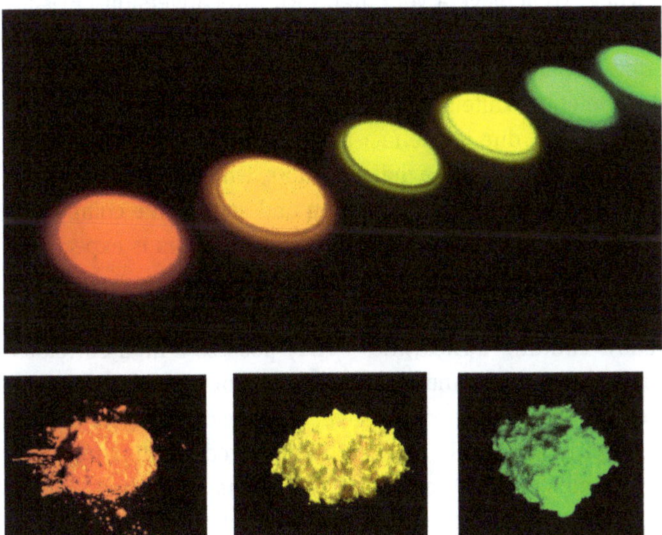

Figure 8.1.5: Photographic images of selected LED phosphors under 365 nm excitation (top row, from left to right): $CaAlSiN_3:Eu^{2+}$, $Sr_2Si_5N_8:Eu^{2+}$, $Lu_3Al_5O_{12}:Ce^{3+}$, $Y_3Al_5O_{12}:Ce^{3+}$, $Sr_2SiO_4:Eu^{2+}$, $(Ba,Sr)_2SiO_4:Eu^{2+}$, and (bottom row, from left to right): $Ba_2Si_5N_8:Eu^{2+}$, $Y_3Al_5O_{12}:Ce^{3+}$, and $Lu_3Al_5O_{12}:Ce^{3+}$. (Photographs: Dr. Danuta Dutczak).

centered at 610 nm. Sulfides are very hygroscopic, a drawback that prevented commercial application of these phosphors. This led to the development of Eu^{2+} activated nitrides that often show red emission. Nowadays, the nitrides $(Ba,Sr)_2Si_5N_8$:Eu^{2+} and specifically $(Sr,Ca)AlSiN_3$:Eu^{2+} are found as the red-emitting component in most pcLEDs for general lighting [47].

8.1.5 White LEDs

Recently, white-emitting LEDs became the dominant light source in general lighting. They consist of a blue-emitting (In,Ga)N semiconductor that is coated with a single or a blend of phosphors dispersed in epoxy or silicone resin. Typically, 10–15 wt% of phosphor are present in the coating. The primary radiation from the LED chip is partially absorbed by the phosphor layer upon which it exhibits luminescence, i.e. it emits light. The combined emission from the semiconductor and the phosphor layer yields a continuous spectrum that is perceived as white by the human eye. By using more than one conversion phosphor, the CRI of the white LED can be increased. However, due to the human eye sensitivity curve, this results in a decreased luminous efficacy. To achieve a very high CRI, full-conversion LEDs are employed. In that case, a 400–420 nm emitting semiconductor chip is used, whose radiation is fully converted by phosphors. Thus, at least blue, green and red-emitting phosphors are required to generate white light with a high CRI. The advantage of this approach is the higher versatility, due to the fact that the primary radiation of the semiconductor is only responsible for exciting the phosphors, whereas in conventional white LEDs, the primary radiation is used for excitation and as the blue component. Furthermore, the quantum efficiency of 400–420 nm semiconductors is expected to surpass that of the blue-emitting chips shortly [48].

Early white LEDs consisted of a 450–465 nm blue-emitting (In,Ga)N semiconductor chip coated with yellow-emitting $Y_3Al_5O_{12}$:Ce^{3+}. They possess a high correlated color temperature (CCT) beyond 4000 K and a rather low CRI of around 70–80, due to a lack of emission intensity in the red spectral range, as depicted in Figure 8.1.6 a). To enable a lower CCT, i.e. warm-white LEDs, higher CRI and a red-emitting phosphor are required. Fluorescent lamps use Eu^{3+}-activated phosphors as the red emitter; however, the absorption of these phosphors in the blue spectral region is too low for efficient excitation via an (In,Ga)N semiconductor. Therefore, new red emitters on the basis of Eu^{2+} were developed that exhibit strong absorption around 450 nm. The sulfides $SrGa_2S_4$:Eu^{2+} and SrS:Eu^{2+} are excellent green and red-emitting materials, respectively, and LEDs with very high CRI can be constructed by a blend of these phosphors (see Figure 8.1.6 b)) [41]. They are hygroscopic, though, and thus suffer from stability issues. It was found that various nitrides, such as $Sr_2Si_5N_8$:Eu^{2+} and $CaAlSiN_3$:Eu^{2+}, exhibit similar optical properties, but possess a much higher chemical and physical stability. Today, the majority of LEDs for general lighting employ such

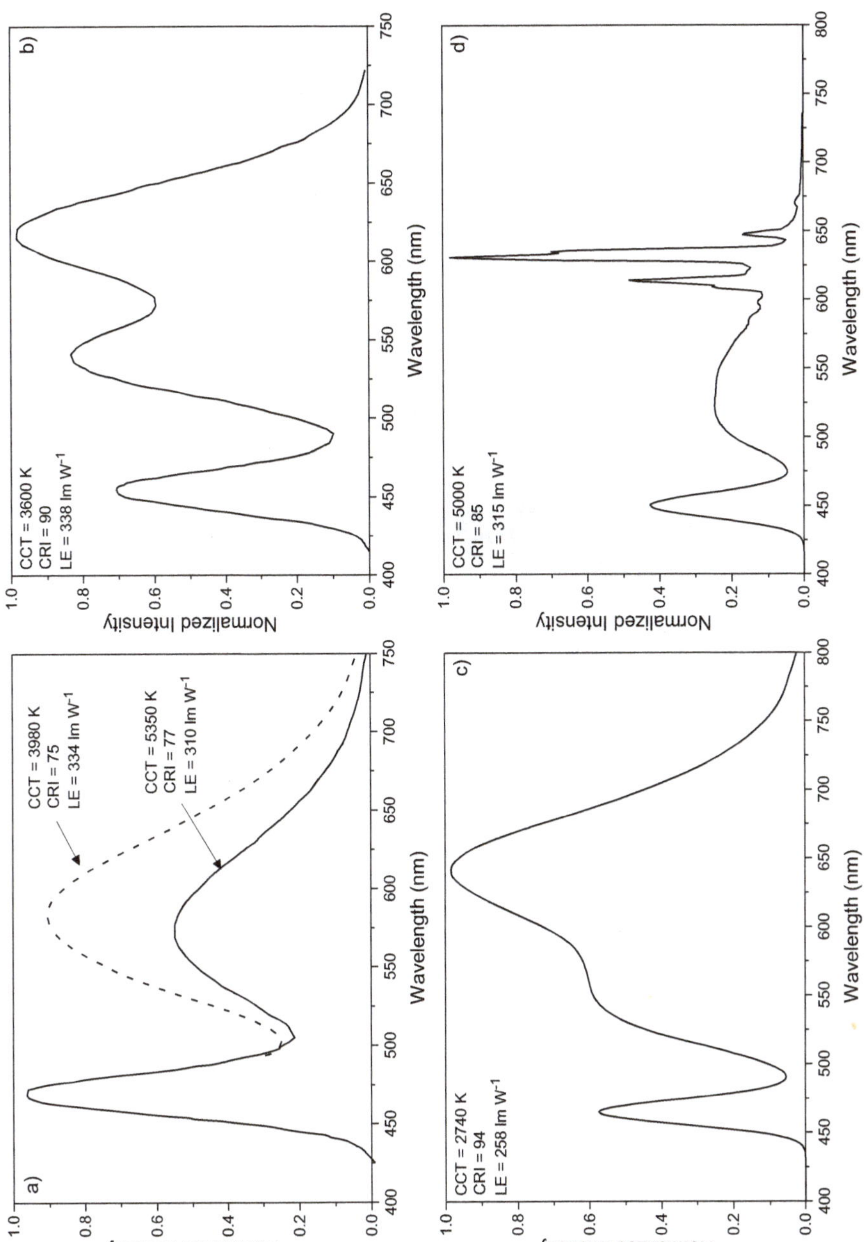

Figure 8.1.6: Emission spectra of a blue-emitting (In,Ga)N semiconductor chip coated with a) $Y_3Al_5O_{12}:Ce^{3+}$ with lower (solid line) and higher (dashed line) phosphor concentration, b) $SrGa_2S_4:Eu^{2+}$ and $SrS:Eu^{2+}$, c) $Y_3Al_5O_{12}:Ce^{3+}$ and $CaAlSiN_3:Eu^{2+}$, d) $Lu_3Al_5O_{12}:Ce^{3+}$ and $K_2SiF_6:Mn^{4+}$. The correlated color temperature (CCT), color-rendering index (CRI) and luminous efficacy (LE) of the respective spectrum is listed as well.

Eu^{2+}-activated nitrides as the red-emitting component [48]. However, they show also emission in the deep-red to NIR spectral region, as depicted in Figure 8.1.6 c), which decreases the luminous efficacy as the human eye sensitivity in that region is very low [49]. Therefore, many research activities focused on the development of narrow band or line-emitting phosphors. The red line emitter K_2SiF_6:Mn^{4+} can be excited in the blue spectral region and has a narrow emission spectrum centered around 625 nm without any spill-over to the deep-red or NIR region (see Figure 8.1.6 d)) [50]. K_2SiF_6:Mn^{4+} has some drawbacks, such as the absorption cross-section of Mn^{4+} being much lower than that of Eu^{2+}, due to the parity-forbidden nature of the respective transition. Furthermore, the phosphor is hygroscopic and decomposes slowly due to a redox reaction between Mn^{4+} and F^- which leads to the formation of MnF_3. The stability issues could be solved by depletion of Mn^{4+} close to the surface of the individual particles, preventing direct contact of Mn^{4+} with moisture from the environment.

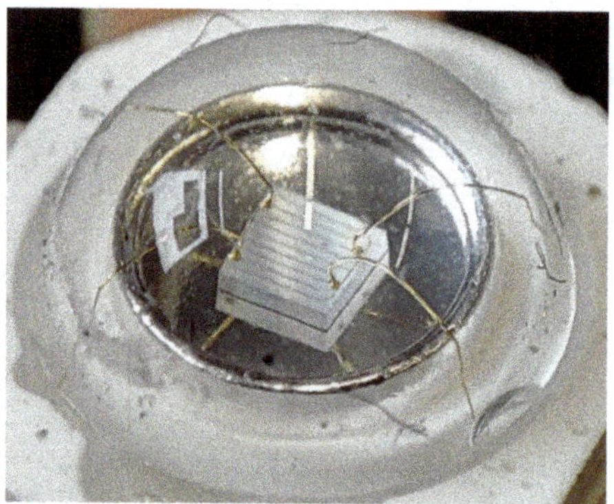

Figure 8.1.7: Photographic image of an LED comprised of a semiconductor chip contacted with gold wires and a clear epoxy dome. (Photograph: Patrick Pues).

8.1.6 Application areas of LEDs

LEDs widely replaced the traditional incandescent and gas-discharge lamps in indoor lighting. While initially LEDs were only used in lamps for signaling, their market share in many areas further increases. The following list summarizes application areas where inorganic LEDs heavily penetrated the established lighting markets:
- Advertisement and decoration lighting
- Office and residential lighting
- Automotive lighting
- LCD backlighting
- Flood-lighting
- Architecture lighting
- Street lighting

In other words, the high energy efficiency and long lifetime of inorganic LEDs makes them highly useful for all kind of lighting applications. Due to the development of special LED phosphors, the initial problems such as low CRI and high CCT were eliminated. For instance, in automotive lighting, the advantage of the LED is mainly the high illuminance that can be achieved, the long lifetime and the possibility to selectively illuminate specific segments of the road. Backlights for liquid crystal displays (LCDs) require narrow emission bands to obtain primaries with a high color purity and thus a large color space. Therefore, today all kinds of LCDs employ LEDs for backlighting, while discharge lamps completely disappeared here!

Even in special applications such as flood-lights, the LED steadily increases its market share. The problem of a comparatively low luminous flux of a single LED has been overcome by the arrangement of a large number of LEDs in an array.

8.1.7 The future of LEDs

For general lighting, the newest generation of inorganic LEDs already exhibit a very good performance in terms of CRI, energy efficiency, luminous flux and lifetime. To increase the luminous efficacy even further upon maintaining a high CRI, the broad-band red-emitting component would need to be substituted by narrow band or even line-emitting phosphors. The Eu^{2+}-activated nitrides show relatively strong emission in the deep-red to NIR spectral region, where the human eye sensitivity is low. These photons can be considered as loss, as they do not significantly enhance the luminous flux or perceived brightness of the respective light source [51]. Potential activators for red line emission are Eu^{3+} and Mn^{4+}; however, both possess disadvantages compared to Eu^{2+}. This is mainly a low absorption cross-section in the blue spectral region in case of Eu^{3+} and emission that is mostly slightly too far in the deep-red spectral range in the case of Mn^{4+}. Presently, many research activities are dedicated to the sensitization of Eu^{3+} in order to achieve sufficient absorption strength in the blue spectral range. While it can show strong absorption in that region, Ce^{3+} cannot be used as a sensitizer, as it will cause quenching of the photoluminescence due to metal-to-metal charge transfer between Ce^{3+} and Eu^{3+}. Various means of sensitization were investigated, such as separating Ce^{3+} and Eu^{3+}, by employing a core-shell particle with energy migration via a Tb^{3+} or Gd^{3+} substructure [52], or sensitization via the uranyl cation $[UO_2]^{2+}$ [53]. Although promising results were obtained, there is no commercial application so far.

However, Mn^{4+} is used commercially, as it exhibits an emission multiplet around 625-630 nm, while the ion shows sufficient absorption strength around 450 nm if it is located in a fluoride host material. These properties of $K_2SiF_6:Mn^{4+}$ were published in 1973 by Paulusz [54], but the compound shows fast degradation, due to a redox reaction between 2 F^- and 2 Mn^{4+} ions to form F_2 and 2 Mn^{3+} that is accelerated by moisture. In 2010, General Electric patented a method to stabilize Mn^{4+} doped fluorides by the depletion of Mn^{4+} close to the surface of the phosphor particles [55].

Broad-band, continuous IR sources are required for applications, such as IR sensors and spectroscopy, where the replacement of incandescent lamps is required due to their phase-out. LEDs without a phosphor conversion layer show narrow emission bands and are thus not suitable in this respect. Therefore, highly efficient IR-emitting phosphors are under investigation that can be used in combination with a blue-emitting LED to generate broadband IR emission. Cr^{3+} is a useful activator, as it possesses a sufficiently large absorption cross-section in the blue spectral region and can exhibit band emission in hosts where the emission originates from a $^4T_2 \rightarrow {}^4A_2$ transition [56]. Additionally, Cu^{2+}, Cr^{4+}, Ni^{2+} or Co^{2+} are potential candidates for broadband NIR emission. Table 8.1.2 lists the most red-shifted documented emission wavelength and exemplary host materials for these activators.

Table 8.1.2: Broad-band activators that can show emission in the NIR spectral region with exemplary host materials.

Activator ion	Exemplary host materials	Longest known emission wavelength at about (nm)
Mn^{4+}	$SrLaAlO_4$, $CaLaMgSbO_6$	730
Eu^{2+}	$Ca_3Sc_2Si_3O_{12}$	840
Cr^{3+}	$Sr_8MgLa(PO_4)_7$	850
Cu^{2+}	$CaCuSi_4O_{10}$	910
Mn^{6+}	$BaSO_4$	1100
Cr^{4+}	Mg_2SiO_4	1250
Ni^{2+}	$KMgF_3$	1600
Co^{2+}	ZnSe	3200

As the production of Hg discharge lamps will likely come to an end in the near future, there is also strong interest in UV radiation sources based on the LED technology. The theoretical limit for III–V nitride semiconductors is 210 nm upon the use of single phase AlN. To date, such LEDs are only available in the form of prototypes and their efficiency is very low. Nonetheless, the efficiency of UV-C emitting LEDs increases continuously while prices decrease at a fast pace and 265 nm emitting (Al,Ga)N LEDs are already commercially available. To generate a broadband continuous UV emission spectrum, phosphor-converted LEDs will have to be employed. For this purpose, Pr^{3+} appears to be the most promising activator ion, since the interconfigurational $[Xe]5d^14f^1$-$[Xe]4f^2$ radiative transitions are mostly located in the UV-C to UV-A range, while its excitation is possible either via the allowed $[Xe]4f^2$-$[Xe]5d^14f^1$ transitions or via the band gap of the host material. $Lu_3Al_5O_{12}:Pr^{3+}$ has been demonstrated to exhibit broad-band emission in the UV-B to UV-A range and could be used in combination with a UV-C emitting LED to provide a continuous UV emission spectrum [57]. Table 8.1.3 lists various Pr^{3+} activated phosphors with their peak emission wavelength.

Table 8.1.3: Overview of Pr^{3+} activated phosphors with the peak emission wavelength of the 5d-4f radiative transition.

Phosphor	Peak emission wavelength of the 5d-4f transition (nm)
$LaPO_4:Pr^{3+}$	225
$LuPO_4:Pr^{3+}$	233
$YPO_4:Pr^{3+}$	235
$YAlO_3:Pr^{3+}$	245
$La_2Si_2O_7:Pr^{3+}$	247
$CaLi_2SiO_4:Pr^{3+}$	253
$YBO_3:Pr^{3+}$	263
$Lu_2Si_2O_7:Pr^{3+}$	266
$Y_2SiO_5:Pr^{3+}$	270
$Lu_2SiO_5:Pr^{3+}$	272
$Lu_3Ga_2Al_3O_{12}:Pr^{3+}$	300
$Lu_3Al_5O_{12}:Pr^{3+}$	310
$Y_3Al_5O_{12}:Pr^{3+}$	320

References

[1] Round HJ. A note on carborundum. Elec World, 1907, 19, 309.

[2] Bay Z, Szigeti G. US2254957A, 1940.

[3] Carr WN, Pittman GE. One-Watt GaAs p-n junction infrared source. Appl Phys Lett, 1963, 3, 173–75. DOI: https://doi.org/10.1063/1.1753837.

[4] Allen JW, Cherry RJ. Some properties of copper-doped gallium phosphide. J Phys Chem Solids, 1962, 23, 509–11. DOI: https://doi.org/10.1016/0022-3697(62)90090-2.

[5] Holonyak N, Bevacqua SF. Coherent (Visible) light emission from $Ga(As_{1-x}P_x)$ junctions. Appl Phys Lett, 1962, 1, 82–83. DOI: https://doi.org/10.1063/1.1753706.

[6] Schubert EF. Light-emitting Diodes, 2nd edition, 3rd print with corr, Cambridge Univ. Press, Cambridge, 2010.

[7] Nakamura S. Nobel lecture: Background story of the invention of efficient blue InGaN light emitting diodes. Rev Mod Phys, 2015, 87, 1139–51. DOI: https://doi.org/10.1103/RevModPhys.87.1139.

[8] Shimizu Y, Sakano K, Noguchi Y, Moriguchi T. JP5610056B2, 1996.

[9] Hintzen HT, van Krevel JWH, Botty GI. EP1104799A1, 1999.

[10] Hirosaki N, Ueda K, Yamamoto H. WO2005052087A1, 2004.
[11] Ben Y, Liang F, Zhao D, Wang X, Yang J, Liu Z, Chen P. Anomalous temperature dependence of photoluminescence caused by non-equilibrium distributed carriers in InGaN/(In)GaN multiple quantum wells. Nanomater, 2021, 11, 1023 DOI: https://doi.org/10.3390/nano11041023.
[12] Shim J-I, Shin D-S. Measuring the internal quantum efficiency of light-emitting diodes: Towards accurate and reliable room-temperature characterization. Nanophotonics, 2018, 7, 1601–15. DOI: https://doi.org/10.1515/nanoph-2018-0094.
[13] Žukauskas A, Shur M, Gaska R. Introduction to Solid-state Lighting, Wiley, New York, 2002.
[14] Cheng KY. III–V Compound Semiconductors and Devices: An Introduction to Fundamentals, Springer International Publishing, Basel, 2020.
[15] Chu J, Sher A. Physics and Properties of Narrow Gap Semiconductors, Springer, New York, NY, 2008.
[16] Irvine S, Capper P. (Eds.) Metalorganic Vapor Phase Epitaxy (MOVPE): Growth, Materials Properties, and Applications, Wiley, Hoboken, NJ, USA, 2019.
[17] Fischer S, Baur F, Jüstel T. Influence of Ga^{3+} substitution on the spectroscopic properties of Ce^{3+} doped $Tb_3(Al,Ga)_5O_{12}$ garnet phosphors. ECS J Solid State Sci Technol, 2018, 7, R142–8. DOI: https://doi.org/10.1149/2.0131809jss.
[18] Bachmann V, Ronda C, Oeckler O, Schnick W, Meijerink A. Color point tuning for (Sr,Ca,Ba) $Si_2O_2N_2$: $Eu^{2+}$for white light LEDs. Chem Mater, 2009, 21, 316–25. DOI: https://doi.org/10.1021/cm802394w.
[19] Yu R, Luan R, Wang C, Chen J, Wang Z, Moon BK, Jeong JH. Photoluminescence properties of green-emitting $ZnGa_2S_4$: Eu^{2+}Phosphor. J Electrochem Soc, 2012, 159, J188–92. DOI: https://doi.org/10.1149/2.099205jes.
[20] Lim MA, Park JK, Kim CH, Park HD, Han MW. Luminescence characteristics of green light emitting Ba_2SiO_4: Eu^{2+}phosphor. J Mater Sci Lett, 2003, 22, 1351–53. https://doi.org/10.1023/A:1025739412154.
[21] Shimomura Y, Kurushima T, Shigeiwa M, Kijima N. Redshift of green photoluminescence of $Ca_3Sc_2Si_3O_{12}$: Ce^{3+}Phosphor by charge compensatory additives. J Electrochem Soc, 2008, 155, J45. DOI: https://doi.org/10.1149/1.2814144.
[22] Yu R, Wang J, Zhang J, Yuan H, Su Q. Luminescence properties of Eu^{2+}- and Ce^{3+}-doped $CaAl_2S_4$ and application in white LEDs. J Solid State Chem, 2008, 181, 658–3. DOI: https://doi.org/10.1016/j.jssc.2007.12.038.
[23] Katelnikovas A, Plewa J, Dutczak D, Möller S, Enseling D, Winkler H, Kareiva A, Jüstel T. Synthesis and optical properties of green emitting garnet phosphors for phosphor-converted light emitting diodes. Opt Mater, 2012, 34, 1195–201. DOI: https://doi.org/10.1016/j.optmat.2012.01.034.
[24] Tang J, Chen J, Hao L, Xu X, Xie W, Li Q. Green Eu^{2+}-doped $Ba_3Si_6O_{12}N_2$ phosphor for white light-emitting diodes: Synthesis, characterization and theoretical simulation. J Lumin, 2011, 131, 1101–06. DOI: https://doi.org/10.1016/j.jlumin.2011.02.007.
[25] Li YQ, de With G, Hintzen HT. Synthesis, structure, and luminescence properties of Eu^{2+} and Ce^{3+} activated $BaYSi_4N_7$. J Alloys Compd, 2004, 385, 1–11. DOI: https://doi.org/10.1016/j.jallcom.2004.04.134.
[26] Xie R-J, Hirosaki N, Mitomo M, Uheda K, Suehiro T, Xu X, Yamamoto Y, Sekiguchi T. Strong green emission from alpha-SiAlON activated by divalent ytterbium under blue light irradiation. J Phys Chem B, 2005, 109, 9490–94. DOI: https://doi.org/10.1021/jp050580s.
[27] Li YQ, Fang CM, de With G, Hintzen HT. Preparation, structure and photoluminescence properties of Eu^{2+} and Ce^{3+}-doped $SrYSi_4N_7$. J Solid State Chem, 2004, 177, 4687–94. DOI: https://doi.org/10.1016/j.jssc.2004.07.054.

[28] Yu R, Li H, Ma H, Wang C, Wang H, Moon BK, Jeong JH. Photoluminescence properties of a new Eu^{2+}-activated $CaLaGa_3S_7$ yellowish-green phosphor for white LED applications. J Lumin, 2012, 132, 2783–87. DOI: https://doi.org/10.1016/j.jlumin.2012.05.004.

[29] Katelnikovas A, Winkler H, Kareiva A, Jüstel T. Synthesis and optical properties of green to orange tunable garnet phosphors for pcLEDs. Opt Mater, 2011, 33, 992–95. DOI: https://doi.org/10.1016/j.optmat.2010.11.023.

[30] Chiang -C-C, Tsai M-S, Hon M-H. Luminescent properties of cerium-activated garnet series phosphor: structure and temperature effects. J Opt Soc Am, 2008, 155, B517. DOI: https://doi.org/10.1149/1.2898093.

[31] Saradhi MP, Varadaraju UV. Photoluminescence studies on Eu^{2+}-Activated Li_2SrSiO_4 – a potential orange-yellow phosphor for solid-state lighting. Chem Mater, 2006, 18, 5267–72. DOI: https://doi.org/10.1021/cm061362u.

[32] Zhang X, Zhang J, Dong Z, Shi J, Gong M. Concentration quenching of Eu^{2+} in a thermal-stable yellow phosphor Ca_2BO_3Cl: $Eu^{2+}$for LED application. J Lumin, 2012, 132, 914–18. DOI: https://doi.org/10.1016/j.jlumin.2011.11.001.

[33] Park JK, Kim CH, Park SH, Park HD, Choi SY. Application of strontium silicate yellow phosphor for white light-emitting diodes. Appl Phys Lett, 2004, 84, 1647–49. DOI: https://doi.org/10.1063/1.1667620.

[34] Piao X, Machida K-I, Horikawa T, Hanzawa H. Self-propagating high temperature synthesis of yellow-emitting $Ba_2Si_5N_8$: Eu^{2+}phosphors for white light-emitting diodes. Appl Phys Lett, 2007, 91, 41908. DOI: https://doi.org/10.1063/1.2760038.

[35] Im WB, Fourré Y, Brinkley S, Sonoda J, Nakamura S, DenBaars SP, Seshadri R. Substitution of oxygen by fluorine in the $GdSr_2AlO_5$: Ce^{3+}phosphors: $Gd_{1-x}Sr_{2+x}AlO_{5-x}F_x$ solid solutions for solid state white lighting. Opt Expr, 2009, 17, 22673–79. DOI: https://doi.org/10.1364/OE.17.022673.

[36] Li YQ, Hirosaki N, Xie RJ, Takeda T, Mitomo M. Yellow-orange-emitting $CaAlSiN_3$:Ce^{3+} Phosphor: structure, photoluminescence, and application in white LEDs. Chem Mater, 2008, 20, 6704–14. DOI: https://doi.org/10.1021/cm801669x.

[37] Kechele JA, Hecht C, Oeckler O, Schmedt auf der Günne J, Schmidt PJ, Schnick W. $Ba_2AlSi_5N_9$ – A new host lattice for Eu^{2+} -doped luminescent materials comprising a nitridoalumosilicate framework with corner- and edge-sharing tetrahedra. Chem Mater, 2009, 21, 1288–95. DOI: https://doi.org/10.1021/cm803233d.

[38] Katelnikovas A, Bettentrup H, Uhlich D, Sakirzanovas S, Jüstel T, Kareiva A. Synthesis and optical properties of Ce^{3+}-doped $Y_3Mg_2AlSi_2O_{12}$ phosphors. J Lumin, 2009, 129, 1356–61. DOI: https://doi.org/10.1016/j.jlumin.2009.07.006.

[39] Xie R-J, Hirosaki N, Sakuma K, Yamamoto Y, Mitomo M. Eu^{2+}-doped Ca-α-SiAlON: A yellow phosphor for white light-emitting diodes. Appl Phys Lett, 2004, 84, 5404–06. DOI: https://doi.org/10.1063/1.1767596.

[40] Li YQ, van Steen JEJ, van Krevel JWH, Botty G, Delsing ACA, DiSalvo FJ, de With G, Hintzen HT. Luminescence properties of red-emitting $M_2Si_5N_8$: Eu^{2+}(M = Ca, Sr, Ba) LED conversion phosphors. J Alloys Compd, 2006, 417, 273–79. DOI: https://doi.org/10.1016/j.jallcom.2005.09.041.

[41] Jia D, Wang X-J. Alkali earth sulfide phosphors doped with Eu^{2+} and Ce^{3+} for LEDs. Opt Mater, 2007, 30, 375–79. DOI: https://doi.org/10.1016/j.optmat.2006.11.061.

[42] Watanabe H, Yamane H, Kijima N. Crystal structure and luminescence of $Sr_{0.99}Eu_{0.01}AlSiN_3$. J Solid State Chem, 2008, 181, 1848–52. DOI: https://doi.org/10.1016/j.jssc.2008.04.017.

[43] Piao X, Horikawa T, Hanzawa H, Machida K-I. Characterization and luminescence properties of $Sr_2Si_5N_8$: Eu^{2+}phosphor for white light-emitting-diode illumination. Appl Phys Lett, 2006, 88, 161908. DOI: https://doi.org/10.1063/1.2196064.

[44] Hecht C, Stadler F, Schmidt PJ, Auf der Günne JS, Baumann V, Schnick W. $SrAlSi_4N_7$: Eu^{2+}– A nitridoalumosilicate phosphor for warm white light (pc)LEDs with edge-sharing tetrahedra. Chem Mater, 2009, 21, 1595–601. DOI: https://doi.org/10.1021/cm803231h.

[45] Piao X, Machida K-I, Horikawa T, Hanzawa H, Shimomura Y, Kijima N. Preparation of $CaAlSiN_3$: Eu^{2+}Phosphors by the self-propagating high-temperature synthesis and their luminescent properties. Chem Mater, 2007, 19, 4592–99. DOI: https://doi.org/10.1021/cm070623c.

[46] Mueller-Mach R, Mueller GO, Krames MR, Höppe HA, Stadler F, Schnick W, Jüstel T, Schmidt P. Highly efficient all-nitride phosphor-converted white light emitting diode. Phys Status Solidi (A), 2005, 202, 1727–32. DOI: https://doi.org/10.1002/pssa.200520045.

[47] Uheda K, Hirosaki N, Yamamoto H. Host lattice materials in the system Ca_3N_2-AlN-Si_3N_4 for white light emitting diode. Phys Status Solidi (A), 2006, 203, 2712–17. DOI: https://doi.org/10.1002/pssa.200669576.

[48] Krames MR. Light emitting diode materials and devices. In: Kitai A. (Ed.) Materials for Solid State Lighting and Displays, John Wiley & Sons Inc, Chichester, West Sussex, UK, Hoboken, NJ, USA, 2017, 273–312.

[49] Baur F, Glocker F, Jüstel T. Photoluminescence and energy transfer rates and efficiencies in Eu^{3+} Activated $Tb_2Mo_3O_{12}$. J Mater Chem C, 2015, 3, 2054–64. DOI: https://doi.org/10.1039/C4TC02588A.

[50] Adachi S, Takahashi T. Direct synthesis and properties of K_2SiF_6: Mn^{4+}phosphor by wet chemical etching of Si wafer. J Appl Phys, 2008, 104, 23512. DOI: https://doi.org/10.1063/1.2956330.

[51] Žukauskas A, Vaicekauskas R, Ivanauskas F, Vaitkevičius H, Shur MS. Spectral optimization of phosphor-conversion light-emitting diodes for ultimate color rendering. Appl Phys Lett, 2008, 93, 51115. DOI: https://doi.org/10.1063/1.2966150.

[52] Fischer S, Baur F, Jüstel T. Suppression of metal-to-metal charge transfer quenching in Ce^{3+} and Eu^{3+} comprising garnets by core-shell structure. J Lumin, 2018, 203, 467–2. DOI: https://doi.org/10.1016/j.jlumin.2018.07.001.

[53] Baur F, Jüstel T. Warm-white LED with ultra high luminous efficacy due to sensitisation of Eu^{3+} photoluminescence by the uranyl moiety in $K_4(UO_2)Eu_2(Ge_2O_7)_2$. J Mater Chem C, 2018, 6, 6966–74. DOI: https://doi.org/10.1039/C8TC01970C.

[54] Paulusz AG. Efficient Mn(IV) emission in fluorine coordination. J Electrochem Soc, 1973, 120, 942–47. DOI: https://doi.org/10.1149/1.2403605.

[55] Setlur AA, Siclovan OP, Lyons RJ, Grigorov LS. General Electric Company. US8057706B1, 2010.

[56] Zhang L, Wang D, Hao Z, Zhang X, Pan G-H, Wu H, Zhang J. Cr^{3+}-Doped broadband NIR garnet phosphor with enhanced luminescence and its application in NIR spectroscopy. Adv Opt Mater, 2019, 7, 1900185. DOI: https://doi.org/10.1002/adom.201900185.

[57] Pues P, Laube M, Fischer S, Schröder F, Schwung S, Rytz D, Fiehler T, Wittrock U, Jüstel T. Luminescence and up-conversion of single crystalline $Lu_3Al_5O_{12}$:Pr^{3+}. J Lumin, 2021, 234, 117987. DOI: https://doi.org/10.1016/j.jlumin.2021.117987.

8.2 Upconverters

Florian Baur, Thomas Jüstel

By definition, upconverters are optically active materials, which convert absorbed photons with a certain energy into photons with a higher energy, i.e. they cause a blue-shift of electromagnetic radiation. This phenomenon is called upconversion, as opposed to downconversion, the much more frequently occurring luminescence process, where the energy of the absorbed photon is higher than that of the emitted photon.

To comply with the conservation of energy, in all but one case upconversion requires two photons to be absorbed per emitted photon, which fundamentally limits the quantum efficiency to 50%. Regardless, upconverters are widely used due to their unique properties.

8.2.1 Historical remarks

A 2-photon process that can result in upconversion luminescence was first discussed as a theoretical concept by Maria Göppert-Mayer in her 1931 doctoral thesis [1]. The first experimental proof of this mechanism was published by Kaiser and Garrett in 1961 for $CaF_2:Eu^{2+}$ [2]. Later in the 1960s, François Auzel, an important figure in modern upconversion, began his research on the topic. He was introduced to the idea by a 1959 paper from Bloembergen, who discussed the theoretical possibility of IR detection via simultaneous excitation of an ion by an IR photon and a visible light photon [3], and Auzel chose upconversion as the topic for his thesis at CNET [4]. In 1966 he published the first report on upconversion between trivalent lanthanides, namely green upconversion emission due to Yb^{3+} and Er^{3+} in Na(La,Yb) $W_2O_8:Er^{3+}$ [5] and shortly afterwards on blue upconversion emission due to Yb^{3+} and Tm^{3+}. In both cases, the compounds are excited with IR radiation.

Since then, there has been steady research on upconversion phosphors, which resulted in the discoveries of various different upconversion mechanisms with different emission and excitation wavelengths. Upconversion phosphors with emission in the spectral range from red to UV-C have been found [6, 7] with excitation ranging from the IR to the blue spectral region.

8.2.2 Upconversion mechanisms

While the initially discovered upconversion phosphors relied on 2-photon excitation and energy-transfer upconversion (ETU), it was found over the next decades, that various other mechanisms can be responsible for upconversion luminescence.

https://doi.org/10.1515/9783110798890-021

With the exception of anti-Stokes emission, they are all based on the principle of simultaneous (within a certain timeframe) absorption of two photons. Therefore, the theoretical limit for the quantum efficiency, i.e. ratio of the number of emitted photons and the number of absorbed photons, amounts to 50%.

In practice, downconversion luminescence and non-radiative return to the ground state decrease the upconversion efficiency significantly. Upconversion efficiencies of 10^{-2} or 10^{-1} are the largest reported values. The efficiency depends strongly on the upconversion mechanism and on the excitation density. Due to upconversion being a 2-photon process, usually a quadratic dependence of the upconversion emission intensity on the excitation density is observed. The presence of such a dependence serves as a strong indicator of upconversion luminescence. The upconversion efficiencies listed for the mechanisms in this section are the ratios of optical power of the emitted upconversion photons and the *incident* excitation photons – i.e. not only of the absorbed photons.

8.2.2.1 Anti-Stokes

In anti-Stokes luminescence, the luminescent species possesses a low-lying, thermally populated excited state E1, which is usually a vibronic state. The excitation transition occurs between E1 and a higher-lying excited state E2. The radiative transition takes place between E2 and the ground state G. Thus, the energy of the emitted photon is higher than that of the absorbed photon by the amount of energy difference ΔE between E1 and G. Since E1 is thermally populated, ΔE must not be greater than 100–200 cm^{-1} for room-temperature anti-Stokes luminescence. Consequently, only short upconversion shifts can be realized, e.g. about 10 nm in the red spectral range. Anti-Stokes luminescence is a special case of upconversion, insofar that solely one photon is absorbed per emitted photon and the excitation density does not influence the upconversion efficiency. Due to the usually low thermal population density of the excited state, the efficiency is on the order of magnitude of 10^{-13}.

8.2.2.2 Two-photon excitation

This mechanism was first proposed by Maria Göppert-Mayer in her doctoral thesis in 1931 [1] and later demonstrated by laser excitation of $CaF_2:Eu^{2+}$ by Kaiser and Garrett in 1961 [2]. Two photons, whose combined energy amounts to the energy difference between excited state E and ground state G, are absorbed simultaneously. The emitted photon has, therefore, twice the energy of the absorbed photons i.e. a wavelength of 1/2 λ_{abs}. To achieve 2-photon excitation, the excitation density has to be high, to ensure simultaneous absorption of 2 photons with a sufficiently high probability. The upconversion efficiency amounts to 10^{-13}.

8.2.2.3 Cooperative luminescence

Cooperative luminescence can be seen as the reverse process of 2-photon excitation. In this case, two ions are in an excited state E and return to the ground state G simultaneously. One photon is emitted with an energy of the sum of the two individual transitions. Similar to 2-photon excitation, high excitation densities are required. In 1970 Nakazawa and Shionoya reported this phenomenon in $YbPO_4$ for excitation with 998 nm at excitation densities between 10^{17} and 10^{19} photons cm^{-2} s^{-1}, corresponding to energy densities of 2 10^5 to 2 10^7 J cm^{-2} s^{-1}. They observed a "weak" emission at 449 nm, with a maximum efficiency of 10^{-8} [8].

8.2.2.4 Cooperative sensitization

Cooperative sensitization is an extension of cooperative luminescence, where the energy released from the simultaneous return to the ground state by the ion pair is transferred to another ion. This process has been demonstrated by Ostermayer and Van Uitert in 1970 for $YF_3:Yb^{3+},Tb^{3+}$ [9]. Upon excitation with 930 nm radiation, an excited Yb^{3+} ion pair decays cooperatively with subsequent energy transfer to Tb^{3+}, which returns to the ground state radiatively. Ostermayer and Van Uitert reported an upconversion efficiency on the order of 10^{-6}.

8.2.2.5 Excited state absorption (ESA)

Excited state absorption occurs when an ion in an excited state E1 absorbs another photon and transitions to the higher excited state E2, as depicted in Figure 8.2.1. ESA requires a meta-stable excited state E1, so that sufficiently high population ratios can be achieved. To that end, generally trivalent rare-earth cations are used, whose respective transitions are spin- and parity-forbidden, and transition rates are on the order of 10^3 s^{-1}. A common example for this type of upconversion is Er^{3+}, such as in $SrF_2:Er^{3+}$. Specifically, for Er^{3+} the E1 state corresponds to the $^4I_{13/2}$ level with an energy of approximately 6600 cm^{-1} (1500 nm) and the E2 state corresponds to the $^4I_{9/2}$ level at 12,500 cm^{-1}. Upconversion emission does not occur immediately from that level, though, but the slightly lower-lying $^4I_{11/2}$ level is populated via multi-phonon relaxation. Thus, excitation in the NIR at around 1500 nm results in emission at around 950–1000 nm with the energy difference being converted into lattice vibrations, i.e. heat [10, 11].

Additionally, upconversion of 980 nm photons is possible by ESA from the $^4I_{11/2}$ state (approximately 10,300 cm^{-1}) to the $^4F_{7/2}$ state (approximately 20,600 cm^{-1}). Multi-phonon relaxation to a lower lying level takes place and the subsequent radiative transition originates from the $^2H_{11/2}$ or the $^4S_{3/2}$ state (18,400 and 19,200 cm^{-1}, respectively), which results in emission in the green spectral range, or from the $^4F_{9/2}$ state (15,300 cm^{-1}), and results in emission in the red spectral range [12].

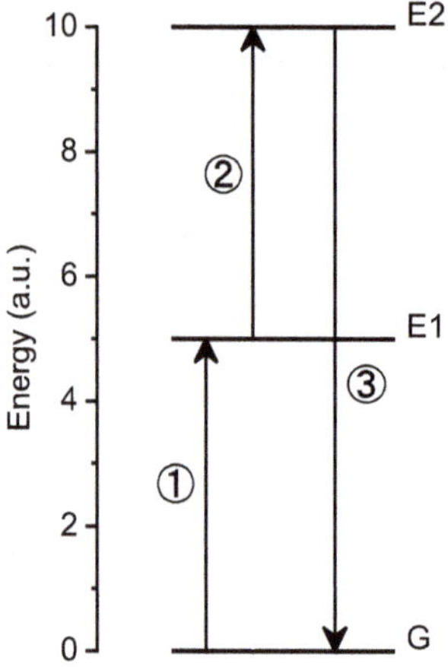

Figure 8.2.1: Schematic depiction of excited state absorption upconversion comprising three steps. Subsequent absorption of two photons, followed by radiative transition to the ground state.

In conclusion, ESA of Er^{3+} can be employed for upconversion luminescence from the NIR to green and red, i.e. for converting invisible to visible radiation. The upconversion efficiency of ESA is on the order of 10^{-5}.

8.2.2.6 Energy transfer upconversion (ETU)

Energy transfer upconversion is similar to ESA. In the first step, a lower-lying excited state E1 is populated via photoexcitation, however, the transition from E1 to the higher excited state E2 occurs via energy transfer from a second excited ion. The process is depicted schematically in Figure 8.2.2.

Both the donor and the acceptor ion are of the same element and ionic charge. Furthermore, ETU and ESA can take place simultaneously in the same host, e.g. in $SrF_2:Er^{3+}$. To discern the two types of upconversion is not simple, but can be attempted via monitoring of the decay times of the involved excited states. Furthermore, ETU has been reported for $YF_3:Er^{3+}$ and can even take place in non-crystalline solids, such as in Er^{3+}-doped fluorozirconate and oxyfluoroaluminate glasses [13]. Generally, in an amorphous environment, the probability for non-radiative transitions increases, which reduces the upconversion efficiency. In highly crystalline solids, upconversion efficiencies of up to 10^{-3} could be achieved [14].

8.2.2.7 Sensitized energy transfer upconversion (sensitized ETU)

The sensitized ETU follows the same basic principle as ESA and ETU: a system is in a lower-lying excited state E1 and further excitation occurs to a higher-lying excited state E2. In the case of sensitized ETU, the donor is a different type of cation than the acceptor. A commonly employed donor or sensitizer is Yb^{3+} with Er^{3+} or Tm^{3+} being the acceptors. The sensitizer is usually substituted in the host in high concentrations of 20–35% to ensure sufficient absorption from the quantum mechanically

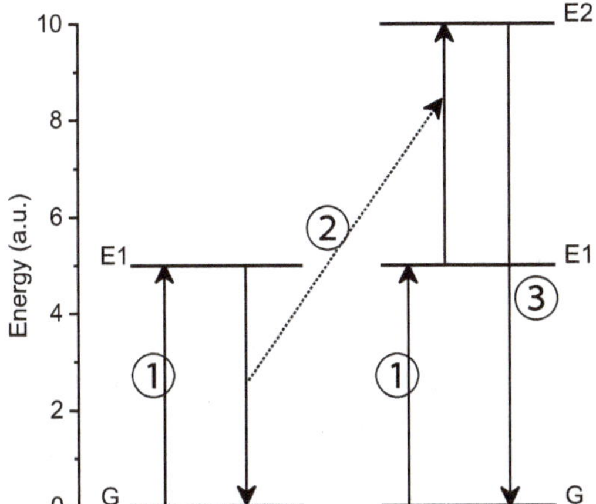

Figure 8.2.2: Schematic depiction of energy transfer upconversion comprising three steps. 1: Photoexcitation of two ions into excited state E1. 2: The first ion returns to the ground state with energy transfer to the second ion, resulting in excitation to E2. 3: The second ion returns radiatively to the ground state.

forbidden transition of Yb^{3+}. As for ESA, the low-lying excited state E1 must be meta-stable with a low transition probability to suppress a radiative return to the ground state in favour of energy transfer. This choice invariably results in a low absorption cross-section.

In these co-substituted systems, the ${}^2F_{7/2}{\rightarrow}{}^2F_{5/2}$ transition of Yb^{3+} is excited with 960–1020 nm photons. Energy transfer originates from the ${}^2F_{5/2}$ level of Yb^{3+} (at approximately 10,000 cm^{-1}) to Er^{3+} or Tm^{3+}. The acceptor transition of Er^{3+} is either from the ${}^4I_{15/2}$ ground state to the ${}^4I_{11/2}$ excited state E1 or from the E1 state to the higher-lying excited E2 state ${}^4F_{7/2}$. The radiative return to the ground state is analogous to the case discussed for ESA and results in green or red luminescence.

For Tm^{3+}, the energy transfer is slightly more complex, as five excited states of Tm^{3+} are involved; in order of increasing energy: ${}^3F_4$(E1a), ${}^3H_5$(E1b), ${}^3H_4(E2a)^3$ $F_{2,3}$(E2b), and ${}^1G_4$(E3). From the Yb^{3+} ${}^2F_{5/2}$ level energy transfer can occur to the E1b level, followed by non-radiative relaxation to the E1a level. Subsequent energy transfer from Yb^{3+} can excite the transition from E1a to E2b, followed by non-radiative relaxation to E2a, and lastly a third energy transfer from E2a to E3. Radiative transitions are generally observed from ${}^1G_4$ (E3) to the ${}^3H_6$ ground state (480 nm) or to the ${}^3F_4$ (E1a) excited state (660 nm), as well as from ${}^3H_4$ (E2a) to the ground state (800 nm). Thus, NIR radiation can be converted into short-wave NIR, deep-red or blue light by the Yb^{3+}–Tm^{3+} dopant pair. The dominant transitions are usually the 480 nm and the 800 nm emissions, with the exact ratios depending on the choice of host material [15, 16].

Sensitized ETU has been reported for other donor/acceptor combinations, such as Yb^{3+}/Ho^{3+} or tri-doped materials such as $Yb^{3+}/Er^{3+}/Tm^{3+}$ or $Nd^{3+}/Yb^{3+}/Er^{3+}$, where in the latter Nd^{3+} acts as a sensitizer for Yb^{3+} to provide excitability at 808 nm. Materials with low phonon-frequencies are used as hosts to decrease the probability of multi-phonon relaxation, e.g. YF_3, $NaYF_4$, BaY_2F_8, or YOCl. The efficiency of sensitized ETU is higher yet than that of non-sensitized ETU and can reach values of 10^{-1}.

8.2.3 Applications areas of upconverters

Despite its relatively low efficiency, upconversion is used in various fields due to its unique properties. Some important, exemplary, fields of application are discussed in this section. Specialized literature is available to grant a more detailed view [17].

In medicine, upconverters are applied in form of nanoparticles to allow their distribution within tissue and even cells. In that respect, the abbreviation UCNP is often used, which stands for *upconverting nanoparticles*.

8.2.3.1 Diagnostics

Tissue exhibits a relatively low absorption coefficient in the deep-red to NIR spectral region (roughly 650 to 1300 nm). This section of the electromagnetic spectrum is called the *biological window* and plays an important role in the usage of upconversion in medicine. UCNP can be excited in vivo from outside the body with NIR radiation and if the resulting upconversion emission is in the deep red spectral region, it can be detected from outside the body as well. Using this concept, it is possible to trace the concentration of UCNP is various parts of the body. By means of surface modification, the UCNP can be constructed to accumulate in specific types of tissue, e.g. tumor cells. UCNP are therefore often used for medical imaging.

Co-doping with Yb^{3+} and Er^{3+} provides upconversion excitable at 960 nm with emission wavelengths of 660, 545, and 525 nm. Since the emission around 525 and 545 nm is strongly absorbed by tissue, UCNP are optimized to show an increased red-to-green emission intensity ratio. This can be achieved by increasing the Yb^{3+} concentration, which facilitates cross-relaxation from the higher excited states of Er^{3+}, thereby increasing the intensity of the 660 nm emission. In addition, excitation at 808 nm, where absorption of tissue is weaker than at 960 nm, can be enabled by employing Nd^{3+} as a sensitizer for Yb^{3+}. This results in two-fold energy transfer, from Nd^{3+} to Yb^{3+} to Er^{3+} [18, 19].

8.2.3.2 Therapy

Upconversion is used in photodynamic therapy to treat cancer. In conventional photodynamic therapy, a photosensitizing compound is administered and excited with radiation in the UV to blue spectral range. Upon photoexcitation, these compounds produce reactive oxygen species that destroy the surrounding cells. To that end, radiation of the adequate wavelength has to reach the photosensitizer from outside the body, which is difficult if the target cells are in deeper regions of the body, due to the high absorption coefficient of tissue in the UV to blue spectral range [20].

By combining the photosensitizer with a suitable upconverter, 808 nm (Nd^{3+}, Yb^{3+}) or 960–980 nm (Yb^{3+}) excitation can be used, which can penetrate much deeper into the tissue. The upconverter emits in the visible spectrum and excites the photosensitizer in its immediate surrounding. Typically, $NaYF_4$ doped with Yb^{3+} as sensitizer (optionally with Nd^{3+} to provide 808 nm excitability) is used for photodynamic therapy in combination with Er^{3+} or Tm^{3+} to provide emission around 460, 540 and 660 nm.

8.2.3.3 Anti-counterfeiting

Documents such as bank notes, ID cards, and passports, or products such as clothing and machinery, are counterfeit in large numbers. Security features help to distinguish the original from the counterfeit. Printable upconverters, so-called upconverter ink, can be used to print invisible codes or markings on the product. To verify the identity, hand-held analytical equipment can be used that excites the ink with NIR radiation and detects the emitted visible light. There are several advantages of using upconverters instead of conventional downconverting phosphors as security markers [21]:

- Upconverters are not as widely available and easy to reproduce as the standard downconverters
- Surfaces often show luminescence upon excitation with UV radiation or blue light, which makes detection of downconversion markers difficult; NIR radiation generally only excites the upconversion marker and not the surface
- Upconversion intensity is strongly dependent on the excitation density, i.e. the counterfeiter would have to reproduce the same dependency to imitate the security marker

8.2.3.4 Photovoltaics

Photovoltaic systems convert electromagnetic radiation into electric energy. Photovoltaic systems that convert solar radiation into electric energy are usually based on silicon. The efficiency of the conversion strongly depends on the energy, viz. the wavelength, of the incident photons. Photons with energy higher than the bandgap are partially thermalized, while photons with energy lower than the bandgap are not absorbed at all. This factor alone is responsible for approximately 50% of the losses in silicon solar cells.

Upconverters can increase the efficiency by converting low-energy, long-wavelength photons into photons that are absorbed by the photovoltaic cell. Silicon solar cells have a maximum efficiency of 31% (Shockley-Queisser limit) which could be increased to nearly 48% with upconversion phosphors [22]. For this application Yb^{3+}/Er^{3+} or Yb^{3+}/Tm^{3+} upconverters (sensitized ETU) are used [23].

8.2.4 Remaining issues

Even though upconverters are widely applied, there are still some tremendous shortcomings. Most annoying is the relatively low efficiency: even upconverters based on sensitized ETU only reach efficiencies of around 10%. Since the fundamental limit for a two-photon process is 50%, the efficiency of upconverters can in theory be strongly increased and thus research activity in this regard is high.

Furthermore, upconversion requires a high excitation density. The upconversion material, therefore, has to be stable even under prolonged irradiation. As discussed previously, fluorides are commonly employed hosts and they can be prone to incorporation of oxygen impurities which decreases the upconversion efficiency. Additional demand is placed on the stability if the upconverters are used in anti-counterfeiting security markers on bank notes or clothing [24]. In the medical field, the upconverters have to be non-toxic and stable in the in vivo environment.

Another open issue is the lack of established upconverters with emission in the UV spectral range. While Tm^{3+} can show emission in the UV, the dominant transitions result in emission in the visible spectral region. To achieve UV upconversion emission, other activators such as Pr^{3+} will have to be used. The 4f-4f transitions of Pr^{3+} in the visible range have a long decay time and can be used for ESA to the fast decaying 5d states. The radiative 5d-4f transition of Pr^{3+} is usually located in the UV region. Thus, Pr^{3+} activated upconversion phosphors are under heavy development [6]. $Lu_3Al_5O_{12}:Pr^{3+}$ was shown to exhibit highly efficient upconversion emission in the UV-A/B region upon excitation at 488 nm [25] und $SrLi_2SiO_4:Pr^{3+}$ reportedly shows emission in the UV-C upon excitation with a 450 nm laser [26].

8.2.5 Outlook

With increasing efficiency and especially with additional emission wavelengths, upconverters can gain further application areas. In the medical field, NIR to NIR upconverters could significantly increase the detection efficiency, since in that case both excitation and emission are in the 700 to 1000 nm region, where the transparency of biological tissue is very high. Such upconverters would allow deep penetration and are useful for in vivo imaging [27].

Night vision can be realized with NIR-to-visible upconverters [28]. Either environmental NIR radiation is converted into visible, e.g. green light, or a NIR-torchlight is used to increase the intensity of the incident NIR radiation. It was even shown in animal experiments that mice could gain night-vision upon injection with specially crafted UCNPs [29].

UV-emitting upconverters can be used for disinfection, photobiology, and photocatalysis. While strong UV sources, such as Hg-gas discharge lamps, LEDs or Xe excimer lamps, are available, upconverters exhibit advantages over them. For example, it can be difficult to direct the UV radiation to a specific spatial position. The penetration depth of UV radiation in water is several orders of magnitude smaller than that of blue to cyan light. By exciting an upconverter with blue light, UV radiation can be generated at the point of care and thus increase the energy efficiency of the disinfection process [30]. Surfaces coated with a visible-to-UV upconverter can be disinfected upon excitation with visible light, without irradiating the surface with UV radiation, which can be harmful to humans and can degrade the material [31]. Figure 8.2.3 shows a Pr^{3+} activated silicate upconverter for UV-C emission, the greenish hue is a result of a set of absorption lines of Pr^{3+} in the blue and red range.

Figure 8.2.3: Photographic image of a Pr^{3+} activated upconverter under standard indoor illumination (Photograph by Patrick Pues).

References

[1] Göppert-Mayer M. Über Elementarakte mit zwei Quantensprüngen. Ann Phys, 1931, 401, 273–94. DOI: https://doi.org/10.1002/andp.19314010303.

[2] Kaiser W, Garrett CGB. Two-photon excitation in CaF_2: Eu^{2+}. Phys Rev Lett, 1961, 7, 229–31. DOI: https://doi.org/10.1103/PhysRevLett.7.229.

[3] Bloembergen N. Solid state infrared quantum counters. Phys Rev Lett, 1959, 2, 84–85. DOI: https://doi.org/10.1103/PhysRevLett.2.84.

[4] Auzel F. History of upconversion discovery and its evolution. J Lumin, 2020, 223, 116900. DOI: https://doi.org/10.1016/j.jlumin.2019.116900.

[5] Auzel F. Compteur quantique par transfer d'énergie entre deux ions de terres rares dans un tungstate mixte et dans un verre. C R Séances Acad Sci, Ser B, 1966, 262, 1016–19.

[6] Cates EL, Wilkinson AP, Kim J-H. Visible-to-UVC upconversion efficiency and mechanisms of $Lu_7O_6F_9Pr^{3+}$ and Y_2SiO_5: Pr^{3+}ceramics. J Lumin, 2015, 160, 202–09. DOI: https://doi.org/10.1016/j.jlumin.2014.11.049.

[7] Heer S, Lehmann O, Haase M, Güdel H-U. Blue, green, and red upconversion emission from lanthanide-doped $LuPO_4$ and $YbPO_4$ nanocrystals in a transparent colloidal solution. Angew Chem Int Ed Engl, 2003, 42, 3179–82. DOI: https://doi.org/10.1002/anie.200351091.

[8] Nakazawa E, Shionoya S. Cooperative luminescence in $YbPO_4$. Phys Rev Lett, 1970, 25, 1710–12. DOI: https://doi.org/10.1103/PhysRevLett.25.1710.

[9] Ostermayer FW, van Uitert LG. Cooperative energy transfer from Yb^{3+} to Tb^{3+} in YF_3. Phys Rev B, 1970, 1, 4208–12. DOI: https://doi.org/10.1103/PhysRevB.1.4208.

[10] Labbé C, Doualan JL, Moncorgé R, Braud A, Camy P. Excited-state absorption and fluorescence dynamics of Er^{3+}:KY_3F_{10}. Opt Mater, 2018, 79, 279–88. DOI: https://doi.org/10.1016/j.optmat.2018.03.058.

[11] Luitel HN, Mizuno S, Tani T, Takeda Y. Broadband-sensitive Ni^{2+}–Er^{3+} based upconverters for crystalline silicon solar cells. RSC Adv, 2016, 6, 55499–506. DOI: https://doi.org/10.1039/C6RA10713C.

[12] Li Y, Song Z, Li C, Wan R, Qiu J, Yang Z, Yin Z, Yang Y, Zhou D, Wang Q. High multi-photon visible upconversion emissions of Er^{3+} singly doped BiOCl microcrystals: A photon avalanche of Er^{3+} induced by 980 nm excitation. Appl Phys Lett, 2013, 103, 231104. DOI: https://doi.org/10.1063/1.4838636.

[13] Huang F, Liu X, Hu L, Chen D. Spectroscopic properties and energy transfer parameters of Er^{3+}-doped fluorozirconate and oxyfluoroaluminate glasses. Sci Rep, 2014, 4, 5053. DOI: https://doi.org/10.1038/srep05053.

[14] Auzel F. Upconversion and anti-Stokes processes with f and d ions in solids. Chem Rev, 2004, 104, 139–73. DOI: https://doi.org/10.1021/cr020357g.

[15] Chen G, Ohulchanskyy TY, Kumar R, Agren H, Prasad PN. Ultrasmall monodisperse $NaYF_4$: Yb^{3+}/Tm^{3+}nanocrystals with enhanced near-infrared to near-infrared upconversion photoluminescence. ACS Nano, 2010, 4, 3163–68. DOI: https://doi.org/10.1021/nn100457j.

[16] Ma M, Xu C, Yang L, Ren G, Lin J, Yang Q. Intense ultraviolet and blue upconversion emissions in Yb^{3+}–Tm^{3+} codoped stoichiometric $Y_7O_6F_9$ powder. Physica B Condens Matter, 2011, 406, 3256–60. DOI: https://doi.org/10.1016/j.physb.2011.05.035.

[17] Yang R Ed. Principles and Applications of Up-converting Phosphor Technology, 1st edition, Springer Singapore, Singapore, 2019.

[18] Ju Q, Chen X, Ai F, Peng D, Lin X, Kong W, Shi P, Zhu G, Wang F. An upconversion nanoprobe operating in the first biological window. J Mater Chem B, 2015, 3, 3548–55. DOI: https://doi.org/10.1039/c5tb00025d.

[19] Song N, Zhou B, Yan L, Huang J, Zhang Q. Understanding the role of Yb^{3+} in the Nd/Yb Coupled 808-nm-Responsive upconversion. Front Chem, 2018, 6, 673. DOI: https://doi.org/10.3389/fchem.2018.00673.

[20] Qiu H, Tan M, Ohulchanskyy TY, Lovell JF, Chen G. Recent progress in upconversion photodynamic therapy. Nanomaterials, 2018, 8, 344. DOI: https://doi.org/10.3390/nano8050344.

[21] Meruga JM, Cross WM, Stanley May P, Luu Q, Crawford GA, Kellar JJ. Security printing of covert quick response codes using upconverting nanoparticle inks. Nanotechnology, 2012, 23, 395201. DOI: https://doi.org/10.1088/0957-4484/23/39/395201.

[22] van Sark WG, de Wild J, Rath JK, Meijerink A, Schropp RE. Upconversion in solar cells. Nanoscale Res Lett, 2013, 8, 81. DOI: https://doi.org/10.1186/1556-276X-8-81.

[23] de Wild J, Meijerink A, Rath JK, van Sark WGJHM, Schropp REI. Upconverter solar cells: Materials and applications. Energy Environ Sci, 2011, 4, 4835. DOI: https://doi.org/10.1039/c1ee01659h.

[24] Li W, Xu J, He Q, Sun Y, Sun S, Chen W, Guzik M, Boulon G, Hu L. Highly stable green and red up-conversion of $LiYF_4$: Yb^{3+},Ho^{3+} for potential application in fluorescent labeling. J Alloys Compd, 2020, 845, 155820. DOI: https://doi.org/10.1016/j.jallcom.2020.155820.

[25] Pues P, Laube M, Fischer S, Schröder F, Schwung S, Rytz D, Fiehler T, Wittrock U, Jüstel T. Luminescence and up-conversion of single crystalline $Lu_3Al_5O_{12}$:Pr^{3+}. J Lumin, 2021, 234, 117987. DOI: https://doi.org/10.1016/j.jlumin.2021.117987.

[26] Yin Z, Yuan P, Zhu Z, Li T, Yang Y. Pr^{3+} doped Li_2SrSiO_4: An efficient visible-ultraviolet C up-conversion phosphor. Ceram Int, 2021, 47, 4858–63. DOI: https://doi.org/10.1016/j.ceramint.2020.10.058.

[27] Li Z, Zhang Y, La H, Zhu R, El-Banna G, Wei Y, Han G. Upconverting NIR Photons for Bioimaging. Nanomaterials, 2015, 5, 2148–68. DOI: https://doi.org/10.3390/nano5042148.
[28] Li N, Lau YS, Xiao Z, Ding L, Zhu F. NIR to visible light upconversion devices comprising an NIR charge generation layer and a perovskite emitter. Adv Opt Mater, 2018, 6, 1801084. DOI: https://doi.org/10.1002/adom.201801084.
[29] Ma Y, Bao J, Zhang Y, Li Z, Zhou X, Wan C, Huang L, Zhao Y, Han G, Xue T. Mammalian near-infrared image vision through injectable and self-powered retinal nanoantennae. Cell, 2019, 177, 243–55. DOI: https://doi.org/10.1016/j.cell.2019.01.038.
[30] Cates EL, Kim J-H. Bench-scale evaluation of water disinfection by visible-to-UVC upconversion under high-intensity irradiation. J Photochem Photobiol B, 2015, 153, 405–11. DOI: https://doi.org/10.1016/j.jphotobiol.2015.10.021.
[31] Cates EL, Cho M, Kim J-H. Converting visible light into UVC: Microbial inactivation by Pr^{3+}-activated upconversion materials. Environ Sci Technol, 2011, 45, 3680–86. DOI: https://doi.org/10.1021/es200196c.

8.3 Organometallic Ir(III) and Pt(II) complexes in phosphorescent OLEDs: an industrial perspective

Ilona Stengel, Stefan Schramm

8.3.1 Introduction

Thin, efficient, flexible and infinite contrast – all these are the attributes of a technology that has shaped the last 10 years' development of disruptive innovations (such as smartphones) like no other, and is now becoming ever more an increasingly important topic for large displays and TVs. The name of this technology is organic light-emitting diodes – short OLED – and organometallic complexes containing Ir(III) and Pt(II) are at the heart of it.

In this chapter, we want to give the reader an overview on this topic from an industrial point of view in order to enrich the fundamental academic knowledge from an application-specific perspective.

Before we explain the precise role of Ir(III) and Pt(II) complexes in OLEDs in further detail, we would like to take time and give a general introduction into the topic.

Then, we will discuss in detail a selection of Ir(III) and Pt(II) complexes sorted by their emission color, and we will use some relevant examples to illustrate key factors in the development of these molecular species.

Thereafter, we will conclude our overview by highlighting some of the corner stones that, in our view, will drive the development of OLEDs in the future.

Fiat lux!

8.3.1.1 The OLED working principle

Figure 8.3.1 shows an RGB OLED in a side-by-side configuration. This means that each pixel has three subpixels: **r**ed, **g**reen and **b**lue. Every subpixel is an OLED in itself and therefore comprised of many functional layers. Between an anode and a cathode there are typically charge transporting layers. Close to the anode side there are the hole injection and transport layers (HIL & HTL), whereas on the cathode side are electron injection and transport layers (EIL & ETL). In the middle of the OLED stack, one can typically find the EML – the emissive layer. This is the place where the actual light emission happens.

In the OLED context, positive charges that are injected on the side of the anode are typically called "holes". Negative charges that are injected on the cathode side are called "electrons". Both meet in an ideal case only within the EML. Here, they recombine forming an exciton, a quasiparticle that describes the excited hole – electron pair. When the exciton relaxes to its ground state, the emission of light can happen. In most OLEDs, the light exits the system through one transparent electrode; in

https://doi.org/10.1515/9783110798890-022

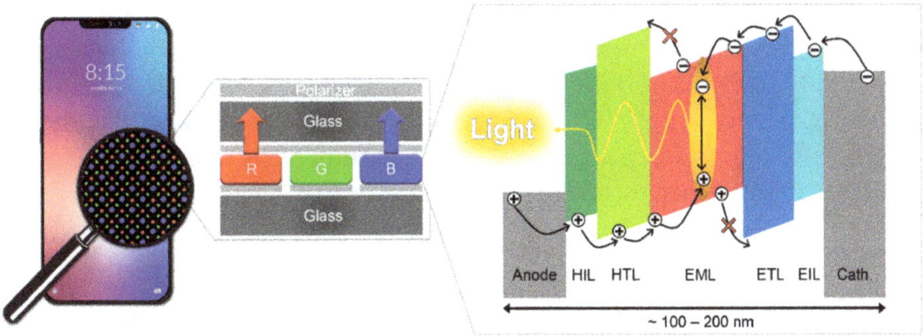

Figure 8.3.1: The OLED display of a smartphone (left) is comprised of many RGB (red, green, blue) subpixels in a side-by-side configuration (middle). Each of these subpixels is by itself constructed of an entire OLED stack, and is made of several functional layers such as HIL, HTL, EML, ETL or EIL deposited between the cathode (abbreviated as Cath.) and the anode (right).

many cases, this is the anode built of indium-tin-oxide (ITO, a conductive transparent material deposited onto a suitable substrate).

The whole OLED stack is about 100–200 nm in height, which means that it is about 500 times thinner than a human hair. In some cases, the layers are only a few nanometers thick. Nevertheless, a modern OLED is made of at least a dozen different molecular species that are organized in the various layers, which illustrates the enormous technical understanding and expertise that goes into the design and the construction of such a complex system.

Looking at the OLED stack from an architectural point of view, it is important to realize that the energy levels of the materials that constitute each functional layer have to be perfectly tuned and aligned towards each other. The right part of Figure 8.3.1 can be interpreted as an energy diagram where the lower edge of each functional layer marks the highest occupied molecular orbital (HOMO) and the upper edge corresponds to the lowest unoccupied molecular orbital (LUMO) of the material used therein. Optimizing the energetic matching of levels to fit adjacent layers is a fundamental requirement for efficient charge transport and recombination. In general, there are no “good” or “bad” HOMO/LUMO levels for a given material; however, they have to be well-adjusted to the other compounds used in the device.

8.3.1.2 Organometallic triplet emitters: a milestone in the history of OLED

There are several mechanisms which can lead to the emission of light in the EML. The two basic ones are fluorescence and phosphorescence.

Earliest experiments and reports on organic-material-based electroluminescence were related to the emission from purely organic fluorescent molecules, such as anthracene crystals [1, 2]. The first ever made OLED, which was invented in the laboratories of Eastman Kodak in 1987 [3], used a fluorescent emitter based on three 8-hydroxyquinolinato ligands linked to an aluminum(III) center in a quasi-octahedral coordination environment; the name of this complex is usually abbreviated as Alq_3. This first OLED was able to emit green light at 550 nm with less than 1% external quantum efficiency (EQE) and enormous driving voltages of *ca.* 10 V.

The EQE can be described as ratio of extracted photons from a device by the number of injected charges, and can be expressed by the following formula:

$$\eta_{EQE} = \gamma \times \Phi_L \times \eta_r \times \eta_{out} = \eta_{int} \times \eta_{out}$$

where η_{EQE} is the EQE, γ the charge carrier balance factor, Φ_L the intrinsic photoluminescence quantum yield of the emitter, η_r the radiative exciton utilization ratio, η_{int} and η_{out} the overall internal quantum efficiency and light outcoupling efficiency [4].

Reflected in there is the central point that the photoluminescence quantum yield of the emitting species should be large, and the excitons should be utilized to a maximum.

Today, only blue OLEDs rely on fluorescent emitters in which the maximum EQEs lie around 10% [5]. Modern green and red OLEDs make use of phosphorescence, which can achieve typically EQEs of 25–30% [6].

When two uncorrelated charges (hole and electron, each with a spin angular momentum characterized by $s = ½$) meet and recombine to an exciton in an OLED stack, the chance that they form a non-degenerate singlet state ($S = 0$) is 25% and 75% for threefold-degenerate triplet state ($S = 1$), due to spin statistics (Figure 8.3.2). In the case of fluorescent devices, the internal quantum efficiency (IQE) is limited to 25%, since the triple excitons relax via radiationless deactivation paths as the phosphorescence (i.e. radiative deactivation from the triplet state into the ground state) is spin-forbidden. Thus, only excitons corresponding to the lowest excited singlet state of the molecular species lead to the emission of light, meaning that 75% of the excitons are lost as heat (even if the intrinsic fluorescence quantum yield of the emitter reaches 100%). Since light outcoupling from the device also occurs under a significant loss factor, the EQE of the early fluorescent OLEDs is typically low.

The utilization of diamagnetic iridium(III) and platinum(II) complexes as triplet emitters in OLEDs was pioneered by the research groups of Forrest and Thompson, giving the EQE and OLED technology in general a great boost. They found a way of harvesting the remaining 75% of excitons to get a maximum possible IQE of 100% [7], and realized that phosphorescent molecules could be employed to also access the so-far lost triplet excitons. Unfortunately, phosphorescence of pure organic molecules rarely occurs at room temperature, so they needed to find a way to accelerate the radiative relaxation of the excited triplet states by strong spin-orbit coupling (SOC). This can be realized by means of the heavy-atom effect, which mixes singlet character into the lowest triplet state while making the radiative relaxation feasible.

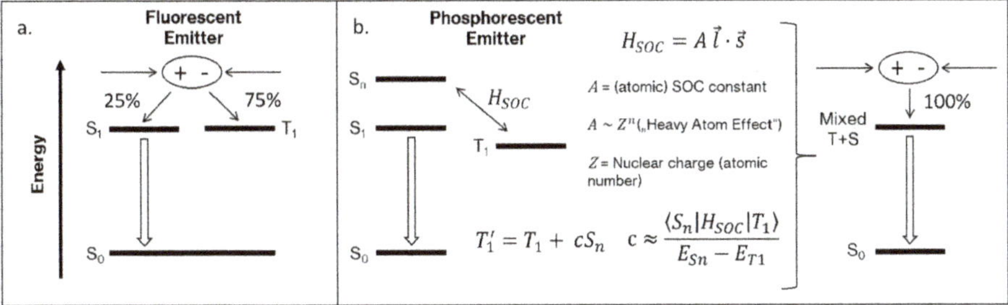

Figure 8.3.2: Exciton utilization in an OLED for a) a fluorescent emitter and b) a phosphorescent emitter.

The singlets, on the other hand (and supported by the aforementioned mixing of states) rapidly intersystem-cross into the triplet manifold, from where the phosphorescence can occur; hence, both classes of excitons are harvested. Consequently, luminescent metal-organic complexes of Ir(III) and Pt(II) became of large interest.

These triplet emitters are typically applied in conjunction with suitable host materials. Their purpose is to disperse the emitter in the EML and therefore to attenuate several detrimental effects, such as aggregation-caused emission quenching and exciton annihilation processes [8].

The underlying process of the phenomenon comprising a molecular entity that emits light upon application of an electric field is called electroluminescence, and was already discovered in the 1950s [9]. Nevertheless, it took about 50 years from initial conception to the first real-world application of an OLED device (Figure 8.3.3).

The discovery of organometallic triplet emitters marked a milestone in the technological development of OLEDs and soon after it enabled early display applications, at first in the high-end consumer market. An early product, with rather prototype character, was the Sony XEL-1 released in 2007 [10]. In the following years, OLED technology matured further by incorporating new ideas and concepts to finally yield significant improvements in EQE, lifetime, color accuracy and maximal luminance.

In the year 2013, the production of active-matrix OLED (AMOLED) smartphone displays by Samsung Display (SDC) hit the mark of 500.000.000, making small OLED panels a mass market product [11]. Today, large-area OLED TVs by LG Display are getting increasingly popular and are not a niche product anymore.

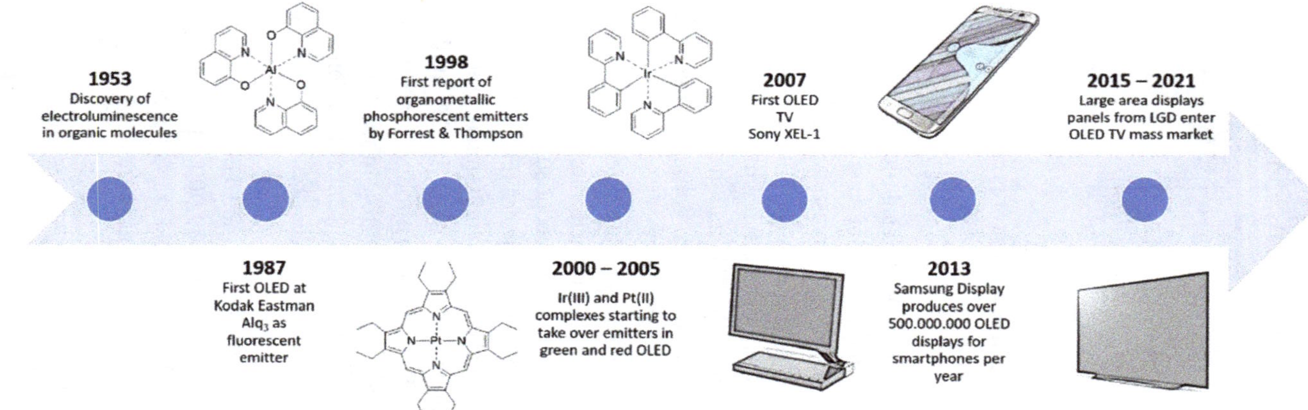

Figure 8.3.3: Historic development of OLEDs. A progression from basic research in the latter half of the twentieth century to modern mass market applications in the twenty-first century.

8.3.2 OLED emitter materials based on organometallic complexes

8.3.2.1 Highly efficient green and red complexes as phosphorescent emitters comprising Ir(III) or Pt(II) centers

While the first reported coordination compound utilizing singlets and triplets in an OLED was a Pt(II) tetrapyrrolate, namely PtOEt [7], the Ir(III)-based complexes have been more relevant in industrial OLED applications so far. Nonetheless, in recent years, Pt(II) complexes have become increasingly prominent, due to their perfectly competitive emission properties, if compared to Ir(III) emitters, mainly due to constantly evolving and optimization-driven ligand sophistication.

Generally speaking, it is necessary to use chromophoric ligands that ensure (i) a high participation of the metal center on the excited state configuration to maximize SOC and the radiative deactivation rate k_r, (ii) a high rigidity to suppress radiationless deactivation channels characterized by the rate constant k_{nr} and (iii) a strong ligand-field splitting to destabilize dissociative metal-centered *p/d-d**-configurations to avoid their thermal population from the emissive state of interest (which otherwise leads to radiationless deactivation or photochemical decomposition). These measures maximize the intrinsic quantum yield $\Phi_L = k_{nr}/(k_r + k_{nr})$ and shorten the excited state lifetime $\tau = 1/(k_r + k_{nr})$.

The first highly efficient (and in detail studied) green-emitting Ir(III) complex was *fac*-Ir(ppy)$_3$ [12, 13]. This homoleptic complex can be synthesized in one or two steps, starting from Ir(acac)$_3$ or IrCl$_3$ · xH$_2$O, respectively [14, 15]. *fac*-Ir(ppy)$_3$ is the most widely used green OLED emitter, both in academia and in industry, due to its elegantly simple structure, easy access and bright emissive properties [16–18]. Starting from this basic emitter structure, several series of derivatives have been developed finding applications in green and red OLEDs. An overview of commonly utilized structures can be found in Figure 8.3.4.

All commercially relevant iridium-based OLED emitters contain low-spin Ir(III) centers (with six paired 5*d* electrons) in a quasi-octahedral coordination environment. Ir(III) emitters are generally neutral coordination compounds with three monoanionic ligands. Although the literature also reports on tridentate ligands, mainly Ir(III) complexes bearing three bidentate ligands carrying evenly distributed negative charges are commercially relevant for display applications. Thus, all Ir(III) emitters have a HOMO located sizeably on the metal center, whereas the LUMO located either solely on the ligands, or partly on the ligands and partially on the metal center. As a result, the excited electronic configurations involve one unpaired electron in each frontier orbital, namely in the HOMO and in the LUMO, whereas the energy gap between these orbitals can be inversely correlated with the wavelength emitted from the excited state. The excited state configurations generally have a metal-to-ligand charge transfer (MLCT) character. These MLCT states produce relatively broad emission bands (due to a shallower potential energy hypersurface with an increased density of vibrational states), but also giving

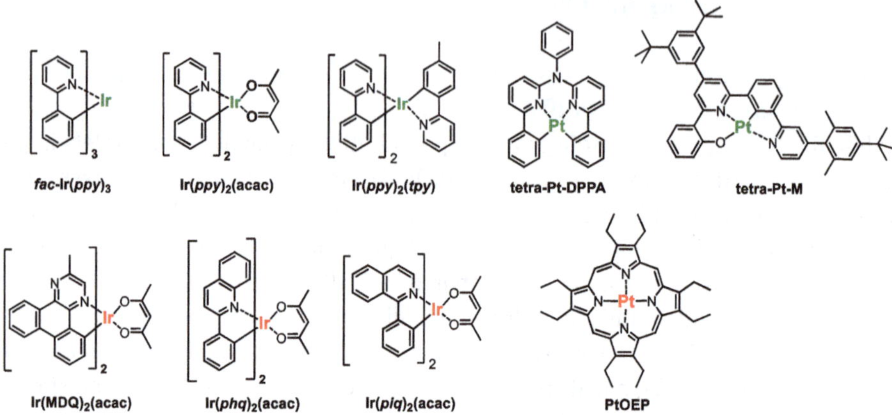

Figure 8.3.4: Overview of commonly utilized Ir(III) and Pt(II) structures in green and red OLED display applications.

versatile options for tuning of the luminescence properties by ligand design, as we will see in the following examples.

Favorable luminescence properties comprise narrow emission spectra (defined by the maximum emission peak wavelength λ_{max}[nm] and the full-width-at-half-maximum (FWHM) in [eV] or [nm]), high photoluminescence quantum yields (Φ_L) in films and in solution, and shortest possible excited state decay times τ [µs] to avoid triplet-triplet annihilation. Quite often, it can be observed in the literature that for one complex and one particular photoluminescence (PL) parameter, different values are reported. The nature of the solvent or the matrix, the concentration, the partial pressure of oxygen in the sample, the temperature and other possible variations resulting from experiments in different laboratories have a drastic influence on the measured values, as reported in Table 8.3.1. For instance, reference [17] reports the Φ_L and τ of *fac*-Ir(ppy)$_3$ in different solvents and film-forming matrices. These values stand in strong contrast to previously reported Φ_L and τ values of *fac*-Ir(ppy)$_3$ using the same medium [16, 17].

After the discovery that homoleptic *fac*-Ir(ppy)$_3$ is a very useful phosphorescent emitter for OLED applications, mainly heteroleptic Ir(III) complexes followed the developments and succeeded in commercial display applications up to date. In such compounds, typically two identical aromatic and cyclometalated ligands are present, which define and carry the LUMO, and one aromatic or non-aromatic ancillary ligand.

An early example of such a compound is Ir(ppy)$_2$(acac), which was one of the first examples that pushed the maximum external quantum efficiencies to a new high level [19]. If the ligands are available, the heteroleptic complexes can be synthesized in two steps. First, two cyclometalating ligands are introduced. Second, the ancillary ligands, such as acetylacetone (acac), are attached [20–22]. If substituted phenyl pyridines are introduced instead of ppy, or other ancillary ligands are utilized, these ligands have to be prepared upfront by additional synthesis.

Comparing *fac*-Ir(ppy)$_3$ and Ir(ppy)$_2$(acac), one can see that there is little difference in the synthetic effort to produce them, as well as in their photoluminescence properties. The emission spectra and also the Φ_L and decay times are very similar, which is due to the fact that the HOMO is located mostly on the iridium center and only partly on the ppy ligands, while the LUMO is predominantly located on the ppy units.

There are also heteroleptic complexes which comprise three cyclometalated ppy ligands with one of them being differently substituted, e.g. Ir(ppy)$_2$(tpy) [23]. Depending on how electron poor or rich the ancillary ligand is (compared to the other two cyclometalated ligands), it might contribute more or less to the LUMO. In the case of Ir(ppy)$_2$(tpy), the additional methyl group has little influence on the spectral profiles, Φ_L and decay times, when compared to homoleptic Ir(ppy)$_3$ (ref. [23] took the Φ_L of 40% for Ir(ppy)$_3$ [17] as a reference when investigating the Φ_L of Ir(ppy)$_2$(tpy) and thus, came to a similarly low value of 41%).

Nevertheless, there is an important effect that makes heteroleptic Ir(III) complexes more favorable compared to their homoleptic analogs: They are able to show a preferred orientation upon evaporation into the host of the emissive layer. This holds true for all sorts of heteroleptic Ir(III) emitters and all emission colors while determining the relative orientation of the transition dipole moments, hence affecting the outcoupling efficiency.

Increasing the EQE of OLEDs via emitter orientation has become a topic of big interest since it was first observed in 2011 [24, 25]. If *fac*-Ir(ppy)$_3$ and Ir(ppy)$_2$(acac) are compared directly in the same device stack architecture, an increased EQE of 21.7% for Ir(ppy)$_2$(acac) is observed, compared to 18.3% for *fac*-Ir(ppy)$_3$ [26].

By making use of this effect, the outcoupling efficiency (η_{out}), and therefore the EQE, is systematically addressed by structural design of the emitter molecules. If the transition dipole moments of the emitters are oriented randomly, such as in the case of homoleptic *fac*-Ir(ppy)$_3$, the light is emitted isotropically from the EML. This results in a sizeable amount of light losses from side and backward-directed emission as well as multiple reflections. If the optical transition dipole moment vectors are horizontally oriented with respect to the substrate plane, the light emission occurs perpendicularly towards them and hence becomes strongly anisotropic, raising the EQE [27]. An ideal triplet emitter should therefore not only be designed in a way that its Φ_L approaches unity, but that it also aligns itself horizontally in evaporated films.

The initial observation of an oriented Ir(III) complex was made with Ir(MDQ)$_2$(acac) [25], which is a well-studied red phosphorescent emitter. Its synthesis was described in analogy to the green phenyl pyridine derivative [28], and its photophysical properties are similarly suited for OLED applications. It shows an intense unstructured and relatively broad emission band from an MLCT state peaking at 609 nm.

Table 8.3.1: Relevant and representative PL properties of green and red Ir(III) and Pt(II) emitters.

Complex	Matrix	λ_{max} [nm]	FWHM [nm]/[eV]	Φ_L [%]	τ [µs]	Ref.
green						
***fac*-Ir(*ppy*)$_3$**	toluene	508	66 / 0.30	40	2.0	[17]
	toluene			73	1.1	[16]
	CBP (8 wt%)			61	1.2	[16]
	PS (8 wt%)			82	1.1	[16]
	MeTHF	508	41 / 0.19	97	1.6	[72]
	DCM	519	84 / 0.36	90	1.6	[18]
	DCM	510			1.9	[15]
	DCE	517	70 / 0.30	89	1.4	[73]
	CBP	512		90	1.3	[73]
Ir(*ppy*)$_2$(*acac*)	toluene	520		46	1.6	[74]
	toluene	516	57 / 0.25	36	1.7	[21]
	DCM	520				[74]
	MeTHF	516		34	1.6	[20]
	DCM	515	70 / 0.31			[20]
	MeTHF	516	77 / 0.36	34	1.6	[22]
Ir(*ppy*)$_2$(*tpy*)	DCM	517	84 / 0.36	41	3.3	[23]
	MeTHF	513		36		[75]
tetra-Pt-DPPA	MeTHF	512	56 / 0.24	74	7.6	[76]
tetra-Pt-M	DCM	515		90	3.7	[38]
	PMMA	510	70 / 0.30	62	1.9	[38]
red						
Ir(MDQ)$_2$(*acac*)	DCM	609	84 / 0.27	48		[28]
Ir(*piq*)$_2$(*acac*)	DCM	611	107 / 0.35			[30]
	DCM	622	65 / 0.20	20	1.7	[31, 32]
	DCM	618	44 / 0.14			[33]
Ir(*phq*)$_2$(*acac*)	DCM	581				[29]
	PMMA			68	1.3	[29]
	MeTHF	597		10	2.0	[20]
	DCE	598	71 /0.24	70	1.6	[73]
	CBP	600		80	2.4	[73]
PtOEP	PS	647		50	91	[7]
	PS	647	19 / 0.06		52	[37]
	PS	648	21 / 0.06			[36]
	toluene/DMF			45	83	[36]
	toluene	647		42	81	[37]
	THF	646		37	100	[37]

CBP = 4,4′-bis(9-carbazolyl)-1,1′-biphenyl; PS = polystyrene; MeTHF = 2-methyl-tetrahydrofuran; DCM = dichloromethane; DCE = dichloroethane; PMMA = poly(methyl methacrylate); DMF = dimethylformamide; THF = tetrahydrofuran; λ_{max} = maximum of emission spectrum in nm; FWHM = full-width-at-half-maximum of emission spectrum in nm and eV was extracted from the pdf of the respective reference by the digitalization function of OriginPro 2021b;
$\Phi_L = k_r / (k_r + k_{nr})$ = photoluminescence quantum yield;
$\tau = 1/(k_r + k_{nr})$ = observed excited state decay time in μs.

Further design improvements were mostly achieved around heteroleptic phenyl quinoline (phq) or phenyl isoquinoline (piq) acac complexes, such as $Ir(phq)_2(acac)$ [20, 22, 29] and $Ir(piq)_2(acac)$ in order to optimize the emission properties [30–33]. At the same time, as their green ppy counterpart became famous, also phq ligands for Ir(III) derivatives gained relevance. The synthesis of these derivatives shows strong similarities.

In order to emphasize the color purity and thus, the narrowness of emission spectra, we extracted and collected the FWHM from spectra given in literature and included the data in Table 8.3.1. Once more, it becomes obvious that literature data on these parameters is pretty heterogeneous and very much dependent on the experimental conditions (*vide supra*).

Besides the fact that literature data on photophysical properties of phosphorescent complexes is subject to large deviations, another important trend becomes visible when analyzing different literature reports: Emission properties measured in solution often differ a lot from the emission characteristics obtained from doped films, in which the emitter is applied in low concentrations (<10 wt%) into a film-forming solid matrix. Even the nature of the matrix material can have a significant influence. Therefore, the host material design carried out to match a particular emitter is equally important for OLED applications.

Additional aryl and alkyl-substitutions are helpful tools in order to tune the emission properties of green and red Ir(III) complexes for deeper color purity and higher efficiency. Several such series of emitter compounds have been successfully synthesized and investigated [29, 34, 35].

The search for new and optimized emitter structures brought the attention back to Pt(II) complexes. Already PtOEP, with an emission peak of 648 nm and a FWHM of 26 nm in a polystyrene matrix, was one of the first examples to show that phosphorescent Pt(II) emitters can have very narrow emission spectra [36], a particularly favorable attribute for OLED display applications. However, PtOEP also presents the weak point of square planar complexes, which is their tendency to form aggregates *via* stacking interactions. This phenomenon is usually observed when spectra of films with increasing concentrations are measured [37]. As a result, broadened and red-shifted emission profiles are obtained, which are problematic for industrial applications. The incorporation of sterically bulky substituents, such as twisted *ortho*-alkyl-phenyl rings or *tert*-butyl groups (as comprised by the green emitter tetra-Pt-M [38]) is an effective countermeasure in this respect. Many Pt(II) complexes have a reduced MLCT character in their excited states, and the enhanced ligand-centered (LC) character provides a sharp vibrational progression in the emission profile with higher color purity. In turn, the lowered metal participation in the excited state goes along with a reduced radiative deactivation rate and with a prolonged excited state lifetime.

Another downside of Pt(II) complexes is their high susceptibility towards radiationless decay. Due to the square-planar configuration, distorted excited states geometries are generally more accessible compared to octahedral Ir(III) complexes, opening up channels for internal conversion and thus, lowering the Φ_L. Rigid ligands which are able to reduce distortions and to cause a strong splitting of *d*-orbitals are required in order to realize efficient Pt(II) emitters (typically via cyclometallation of polydentate ligands to push the metal-centered π/d-d^* states up in energy, making them less thermally accessible) [39].

As a result of the aggregation and radiationless decay tendencies, the early investigated Pt(II) emitters, which were designed in analogy to the Ir(III) emitters while bearing bidentate cyclometalated ppy ligands and acac, could not compete their Ir(III)-based counterparts [40]. The first examples of Pt(II) complexes showing a stronger ligand field splitting and more efficient emission typically comprised tridentate and cyclometalated 1,3-di(pyridylbenzene) ligands [41].

Along with a stronger ligand field splitting, also the luminescence decay times of such Pt(II) compounds became shorter by increasing the participation of the metal on the excited state. While for early bidentate chromophoric units the lifetimes were commonly above 8–10 μs [40], tridentate 1,3-di(pyridylbenzene) ligands provide decay times of 7 μs while the latest examples of Pt(II) emitters are able to go well below 4 μs (as it is the case for tetra-Pt-M, listed in Table 8.3.1 [38]).

Modern green Pt(II) emitters, such as tetra-Pt-M, are well-armed against aggregation and possess tetradentate ligands providing a strong ligand field splitting. Thus, they are able to compete with green Ir(III) emitters in OLED display applications in every respect.

For red phosphorescent emitters, Ir(III) complexes hold a strong position, because they offer very high efficiency and long durability, creating little need for substantial improvement. There has been little effort in Pt(II) chemistry attempting to keep up with red Ir(III) emitters. Since the situation is different in the green, where more efficient, more color pure and more stable green phosphorescent OLED emitters are needed, Pt(II) complexes can provide extra levers in order to narrow down the FWHM [42], pushing the efficiency and making green emitter materials more stable.

It is an ongoing endeavor in OLED material research and development to find the right ligand and substitution pattern for deepest color, highest efficiency, shortest luminescence decay time and longest device durability.

8.3.2.2 Stable and blue phosphorescent Ir(III) and Pt(II) emitters

In analogy to green and red bis or tris-cyclometalated Ir(III) complexes, there is a broad variety of blue Ir(III) emitters reported in the literature since their discovery almost 20 years ago. Besides *fac*-Ir(ppy)$_3$, the most widely studied Ir(III) emitter is FIrpic [43, 44]. It shows sky-blue emission with a CIE-y coordinate of around 0.3, which does not fulfill the deep-blue color requirements of display applications (a CIE-y coordinate below 0.1 is needed). Nevertheless, it has often been used as a model compound to investigate blue phosphorescent OLEDs [45–48]. Synthetically, it is obtained in a similar fashion as its green and red counterparts by a two-step approach [48, 49]. There is much literature data available on the photophysical properties of FIrpic, a representative collection is listed in Table 8.3.2. Here, a scattering of experimental data can also be observed, although to a lesser extent if compared with the green and red emitter examples given in Table 8.3.1. While FIrpic is brightly luminescent, a deeper blue emission color is required for display applications. Thus, further ligand systems have been investigated over the years, such as phenyl-pyridine derivatives with additional electron-withdrawing substituents [50, 51], or completely different core structures, such as imidazole or carbene-based ligands.

An example of a deep-blue carbene-based Ir(III) emitter is shown in Figure 8.3.5. Due to the asymmetric ligand, isomeric mixtures are obtained in the synthesis of $Ir(cb)_3$. However, all facial isomers have similar photophysical properties [52]. While the synthesis and isolation of the derivative without the *tert*-butyl group is more straightforward, the unsubstituted version gives a rather light-blue emission [53], which does not meet deep-blue color requirements.

Homoleptic Ir(III) complexes with phenyl-imidazole ligands, such as CNImIr, also show bright and deep blue emission [54]. Nevertheless, deep-blue phosphorescence has not been realized in OLED display applications so far. Besides reaching the required deep-blue color with phosphorescent emitters, the most fundamental obstacle to make such phosphorescence fit for industrial OLED applications is their durability. Up to date, blue OLEDs utilizing phosphorescence suffer from short device lifetime [55]. This is mainly related to the higher energy content of the emissive triplet state, which can lead to thermal population of dissociative states promoting the breakup of metal-ligand bonds.

Tackling the challenge of blue OLED durability can be addressed from a device and a material point of view. The intrinsic material stability has been investigated intensively during the past 20 years. Although this topic is still relevant for all colors and all sort of OLED materials, FIrpic is a particularly popular model compound for degradation studies [56–59]. The weakest bonds in OLED materials can be identified by investigating the structures of decomposition products after heavily stressing the materials, or by computational methods calculating bond dissociation energies (BDE). Ligand dissociation reactions such as degradation pathways are a weak point to consider. Therefore, the idea of using multidentate ligands is especially interesting for blue emitters. Pt(II) complexes with tetradentate ligands show bright, and most importantly also deep-blue emission properties. Similar to their green analogues, they have raised an increasing attention in recent years [60, 61]. Especially deep-blue Pt(II) emitters, such as PtON7-dtb [61], are able to show highly favored emission spectra with narrow maxima. Despite sophisticated ligand design for many years, the search for a stable phosphorescent deep-blue OLED emitter is still an ongoing challenge.

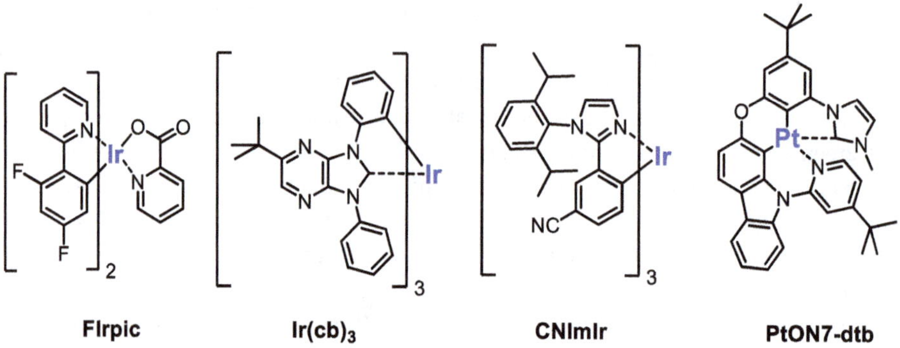

Figure 8.3.5: Overview of commonly utilized Ir(III) and Pt(II) structures in research of blue OLEDs.

Table 8.3.2: Relevant and representative PL properties of blue Ir(III) and Pt(II) emitters.

Complex	Matrix	λ_{max} [nm]	FWHM [nm]/[eV]	Φ_L [%]	τ [µs]	Ref.
FIrpic	DCE	471	57/0.29	89	1.7	[73]
	mCP	476		85	1.2	[73]
	DCM	468	57/0.29			[43]
	DCM	468		62	1.7	[49]
	ACN	471	60/0.31	68		[44]
	DCM	470	58/0.30	83	1.9	[77]
	THF	471		84	1.8	[77]
	PMMA	468		89	1.7	[77]
Ir(cb)$_3$	toluene	456	53/0.30			[78]
	PS (1%)			88	0.5	[78]
	PMMA (2 wt%)	466–468		89–99	2.8–2.9	[52]
CNImIr	toluene	462	75/0.39			[54]
	mCP (5 wt%)			99	1.8	[54]
PtON7	DCM	452	66/0.36	78	4.2	[60]
	PMMA	542		89	4.1	[60]
PtON7-dtb	PMMA (5%)			91	4.7	[61]
	DCM	446	19/0.12			[61]

DCE = dichloroethane; mCP = 1,3-bis(*N*-carbazolyl)benzene; DCM = dichloromethane; ACN = acetonitrile; THF = tetrahydrofuran; PMMA = poly(methyl methacrylate); PS = polystyrene; λ_{max} = maximum of emission spectrum in nm; FWHM = full-width-at-half-maximum of emission spectrum in nm and eV was extracted from the pdf of the respective reference by the digitalization function of OriginPro 2021b;

$\Phi_L = k_r/(k_r + k_{nr})$ = photoluminescence quantum yield;

$\tau = 1/(k_r + k_{nr})$ = observed excited state decay time in µs.

8.3.3 Industrial relevance and next generation approaches

In the previous part, we discussed a series of relevant Ir(III) and Pt(II) complexes that are commonly used as triplet emitters in OLEDs. These compounds are a key ingredient for today's success of OLED displays. Much has been achieved for green and red emitters to improve their performance in devices, such as higher efficiency by making use of emitter orientation (e.g. with heteroleptic complexes), higher color purity (narrowing the emission spectra, e.g. by sterically demanding substituents) and increasing the durability through enhanced chemical stability (e.g. with polydentate ligands).

Nevertheless, to this date, there is no commercial blue OLED that uses a phosphorescent triplet emitter in its EML [62]. This can be explained by the still too short operational lifetime of phosphorescent blue OLED. For the emission of deep blue light (typically around 450 nm), a wide HOMO-LUMO gap of about 3 eV is required. Long exciton lifetimes of the phosphorescent emitters that are in the order of microseconds (compared to nanoseconds for fluorescence emitters) and exciton-polaron interactions can create “hot” excited states that have needlessly high energies [63], lying beyond typical bond dissociation energies, and leading to degradation of the material.

While the search for stable and efficient phosphorescent blue emitters has been ongoing for about 20 years now, there are several alternative materials and technologies which are being developed as a potential solution for the blue range, but which could possibly also find applications in green and red devices. Phosphor-sensitized fluorescence (PSF) has long been known [64], but generated more interest in recent years [53]. It can be regarded as a follow-up technology to classical phosphorescence, even though it still relies on Ir(III) or Pt(II) complexes. So far, most widely investigated alternatives without organometallic compounds are high triplet-triplet annihilation (TTA) host materials for fluorescent emitters [65], and thermally activated delayed fluorescence (TADF) materials [66], as well as related approaches such as hyperfluorescence [62, 67, 68].

All these follow-up and alternative technologies proved to have their advantages and disadvantages, if compared to classical phosphorescence. It is currently vividly discussed among experts which technology might be the most promising one to realize efficient and stable blue OLEDs. Common ground in these discussions is the thesis that the excited state decay time has to be shortened with respect to state-of-the-art phosphorescent decay times [69]. Similarly to efficiency enhancement via either molecular orientation and/or device-engineered outcoupling techniques, reducing the excited state decay time in phosphorescent devices may be done by means of judicious molecular design [70] and/or additionally engineered outcoupling layers in the device [71].

Finally, we can conclude that Ir(III) and Pt(II) complexes are the key compounds that made OLED display applications successful during the past 20 years, and yet research on new materials still remains a challenge in order to realize efficient and stable blue OLEDs, for which organometallic Ir(III) and Pt(II) compounds remain promising candidates.

References

[1] Hong G, Gan X, Leonhardt C, Zhang Z, Seibert J, Busch JM, Bräse S. Adv Mater, 2021, 33, 2005630.
[2] Pope M, Kallmann HP, Magnante P. J Chem Phys, 1963, 38, 2042–43.
[3] Tang CW, VanSlyke SA. Appl Phys Lett, 1987, 51, 913–15.
[4] Cai X, Su S-J. Adv Funct Mater, 2018, 28, 1802558.
[5] Lee KH, Lee JY, Oh HY. Dig Tech Pap SID Int Symp, 2019, 50, 1924–27.
[6] Kim S-Y, Jeong W-I, Mayr C, Park Y-S, Kim K-H, Lee J-H, Moon C-K, Brütting W, Kim J-J. Adv Funct Mater, 2013, 23, 3896–900.
[7] Baldo MA, O'Brien DF, You Y, Shoustikov A, Sibley S, Thompson ME, Forrest SR. Nature, 1998, 395, 151–54.
[8] Wang Y, Yun JH, Wang L, Lee JY. Adv Funct Mater, 2021, 31, 2008332.
[9] Bernanose A, Comte M, Vouaux P. J Chim Phys, 1953, 50, 64–68.
[10] https://en.wikipedia.org/wiki/Sony_XEL-1 accessed on July 22nd 2021.
[11] https://www.oled-info.com/files/samsung-display-2013-analyst-day.pdf accessed on July 22nd 2021.
[12] Baldo MA, Lamansky S, Burrows PE, Thompson ME, Forrest SR. Appl Phys Lett, 1999, 75, 4–6.
[13] Adachi C, Baldo MA, Forrest SR, Thompson ME. Appl Phys Lett, 2000, 77, 904–06.
[14] Dedeian K, Djurovich PI, Garces FO, Carlson G, Watts RJ. Inorg Chem, 1991, 30, 1685–87.
[15] Tamayo AB, Alleyne BD, Djurovich PI, Lamansky S, Tsyba I, Ho NN, Bau R, Thompson ME. JACS, 2003, 125, 7377–87.
[16] Holzer W, Penzkofer A, Tsuboi T. Chem Phys, 2005, 308, 93–102.
[17] King KA, Spellane PJ, Watts RJ. J Am Chem Soc, 1985, 107, 1431–32.
[18] Hofbeck T, Yersin H. Inorg Chem, 2010, 49, 9290–99.
[19] Adachi C, Baldo MA, Thompson ME, Forrest SR. J Appl Phys, 2001, 90, 5048–51.
[20] Lamansky S, Djurovich P, Murphy D, Abdel-Razzaq F, Kwong R, Tsyba I, Bortz M, Mui B, Bau R, Thompson ME. Inorg Chem, 2001, 40, 1704–11.
[21] Coughlin FJ, Westrol MS, Oyler KD, Byrne N, Kraml C, Zysman-Colman E, Lowry MS, Bernhard S. Inorg Chem, 2008, 47, 2039–48.
[22] Lamansky S, Djurovich P, Murphy D, Abdel-Razzaq F, Lee H-E, Adachi C, Burrows PE, Forrest SR, Thompson ME. J Am Chem Soc, 2001, 123, 4304–12.
[23] Lu J, Liu Q, Ding J, Tao Y. Syn Met, 2008, 158, 95–103.
[24] Flämmich M, Frischeisen J, Setz DS, Michaelis D, Krummacher BC, Schmidt TD, Brütting W, Danz N. Org El, 2011, 12, 1663–68.
[25] Schmidt TD, Setz DS, Flämmich M, Frischeisen J, Michaelis D, Krummacher BC, Danz N, Brütting W. Appl Phys Lett, 2011, 99, 163302.
[26] Liehm P, Murawski C, Furno M, Lüssem B, Leo K, Gather MC. Appl Phys Lett, 2012, 101, 253304.
[27] Schmidt TD, Lampe T, Sylvinson MRD, Djurovich PI, Thompson ME, Brütting W. Phys Rev Appl, 2017, 8, 037001.
[28] Duan JP, Sun PP, Cheng CH. Adv Mater, 2003, 15, 224–28.

[29] Kim DH, Cho NS, Oh H-Y, Yang JH, Jeon WS, Park JS, Suh MC, Kwon JH. Adv Mater, 2011, 23, 2721–26.
[30] Lepeltier M, Dumur F, Wantz G, Vila N, Mbomekallé I, Bertin D, Gigmes D, Mayer CR. J Lumi, 2013, 143, 145–49.
[31] Su Y-J, Huang H-L, Li C-L, Chien C-H, Tao Y-T, Chou P-T, Datta S, Liu R-S. Adv Mater, 2003, 15, 884–88.
[32] Li C-L, Su Y-J, Tao Y-T, Chou P-T, Chien C-H, Cheng -C-C, Liu R-S. Adv Funct Mater, 2005, 15, 387–95.
[33] Yang C-H, Tai -C-C, Sun IW. J Mater Chem, 2004, 14, 947–50.
[34] Kim J, Lee KH, Lee SJ, Lee HW, Kim YK, Kim YS, Yoon SS. Chem Eur J, 2016, 22, 4036–45.
[35] Kim K-H, Ahn ES, Huh J-S, Kim Y-H, Kim J-J. Chem Mater, 2016, 28, 7505–10.
[36] Kwong RC, Sibley S, Dubovoy T, Baldo M, Forrest SR, Thompson ME. Chem Mater, 1999, 11, 3709–13.
[37] Nifiatis F, Su W, Haley JE, Slagle JE, Cooper TM. J Phys Chem A, 2011, 115, 13764–72.
[38] Mao M, Peng J, Lam T-L, Ang W-H, Li H, Cheng G, Che C-M. J Mater Chem C, 2019, 7, 7230–36.
[39] Williams JAG. Photochemistry and Photophysics of Coordination Compounds II, Springer, Berlin, Heidelberg, 2007, 205–68.
[40] Brooks J, Babayan Y, Lamansky S, Djurovich PI, Tsyba I, Bau R, Thompson ME. Inorg Chem, 2002, 41, 3055–66.
[41] Rausch AF, Murphy L, Williams JAG, Yersin H. Inorg Chem, 2009, 48, 11407–14.
[42] Fukagawa H, Oono T, Iwasaki Y, Hatakeyama T, Shimizu T. Mater Chem Front, 2018, 2, 704–09.
[43] Baranoff E, Curchod BFE. Dalton Trans, 2015, 44, 8318–29.
[44] Zhou Y, Li W, Liu Y, Zeng L, Su W, Zhou M. Dalton Trans, 2012, 41, 9373–81.
[45] Adachi C, Kwong RC, Djurovich P, Adamovich V, Baldo MA, Thompson ME, Forrest SR. Appl Phys Lett, 2001, 79, 2082–84.
[46] Tokito S, Iijima T, Suzuri Y, Kita H, Tsuzuki T, Sato F. Appl Phys Lett, 2003, 83, 569–71.
[47] Holmes RJ, Forrest SR, Tung Y-J, Kwong RC, Brown JJ, Garon S, Thompson ME. Appl Phys Lett, 2003, 82, 2422–24.
[48] Yeh S-J, Wu M-F, Chen C-T, Song Y-H, Chi Y, Ho M-H, Hsu S-F, Chen CH. Adv Mater, 2005, 17, 285–89.
[49] Baranoff E, Curchod BFE, Monti F, Steimer F, Accorsi G, Tavernelli I, Röthlisberger U, Scopelliti R, Grätzel M, Nazeeruddin MK. Inorg Chem, 2012, 51, 799–811.
[50] Li J, Djurovich PI, Alleyne BD, Yousufuddin M, Ho NN, Thomas JC, Peters JC, Bau R, Thompson ME. Inorg Chem, 2005, 44, 1713–27.
[51] Jeon SO, Jang SE, Son HS, Lee JY. Adv Mater, 2011, 23, 1436–41.
[52] Murer P, Dormann K, Benedito FL, B G, Metz S, Heinemeyer U, Lennartz C, Wagenblast G, Watanabe S, Gessner T. WO15000955A1.
[53] Heimel P, Mondal A, May F, Kowalsky W, Lennartz C, Andrienko D, Lovrincic R. Nat Commun, 2018, 9, 4990.
[54] Kwon Y, Han SH, Yu S, Lee JY, Lee KM. J Mater Chem C, 2018, 6, 4565–72.
[55] Kim S, Bae HJ, Park S, Kim W, Kim J, Kim JS, Jung Y, Sul S, Ihn S-G, Noh C, Kim S, You Y. Nat Commun, 2018, 9, 1211.
[56] Moraes IRD, Scholz S, Lüssem B, Leo K. Org El, 2011, 12, 341–47.
[57] Seifert R, Rabelo de Moraes I, Scholz S, Gather MC, Lüssem B, Leo K. Org El, 2013, 14, 115–23.
[58] Penconi M, Cazzaniga M, Panzeri W, Mele A, Cargnoni F, Ceresoli D, Bossi A. Chem Mater, 2019, 31, 2277–85.
[59] Wang D, Cheng C, Tsuboi T, Zhang Q. CCS Chem, 2020, 2, 1278–96.

[60] Hang X-C, Fleetham T, Turner E, Brooks J, Li J. Angew Chem Int Ed, 2013, 52, 6753–56.
[61] Fleetham T, Li G, Wen L, Li J. Adv Mater, 2014, 26, 7116–21.
[62] Lee J-H, Chen C-H, Lee P-H, Lin H-Y, Leung M-K, Chiu T-L, Lin C-F. J Mater Chem C, 2019, 7, 5874–88.
[63] Lee J, Jeong C, Batagoda T, Coburn C, Thompson ME, Forrest SR. Nat Commun, 2017, 8, 15566.
[64] D'Andrade BW, Baldo MA, Adachi C, Brooks J, Thompson ME, Forrest SR. Appl Phys Lett, 2001, 79, 1045–47.
[65] Kukhta NA, Matulaitis T, Volyniuk D, Ivaniuk K, Turyk P, Stakhira P, Grazulevicius JV, Monkman AP. J Phys Chem Lett, 2017, 8, 6199–205.
[66] Uoyama H, Goushi K, Shizu K, Nomura H, Adachi C. Nature, 2012, 492, 234–38.
[67] Chan C-Y, Tanaka M, Lee Y-T, Wong Y-W, Nakanotani H, Hatakeyama T, Adachi C. Nat Photon, 2021, 15, 203–07.
[68] Nakanotani H, Higuchi T, Furukawa T, Masui K, Morimoto K, Numata M, Tanaka H, Sagara Y, Yasuda T, Adachi C. Nat Commun, 2014, 5, 4016.
[69] https://www.youtube.com/watch?v=o3jiwMP_xdU accessed on July 22nd 2021.
[70] Paterson L, May F, Andrienko D. J Appl Phys, 2020, 128, 160901.
[71] Fusella MA, Saramak R, Bushati R, Menon VM, Weaver MS, Thompson NJ, Brown JJ. Nature, 2020, 585, 379–82.
[72] Sajoto T, Djurovich PI, Tamayo AB, Oxgaard J, Goddard WA, Thompson ME. JACS, 2009, 131, 9813–22.
[73] Endo A, Suzuki K, Yoshihara T, Tobita S, Yahiro M, Adachi C. Chem Phys Lett, 2008, 460, 155–57.
[74] Edkins RM, Wriglesworth A, Fucke K, Bettington SL, Beeby A. Dalton Trans, 2011, 40, 9672–78.
[75] McDonald AR, Lutz M, von Chrzanowski LS, van Klink GP, Spek AL, van Koten G. Inorg Chem, 2008, 47, 6681–91.
[76] Vezzu DAK, Deaton JC, Jones JS, Bartolotti L, Harris CF, Marchetti AP, Kondakova M, Pike RD, Huo S. Inorg Chem, 2010, 49, 5107–19.
[77] Rausch AF, Thompson ME, Yersin H. The J Phys Chem A, 2009, 113, 5927–32.
[78] Maheshwaran A, Sree VG, Park H-Y, Kim H, Han SH, Lee JY, Jin S-H. Adv Funct Mater, 2018, 28, 1802945.

8.4 Luminescent thermometry materials

Miroslav Dramicanin

Luminescence thermometry is a semi-invasive method for temperature measurements [1]. In this method, a luminescent probe material is placed in a contact with an object where the temperature should be measured. After the probe and object reach thermal equilibrium, the temperature is determined from temperature-induced changes in the luminescent features of the probe. These changes are the fluctuations in emission intensity, the shape of the emission spectrum, spectral position and width of the emission band and a lifetime of the excited state, as illustrated in Figure 8.4.1. Temperature readings from changes in the shape of the emission spectrum and excited state lifetimes are self-calibrated methods, meaning that they do not require calibration with a reference, and are the preferable choice. Luminescence thermometry is used for many applications where conventional thermometry is not reliable or completely fails, such as, for example, for temperature measurements at the nanoscale, in biomedicine and in optoelectronics [2]. It is useful also for measurements in harsh environments, such as high-energy radiation fields [3], and on rotating or moving objects.

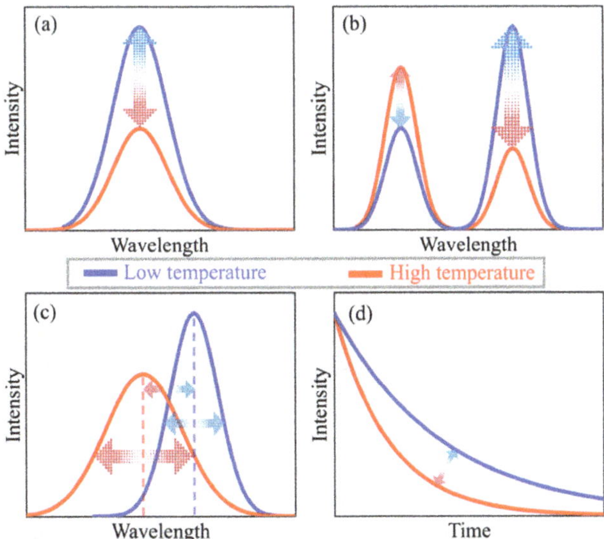

Figure 8.4.1: Changes in luminescence of materials that can be used for measurements of temperature: (a) emission intensity, (b) shape of the emission spectrum, (c) spectral position and width of the emission band and (d) a lifetime of the excited state.

https://doi.org/10.1515/9783110798890-023

Since luminescence of all materials depends on temperature, there are endless possibilities in selecting a material for the luminescence thermometry probe. The selection is usually done considering application requirements and the measuring environment. As a rule of thumb, the probe material should be chemically stable and have strong sensitivity of luminescence on temperature (>1%/K) in the measurement range of interest. So far, lanthanide and transition metal ion-activated phosphors and semiconductor quantum dots have been the most frequently selected materials for luminescence thermometry probes. Other materials include organic dyes, metal-organic frameworks, luminescent polymers and carbon materials.

8.4.1 Phosphors activated with trivalent lanthanide ions

Lanthanide luminescence centers in solids are created from the triply or doubly charged lanthanide ions. These two types of lanthanide centers provide distinct luminescence appearances, even though they are formed from the same elements. Luminescence of trivalent lanthanide ions (Ln^{3+}) mainly occurs from 4f-4f electronic transitions, and it is characterized by narrow emission bands and long excited state lifetimes (>μs). The abundance of $4f^n$ energy levels and transitions between them provides emissions that cover a vast energy range of about 40000 cm^{-1}, from UV through VIS into NIR. Ln^{3+}-activated phosphors generally show excellent chemical stabilities and can be prepared in many forms: micro and nano-powders, crystals, composites, glasses, glass ceramics, thin films and coatings. They are also the class of materials that up to now has received, by far, the greatest attention in luminescence thermometry [4]. Transition energies of Ln^{3+} centers show weak sensitivity on temperature fluctuations (<0.1 cm^{-1}/K), so the thermally induced shifts in the energy of their emission bands are rarely exploited for thermometry. The same stands for thermally induced changes in Ln^{3+} emission bandwidths.

When using Ln^{3+}-activated phosphors, the temperature is determined either from the ratio of emission intensities that commence from the adjacent thermally coupled excited levels (LIR) or from the excited state lifetime. In both cases, one can exploit both downshifting and upconversion emissions, as illustrated in Figure 8.4.2 for Dy^{3+} and Yb^{3+}/Er^{3+} LIRs, respectively.

The population of the higher energy excited state increases with temperature, in accordance with the Boltzmann statistical distribution when two excited states are in thermal equilibrium. Then, the ratio of emission intensities from thermally coupled levels is expressed as

$$LIR(T) = B \cdot \exp\left(-\frac{\Delta E}{kT}\right),$$

where B is a constant, ΔE is the energy difference between thermally coupled levels, k is the Boltzmann's constant and T is the temperature. The LIR sensitivity to

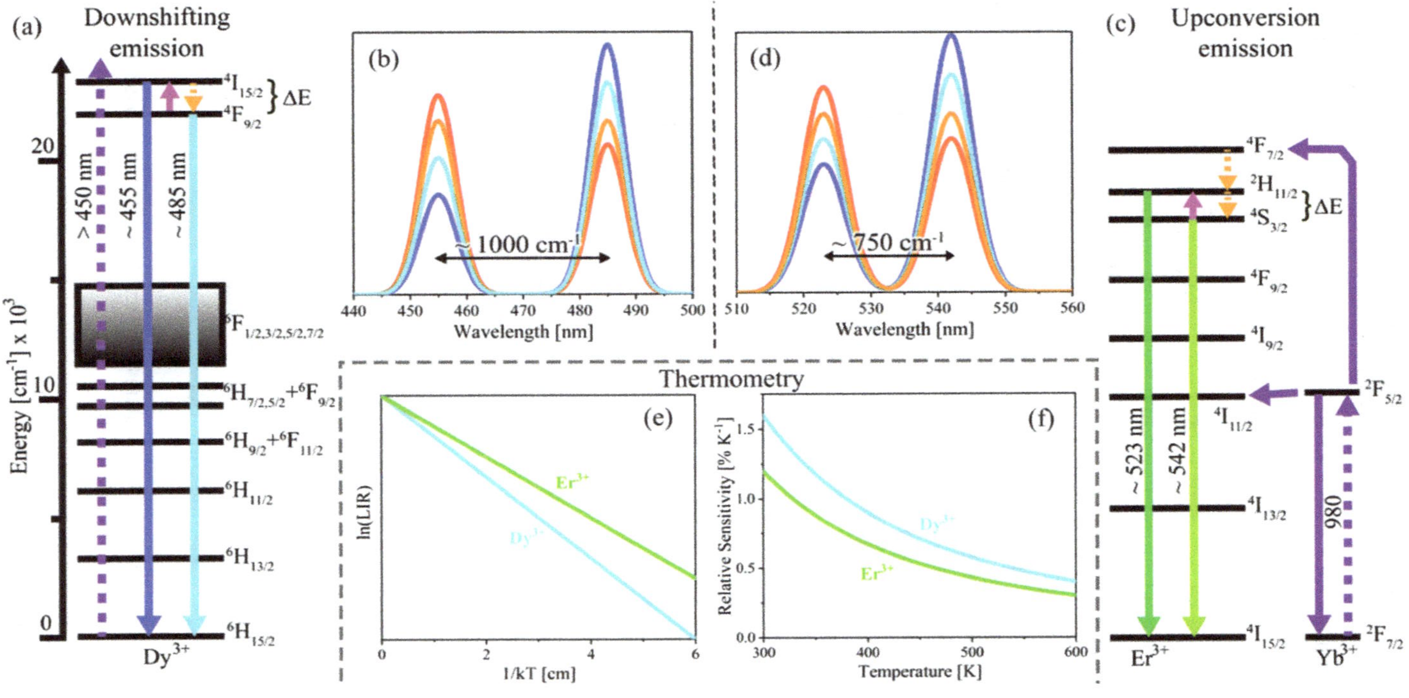

Figure 8.4.2: Energy level diagrams of Dy^{3+} (a) and Yb^{3+}/Er^{3+} (c), changes of their respective emission intensities with temperature (b, d), dependence of ratios of emission intensities (LIRs) with temperature (e) and LIRs sensitivities on temperature (f).

temperature changes is greater for larger ΔE values (the largest difference of approximately 1750 cm^{-1} is between Eu^{3+} $^5D_0$ and $^5D_1$ excited states) and decreases with an increase in temperature. Emission and thermometric properties of Ln^{3+} centers are easily estimated using the famous Judd-Ofelt theory [5].

Measurements of excited state lifetimes are not affected by the concentration and distribution of probe material and are not influenced by fluctuations in excitation power. For these reasons, lifetime-based luminescence thermometry can be considered a robust method. It is a convenient technique suitable for luminescence thermal imaging and mapping, since it requires monitoring of just one emission band. On the other hand, this technique requires more complex equipment and provides smaller temporal resolution than intensity-based LIR thermometry. When used with Ln^{3+} probes, the major problem of this technique is its measurement range. A Ln^{3+} excited state lifetime shows negligible change with an increase in temperature over a wide range, and starts to decrease only at very high temperatures, as shown in Figure 8.4.3 for several Eu^{3+}-activated phosphors. For this reason, the method cannot be used in a physiologically relevant temperature range (from 20 °C to 50 °C), thus, it is not useful for most biomedical applications. However, the technique is appropriate for high-temperature thermometry.

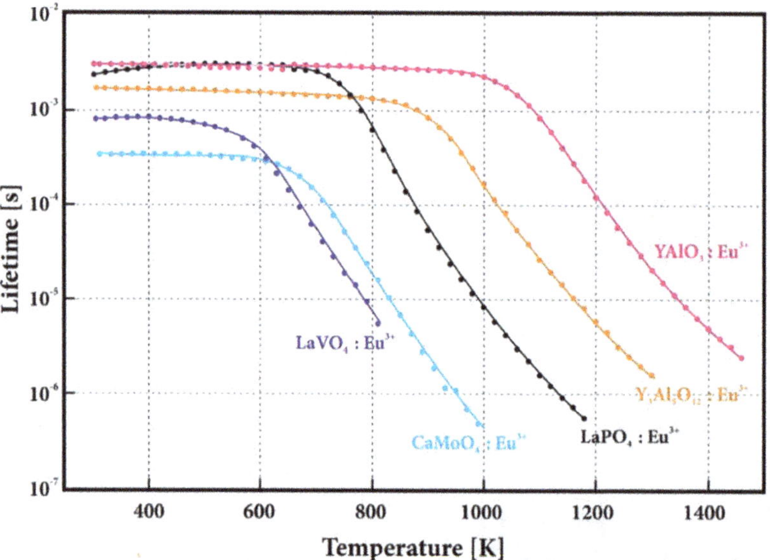

Figure 8.4.3: Changes in lifetime values with temperature for several Eu^{3+}-activated phosphors (Adapted from Ref. [6], with permission from Elsevier).

LIR thermometry with Ln^{3+}-activated phosphors are used for the thermal mapping of surfaces with a high spatial resolution in a scanning thermal microscope (SThM)

[7]. In this instrument, the luminescent temperature probe is attached to the end of the tip of a scanning probe microscope (SPM) and the measuring object is placed on a piezoelectric scanning stage, Figure 8.4.4. During scanning, the luminescent probe that contains Yb^{3+} and Er^{3+} ions is excited with a 980 nm laser, the upconversion radiation from the tip is passed through an objective, then divided into two beams by a beam splitter, and finally passed to two photodetectors through interference filters (one centered at 520 nm and a second at 550 nm). After the scan, a thermal image is obtained as the ratio of respective pixel intensities in 520 nm and 550 nm images.

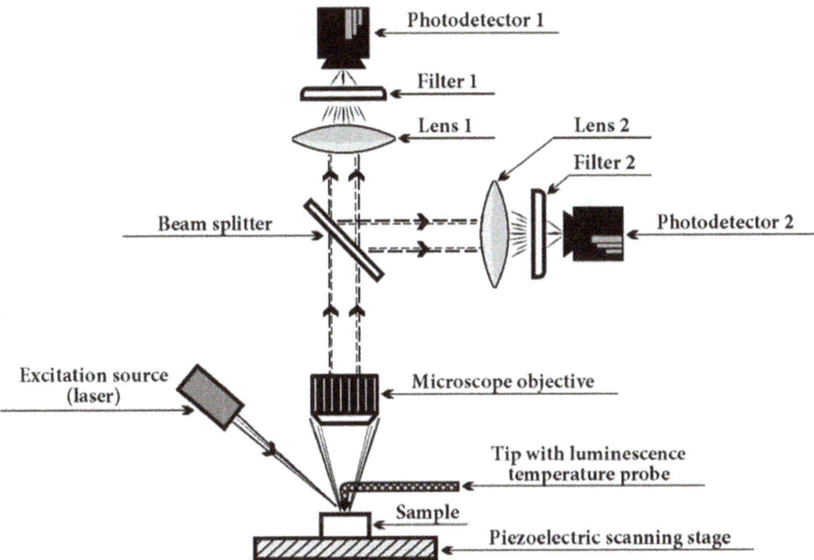

Figure 8.4.4: Schematic of the luminescence scanning thermal microscope (SThM) (reprinted from Ref. [1], with permission from Elsevier).

Biomedical subcutaneous in vivo thermal imaging is difficult to perform with thermographic cameras since they record only tissues' surface temperature. However, it can be done by using luminescence thermometry with near-infrared-emitting nanoparticles, whose emission is strongly temperature-dependent in the physiological range of temperatures (20–50 °C). Er^{3+}, Tm^{3+} and Yb^{3+}-activated core-shell LaF_3 nanoparticles emit in the near-infrared spectral range under 690 nm excitation with emission bands centered at 1000, 1230 and 1550 nm, due to electronic f–f transitions in Yb^{3+}, Tm^{3+} and Er^{3+} ions, respectively [8]. The ratio between emission intensity at 1000 nm (Yb^{3+} emission) and 1230 nm (Tm^{3+} emission) is highly sensitive to temperature changes, and it is suitable for the in vivo acquisition of two-dimensional thermal images at the subcutaneous level. The thermal subcutaneous imaging during laser heating of the mouse is illustrated in Figure 8.4.5 [8]. It involves the injection of a phosphate-buffered saline dispersion of nanoparticles into a mouse, excitation of the nanoparticles at 690 nm laser and the collection of NIR emission through two bandpass filters (1000 and 1200 nm) placed in a motorized wheel (with a 0.3 Hz switching frequency).

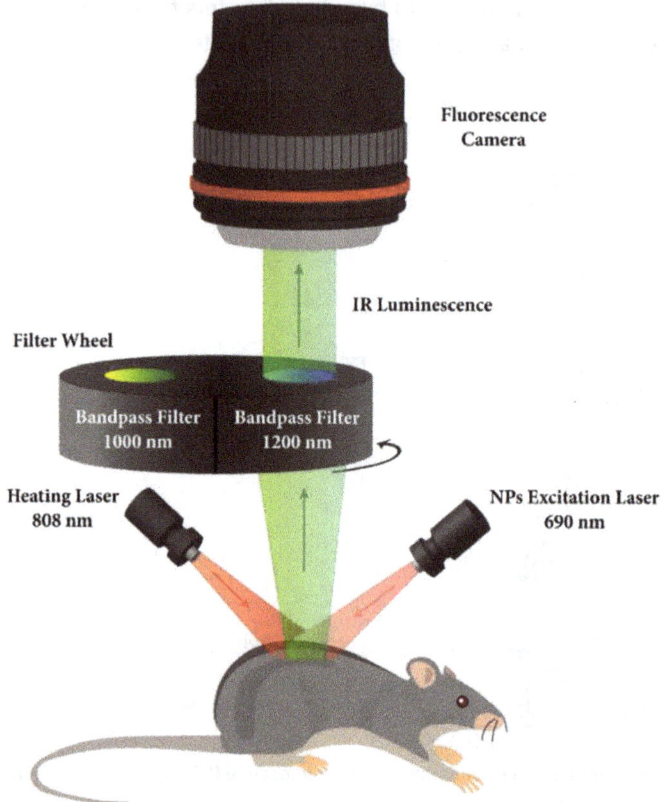

Figure 8.4.5: The two-dimensional subcutaneous in vivo thermal imaging on a mouse, acquired from the ratio of intensities of emissions at 1000 and 1230 nm from Er/Yb@Yb/Tm LaF_3 core-shell nanoparticles. (Reprinted from Ref. [8], with permission from Wiley.).

Ln^{3+}-activated phosphors can be used to give forensic evidence of the temperatures that they have experienced during thermal events [9]. In such cases, they act as off-line thermometers, i.e. thermal history sensors, usually in high-temperature environments for the development and safe use of chemical reactors, engines and gas turbines. Furthermore, off-line temperature measurements are sometimes the only option to assess the temperature distribution on an object placed in harsh environments, such as during a laser annealing or an explosion. In recent years, thermal history sensors have been used also for monitoring the integrity of pharmaceutical products, ensuring the quality of foods and beverages, and monitoring the degradation of consumer electronics and batteries. The operation of these sensors is based on an irreversible and repeatable change of sensor properties with temperature, most frequently the phase transformation, crystallization of phosphor, diffusion of activator/sensitizer ions or emission killer center ions as well as oxidation of activator ions.

8.4.2 Phosphors activated with divalent lanthanide ions

In some host materials that have divalent constituent ions, Sm^{2+}, Eu^{2+}, Tm^{2+} and Yb^{2+} are stable and can luminesce. These luminescent centers provide a broadband emission due to 5d → 4f transitions with a wavelength of the emission band that significantly depends on the host material, and a line emission due to 4f → 4f transitions. The energy difference between 5d and 4f levels is relatively high, much higher than the difference between adjacent 4f levels in trivalent lanthanide centers. For this reason, luminescence thermometry based on the temperature dependence of the ratio of intensities between 5d → 4f and 4f → 4f transitions is more sensitive to temperature fluctuations than conventional LIR with Ln^{3+} centers. However, thermal coupling between d and f centered configurations is only possible when there is a strong electron-phonon coupling.

Thermal image of operating printed circuit boards (PCB) can be acquired using fine Sm^{2+}-doped SrB_4O_7 luminescent powder and a luminescence microscope [10], as shown in Figure 8.4.6. To make a thermal image, the part of the operating PCB is covered with the powder, illuminated with 365 nm radiation, and the luminescence is recorded with two monochrome CMOS cameras after splitting the emission beam and passing one of them through a short-wave pass-filter (590 nm), and the second through a long-wave pass-filter (650 nm). The thermal image is then created by rationing the respective pixel intensities of two images and calculating the temperature with calibration data. These two emission windows correspond to the 5d → 4f and 4f → 4f transitions, respectively.

8.4.3 Phosphors activated with transition metal ions

Phosphors activated by Cr^{3+} (in a strong crystal field) and Mn^{4+} are frequently used as luminescence thermometry probes both in LIR and in lifetime-based measurements. Thermally induced shifts in the energy of their emission bands and bandwidths can be exploited for thermometry since they are much larger than in Ln^{3+}. These materials are efficient emitters in the deep-red and NIR spectral regions, which make them attractive for many applications, for example, in biomedicine [11] due to low interference between biological media and radiation of long wavelengths. In the LIR approach, the temperature can be determined from the ratio of emission intensities from the $^4T_2 \rightarrow {}^4A_2$ and $^2E \rightarrow {}^4A_2$ transitions in Mn^{4+} and Cr^{3+},

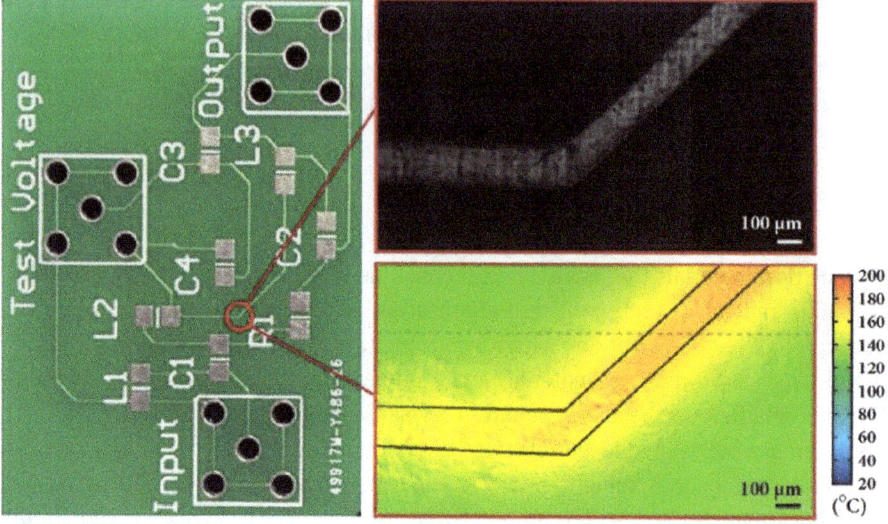

Figure 8.4.6: (left) Image of the printed circuit board with a *circle* marking the imaged area; (upper right) the optical monochrome image of the thermally imaged area; (lower right) thermal image of the marked circuit area operating with a DC current of 5.3 A; the scale bars represent 100 µm. (Reprinted from Ref. [10] under a Creative Commons Attribution 4.0 license.).

Figure 8.4.7. However, these emissions overlap, and data processing requires deconvolution of the emission spectrum [12].

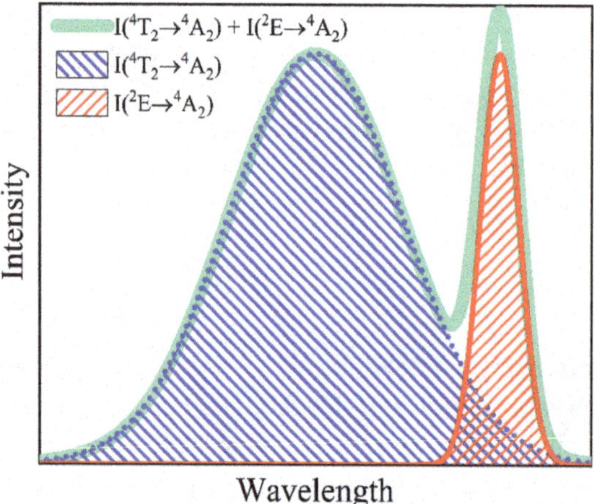

Figure 8.4.7: Overlapping emissions from the $^4T_2 \rightarrow {}^4A_2$ and $^2E \rightarrow {}^4A_2$ transitions in Cr^{3+} and Mn^{4+} used for measurements of temperature.

Both Cr^{3+} and Mn^{4+} have long excited state lifetimes, ranging from a few hundred µs to a few ms, depending on the host material, and are strongly affected by temperature variations. Temperature quenching of their emission occurs by a thermally-activated crossover through the $^4T_2$ level or by thermally-activated auto-ionization in the case of hosts with small energy band gaps. These processes lead to temperature-sensitive lifetimes in the 0.5–3%/K range. As a rule of thumb, in phosphors with a small $^4T_2$ energy, the lifetime sensitivity on temperature is large. In recent times, Ti^{3+} luminescence centers have been demonstrated as efficient probes for both LIR and lifetime luminescence thermometry [13]. The ^2E level lifetime from Ti^{3+} centers in $SrTiO_3$ shows a sensitivity to temperature as high as 5.5%/K.

8.4.4 Phosphors with two luminescent activator ions

The main limitation for applications of LIR and lifetime-based luminescent thermometers is their relatively low sensitivity on temperature, which affects the temperature resolution of measurements. To overcome the sensitivity limitation, it is useful to use probes that contain two types of luminescence centers that have different sensitivity on temperature. They can be sought among different combinations of Ln^{3+} and transition metal centers. These may be co-doped into a single host material or separated into different hosts comprising a mixture. With such probes, the LIR thermometry usually utilizes transition metal centers for rapidly decaying emissions and Ln^{3+} centers for temperature-independent or slow-changing emissions. Some exemplary dual-activated probes are a $Ho^{3+}:Y_2O_3$ and $Mn^{4+}:Mg_2TiO_4$ binary mixture, $Dy^{3+}/Mn^{4+}:BaLaMgNbO_6$, $Er^{3+}/Ni^{2+}:SrTiO_3$, $Nd^{3+}/Cr^{3+}:Gd_5Al_{5-x}Ga_xO_{12}$, $Ce^{3+}/Mn^{4+}:Lu_3Al_5O_{12}$, $Eu^{3+}/Mn^{4+}:Y_3Al_5O_{12}$ and $Eu^{3+}:GdVO_4$ + $Cr^{3+}:Al_2O_3$ hybrid particles.

A high sensitivity of luminescence thermometry can be achieved with several probes activated by two Ln^{3+} centers (for example, $Nd^{3+}/Eu^{3+}:YVO_4$ and $Ce^{3+}/Tb^{3+}:YBO_3$ phosphors) and a combination of Ln^{2+} and Ln^{3+} centers, such as $Eu^{2+}/Eu^{3+}:Sc_2O_3$ nanoparticles [14]. Different quenching rates for Eu^{2+} and Eu^{3+} emissions by temperature is easily observed from changes in the color of $Eu^{2+}/Eu^{3+}:Sc_2O_3$ nanoparticles in Figure 8.4.8. The use of defect emissions from the host material and Ln^{3+} luminescence is another possibility for the creation of highly sensitive thermometry probes. The typical examples are anatase TiO_2 nanoparticles doped with Eu^{3+} or Sm^{3+}.

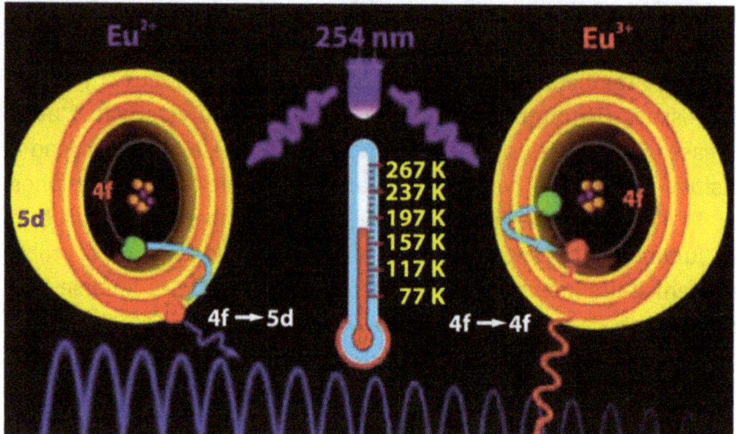

Figure 8.4.8: $Eu^{2+}/Eu^{3+}:Sc_2O_3$ luminescence thermometry nanoprobes (reprinted from Ref. [14], with permission from Wiley).

8.4.5 Semiconductor quantum dots

In addition to the well-known set of the size-dependent properties of semiconductor quantum dots (QDs), their luminescence also displays sensitivity to temperature, which is dependent on nanoparticle size. This feature boosted the use of QDs in luminescence thermometry, especially at nanoscale and in biomedical environments. Temperature fluctuations affect QDs emission intensity, the spectral position of the emission peak, the width of the band and the lifetime. All these luminescence features usually change linearly with temperature, which simplifies calibration of luminescence thermometry sensors and provides an invariable sensitivity across the complete measurement range. The extent of temperature-induced changes in luminescence is different for different types of QDs, it depends on their size and surface chemistry. The emission intensity of QDs demonstrates the strongest variation with temperature changes (it rapidly decreases with temperature increase, i.e. emission quenching). However, in some cases, the emission of QDs may exhibit an anti-quenching effect, i.e. the increase in the intensity of emission with a rise in temperature [15].

The sensitivity to temperature of QDs emission maxima can be either positive or negative. The emission peak redshifts with a negative temperature sensitivity in cases of QDs with a strong confinement of electrons and holes, while the QDs emission peaks show blueshift with temperature increase when there is a weak electron and hole confinement. For example, in PbS QDs with a diameter <5 nm, the emission peak redshifts with a rise in temperature, but for larger dots (diameter >5 nm) the blueshift occurs and becomes more pronounced with an increase in dot size.

Highly sensitive luminescence thermometry probes can be constructed by doping QDs with Ln^{3+} or transition metals (for example, Eu^{3+}-doped ZnS, Mn^{2+}-doped ZnSe or Mn^{2+}-doped $CsPbCl_3$). In such probes, intensities and shapes of spectra are determined by the QDs' exciton energy, energies of excited states from impurities and the energy difference between them; see Figure 8.4.9.

For luminescence thermometry, the interest focusses on in QDs where impurity levels are located within the semiconductor energy gap, close to exciton energy states (Figure 8.4.9b; illustrated for the Mn^{2+} $^4T_1$ level). Then, the excited populations are shared between excitonic states and impurity energy levels. The ratio of intensities of the exciton recombination emission and impurity emission is exceptionally sensitive to temperature fluctuations and can be exploited for thermometry.

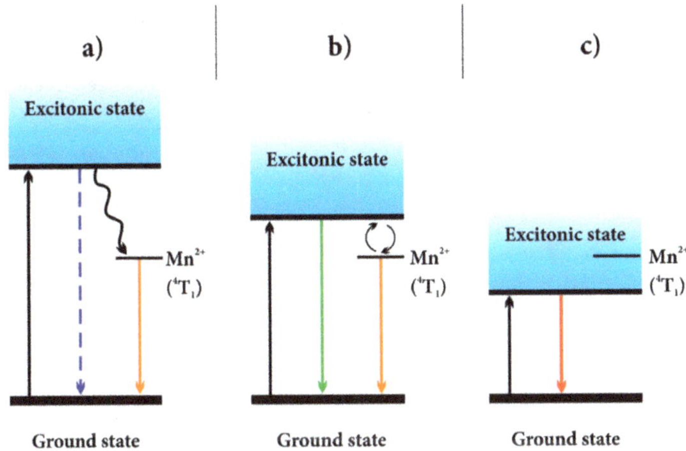

Figure 8.4.9: Energy levels, configurations and optical transitions for Mn^{2+}-doped QDs: the Mn^{2+}-centered $^4T_1$ level (A) inside the energy gap, (B) close to the exciton energy and (C) outside the energy gap. (Reprinted from Ref. [1], with permission from Elsevier.).

References

[1] Dramićanin MD. Luminescence Thermometry, Methods, Materials, and Applications, Elsevier, Woodhead Publishing, Sawston, United Kingdom, 2018. ISBN 978-0-08-102029-6

[2] Carlos LD, Palacio F. (Eds.) Thermometry at the Nanoscale: Techniques and Selected Applications, Royal Society of Chemistry, United Kingdom, Cambridge, 2016. ISBN 978-1-84973-904-7.

[3] Antić Ž, Dramićanin MD, Prashanthi K, Jovanović D, Kuzman S, Thundat T. Adv Mater, 2016, 28, 7745–52. DOI: 10.1002/adma.201601176.

[4] Dramićanin MD. Methods Appl Fluoresc, 2016, 4(042001), 23. DOI: 10.1088/2050-6120/4/4/04200.

[5] Ćirić A, Stojadinović S, Dramićanin MD. J Lumines, 2019, 216(116749), 8p. DOI: 10.1016/j.jlumin.2019.116749.

[6] Brübach J, Pflitsch C, Dreizler A, Atakan B. Progr Energy Combustion Sci, 2013, 39, 37–60. DOI: 10.1016/j.pecs.2012.06.001.

[7] Aigouy L, De Wilde Y, Mortier M, Giérak J, Bourhis E. Appl Optics, 2004, 43, 3829–37. DOI: 10.1364/AO.43.003829.

[8] Ximendes EC, Rocha U, Sales TO, Fernández N, Sanz-Rodríguez F, Martín IR, Jacinto C, Jaque D. Adv Funct Mater, 2017, 27(1702249), 10. DOI: 10.1002/adfm.201702249.

[9] Dramićanin MD. J Appl Phys, 2020, 128(040902), 18. DOI: 10.1063/5.0014825.

[10] Xiong J, Zhao M, Han X, Cao Z, Wei X, Chen Y, Duan C, Yin M. Sci Rep, 2017, 7(41311), 8. DOI: 10.1038/srep41311.

[11] Bednarkiewicz A, Marciniak L, Carlos LD, Jaque D. Nanoscale, 2020, 12, 14405–21. DOI: 10.1039/D0NR03568H.

[12] Ćirić A, Ristić Z, Antić Ž, Dramićanin MD. Physica B: Condens Matter, 2022, 624(413454), 6. DOI: 10.1016/j.physb.2021.413454.

[13] Piotrowski W, Kuchowicz M, Dramićanin M, Marciniak L. Chem Eng J, 2021, 428(131165), 7. DOI: 10.1016/j.cej.2021.131165.

[14] Pan Y, Xie X, Huang Q, Gao C, Wang Y, Wang L, Yang B, Su H, Huang L, Huang W. Adv Mater, 2018, 30(1705256), 6. DOI: 10.1002/adma.201705256.

[15] Wuister SF, van Houselt A, de Mello Donegá C, Vanmaekelbergh D, Meijerink A. Angew Chem Int Ed, 2004, 43, 3029–33. DOI: 10.1002/anie.200353532.

8.5 Crystals for solid-state lasers

Christian Kränkel

8.5.1 Introduction

Lasers based on solid-state gain materials can be roughly classified into three groups: semiconductor-based lasers, fiber lasers and crystal lasers. Only the last are typically referred to as solid-state lasers. They usually utilize single crystalline host materials, which are optically inactive in the region of interest and doped with few per mill to few percent of an optically active ion, in most cases from the transition metal or rare-earth group. In recent years, also polycrystalline host materials, so-called ceramic laser materials, raised the interest of researchers. This chapter is focused on single crystals, but many of the considerations made here are valid for ceramics, too. The brevity of this chapter requires to present several correlations without supplying proper physical or chemical background and some are just useful rule-of-thumb guidelines.

Figure 8.5.1 highlights the multitude of parameters to be considered for the proper choice of a host crystal for a solid-state laser application. In most cases compromises have to be made, because some desirable properties are even contradictory. As an example, a large gain bandwidth required for the generation of ultrashort laser pulses often goes along with a disordered structure, which is detrimental for the

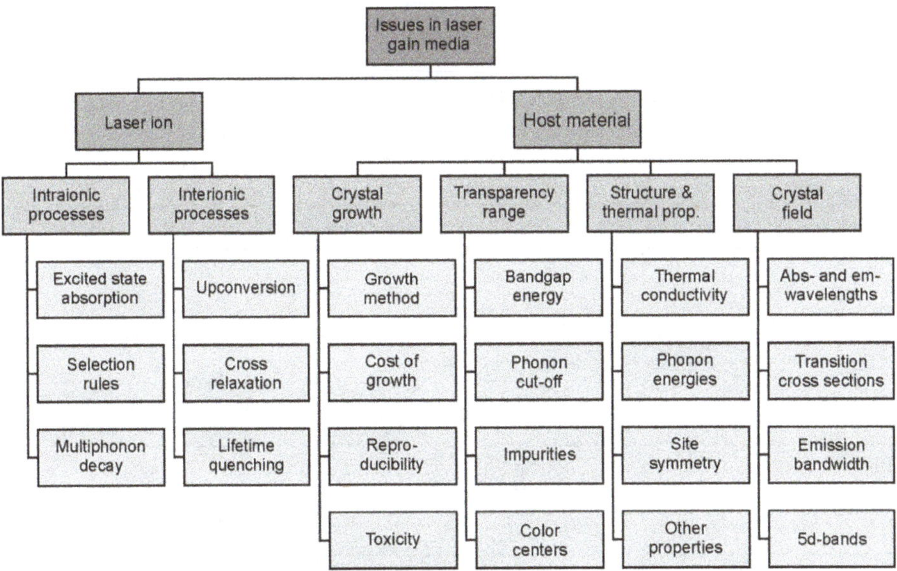

Figure 8.5.1: Aspects and properties to be considered for the choice of a suitable laser host crystal.

https://doi.org/10.1515/9783110798890-024

thermal conductivity. Thus, when targeting high average output powers in ultrashort pulses, there is always a trade-off between highest thermal conductivity and largest gain bandwidth.

In the following, several of the parameters listed in Figure 8.5.1 and their relevance will be introduced briefly, before some aspects on crystal growth of laser materials are highlighted. Finally, the most common host crystals for solid-state lasers with their advantages and disadvantages are introduced.

8.5.2 Crystal properties relevant for solid-state lasers

In this section, the most relevant properties for the choice of a proper host material for a rare-earth-doped laser are introduced. Some of the parameters may appear obvious. However, in order to get a complete picture for the interrelations between important parameters, it is essential to mention them as well.

8.5.2.1 The doping ion site

With only a few exceptions, most solid-state lasers are based on intra-shell transitions. For rare-earth laser active ions these transitions take place within the $4f$ shell, while for transition metals the $3d$ shell is relevant. According to the Laporte rule, such transitions are parity forbidden and an inversion-center-free site symmetry is required to promote relevant transition cross sections for active ions incorporated on this site. The 11 Laue groups among the 32 point groups exhibit inversion centers and are thus not suited for the incorporation of laser active ions (Table 8.5.1).

Table 8.5.1: Overview of the 32 crystallographic point groups in Hermann-Mauguin (Schoenflies) notation. Dark grey background indicates sites without inversion symmetry, light grey background indicates the presence of inversion symmetry, making these sites unsuited for laser ions.

Symmetry	Point group						
Triclinic	1 (C_1)	$\bar{1}$ (C_i)					
Monoclinic	2 (C_2)	m (C_s)	$2/m$ (C_{2h})				
Orthorhombic	222 (D_2)	$mm2$ (C_{2v})	mmm (D_{2h})				
Tetragonal	4 (C_4)	$\bar{4}$ (S_4)	$4/m$ (C_{4h})	422 (D_4)	$4mm$ (C_{4v})	$\bar{4}2m$ (D_{2d})	$4/mmm$ (D_{4h})
Trigonal	3 (C_3)	$\bar{3}$ (C_{3i})	32 (D_3)	$3m$ (C_{3v})	$\bar{3}m$ (D_{3d})		
Hexagonal	6 (C_6)	$\bar{6}$ (C_{6h})	$6/m$ (C_{6h})	622 (D_6)	$6mm$ (C_{6v})	$\bar{6}m2$ (D_{3h})	$6/mmm$ (D_{6h})
Cubic	23 (T)	$m\bar{3}$ (T_h)	432 (O)	$\bar{4}3m$ (T_d)	$m\bar{3}m$ (O_h)		

For the actual incorporation of a dopant into the structure, the charge and the size of the dopant are relevant. Most rare earths ions are only incorporated in their trivalent state, which is usually desired for laser applications. However, Ce^{3+} with only one electron in the 4*f* shell and Tb^{3+} with a half-plus-one filled 4*f* shell are prone to be incorporated in the tetravalent state. In turn, ytterbium, located at the end of the lanthanide row, is sometimes found to be incorporated in the divalent state to completely fill its 4*f* shell; also samarium exists as a divalent ion. The situation is even more complicated for transition metal ions, many of which can exist in various oxidation states from II to VII.

Both criteria, size and charge, have to be considered when selecting a suitable host material to avoid the incorporation in an unwanted oxidation state or a strong segregation of the doping ion in the host crystal. Even possible clustering of the doping ions must be considered when rare-earth ions are incorporated on divalent sites, e.g. in CaF_2 or SrF_2 [1]. To avoid this, monovalent co-doping (mostly by Na^+) is often applied.

The ionic radius of trivalent rare-earth ions is on the order of 1 Å, while the ionic radius of transition metals varies strongly between ~0.3 Å and ~0.9 Å depending on element and oxidation state. Further details can be found in Table 8.5.2. From this aspect, it is straightforward to assume that a rare-earth doping ion replacing another rare-earth ion in the host crystal is a good choice. However, most rare-earth ions are optically active and their energy levels are likely to interact with the dopants'. Only Sc^{3+}, Y^{3+} with their empty 4*f* shell and Lu^{3+} with its completely filled 4*f* shell are usually good choices for cations to be replaced in the host – though the effective radius of Sc^{3+} of 0.87 Å is somewhat smaller than for the optically active rare-earth ions (Table 8.5.2), possibly leading to low segregation coefficients. Gd^{3+} is another possible choice, as its half-filled 4*f* shell is strongly bound and the lowest excited states are found in the ultraviolet (UV) with a low probability to interact with visible or infrared lasers.

Regarding the size of a site, the coordination number of a doping ion site can also serve as a rough guideline. As an example, fourfold coordinated sites with tetrahedral coordination spheres are too small to incorporate rare-earth ions. Figure 8.5.2 shows which sites are occupied by common doping ions for the example of the popular laser host material YAG ($Y_3Al_5O_{12}$). Due to the lanthanide contraction, lighter rare-earth ions including e.g. Nd^{3+} exhibit a larger ionic radius and can thus only incorporated in small amounts, typically below 2%, into YAG, whereas from a crystallographic point of view, the growth of YbAG, i.e. YAG doped with 100% of Yb^{3+}, is possible.

In most cases, there is an upper limit for the doping ion concentration useful for efficient laser operation. Electric dipole-dipole interaction probabilities follow a strong R^{-6} dependency on the distance R. Thus, with increasing doping level and reduced average distance of the doping ions, a quite drastic onset of – usually detrimental – processes such as cross-relaxation and upconversion can be observed. To increase the doping limit for the onset of such processes, host materials with a larger distance between the doping ion sites are useful in some cases.

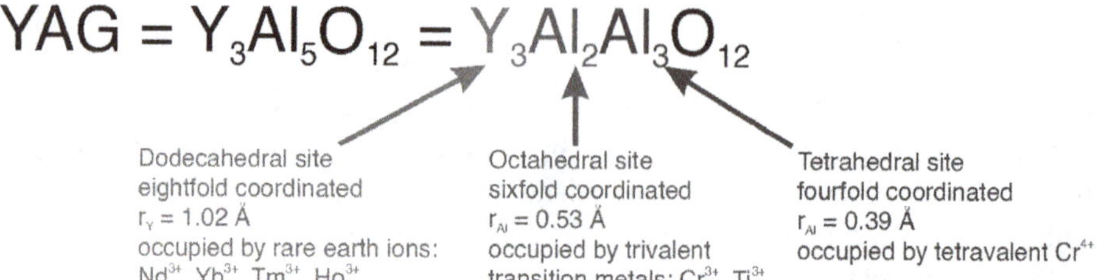

Figure 8.5.2: Preferred doping sites of rare-earth and transition metal ions for the example of yttrium aluminum garnet (YAG).

Table 8.5.2: Effective ionic radii and masses of typical doping ions in oxide *[fluoride]* configuration with data from [2]. Values in (parentheses) are from [3] for halide and chalcogenide configurations. In [3] also a value of 1.27 Å for Nd^{3+} in twelvefold coordination is found. "–" indicates that the ion is usually not found in the respective coordination. For missing values, no data are available. Diameters of respective cations in the host structure are found in Table 8.5.3 a-d (*vide infra*).

Coordination / Ion	Fourfold	Sixfold	Eightfold	Ninefold	Atomic mass [u]
	Ionic radius in Å				
Pr^{3+}	–	1.00 [*1.14*]	1.14 [*1.28*]	(1.18)	140.9
Nd^{3+}	–	0.98 [*1.14*]	1.12 [*1.26*]	1.09 [*1.23*]	144.2
Tb^{3+}	–	0.92 [*1.08*]	1.04 [*1.18*]		158.9
Ho^{3+}	–	0.90 [*1.04*]	1.02 [*1.16*]	(1.07)	164.9
Er^{3+}	–	0.89 [*1.03*]	1.00 [*1.14*]	(1.06)	167.3
Tm^{3+}	–	0.88 [*1.02*]	0.99 [*1.13*]	(1.05)	168.9
Yb^{3+}	–	0.87 [*1.01*]	0.98 [*1.12*]	(1.04)	173.1
Cr^{2+}	–	0.67 [*0.81*]	0.82 [*0.96*]	–	52.0
Cr^{3+}	–	0.62 [*0.76*]	–	–	52.0
Cr^{4+}	0.44 [*0.58*]	0.55 [*0.69*]	–	–	52.0
Ti^{3+}	–	0.67 [*0.81*]	–	–	47.9

8.5.2.2 Transparency

A laser host crystal should be optically inactive and transparent for the pump and laser wavelength. This requires the absence of optically active ions in the host structure, which would be prone to interact with the energy level scheme of the desired dopant. Besides this aspect, the short wavelength transmittance range is ultimately limited by the band-gap energy and the long wavelength limit is determined by the phonon energies of the host materials, leading to the so-called phonon cut-off. However, this definition of a transmission window covers the whole range with a transmittance above zero, while for efficient laser operation the absorption losses in the host crystal should be significantly below the output coupler transmission, i.e. on the order of 1% or lower.

As an example, the transmission of YAG is usually stated to range from ~0.2 µm to >6 µm (Figure 8.5.3). Still, the range with a transmittance high enough to warrant efficient laser operation is much narrower and extends only from ~0.4 µm to ~3.0 µm. To account for this fact, in Table 8.5.3 at the end of this chapter the minimum pump and laser wavelength are defined by the energy at which two photons can bridge the band gap, either by two-photon absorption or by excited state absorption (ESA) from the upper laser level. The long wavelength limit of the transmission is the wavelength corresponding to photon energies of four times the maximum phonon energy of the host material. The possibility of a fourth-order multiphonon decay process was shown to decrease the radiative transition probability enough to be detrimental for the laser process [4]. Even though other factors are relevant for the actual wavelength limits of laser transitions in a host material, these simple approximations can serve as useful rules of thumb to compare different host materials. It should, however, be noted, that the ground state of rare-earth ions can be energetically located within the band gap, i.e. even lower photon energies can be sufficient to reach the conduction band by two-photon processes.

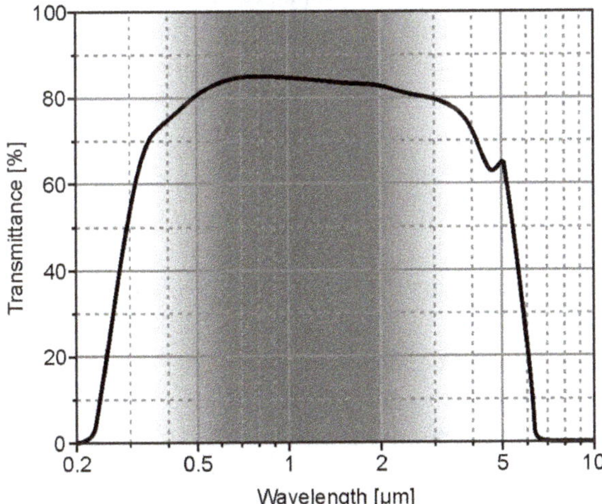

Figure 8.5.3: Transmittance of an undoped 10 mm long YAG sample. The curve shows the transmission range, the grey area roughly indicates the range useful for efficient laser operation. Note that the transmission does not reach 100% due to Fresnel reflection at both surfaces of the sample, which amounts to ~8% e.g. at 1 µm.

Notably, in most cases, transmittance curves are not corrected by (wavelength dependent) Fresnel reflections at the surfaces of the sample. This is also seen in Figure 8.5.3, where the transmittance of YAG appears to be below 90% at any wavelength, caused by the total Fresnel reflection of more than 10% at any wavelength. In a laser, Fresnel losses can be avoided e.g. by respective surface coatings or Brewster-angle arrangement of the laser crystal.

The transparency range can further be affected by impurities, in particular broadband absorbing transition metal ions like the omnipresent iron and by the presence of color centers in the host crystals. In oxides, color centers are caused by a lack of oxygen atoms in the crystal, which can result from a very reducing growth atmosphere sometimes required to protect the crucible from damage. The empty O^{2-}-sites are filled by electrons (e^-), causing characteristic broad absorption bands in the visible, which gave color centers their name.

8.5.2.3 Thermal conductivity and phonon energies

Besides influencing the transparency of laser host crystals, phonons are also relevant for the thermal conductivity of a crystalline material. The thermal conductivity κ of a solid-state material is given by the relation

$$\kappa = c_p \cdot \rho \cdot \alpha \tag{8.5.1}$$

with the specific heat capacity c_p, the density ρ and the thermal diffusivity α. In insulating laser crystals, α is mainly driven by photon propagation due to the lack of mobile conduction band electrons, which provide the main contribution to heat transport in metals. In eq. (8.5.1), it may appear as if a high density of a material promotes high thermal conductivity. However, this effect is partially counterbalanced in the phonon energy, which decreases with the mass m of the oscillating atoms and is proportional to $m^{-1/2}$. As a consequence, e.g. among the isostructural cubic sesquioxide crystals scandia (Sc_2O_3), yttria (Y_2O_3) and lutetia (Lu_2O_3), yttria possesses the highest phonon energies and also the highest thermal conductivity in the un-doped state (Figure 8.5.4) despite having the lowest density among the three materials.

As a general trend, oxide crystals exhibit higher phonon energies than fluoride crystals. However, besides high phonon energies, also an efficient propagation of the phonons is required. In the context of ion-doped laser crystals, mass difference phonon scattering must be considered: It is intuitive that a heavy doping ion replacing a light cation in a crystal structure is detrimental for the propagation of phonons. Therefore, thermal conductivity often drops significantly with doping concentration (Figure 8.5.4). This effect can be reduced in materials with a low mass difference between the doping ion and the replaced cation. For example, the thermal conductivities of Lu_2O_3 and LuAG ($Lu_3Al_5O_{12}$), both shown in blue in Figure 8.5.4, drop only marginally with increasing Yb^{3+}-doping concentration, because lutetium and ytterbium are neighbors in the Periodic Table with very similar atomic masses, whereas a significant decrease of the thermal conductivity with Yb^{3+}-doping can be observed in Y_2O_3 and YAG, both shown in green in Figure 8.5.4. As a consequence, due to the strong influence of the doping ions on the thermal conductivity, the value for the undoped crystal is in many cases not sufficient for the thermal design of a laser.

The average phonon propagation length further reduces in materials with a disordered structure incorporating different ions statistically distributed on the same lattice site – and obviously in glasses with an amorphous structure. Such materials possess low thermal conductivities even when undoped, which, however, does not significantly decrease with doping ion density. This is also seen in Figure 8.5.4 for the disordered mixed sesquioxide crystal $(Lu,Sc)_2O_3$ (red curve), where Lu^{3+} and Sc^{3+} share the same crystallographic sites.

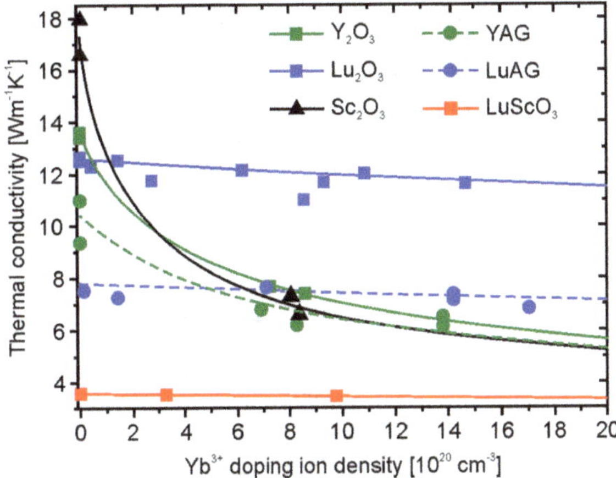

Figure 8.5.4: Thermal conductivity of different garnet and sesquioxide laser host materials vs. Yb^{3+} doping concentration. The *x*-axis extends to ~14 at% doping in YAG or ~7 at% doping in Lu_2O_3, which are realistic upper limits for the use of Yb^{3+} in efficient lasers.

Lattice phonon energies are also relevant for other aspects of laser gain media. As briefly mentioned before, a too low energetic difference between the excited level and the next energetically lower level (not necessarily the laser terminal level), imposes a risk of a multiphonon decay of the excited state, creating phonons (i.e. heat) instead of photons. This risk exists for laser transitions with an energetic gap, which is smaller than four to five times the maximum phonon energy. As the phonon energies in laser crystals typically do not exceed 1000 cm^{-1} (the unit cm^{-1} chosen here for convenience is a unit of wavenumbers; multiplication with hc yields the actual energy), this has in most cases low implications for visible and near-infrared lasers. In contrast, for mid-infrared lasers in the 3 µm range, corresponding to ~3000 cm^{-1}, the phonon energies are crucial for the choice of a host material. It should be noted that not only nonradiative decay directly on the laser transition is detrimental for the laser efficiency. The energy level scheme of Pr^{3+} is an example for a highly energetic laser transition, which is prone to detrimental nonradiative decay from the upper laser level $^3P_0$ into an intermediate $^1D_2$ energy level in host materials with high phonon energies.

The Dieke diagram (Figure 8.5.5) clearly reveals that the presence of phonons with hundreds of cm^{-1} energy in typical laser host crystals alone is responsible for most of the energy levels not being useful upper laser levels, because their lifetimes are strongly quenched by (multi)phonon decay. While this process is always detrimental for the upper laser level, it is required to efficiently depopulate the laser terminal level back into the ground state and, also for a fast energy transfer from the pump acceptor level into the upper laser level, if not in-band pumping is applied.

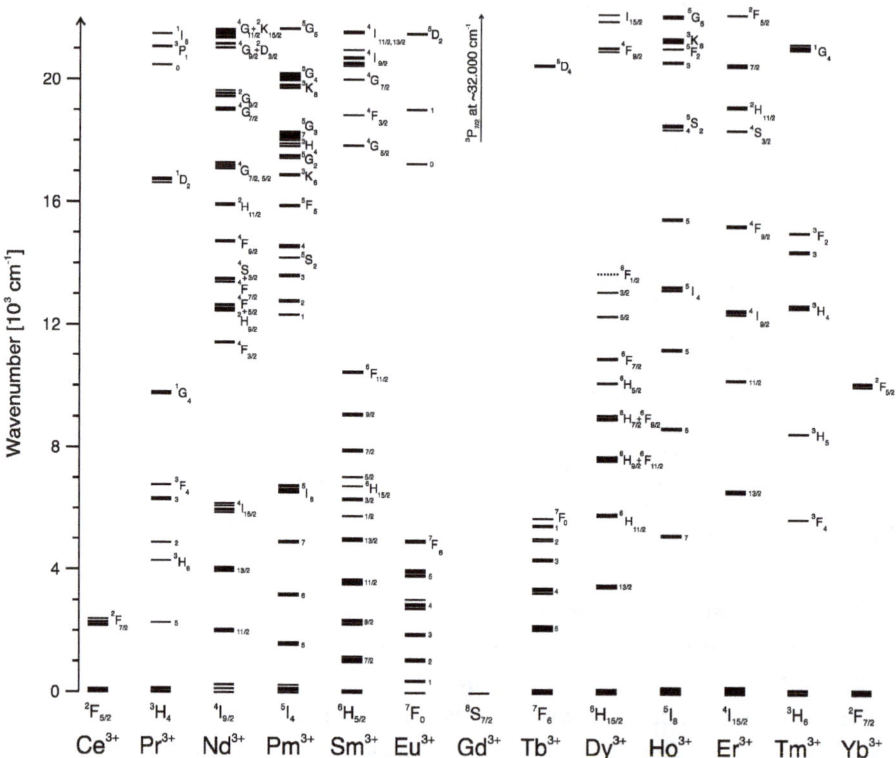

Figure 8.5.5: The Dieke diagram after [5] updated with data from [2, 6] shows the energy levels of rare-earth ions in $LaCl_3$. As the influence of the crystal field on the energy levels is low, this diagram is also useful for all other rare-earth-doped laser crystals. It should be noted that various further energy levels exist at higher energies, while most of them are not relevant for the respective laser schemes (though ESA into these levels may be relevant).

8.5.2.4 Crystal field strength

It goes beyond the scope of this chapter to introduce all peculiarities of possible laser ions. Rare-earth ions possess a shielded $4f$ shell, making their energy levels widely insensitive to a change of the host matrix. Therefore, the Dieke diagram in Figure 8.5.5 is very useful, as the characteristic spectroscopic fingerprint of any rare-earth doping ion can be recognized in all host materials. In contrast, for transition metals as the second class of prominent laser ions, the laser transitions take place in the unshielded $3d$ shell, which makes this class of ions highly sensitive to changes of the crystal field. For the case of transition metals, thus, the less intuitive Tanabe-Sugano diagrams [7] are required, which indicate the level energies versus a normalized relative crystal field strength.

Despite the low influence of the crystal field on rare-earth ions, depending on the laser ion, a high or a low crystal field strength can be desirable. A low crystal field strength decreases the splitting of energy levels of the $4f^{n-1}5d^1$ configuration of rare-earth ions, which is beneficial e.g. for visible lasers based on Pr^{3+} and Tb^{3+}, which are prone to ESA from the upper laser level into $4f^{n-1}5d^1$ energy levels. Even though the influence of the crystal field on the energetic positions of the energy levels is low, the transition cross sections can vary by orders of magnitude, depending on the symmetry of the doping ion site. The combination of both effects shifts the spectral emission peak positions only slightly, but it may change the ratio of the intensity of different emission peaks so that another peak becomes the dominant laser transition. Moreover, effects like ESA or cross relaxation may become more (or less) resonant, which can also strongly influence the laser performance. It should again be noted, that many of these effects – namely the interionic processes – also strongly depend on the (average) distance between the doping ions. This quantity depends stronger on the chosen doping ion density than on the doping ion site distance of the host.

Figure 8.5.6 shows the redshift of the luminescence from Ce^{3+} doped into different host materials compared to the free Ce^{3+} ion. Due to the strong effect of the crystal field on the unshielded $5d$ levels, an increasing crystal field strength shifts the lowest energy levels of the $5d$ configuration downwards. This reduces the energy of $5d \rightarrow 4f$ transitions and yields a red-shift. Thus, the crystal field depression of the $5d$ fluorescence of Ce^{3+} can serve as an indicator for the strength of the crystal field of a given host material. P. Dorenbos performed such measurements for a vast number of materials [8–12], and deduced some general trends. As can be seen in Figure 8.5.6, the crystal field depression of oxide host materials is usually stronger than that of fluoride host materials. This can be easily understood, because in the same coordination with similar distances, a divalent O^{2-} ligand provides a stronger Coulomb field than a singly charged F^-. In the same manner, a highly coordinated site provides a lower crystal field strength than a site with a lower coordination number, because usually the distance between the ligands and the doping ion increases with coordination number and the Coulomb field decreases.

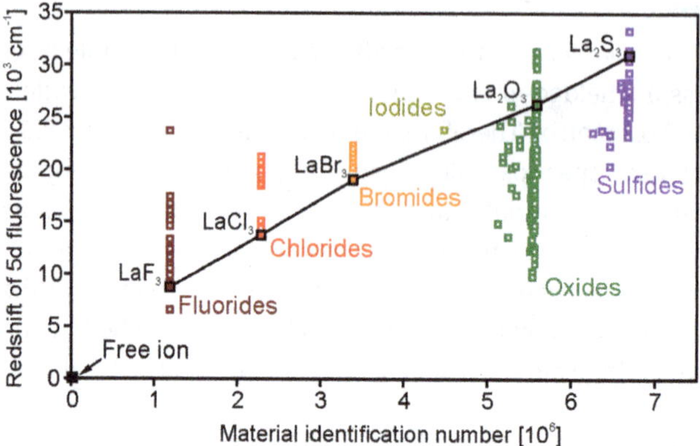

Figure 8.5.6: Crystal field depression represented by the redshift of the $5d \rightarrow 4f$ fluorescence of Ce^{3+} in various host materials over their material identification number modified from [8]. The material identification number is somewhat random, but it clearly separates fluorides, oxides and other classes of materials. The solid line connects the respective La compositions and serves as a guide to the eye to indicate the increasing trend of redshift from left to right.

For this reason, for visible lasers which are prone to *4f-5d* ESA as previously mentioned, fluoride crystals are often preferred. Nevertheless, as already mentioned, too, fluorides provide a lower mechanical strength and lower thermal conductivities, thus they are not the first choice when it comes to high power laser operation. Here, highly coordinated materials, such as the twelvefold coordinated SRA ($SrAl_{12}O_{19}$) from the hexaaluminate family, may be favorable. However, as a trade-off, a high coordination number intrinsically brings the coordination sphere closer to a spherical symmetry, which often goes along with significantly reduced transition cross sections.

8.5.2.5 Further thermal and mechanical properties

Though not of paramount importance in many cases, there are various further properties which can make a laser host crystal more or less useful for the targeted laser ion and/or application. Among these are the mechanical stability of the crystal including its hardness, brittleness and fracture toughness. These properties can strongly hamper the possibility to prepare and polish samples of the desired orientation and/or geometry. The handling of brittle and soft materials is obviously quite delicate.

Chemical properties such as hygroscopicity and solubility must as well be considered, as they influence storage and cleaning of the samples. In these aspects, in particular most chloride and bromide crystals are not favorable, but also various nonlinear crystals such as ß-BBO (ß-$Ba(BO_2)_2$) are slightly hygroscopic.

When it comes to high power laser operation and/or the layout of fundamental mode resonators, thermal effects such as thermal expansion dL/dT and the temperature coefficient of the refractive index dn/dT have to be considered, as both effects contribute to thermal lensing, i.e. the formation of a lens effect due to temperature gradients between the pumped center of a laser crystal and its colder outer regions.

Mechanical, thermal and chemical properties as well as the refractive index are also relevant, when a coating of the laser crystal is desired. Obviously, the crystal should be inert towards the coating materials, but less obviously a certain heating of the sample during the coating process is unavoidable and strong differences between the thermal expansion of the coating and the sample may lead to stress, resulting in an exfoliation of the coatings or even the formation of cracks in the samples.

Finally, there is a large number of further materials properties, which have to be considered. These include linear and nonlinear refractive indices, the possible formation of transient and permanent color centers by intense and/or high photon energy light, thermal shock resistance, Young modulus and tensile strength, along many others, which can not be discussed in detail here.

8.5.3 Growth of laser crystals

The growth of high-quality laser crystals (see also Chapter 6.7) can be very demanding. The melting points of oxide materials can exceed 2000 °C, imposing harsh conditions for all materials involved. On the other hand, the success of a laser material is largely determined by the possibility to grow it in sufficient size and quality by a cost-efficient growth method. Consequently, only for few established laser crystals, such as YAG, YVO_4 and Ti:sapphire, a steady commercial supply is available. In contrast, for many other laser crystals, further research is required to adapt existing growth techniques for this purpose. The details of laser crystal growth methods extend far beyond the scope of this chapter and can be found e.g. in [13, 14]. However, few aspects are mentioned here to explain the difficulties for the commercial availability of laser crystals and existing variations in their quality.

One important aspect regards the required crucible material. Materials with a phase transition from the solid to the liquid phase without strong evaporation of any component can be melted in a chemically inert crucible with a sufficiently high melting point. For oxides, only few suitable crucible materials exist. Among them are molybdenum, tungsten, tantalum and rhenium as well as the noble metals osmium, iridium and platinum. The material price and its volatility as well as the fabrication costs of a crucible can be limiting factors for the availability of a laser crystal. As an example, sesquioxide crystals like Lu_2O_3 are typically grown from expensive to fabricate rhenium crucibles [15], which are prone to damage in insufficient growth atmospheres. On the other hand, fluoride crystals possess typically much lower melting temperatures below 1500 °C. Such temperatures allow for the growth from comparably cheap to fabricate platinum crucibles, but also affordable vitreous carbon crucibles are feasible due to the required oxygen free growth atmosphere.

Besides the crucible materials, also the cost of the starting materials should be considered. As an example, starting powders of the widely applied yttria are easily available in very high purities of 6 N (99.9999%) for few hundred euros per kg, while lutetia powder is usually an order of magnitude more expensive. In addition, due to lutetia's nearly twice molecular weight compared to yttria, the actual mass required for a specified crystal volume is higher. Therefore, despite the good match of weight and size of Lu^{3+} and the heavy rare-earth ions Yb^{3+}, Tm^{3+}, Er^{3+} and Ho^{3+} and the corresponding advantages, the commercial availability of Lu-containing crystals is currently lower than for those incorporating yttrium (Y).

The Czochralski technique [16] is widely applied to the growth of laser crystals including garnet crystals such as YAG, vanadates, CALGO ($CaAlGdO_4$) and fluoride crystals, such as YLF ($LiYF_4$) or LiCAF ($LiCaAlF_6$). Recently, there are also first reports on the growth of mixed sesquioxides by this method [17]. In the Czochralski technique, the crystals are grown on a seed from the crucible. Even without detailed knowledge on phase diagrams, it is evident that the composition of a melt steadily changes if the composition of the growing crystal differs from that of the melt. For such incongruent melts, where the segregation coefficients strongly deviate from unity, not all of the melt can be used, limiting the achievable crystal size. For Pr^{3+} in YLF, the segregation coefficient is ~0.2 [18], which leads to strong and often undesired gradients of the doping ion in the grown crystal. Some materials, including most double tungstates such as KYW ($KY(WO_4)_2$) and many borate crystals, do not crystallize in the desired phase at their melting temperature. Successful growth is realized by adding a solvent to the melt, which solves the material at temperatures significantly below the actual melting point. This so-called flux growth method usually significantly decreases the growth velocity and bears the risk for undesired incorporation of the flux into the grown crystal, which explains strong quality fluctuations in commercially available tungstate crystals.

Others like the Bridgman technique [19], the Bagdasarov method [20] and heat exchanger growth (HEM) [21] and their modifications are based on a directional solidification in the crucible. Such methods allow for the growth of large crystals, e.g. used as amplifier crystals for petawatt Ti:sapphire based laser facilities, but they are also successfully applied to some fluoride crystals such as LaF_3 and CaF_2. In particular, HEM proved useful for the growth of sesquioxide laser crystals [15, 22]. For directional solidification methods, a (nearly) congruent melting behavior is a prerequisite for homogeneous crystals of useful size. It should also be noted that in some cases it is not straightforward to release the crystal as a whole from the crucible.

Crucible-free crystal growth methods like the optical floating zone growth (OFZ) [23] and laser heated pedestal growth (LHPG) [24] do exist. However, due to the low yield, these methods are of low practical relevance for the growth of laser crystals.

Chalcogenides like ZnSe and ZnS sublimate well below their melting points. These crystals need to be grown by chemical or physical vapor transport growth (CVT or PVT) in high pressure ampules [25, 26]. These methods exhibit very low growth rates and yield limited crystal sizes. Both growth methods do not allow for an in-situ doping during the growth process and successive diffusion doping is required [27]. As an alternative, more recently the growth of polycrystalline chalcogenide ceramics by chemical vapor deposition (CVD) gained increasing attention [28].

The feasibility and yield of a growth method are determined by many further effects. Evaporation of one component of the melt – often observed in gallium containing melts, e.g. the garnet YGG ($Y_3Ga_5O_{12}$) [29] – can be another serious issue as it changes the composition of the melt and pollutes the growth chamber.

Finally, in some cases, very sophisticated (and possibly toxic) growth atmospheres including H_2 and CO are required to keep the oxygen partial pressure low enough to prevent damage of the crucible. In particular, for fluorides, very aggressive growth atmospheres (e.g. CF_4 or even HF to fluorinate the starting material and binding oxygen atoms in the growth atmosphere) yield best growth results. In some cases, e.g. for ZnSe, but also for the beryllium containing alexandrite ($BeAl_2O_4$), even the starting materials are highly toxic, which further complicates safe crystal growth.

8.5.4 Common laser host materials

There is a vast variety of host materials for lasers. In combination with about 20 laser-active rare-earth and transition metal-doping ions, the number of actual laser crystals, i.e. combinations of doping ion and host crystal, becomes innumerable. It can thus not be the purpose of this section to provide a complete list of laser crystals and much more extensive lists can be found e.g. in the work of A. Kaminskii [2, 30]. Instead, the materials presented here are chosen with a focus on a commercial relevance and/or outstanding laser properties.

8.5.4.1 Oxide host materials

Oxide crystals are the most established host materials for laser applications. They often possess a good mechanical stability, they are very often inert to usual atmospheric conditions and their thermal conductivity is in many cases higher than that of fluoride crystals. Sophisticated cooling concepts allow for several kW of continuous wave output power in the 1 µm spectral range, based on oxide host materials [31]. There is a broad variety of oxide crystals that have been established as laser host materials and in the following only the most prominent oxide laser host materials are shortly introduced hereafter.

8.5.4.1.1 Garnets

Garnets are the most prominent class of laser host crystals, and Yb:YAG and Yb:LuAG crystals deliver the highest solid-state laser output powers up to now. Garnets are composed according to $A_3B_2C_3O_{12}$, where for laser crystals A is mainly Y, Lu or Gd. The octahedral B-site exhibits the optically inactive C_{3i} symmetry (Table 8.5.1) and is thus of low relevance for doping ions. In laser host materials, the B-site is usually occupied by Al, Ga and Sc. The tetrahedral C-site (S_4 symmetry) is too small for Sc and only occupied by Al and Ga. Rare-earth doping ions are incorporated on the dodecahedral A-site. Large amounts of heavier rare-earth ions such as Ho, Er, Tm and Yb can be incorporated on this site, but only few percent of lighter and larger rare-earth ions such as Nd fit on this site. Transition metals such as Cr and Ti are typically incorporated on the C sites, but lasers based on transition metal-doped garnets do not have a strong market. However, Cr is a widely used sensitizer co-dopant to enhance the absorption efficiency in flash-lamp-pumped lasers, e.g. for Cr,Tm,Ho:YAG (CTH:YAG) emitting around 2 µm. In addition, Cr^{4+}:YAG and V^{3+}:YAG are widely applied passive saturable absorbers for Q-switched pulse-lasers in the 1 and 1.5 µm spectral range, respectively.

Garnets possess a cubic crystal structure. They are very hard and show a comparably high thermal conductivity. Despite a high melting point of ~1930 °C, very large YAG crystals of more than 10 cm diameter can be easily grown by the Czochralski method from iridium or molybdenum crucibles [32]. Due to its advantageous properties, YAG is one of the most common laser host materials and commercially available in large quantities, sizes and with arbitrary doping. In particular, Nd:YAG and Yb:YAG are of very high commercial relevance. As a disadvantage, rare-earth-doped garnets often exhibit lower transition cross sections than for rare-earth ions in other materials and their gain bandwidth is lower. As previously mentioned, the thermal conductivity drops strongly when doped with larger amounts of heavy rare-earth ions (Figure 8.5.4), but this issue can be circumvented by the use of LuAG as a host material [31]. Ga-garnets like YSGG ($Y_3Sc_2Ga_3O_{12}$) or GGG ($Gd_3Ga_5O_{12}$) possess somewhat lower phonon energies and – though more difficult to grow – are better suited for 3 µm lasers based on Er^{3+}-doping than YAG [33].

8.5.4.1.2 Perovskites

The perovskites represent a vast range of crystals. Among these, only YAP ($YAlO_3$) and LuAP ($LuAlO_3$) are of relevance as laser hosts. The twelvefold coordinated rare-earth site in these orthorhombic crystals provides a particularly low crystal field, which enables laser operation at unusual wavelengths, e.g. at 999.4 nm in Yb:LuAP [34], and avoids a strong decrease of the 5*d* energy levels, but on the down-side it does not feature strong transition cross sections. Nevertheless, Pr:YAP is one of the few oxide host materials suitable for efficient visible laser operation [35]. Due to the orthorhombic crystal structure, attention has to be paid to proper orientation of the samples. YAP and LuAP can be grown by the conventional Czochralski method.

8.5.4.1.3 Double tungstates

The tungstate crystals KGW ($KGd(WO_4)_2$) and KYW are very attractive laser host materials. The low level of symmetry of the monoclinically coordinated C_2 cation sites yields very high cross sections, making tungstates attractive host materials for compact and microchip lasers. Moreover, their emission spectra are comparably broad, Yb:KGW and Yb:KYW routinely supply sub-150-fs pulse durations [36] and many commercial fs-pulse seed lasers are equipped with these gain crystals.

On the negative side, these crystals exhibit a strongly incongruent melting behavior and need to be grown by flux methods. Due to the difficulties of this growth method, the quality of commercially available double tungstate crystals may strongly vary. Moreover, their thermal conductivity below 3 $Wm^{-1}K^{-1}$ only slightly exceeds that of glasses. Finally, when doped with RE = Tb or Pr, tungstates are prone to an intervalence charge transfer RE^{3+}, $W^{6+} \rightarrow RE^{4+}$, W^{5+} [37] at comparably low photon energies, which makes them unsuitable for visible lasers based on these ions.

8.5.4.1.4 Vanadates

Despite a high symmetry of the D_{2d} doping ion sites, the orthovanadate crystals YVO_4 and $GdVO_4$ exhibit very high transition cross sections, e.g. when doped with Nd^{3+}. In combination with their tetragonal crystal structure, featuring polarized emission, they are first choice gain materials for frequency-doubled lasers. Most green laser pointers incorporate a Nd:YVO_4 laser crystal, but due to their high thermal conductivity, vanadates are also useful for high-power green emitting laser sources at 532 nm. However, their crystal field splitting is too low to qualify them as host materials for Yb-doping, and their low band-gap energy makes them unsuitable for visible lasers. Vanadate crystals can be grown by the Czochralski method and are readily commercially available.

8.5.4.1.5 Sesquioxides

The sesquioxide crystals Sc_2O_3, Y_2O_3, and Lu_2O_3 as well as their solid solutions are very difficult to synthesize, due to their high melting points. The Czochralski method does not yield good results and HEM growth requires very expensive and sensitive crucible and isolation materials, which hindered commercialization up to now. Recently, the growth of mixed sesquioxide crystals by the Czochralski method from iridium crucibles has been reported [17], paving the way for commercial applications. Sesquioxides possess quite high thermal conductivities, which in particular for Lu_2O_3 does not drop for doping with the heavy rare-earth ions Ho, Er, Tm and Yb and enables efficient high-power laser operation [38]. Depending on the substituted cation

(Sc, Lu or Y), the crystal field varies strongly, enabling for a shift of the gain peak in a wide range for many doping ions. Moreover, though at the expense of a reduced thermal conductivity (see Figure 8.5.4), doping of mixed sesquioxide crystals combines the gain spectra of the constituents, enabling broad emission bandwidths suitable for wavelength tuning or ultrafast pulse generation. It should be noted that sesquioxides provide 32 cation sites per unit cell, of which only 24 are optically active, while doping ions incorporated on the remaining C_{3i} sites with inversion symmetry are considered to be optically inactive, i.e. they do not strongly contribute to the laser process. Despite being out of the scope of this chapter, it should be noted that ceramic synthesis methods are well suited for these cubic host materials [39], and sesquioxide laser ceramics are commercially available.

8.5.4.1.6 Aluminates

CALGO ($CaGdAlO_4$) is exceptionally well suited for ultrashort pulse generation. Yb:CALGO has a much higher thermal conductivity than other Yb-doped materials with a comparably broad emission bandwidth, despite a high level of disorder that should go along with a statistical distribution of Ca^{2+} and Gd^{3+} ions on the same site. This is attributed to the formation of a superstructure [40]. In addition, the negative dn/dT compensates for the positive dL/dT, avoiding a strong thermal lensing even at high pump power levels [41]. CALGO can be grown by the Czochralski method. A slight yellow color observed in CALGO crystals with stoichiometric composition is usually not detrimental for infrared laser performance and can be avoided by growing with an excess of Gd in the melt [42].

SRA ($SrAl_{12}O_{19}$) and other members of the hexaaluminate family exhibit a ratio of Al and O close to 2:3, similar to the composition of sapphire. They are hard and possess a high thermal conductivity. The only suitable rare-earth doping site in SRA is the D_{3h} coordinated Sr^{2+}-site. Thus, e.g. Pr:SRA is often co-doped with the same amount of divalent Mg^{2+} to be incorporated on trivalent Al^{3+}-sites for charge compensation. Due to its twelvefold coordinated doping ion site, SRA exhibits one of the lowest crystal field depression values of all oxides. Therefore, it is well suited for visible lasers based on Pr^{3+}-doping, even though the high symmetry of such a highly coordinated site yields low transition cross sections. SRA can be grown by the Czochralski method.

8.5.4.2 Fluoride host materials

Fluorides possess in general lower phonon energies than their oxide counterparts. The corresponding wide mid-infrared transmission window makes them suitable host materials for 3 µm lasers based on Er^{3+}, Ho^{3+} and Dy^{3+} as the doping ion. On the other hand, they exhibit large band-gap energies and often a transparency far into the UV. The monovalent F^- ligands introduce a weaker crystal field than O^{2-} ligands. The 5*d* energy levels thus remain at high energies, reducing the risk of ESA for visible lasers, e.g. based on Pr^{3+} or Tb^{3+}-doping [43]. As a disadvantage, fluoride crystals often exhibit a low thermal conductivity (associated with their low phonon energies) and are in general softer than oxide materials. In addition, the growth of fluorides requires oxygen-free and dry starting materials and growth atmosphere, while many as-grown fluorides are neither hygroscopic nor sensitive to air atmosphere.

8.5.4.2.1 Scheelites

The isostructural scheelite crystals YLF ($LiYF_4$) and LLF ($LiLuF_4$) exhibit a tetragonal crystal structure and thus polarization-dependent spectroscopic features when doped with active ions. Both crystals feature very similar spectroscopic and thermo-mechanical properties. They are successfully grown by the Czochralski growth method, even though some researchers report a slightly incongruent melting behavior for YLF [44]. A minor drawback of LLF might be the higher cost of the required LuF_3 starting materials. Due to their wide transmission window, both materials are suited for laser transitions from the UV when doped with Ce^{3+} [45], to the mid-infrared when doped with Er^{3+} [46]. They are the most successful host materials for visible Pr^{3+} or Tb^{3+} lasers [43], while it should be noted that the segregation coefficients of <0.2 and ~0.7, respectively, are rather low and may lead to strong doping concentration gradients. Both crystals exhibit a negative dn/dT, but a positive thermal expansion, which – depending on laser wavelength and orientation – can cancel out each other and avoid the formation of a strong thermal lens. For this reason, Nd:YLF is also a common gain material for frequency-doubled low repetition rate/highly energetic/nanosecond-pulsed green lasers emitting at a characteristic wavelength of 527 nm and used as pump sources for Ti:sapphire lasers. It is in related to particular difficulties in avoiding oxygen impurities, that the quality of commercially available scheelite-structure crystals may strongly vary.

8.5.4.2.2 Fluorites

CaF_2 exhibits an exceptionally high thermal conductivity for a fluoride crystal. It crystallizes in the cubic fluorite structure and can be grown in very large boules of more than 30 cm in diameter. Like in the scheelites, its negative dn/dT and positive dL/dT may cancel out and reduce thermal lensing. Doping trivalent rare-earth ions onto the divalent Ca^{2+} site introduces a certain level of disorder, which yields inhomogeneously broadened absorption and emission characteristics, even when co-doped with monovalent Na^+ for charge compensation. All these features make CaF_2 an interesting host material for ultrahigh peak power laser pulses based on the 1 μm transition of Yb^{3+} [47]. However, like for other fluorides, the low crystal field depression and the large transparency range enable laser operation with a variety of rare-earth dopants at wavelengths from the visible to the mid-infrared spectral range.

8.5.4.3 Host materials for transition metals

For transition metal doping, a suitable crystal field strength of the host crystal on the doping ion site is even more important than good thermo-mechanical properties. As previously mentioned, the 3*d* orbitals of transition metal ions are not shielded by outer shells, and thus the position of the energy levels is very sensitive to changes of the host crystal field. Due to the corresponding stronger coupling of the electronic transitions to the host properties, transition metal ions exhibit broad absorption and emission spectra, but are also prone to phonon-assisted nonradiative decay. We have seen that materials with lower phonon energies are often disadvantageous in their thermo-mechanical properties. Therefore, the highest output powers obtained with transition metal-doped lasers are orders of magnitude lower than the kW laser powers obtained with rare-earth doped crystals. In turn, the broad spectra enable widely tunable lasers and ultrashort pulse generation, which are required in a wide range of applications.

8.5.4.3.1 Sapphire

α-Al_2O_3, also known as corundum or sapphire, was the host material for the first laser [48]. However, even though ruby (Cr:Al_2O_3) lasers are still around (e.g. for tattoo removal), ruby has never been an efficient laser material. Sapphire is one of the hardest minerals, exhibits a very high thermal conductivity, a comparably large transparency window and can be grown by the Czochralski method. It could thus be seen as an ideal host material. Unfortunately, it does not exhibit a doping site large enough for rare-earth ions and, though a Nd:sapphire laser has been demonstrated [49], the prospects for the growth of rare-earth doped sapphire crystals with practically relevant dimensions and good optical quality are low. Still, sapphire is one of the most important host materials: Ti:sapphire lasers generate pulses with sub-5-fs pulse duration [50, 51], and large Ti:sapphire amplifier crystals with tens of centimeter diameter amplify short pulses to unprecedented peak powers in the petawatt (10^{15} W) range. Ti:sapphire lasers find many commercial applications and Ti:sapphire crystals are thus readily available. It should, however, be noted, that Ti^{3+} ions are larger and heavier than Al^{3+} (Tables 8.5.2 and 8.5.3c). Ti:sapphire is thus prone to strong gradients in the doping concentration, the maximum achievable doping concentrations are low and the thermal conductivity of sapphire drops drastically with Ti-doping. Moreover, Ti:sapphire is prone to nonradiative decay at elevated temperatures, and therefore requires thorough temperature management.

8.5.4.3.2 Chrysoberyl (Alexandrite)

Chrysoberyl is the mineralogic name of $BeAl_2O_4$, while Cr:$BeAl_2O_4$ is named Alexandrite. Chrysoberyl provides two different Al^{3+} sites, of which the C_i site is optically inactive (see Table 8.5.1). Thus, when doped with Cr^{3+}, about 50% of the ions do not contribute to the laser process. The crystal field for Cr^{3+}-ions on the remaining C_s

sites features a complicated energy level structure, which benefits from operation at elevated temperatures (>50 °C) for efficient broadband lasing from the high-spin $^4T_2$ level. Alexandrite features high slope efficiencies. More than 100 nm wavelength tuning range and sub-100-fs pulse duration can be obtained around the center wavelength of 780 nm [52]. Alexandrite is very hard, features a very high thermal conductivity and can be grown by the Czochralski method. However, the use of toxic beryllium requires strict safety measures for the growth and starting materials preparation, which limits the availability of this crystal.

8.5.4.3.3 Forsterite

Forsterite (Mg_2SiO_4) is the most successful host material for lasers based on tetravalent chromium (Cr^{4+}). The doping ion is incorporated on the Si^{4+}-site with tetrahedral configuration. Despite typical issues owed to the energy level scheme of Cr^{4+} such as ESA and nonradiative decay of the excited state [53] – the latter supported by the high phonon energies – Cr^{4+}:Mg_2SiO_4 enables laser efficiencies of more than 35% in the wavelength range between 1.15 and 1.35 µm [54], which is sparsely covered by other laser materials. The broad emission band also supports the generation of sub-100-fs pulses, but the output powers are currently limited to a few watts [55], which is partially owed to the comparably low thermal conductivity of Cr:forsterite, and its comparably high thermal expansion and dn/dT values. Forsterite is hard and large crystals of high optical quality can be grown by the Czochralski method, even though attention should be paid to suppress the formation of unwanted Cr^{3+}.

8.5.4.3.4 Colquiriites

LiCAF ($LiCaAlF_6$) and LiSAF ($LiSrAlF_6$) are among the materials with the widest UV transparency range, which can extend down to below 110 nm. Their sixfold coordinated D_3 symmetry Al^{3+} site is well-suited for Cr^{3+} doping. A comparably low crystal field strength ensures that the high-spin $^4T_2$ level is the lowest excited state, providing a strong phononic broadening of the absorption and emission spectra. Laser wavelength tuning ranges of more than 300 nm around the center wavelength of 850 nm are obtained in Cr:LiSAF, slightly lower ranges in Cr:LiCAF. Such broad emission bands enabled the generation of 10-fs pulses in both materials. Colquiriites can be grown in high optical quality by the Czochralski method. The main disadvantage is the low thermal conductivity typical for fluorides, further reduced by Cr^{3+}-doping, which hinders power scaling of the output power. Further details on colquiriite crystals can be found in [52].

It should further be noted, that colquiriites are among the most suitable materials for lasers based on the "transition metal-like" broad $5d \rightarrow 4f$ transition of Ce^{3+}, yielding UV lasers at wavelengths between 280 and 320 nm [56]. These are, however, hardly operated in continuous wave mode, due to the lack of suitable pump sources.

8.5.4.3.5 Chalcogenides

The very low phonon energies of chalcogenide host materials like ZnS, ZnSe and CdSe enable efficient laser operation of Cr^{2+} and Fe^{2+} at wavelengths up to 5 µm. Their comparably low crystal field strengths ensure simple laser schemes from the excited ^5E to the $^5T_2$ state with all possible ESA channels being spin forbidden. Cr^{2+}:ZnSe enabled an enormous wavelength tuning range between 1.8 and 3.3 µm [57], for Cr^{2+}:CdSe, even laser wavelengths above 3.6 µm are reported [58]. Such a broad emission bandwidth goes along with the possibility to generate ultrashort pulses, and indeed sub-50-fs pulses are obtained from Cr^{2+}-doped chalcogenides [59]. Even longer laser wavelengths up to 5 µm are realized with Fe^{2+}-doped chalcogenides even at room temperature [60].

Laser output power scaling in chalcogenides is difficult. Even though ZnS and ZnSe possess fairly high thermal conductivities, their dn/dT values are about an order of magnitude higher than in oxides and fluorides, resulting in very strong thermal lensing effects even at moderate pump power levels. Finally, as previously mentioned, chalcogenides are grown from the gas phase and often diffusion doping is required, which limits the availability of high quality, large size laser samples.

8.5.5 Materials properties of laser host crystals

In this section, the most relevant properties for the evaluation of a host material are listed in Tables 8.5.3 a-d. The materials are sorted by "crystal type", where this term can refer to many different meanings, including the material class, a reference structure mineral, a chemical compound or in the case of sesquioxides (Greek *sesqui*: one and a half times) to the ratio between cations and oxygen. Thus, one can argue about the actual meaning of these names, but it is important to state them, as they are commonly used in the solid-state laser community.

Several of the values for the listed properties strongly depend on the experimental conditions, i.e., mainly the temperature and the wavelength. Where possible, the values are thus stated for 1.1 µm test wavelength and room temperature to enable a practical comparison. However, it is important to note that, at the actual laser wavelength, the properties may strongly differ. Additionally, the values for the thermal conductivity depend strongly on the purity and structural quality of the host material. As a consequence, literature values in different references may considerably differ.

As previously mentioned, the usually stated transmission window of a host material is not suitable to evaluate the suitability for laser operation at a particular wavelength. To account for this, the minimum laser wavelength λ_{min} stated in the tables is defined as the wavelength, for which the energy of a photon corresponds to half the band-gap energy, i.e.

$$\lambda_{min} = \frac{2 \cdot h \cdot c}{E_{bandgap}}. \tag{8.5.2}$$

For shorter wavelengths, ESA from the emitting laser level or two-photon absorption into the conduction band is feasible, which may strongly suppress efficient lasing.

Accordingly, an increasing risk for nonradiative multiphonon decay of the upper laser level (and a starting decrease of the transmission of a host material in general) is observed if the laser photon energy becomes so low, that it equals four times the maximum phonon energy $(E)_{phonon,max}$ (N.B.: in cm^{-1}) or less [4]. The corresponding maximum laser wavelength λ_{max} is calculated as

$$\lambda_{max} = \frac{1}{4 \cdot (E)_{phonon,max}}. \tag{8.5.3}$$

Both of these estimations do not set strict limits and are rather to be seen as rules-of-thumb. They are, however, in reasonable agreement with the experimental trends.

Another remark regards the crystal field depression according to Dorenbos [9, 10]. This value is defined as the shift of the $5d \rightarrow 4f$ fluorescence of Ce^{3+} compared to the free ion. Thus, values can only exist for host materials with a suitable site for Ce^{3+}-doping.

All missing literature data is indicated by "–" in the tables. Values marked by "~" represent an average of different literature data. Estimated values are indicated by "(est.)". The tables were created with data from [2, 3, 9, 10, 41, 42, 52, 61–135] and references therein.

In conclusion, though effort was made to present only plausible values, all data presented here are intended to be useful for a first evaluation only. Attention should be paid to find data fitting to the planned experimental conditions.

Table 8.5.3a: Materials properties of garnet and perovskite laser host materials.

Crystal type	Garnets	Garnets	Garnets	Garnets	Garnets	Garnets	Perovskites	Perovskites
Host crystal	YAG	LuAG	YGG	GGG	GSGG	YSAG	YAP	LuAP
Chemical formula	$Y_3Al_5O_{12}$	$Lu_3Al_5O_{12}$	$Y_3Ga_5O_{12}$	$Gd_3Ga_5O_{12}$	$Gd_3Sc_2Ga_3O_{12}$	$Y_3Sc_2Al_3O_{12}$	$YAlO_3$	$LuAlO_3$
Crystal system	cubic	cubic	cubic	cubic	cubic	cubic	orthorhombic	orthorhombic
Space group	$Ia\bar{3}d$	$Ia\bar{3}d$	$Ia\bar{3}d$	$Ia\bar{3}d$	$Ia\bar{3}d$	$Ia\bar{3}d$	*Pnma*	*Pnma*
Main doping site	Y^{3+}	Lu^{3+}	Y^{3+}	Gd^{3+}	Gd^{3+}	Y^{3+}	Y^{3+}	Lu^{3+}
Cation site symmetry	D_2	D_2	D_2	D_2	D_2	D_2	C_s	C_s
Coordination number	8	8	8	8	8	8	8	8
Cation site size [Å]	1.02	0.97	1.02	1.06	1.06	1.02	1.02	1.02
Cation density [10^{22} cm^{-3}]	1.39	1.42	1.30	1.26	1.21	1.25	1.91	2.02
Cation atomic mass [u]	88.9	175.0	88.9	157.3	157.3	88.9	88.9	175.0
Band gap energy [eV]	6.5	5.5	6.6	5.7	6.4	6.0	5.6	5.9
Min. laser wavelength [μm]	0.38	0.46	0.38	0.44	0.38	0.42	0.44	0.42
Max. phonon energy [cm^{-1}]	857	871	754	740	741	~800	710	700
Max. laser wavelength [μm]	2.9	2.9	2.9	3.4	3.4	~3.1	3.5	3.6
Thermal conductivity (undoped) [$Wm^{-1}K^{-1}$]	11	8.3	9	8.2	5.8	8.5	11.7 (a), 10.0 (b), 13.3 (c)	~11
Crystal field depression [cm^{-1}]	26,654	27,019	25,261	~27,000 (est.)	~26,500 (est.)	~26,000 (est.)	16,537	16,873

Growth method	Cz	Cz	Cz	Cz	Cz	Cz	Cz	Cz
Melting point [°C]	1930	1980	1790	1730	1850	1900	1875	1960
Mohs hardness	8.5	8.5	7.5	7	6	~8	8.5	~8.5 (est.)
dn/dT [10^{-6} K^{-1}]	9.9	6.1	–	13	9	–	2.3(a), 8.1 (b), 8.7 (c)	–
dL/dT [10^{-6} K^{-1}]	7.8	9.3	~4.5	7.5	7.3	~5 (est.)	7.7(a), 11.7 (b), 8.3 (c)	~12 (a), ~10 (b), ~5 (c)

Cz: Czochralski.

Table 8.5.3b: Materials properties of tungstate, vanadate and sesquioxide host materials.

Crystal type	(Double-) Tungstates	(Double-) Tungstates	Vanadates	Vanadates	Vanadates	Sesquioxides	Sesquioxides	Sesquioxides
Host crystal	KYW	KGW	YVO	GVO, GdVO	LuVO	YO	ScO	LuO
Chemical formula	$KY(WO_4)_2$	$KGd(WO_4)_2$	YVO_4	$GdVO_4$	$LuVO_4$	Y_2O_3	Sc_2O_3	Lu_2O_3
Crystal system	monoclinic	monoclinic	tetragonal	tetragonal	tetragonal	cubic	cubic	cubic
Space group	$C2/c$	$C2/c$	$I4_1/amd$	$I4_1/amd$	$I4_1/amd$	$Ia\bar{3}$	$Ia\bar{3}$	$Ia\bar{3}$
Main doping site	Y^{3+}	Gd^{3+}	Y^{3+}	Gd^{3+}	Lu^{3+}	Y^{3+}	Sc^{3+}	Lu^{3+}
Cation site symmetry	C_2	C_2	D_{2d}	D_{2d}	D_{2d}	C_2 & C_{3i}	C_2 & C_{3i}	C_2 & C_{3i}
Coordination number	8	8	8	8	8	6	6	6
Cation site size [Å]	1.02	1.06	1.02	1.06	0.97	0.90	0.75	0.86
Cation density [10^{22} cm^{-3}]	0.64	0.63	1.26	1.21	1.31	2.69	3.83	2.85
Cation atomic mass [u]	88.9	157.3	88.9	157.3	175.0	88.9	45.0	175.0
Band gap energy [eV]	4.3	4.3	3.8	3.8	3.8	5.6	5.9	5.5
Min. laser wavelength [μm]	0.58	0.58	0.66	0.66	0.66	0.44	0.42	0.46
Max. phonon energy [cm^{-1}]	933	901	890	880	900	621	618	597
Max. laser wavelength [μm]	2.7	2.8	2.9	2.8	2.8	4.0	4.0	4.2
Thermal conductivity (undoped) [$Wm^{-1}K^{-1}$]	2.7 (b)	2.5 (*p*), 3.0 (*m*), 3.5 (*g*)	~7 (a), ~9.5 (c)	~6.5 (a), ~9.5 (c)	~7.5 (a), ~8 (c)	13.4	18	12.6

Crystal field depression [cm^{-1}]	High	High	Low	Low	Low	26,583 (C_2)	27,786 (C_2)	25,354 (C_2)
Growth method	Flux	Flux	Cz	Cz	Cz	HEM	HEM	HEM
Melting point [°C]	1075[a]	1075[a]	1810	~1780	1800	2430	2430	2450
Mohs hardness	4.5	4.5	5	4.5–5	6	6.5	6.5	6–6.5
dn/dT [10^{-6} K^{-1}]	−14.6 (p), −8.9 (m), −12.4 (g)	−15 (*p*), −10 (*m*), −16 (*g*)	14 (a), 9 (c)	9 (a), 12 (c)	7 (a), 15 (c)	6.3	6.4	5.8
dL/dT [10^{-6} K^{-1}]	1.9 (p), 10.3 (m), 16 (g)	2.4 (*p*), 11 (*m*), 17 (*g*)	~3 (a), ~10 (c)	~2 (a), ~10 (c)	~4 (a), ~11 (c)	6.1	8.1	5.3

Cz: Czochralski, HEM: Heat exchanger method. [a] The required low-temperature phase is grown at <850 °C from the flux.

Table 8.5.3c: Materials properties of sesquioxide, aluminate, corundum, scheelite and fluorite host materials.

Crystal type	Sesquioxides	Sesquioxides	Aluminates	Hexaaluminates	Corundum	Scheelites	Scheelites	Fluorites
Host crystal	LuScO	YScO	CALGO	SRA	Sapphire	YLF	LiLuF, LLF	CaF
Chemical formula	$(Lu_{0.5}, Sc_{0.5})_2O_3$	$(Y_{0.5}Sc_{0.5})_2O_3$	$CaGdAlO_4$	$SrAl_{12}O_{19}$	α-Al_2O_3	$LiYF_4$	$LiLuF_4$	CaF_2
Crystal system	Cubic	Cubic	Tetragonal	Hexagonal	Hexagonal	Tetragonal	Tetragonal	Cubic
Space group	$Ia\bar{3}$	$Ia\bar{3}$	$I4/mmm$	$P6_3/mmc$	$R\bar{3}c$	$I4_1/a$	$I4_1/a$	$Fm\bar{3}m$
Main doping site	Lu^{3+}/Sc^{3+}	Y^{3+}/Sc^{3+}	Ca^{2+}/Gd^{3+}	Sr^{2+}	Al^{3+}	Y^{3+}	Lu^{3+}	Ca^{2+}
Cation site symmetry	C_2 & C_{3i}	C_2 & C_{3i}	C_{4v}	D_{3h}	C_3	S_4	S_4	C_{4v}
Coordination number	6 & 6	6 & 6	9	12	6	8	8	8
Cation site size [Å]	0.86/0.75	0.90/0.75	1.18/~1.08	1.40	0.53	1.16	1.11	1.26
Cation density [10^{22} cm^{-3}]	3.08	2.99	1.25	0.34	4.71	1.40	1.44	2.45
Cation atomic mass [u]	175.0/45.0	88.9/45.0	40.1/157.3	87.6	27.0	88.9	175.0	40.1
Band gap energy [eV]	~5.4	~5.4	4.5	5.8	8.7	10.3	10.4	10
Min. laser wavelength [µm]	0.46	0.46	0.56	0.42	0.28	0.24	0.24	0.24
Max. phonon energy [cm^{-1}]	~630	620 (est.)	650	761	756	~450	~430	470
Max. laser wavelength [µm]	4.0	4.0	3.8	3.3	3.3	5.6	5.8	5.3
Thermal conductivity (undoped) [$Wm^{-1}K^{-1}$]	3.9	4.1	6.5 (a), 6.9 (c)	11	30 (a), 33 (c)	5.3 (a), 7.2 (c)	5.0 (a), 6.3 (c)	10

Crystal field depression [cm^{-1}]	26,500 (est.)	26,500 (est.)	~22,500 (est.)	11,050	(20,533) No Ce^{3+} site	15,262	15,140	16,509
Growth method	HEM	HEM & Cz	Cz	Cz	Cz & HEM	Cz	Cz	Cz & DS
Melting point [°C]	2370	2053	1750	1790	2050	850	794	812
Mohs hardness	6.5	6.5	6	8.5–9	9	4–5	4–5	4
dn/dT [10^{-6} K^{-1}]	~6 (est.)	~6 (est.)	−7.6 (a), −8.6 (c)	–	12.8 (a), 9.8 (c)	−4.6 (a), −6.6 (c)	−3.6 (a), −6.0 (c)	−11.5
dL/dT [10^{-6} K^{-1}]	~7 (est.)	~7 (est.)	10 (a), 16 (c)		5.9 (a), 5.2 (c)	14.3 (a), 10.0 (c)	13.6 (a), 10.8 (c)	18.7

HEM: Heat exchanger method, Cz: Czochralski, DS: Directional solidification

Table 8.5.3d: Materials properties of colquiriite, spinel and chalcogenide host materials.

Crystal type	Colquiriites	Colquiriites	Olivine	Chrysoberyl	Chalcogenides	Chalcogenides	Chalcogenides
Host crystal	LiCAF	LiSAF	Forsterite	Alexandrite[b)]	ZnS	ZnSe	CdSe
Chemical formula	$LiCaAlF_6$	$LiSrAlF_6$	Mg_2SiO_4	$BeAl_2O_4$	ZnS	ZnSe	CdSe
Crystal system	Trigonal	Trigonal	Orthorhombic	Orthorhombic	Cubic	Cubic	Hexagonal
Space group	$P\bar{3}1c$	$P\bar{3}1c$	$Pnma$	$Pnma$	$F\bar{4}3m$	$F\bar{4}3m$	$P6_3mc$
Main doping site	Al^{3+}/Ca^{2+}	Al^{3+}/Sr^{2+}	Si^{4+}	Al^{3+}	Zn^{2+}	Zn^{2+}	Cd^{2+}
Cation site symmetry	D_3/D_{3d}	D_3/D_{3d}	C_s	C_i, C_s	T_d	T_d	C_{3v}
Coordination number	6/6	6/6	4	6	4	4	4
Cation site size [Å]	0.67/1.14	0.67/1.27	0.26	0.53	0.60	0.60	0.78
Cation density [10^{22} cm^{-3}]	0.96	0.875	1.38	3.41	2.51	2.20	4.33
Cation atomic mass [u]	27.0	27.0	28.1	27.0	65.4	65.4	112.4
Band gap energy [eV]	12.2	11.8	8.4	9	3.8	3.0	1.74
Min. laser wavelength [μm]	0.20	0.22	0.30	0.28	0.66	0.82	1.42
Max. phonon energy [cm^{-1}]	581	580	883	933	352	251	213
Max. laser wavelength [μm]	4.3	4.3	2.8	2.7	7.1	10.0	11.7
Thermal conductivity (undoped) [$Wm^{-1}K^{-1}$]	4.6 (a), 5.1 (c)	1.8 (a)3.1 (c)	4.7[b)]	28	27	18	6.2 (a), 6.9 (c)
Crystal field depression [cm^{-1}]	13,344	12,165	22,530	Medium (No Ce^{3+} site)	Low (No Ce^{3+} site)	Low (No Ce^{3+} site)	Very low (No Ce^{3+} site)

Growth method	Cz	Cz	Cz	Cz	GP	GP	GP
Melting point [°C]	765	809	1895	1870	1020	1525	1300
Mohs hardness	4	3–4	7	8.5	4	5	4
dn/dT [10^{-6} K^{-1}]	−4.2 (a), −4.6 (c)	−2.5 (a), −4.0 (c)	5.9 (a), 6.9 (b), 15.2 (c)	5.9 (a), 6.9 (b), 15.2 (c)	46	70	89 (a), 95 (c)
dL/dT [10^{-6} K^{-1}]	22 (a), 3.6 (c)	25 (a), −10 (c)	13	5.9 (a), 6.1 (b), 6.7 (c)	6.8	7.3	4.1 (a), 2.8 (c)

[b] Cr^{4+}-doped.

Cz: Czochralski, GP: Growth from the gas phase

References

[1] Sottile A, Damiano E, Rabe M, Bertram R, Klimm D, Tonelli M. Widely tunable, efficient 2 µm laser in monocrystalline Tm^{3+}:SrF_2. Opt Express, 2018, 26, 5368–80.

[2] Kaminskii AA. Laser Crystals – Their Physics and Properties, 2nd edition, Springer-Verlag, Heidelberg, 1990.

[3] Shannon RD. Revised effective ionic radii and systematic studies of interatomic distances in halides and chalcogenides. Acta Crystallogr, 1976, A32, 751–67.

[4] Moos HW. Spectroscopic relaxation processes of rare earth ions in crystals. J Lumin, 1970, 1–2, 106–21.

[5] Dieke GH, Crosswhite HM. The spectra of the doubly and triply ionized rare earths. Appl Opt, 1963, 2, 675–86.

[6] Morrison CA, Leavitt RP. Properties of Triply Ionized Lanthanides in Transparent Host Crystals, North-Holland Publishing Co, Amsterdam, 1982.

[7] Tanabe Y, Sugano S. On the absorption spectra of complex ions II. J Phys Soc Jpn, 1954, 9, 766–79.

[8] Dorenbos P. 5d-level energies of Ce^{3+} and the crystalline environment. II. Chloride, bromide, and iodide compounds. Phys Rev B, 2000, 62, 15650–59.

[9] Dorenbos P. The 5d level positions of the trivalent lanthanides in inorganic compounds. J Lumin, 2000, 91, 155–76.

[10] Dorenbos P. 5d-level energies of Ce^{3+} and the crystalline environment. I. Fluoride compounds. Phys Rev B, 2000, 62, 15640–48.

[11] Dorenbos P. 5d-level energies of Ce^{3+} and the crystalline environment. III. Oxides containing ionic complexes. Phys Rev B, 2000, 64, 125117.

[12] Dorenbos P. The $4f^n$ <-> $4f^{n-1}5d$ transitions of the trivalent lanthanides in halogenides and chalcogenides. J Lumin, 2000, 91, 91–106.

[13] Dhanaraj G, Byrappa K, Prasad V, Dudley M. Handbook of Crystal Growth, Springer, Berlin, Heidelberg, 2010.

[14] Rudolph P. Handbook of Crystal Growth: Bulk Crystal Growth (Volume 2A-2B), 2nd edition, Elsevier, Amstardam, 2014.

[15] Peters R, Kränkel C, Petermann K, Huber G. Crystal growth by the heat exchanger method, spectroscopic characterization and laser operation of high-purity Yb:Lu_2O_3. J Cryst Growth, 2008, 310, 1934–38.

[16] Czochralski J. Ein neues Verfahren zur Messung der Kristallisationsgeschwindigkeit der Metalle. Z Phys Chem Stoechiom Verwandtschafts, 1917, 92, 219–21.

[17] Kränkel C, Uvarorova A, Haurat E, Hülshoff L, Brützam M, Guguschev C, Kalusniak S, Klimm D. Czochralski growth of mixed cubic sesquioxide crystals in the ternary system Lu_2O_3-Sc_2O_3-Y_2O_3. Acta Crystallogr B, 2021, B77, 550–58.

[18] Metz PW. Visible lasers in rare earth-doped fluoride crystals. Dissertation, Institut für Laser-Physik, Universität Hamburg, 2014.

[19] Bridgman PW. The compressibility of thirty metals as a function of pressure and temperature. Proc Amer Acad, 1923, 58, 165–242.

[20] Arzakantsyan M, Ananyan N, Gevorgyan V, Chanteloup JC. Growth of large 90 mm diameter Yb: YAG single crystals with Bagdasarov method. Opt Mat Express, 2012, 2, 1219–25.

[21] Schmid F, Viechnicki D. Growth of sapphire disks from melt by a gradient furnace technique. J Am Ceram Soc, 1970, 53, 528–29.

[22] Peters R, Petermann K, Huber G. Growth technology and laser properties of Yb-doped sesquioxides. In: Capper P, Rudolph P. (Eds.) Crystal Growth Technology – Semiconductors and Dielectrics, WILEY-VCH Verlag GmbH & Co. KGaA, Weinheim, 2010, 267–82.

[23] Koohpayeh SM, Fort D, Abell JS. The optical floating zone technique: A review of experimental procedures with special reference to oxides. Prog Cryst Growth Charact Mater, 2008, 54, 121–37.
[24] Fejer MM, Nightingale JL, Magel GA, Byer RL. Laser-heated miniature pedestal growth apparatus for single-crystal optical fibers. Rev Scient Inst, 1984, 55, 1791–96.
[25] Cantwell G, Harsch WC, Cotal HL, Markey BG, McKeever SWS, Thomas JE. Growth and characterization of substrate-quality ZnSe single-crystals using seeded physical vapor transport. J Appl Phys, 1992, 71, 2931–36.
[26] Li HY, Jie WQ. Growth and characterizations of bulk ZnSe single crystal by chemical vapor transport. J Cryst Growth, 2003, 257, 110–15.
[27] Demirbas U, Sennaroglu A, Somer M. Synthesis and characterization of diffusion-doped Cr^{2+}:ZnSe and Fe^{2+}:ZnSe. Opt Mater, 2006, 28, 231–40.
[28] Firsov KN, Gavrishchuk EM, Ikonnikov VB, Kazantsev S, Kononov IG, Rodin SA, Savin DV, Sirotkin AA, Timofeeva NA. CVD-grown Fe^{2+}:ZnSe polycrystals for laser applications. Laser Phys Lett, 2017, 14, 055805.
[29] Henderson B, Gallagher HG, Han TPJ, Scott MA. Optical spectroscopy and optimal crystal growth of some Cr^{4+}-doped garnets. J Phys: Condens Matter, 2000, 12, 1927–38.
[30] Kaminskii AA. Laser crystals and ceramics: Recent advances. Laser Phot Rev, 2007, 1, 93–177.
[31] Beil K, Fredrich-Thornton ST, Peters R, Petermann K, Huber G. Yb-doped Thin Disk Laser Materials: A Comparison between Yb:LuAG and Yb:YAG. Advanced Solid-State Photonics (ASSP), Denver, USA, 2009, 28.
[32] Kvapil J, Kvapil J, Kubecek V. Laser properties of YAG:Nd grown from the melt contained in molybdenum crucibles. Czech J Phys B, 1979, 29, 1282–92.
[33] Dinerman BJ, Moulton PF. 3-μm cw laser operations in erbium-doped YSGG, GGG, and YAG. Opt Lett, 1994, 19, 1143–45.
[34] Rudenkov A, Kisel V, Yasukevich A, Hovhannesyan K, Petrosyan A, Kuleshov N. Spectroscopy and continuous wave laser performance of Yb^{3+}:$LuAlO_3$ crystal. Opt Lett, 2016, 41, 5805–08.
[35] Fibrich M, Jelinkova H, Sulc J, Nejezchleb K, Skoda V. Diode-pumped Pr:YA Plasers. Laser Phys Lett, 2011, 8, 559–68.
[36] Paunescu G, Hein J, Sauerbrey R. 100-fs diode-pumped Yb: KGW mode-locked laser. Appl Phys B, 2004, 79, 555–58.
[37] Boutinaud P, Bettinelli M, Diaz F. Intervalence charge transfer in Pr^{3+}- and Tb^{3+}-doped double tungstate crystals $KRE(WO_4)_2$ (RE = Y, Gd, Yb, Lu). Opt Mater, 2010, 32, 1659–63.
[38] Kränkel C. Rare earth-doped sesquioxides for diode-pumped high-power lasers in the 1-, 2-, and 3-μm spectral range. IEEE J Sel Top Quant Electron, 2015, 21, 1602013.
[39] Liu ZY, Ikesue A, Li J. Research progress and prospects of rare earth doped sesquioxide laser ceramic. J Eur Ceram Soc, 2021, 41, 3895–910.
[40] Zvereva I, Smirnov Y, Choisnet J. Demixion of the K_2NiF_4 type aluminate $LaCaAlO_4$: Precursor role of the local ordering of lanthanum and calcium. Mater Chem Phys, 1999, 60, 63–69.
[41] Loiko P, Druon F, Georges P, Viana B, Yumashev KV. Thermo-optic characterization of Yb: $CaGdAlO_4$ laser crystal. Opt Express, 2014, 4, 2241–49.
[42] Liebald C. Yb-dotierte Ultrakurzpuls-Lasermaterialien mit K_2NiF_4-Struktur – Züchtung und Verbesserung der Kristallqualität. Dissertation, Fachbereich Chemie, Pharmazie und Geowissenschaften, Johannes Gutenberg-Universität, Mainz, 2017.
[43] Kränkel C, Marzahl DT, Moglia F, Huber G, Metz P. Out of the blue: Semiconductor-laser-pumped visible rare earth doped lasers. Laser Phot Rev, 2016, 10, 548–68.
[44] Thoma RE, Weaver CF, Friedman HA, Insley H, Harris LA, Yakel HA Jr. Phase equilibria in the system LiF-YF_3. J Phys Chem, 1961, 65, 1096–99.

[45] Okada F, Togawa S, Ohta K, Koda S. Solid-state ultraviolet tunable laser – a Ce^{3+} doped $LiYF_4$ crystal. J Appl Phys, 1994, 75, 49–53.
[46] Jensen T, Diening A, Huber G, Chai BHT. Investigation of diode-pumped 2.8-μm $Er:LiYF_4$ lasers with various doping levels. Opt Lett, 1996, 21, 585–87.
[47] Siebold M, Bock S, Schramm U, Xu B, Doualan JL, Camy P, Moncorgé R. $Yb:CaF_2$ – a new old laser crystal. Appl Phys B, 2009, 97, 327–38.
[48] Maiman TH. Stimulated optical radiation in ruby. Nature, 1960, 187, 493–94.
[49] Waeselmann SH, Heinrich S, Kränkel C, Huber G. Lasing of Nd^{3+} in sapphire. Laser Photon Rev, 2016, 10, 510–16.
[50] Morgner U, Kärtner FX, Cho SH, Chen Y, Haus HA, Fujimoto JG, Ippen EP, Scheuer V, Angelow G, Tschudi T. Sub-two-cycle pulses from a Kerr-lens mode-locked Ti:Sapphire laser. Opt Lett, 1999, 24, 411–13.
[51] Sutter DH, Steinmeyer G, Gallmann L, Matuschek N, Morier-Genoud F, Keller U, Scheuer V, Angelow G, Tschudi T. Semiconductor saturable-absorber mirror-assisted Kerr-lens mode-locked Ti:Sapphire laser producing pulses in the two-cycle regime. Opt Lett, 1999, 24, 631–33.
[52] Demirbas U. Cr:colquiriite lasers: Current status and challenges for further progress. Prog Quant Electron, 2019, 68, 100227.
[53] Kück S. Laser-related spectroscopy of ion-doped crystals for tunable solid-state lasers. Appl Phys B, 2001, 72, 515–62.
[54] Petricevic V, Seas A, Alfano RR. Slope efficiency measurements of chromium-doped forsterite laser. Proc Adv Sol Stat Lasers, 1999, 10, C4L3.
[55] Chia SH, Liu TM, Ivanov AA, Fedotov AB, Zheltikov AM, Tsai MR, Chan MC, Yu CH, Sun CK. A sub-100fs self-starting Cr:Forsterite laser generating 1.4 W output power. Opt Express, 2010, 18, 24085–91.
[56] Marshall CD, Speth JA, Payne SA, Krupke WF, Quarles GJ, Castillo V, Chai BHT. Ultraviolet laser emission properties of Ce^{3+}-doped $LiSrAlF_6$ and $LiCaAlF_6$. J Opt Soc Am B, 1994, 11, 2054–65.
[57] Demirbas U, Sennaroglu A. Intracavity-pumped Cr^{2+}:ZnSe laser with ultrabroad tuning range between 1880 and 3100 nm. Opt Lett, 2006, 31, 22932295.
[58] Akimov VA, Kozovskii VI, Korostelin YV, Landman AI, Podmar'kov Y, Skasyrskii Y, Frolov M. Efficient pulsed Cr^{2+}:CdSe laser continuously tunable in the spectral range from 2.26 to 3.61 μm. Quant Electron, 2008, 38, 205–08.
[59] Vasilyev S, Moskalev I, Mirov M, Mirov S, Gapontsev V. Three optical cycle mid-IR Kerr-lens mode-locked polycrystalline Cr^{2+}:ZnS laser. Opt Lett, 2015, 40, 5054–57.
[60] Kernal J, Fedorov VV, Gallian A, Mirov SB, Badikov VV. 3.9-4.8 μm gain-switched lasing of Fe: ZnSe at room temperature. Opt Express, 2005, 13, 10608–15.
[61] Jain A, Ong SP, Hautier G, Chen W, Davidson Richards W, Dacek S, Cholia S, Gunter D, Skinner D, Ceder G, Persson KA. Commentary: The materials project: A materials genome approach to accelerating materials innovation. APL Materials, 2013, 1, 011002.
[62] Downs RT, Hall-Wallace M. The American mineralogist crystal structure database. Am Miner, 2003, 88, 247–50.
[63] Grazulis S, Chateigner D, Downs RT, Yokochi AFT, Quirós M, Lutterotti L, Manakova E, Butkus J, Moeck P, Le Bail A. Crystallography open database – an open-access collection of crystal structures. J Appl Crystallogr, 2009, 42, 726.
[64] Bermudez-G JC, Pinto-Robledo VJ, Kir'yanov AV, Damzen MJ. The thermo-lensing effect in a grazing incidence, diode-side-pumped $Nd:YVO_4$ laser. Opt Commun, 2002, 210, 75.

[65] Czeranowski C. Resonatorinterne Frequenzverdopplung von diodengepumpten Neodym-Lasern mit hohen Ausgangsleistungen im blauen Spektralbereich. Dissertation, Institut für Laser-Physik, Universität Hamburg, 2002.

[66] Fredrich-Thornton ST. Nonlinear losses in single crystalline and ceramic Yb:YAG thin-disk lasers. Dissertation, Institut für Laser-Physik, Universität Hamburg, 2010.

[67] Kränkel C. Ytterbium-dotierte Borate und Vanadate mit großer Verstärkungsbandbreite als aktive Materialien im Scheibenlaser. Dissertation, Institut für Laser-Physik, Universität Hamburg, 2008.

[68] Zhang HJ, Zhu L, Meng XL, Yang ZH, Wang CQ, Yu WT, Chow YT, Lu MK. Thermal and laser properties of Nd: YVO_4 crystal. Cryst Res Technol, 1999, 34, 1011.

[69] Zhao SR, Zhang HJ, Wang JY, Haikuan K, Xiufeng C, Junhai L, Jing L, Yanting L, Xiaobo H, Xiangang X, Xinqiang W, Zongshu S, Minhua J. Growth and characaterization of the new laser crystal Nd:$LuVO_4$. Opt Mater, 2004, 26, 319.

[70] Mond M. Cr^{2+}-dotierte Chalkogenide – Neue durchstimmbare Festkörperlaser und passive Güteschalter im mittleren infraroten Spektralbereich. Dissertation, Institut für Laser-Physik, Universität Hamburg, 2003.

[71] Wang YC. Passive mode-locking of 2-μm solid-state lasers: towards sub-10 optical cycle pulse generation. Dissertation, Department of Physics, FU Berlin, 2017.

[72] Hu QQ, Jia ZT, Tang C, Lin N, Zhang J, Jia N, Wang S, Zhao X, Tao X. The origin of coloration of $CaGdAlO_4$ crystals and its effect on their physical properties. Cryst Eng Comm, 2016, 19, 537–45.

[73] Loiko PA, Yumashev KV, Schödel R, Peltz M, Liebald C, Mateos X, Deppe B. Thermo-optic properties of Yb:Lu_2O_3 single crystals. Appl Phys B-Lasers Opt, 2015, 120, 601–07.

[74] Reichert F. Praseodymium- and Holmium-doped crystals for lasers emitting in the visible spectral region. Dissertation, Institut für Laser-Physik, Universität Hamburg, 2013.

[75] Marzahl DT. Rare-earth-doped strontium hexaaluminate lasers. Dissertation, Institut für Laser-Physik, Universität Hamburg, 2016.

[76] Ileri B. Lattice matching of epitaxial rare earth-doped dielectric PLD-films. Dissertation, Institut für Laser-Physik, Universität Hamburg, 2007.

[77] Selivanov AG, Denisov IA, Kuleshov NV, Yumashev KV. Nonlinear refractive properties of Yb^{3+}-doped $KY(WO_4)_2$ and YVO_4 laser crystals. Appl Phys B, 2006, 83, 61–65.

[78] Ghirnire K, Iqbal H, Collins R, Podraza N. Optical properties of single-crystal $Gd_3Ga_5O_{12}$ from the infraret to ultraviolet. Phys Stat Sol (B), 2015, 252, 2191–98.

[79] Panchal V, Errandonea D, Segura A, Rodríguez-Hernandez P, Muñoz A, Lopez-Moreno S, Bettinelli M. The electronic structure of zircon-type orthovanadates: Effects of high-pressure and cation substitution. J Appl Phys, 2011, 110, 043723.

[80] Koopmann P. Thulium- and holmium-doped sesquioxides for 2 μm lasers. Dissertation, Institut für Laser-Physik, Universität Hamburg, 2012.

[81] Masubuchi Y, Hata T, Motohashi T, Kikkawa S. Synthesis and crystal structure of K_2NiF_4-type novel $Gd_{1+x}Ca_{1-x}AlO_{4-x}N_x$ oxynitrides. J Alloys Compd, 2014, 582, 823–26.

[82] Luong MV, Sarukura N, Pham MH, Schwerdtfeger P, Cadatal-Raduban M. Numerical investigation of the electronic and optical properties of $LiLuF_4$ vacuum ultraviolet material. Jpn J Appl Phys, 2020, 59, 072001.

[83] Shimizu T, Luong MV, Cadatal-Raduban M, Empizo MJF, Yamanoi K, Arita R, Minami Y, Sarukura N, Mitsuo N, Azechi H, Pham MH, Nguyen HD, Ichiyanagi K, Nozawa S, Fukaya R, Adachi S-I, Nakamura KG, Fukuda K, Kawazoe Y, Steenbergen KG, Schwerdtfeger P. High pressure band gap modification of $LiCaAlF_6$. Appl Phys Lett, 2016, 110, 141902.

[84] Luong MV, Empizo MJF, Cadatal-Raduban M, Arita R, Minami Y, Shimizu T, Sarukura N, Azechi H, Pham MH, Nguyen HD, Kawazoe Y, Steenbergen KG, Schwerdtferger P. First-principles

calculations of electronic and optical properties of $LiCaAlF_6$ and $LiSrAlF_6$ crystals as VUV to UV solid-state laser materials. Opt Mater, 2017, 65, 15–20.
[85] Shankland TJ. Band gap of forsterite. Science, 1968, 161, 51–53.
[86] Loiko P, Ghanbari S, Matrosov V, Yumashev K, Major A. Dispersion and anisotropy of thermo-optical properties of alexandrite laser crystal. Opt Mater Express, 2018, 8, 3000–06.
[87] Togo A. Phonon database at 2018-04-17. http://phonondb.mtl.kyoto-u.ac.jp/ph20180417/index.html 2018. Accessed January 13th 2022.
[88] Hinuma Y, Pizzi G, Kumagai Y, Oba F, Tanaka I. Band structure diagram paths based on crystallography. Comp Mater Sci, 2017, 128, 140.
[89] Fechner M. Seltenerd-dotierte Oxidkristalle für Festkörperlaser im sichtbaren Spektralbereich. Dissertation, Institut für Laser-Physik, Universität Hamburg, 2011.
[90] Dimitrov D, Rafailov P, Marinova V, Babeva T, Goovaerts E, Chen YF, Lee CS, Juang JY. Structural and optical properties of $LuVO_4$ single crystals. J Phys: Conf Ser, 2017, 794, 012029.
[91] Song JJ, Klein PB, Wadsack RL, Selders M, Mroczkowski S, Chang RK. Raman-active phonons in aluminum, gallium, and iron garnets. J Opt Soc Am B, 1973, 63, 1135–40.
[92] Papagelis K, Arvanitidis J, Vinga E, Christofilos D, Kourouklis GA, Kimura H, Ves S. Vibrational properties of $(Gd_{1-x}Y_x)_3Ga_5O_{12}$ solid solutions. J Appl Phys, 2010, 107, 113504.
[93] Adachi S, Taguchi T. Optical properties of ZnSe. Phys Rev B, 1991, 43, 9569–77.
[94] Widulle F, Kramp S, Pyka NM, Göbel A, Ruf T, Debernardi A, Lauck R, Cardona M. The phonon dispersion of wurtzite CdSe. Phys Rev B, 1999, 263–264, 448–51.
[95] Fang ZQ, Sun DL, Luo JQ, Zhang H, Zhao X-Y, Quan C, Hu L, Cheng M, Zhang Q-L, Yin S. Study on the raman spectra of a new type GYSGG radiation resistant crystal. Mater Sci, 2017, 7, 515–22.
[96] Macalik L, Hanuza J, Kaminskii AA. Polarized Raman spectra of the oriented $NaY(WO_4)_2$ and $KY(WO_4)_2$ single crystals. J Mol Struc, 2000, 555, 289–97.
[97] Macalik L, Hanuza J, Kaminskii AA. Polarized infrared and Raman spectra of $KGd(WO_4)_2$ and their interpretation based on normal coordinate analysis. J Raman Spec, 2002, 33, 92–103.
[98] Llamas V, Loiko P, Kifle E, Romero C, Vázquez de Aldana JR, Pan Z, Serres JM, Yuan H, Dai X, Cai H, Wang Y, Zhao Y, Zakharov V, Veniaminov A, Thouroude R, Laroche M, Gilles H, Aguiló M, Díaz F, Griebner U, Petrov V, Camy P, Mateos X. Ultrafast laser inscribed waveguide lasers in Tm:CALGO with depressed-index cladding. Opt Express, 2020, 3, 3528–40.
[99] Kontos AG, Tsaknakis G, Raptis YS, Papayannis A, Landulfo E, Baldochi SL, Barbosa E, Vieira ND Jr. Spectroscopic study of Ce- and Cr-doped $LiSrAlF_6$. J Appl Phys, 2003, 93, 2797–803.
[100] Doualan JL, Moncorgé R. Laser crystals with low phonon frequencies. Ann Chim Sci Mater, 2003, 28, 5–20.
[101] Daniel P, Gesland JY, Rousseau M. Raman scattering study of a new laser crystal: $LiCaAlF_6$. J Raman Spec, 1992, 23, 197.
[102] Calistru DM, Wang WB, Petricevic V, Alfano RR. Resonance Raman scattering in Cr^{4+}-doped forsterite. Phys Rev B, 1995, 51, 14980–86.
[103] Malickova I, Fridrichova J, Bacik P, Milovská S, Škoda R, Illášová Ľ, Štubňa J. Laser effect in the optical luminescence of oxides containing Cr. Acta Geol Slov, 2018, 10, 27–34.
[104] Metz PW. Visible lasers in rare earth-doped fluoride crystals. Dissertation, Institut für Laser-Physik, Universität Hamburg, 2014.
[105] Yu HH, Wu K, Yao B, Zhang H, Wang Z, Wang J, Zhang Y, Wei Z, Zhang Z, Zhang X, Jiang M. Growth and characterization of Yb-doped $Y_3Ga_5O_{12}$ laser crystal. IEEE J Quantum Electron, 2010, 46, 1689–95.

[106] Aggarwal RL, Ripin DJ, Ochoa JR, Fan TY. Measurement of thermo-optic properites of $Y_3Al_5O_{12}$, $Lu_3Al_5O_{12}$, $YAlO_3$, $LiYF_4$, $LiLuF_4$, BaY_2F_8, $KGd(WO_4)_2$, and $KY(WO_4)_2$ laser crystals in the 80–300 K temperature range. J Appl Phys, 2005, 98, 103514.

[107] Biswal S, O'Connor SP, Bowman SR. Thermo-optical parameters measured in ytterbium-doped potassium gadolinium tungstate. Appl Opt, 2005, 44, 3093–97.

[108] Sirota NN, Popov PA, Ivanov IA. The thermal conductivity of monocrystalline gallium garnets doped with rare earth elements and chromium in the range 6–300 K. Cryst Res Technol, 1992, 27, 535–43.

[109] Zagumennyi AI, Lutts GB, Popov PA, Sirota NN, Shecherbakov IA. The thermal conductivity of YAG and YSAG laser crystals. Laser Phys, 1993, 3, 1064–65.

[110] Langenberg E, Ferreiro-Vila E, Leborán V, Fumega AO, Pardo V, Rivadulla F. Analysis of the temperature dependence of the thermal conductivity of insulating single crystal oxides. APL Mater, 2016, 4, 5, 104815.

[111] Vinnik DA, Popov PA, Archugov SA, Mikhailov GG. Heat conductivity of chromium-doped alexandrite single crystals. Dokl Phys, 2009, 54, 449–50.

[112] Peters R. Ytterbium-dotierte Sesquioxide als hocheffiziente Lasermaterialien. Dissertation, Institut für Laser-Physik, Universität Hamburg, 2009.

[113] Silvestre O, Grau J, Pujol MC, Massons J, Aguiló M, Díaz F, Borowiec MT, Szewczyk A, Gutowska MU, Massot M, Salazar A, Petrov V. Thermal properties of monoclinic $KLu(WO_4)_2$ as a promising solid state laser host. Opt Express, 2008, 16(7), 5022–34.

[114] Loiko PA, Yumashev KV, Kuleshov NV, Pavlyuk AA. Thermo-optical properties of pure and Yb-doped monoclinic $KY(WO_4)_2$ crystals. Appl Phys B, 2012, 106, 663–68.

[115] Baryshevski VG, Korzhik MV, Livshitz MG, Tarasov AA, Kimaev AE, Mishkel II, Meilman ML, Minkov BJ, Shkadarevich AP. Properties of forsterite and the performance of forsterite lasers with lasers and flash lamp pumping. Adv Solid State Lasers, 1991, C4L1.

[116] Sirdeshmukh DB, Sirdeshmukh L, Subhadra KG, Kishan Rao K, Bal Laxman S. Systematic hardness measurements on some rare earth garnet crystals. Bull Mater Sci, 2001, 24, 469–73.

[117] Kück S, Hartung S, Hurling S, Peterman K, Huber G. Optical transitions in Mn^{3+}-doped garnets. Phys Rev B, 1998, 57, 2203–16.

[118] Cornacchia F, Simura R, Toncelli A, Tonelli M, Yoshikawa A, Fukuda T. Spectroscopic properties of $Y_3Sc_2Al_3O_{12}$ (YSAG) single crystals grown by µ-PD technique. Opt Mater, 2007, 30, 135–38.

[119] Tang LY, Lin ZB, Zhang LZ, Wang GF. Phase diagram, growth and spectral characteristic of Yb^{3+}:$KY(WO_4)_2$. J Cryst Growth, 2005, 282, 376–82.

[120] Klimm D, Lacayo G, Reiche P. Growth of Cr:$LiCaAlF_6$ and Cr:$LiSrAlF_6$ by the Czochralski method. J Cryst Growth, 2000, 210, 683–93.

[121] Klimm D, Dos Santos IA, Ranieri IM, Baldochi SL. Phase equilibria and crystal growth for $LiREF_4$ scheelite laser crystals. MRS Online Proc Lib, 2008, 111, 107.

[122] Brown DC, Tornegard S, Kolis J, McMillen C, Moore C, Sanjeewa L, Hancock C. The application of cryogenic laser physics to the development of high average power ultra-short pulse lasers. Appl Sci, 2016, 6, 23.

[123] Tian L, Yuan H, Wang SX, Wu K, Pan Z, Cai H, Yu H, Zhang H. Evaluation of growth, thermal and spectroscopic properties of Yb^{3+}-doped GSGG crystals for use in ultrashort pulsed and tunable lasers. Opt Mat Express, 2014, 4, 1953–65.

[124] Petrosyan AG, Shirinyan GO, Pedrini C, Durjardin C, Ovanesyan KL, Manucharyan RG, Butaeva TI, Derzyan MV. Bridgman growth and characterization of $LuAlO_3$-Ce^{3+} scintillator crystals. Cryst Res Technol, 1998, 33, 241–48.

[125] Loiko PA, Yumashev KV, Schödel R, Peltz M, Liebald C, Mateos X, Deppe B, Kränkel C. Thermo-optic properties of Yb:Lu_2O_3 single crystals. Appl Phys B, 2015, 120, 601–07.
[126] Roberts RB, White GK, Sabine TM. Thermal expansion of zinc sulfide: 300–1300 K. Austr J Phys, 1981, 34, 701–06.
[127] Su CH, Feth S, Lehoczky SL. Thermal expansion coefficient of ZnSe crystal between 17 and 1080 °C by interferometry. Mater Lett, 2009, 63, 1475–77.
[128] Iwanaga H, Kunishige A, Takeuchi S. Anisotropic thermal expansion in wurtzite-type crystals. J Mater Sci, 2000, 35, 2451–54.
[129] Hoefer CS, Kirby KW, DeShazer LG. Thermo-optic properties of gadolinium garnet laser crystals. J Opt Soc Am B, 1988, 5, 2327–32.
[130] Fan TY, Ripin DJ, Aggarwal RL, Ochoa JR, Chann B, Tilleman M, Spitzberg J. Cryogenic Yb^{3+}-doped solid-state lasers. IEEE J Sel Top Quant Electron, 2007, 13, 448–59.
[131] Feldman A, Horowitz D, Waxler RM, Dodge J. Optical Materials Characterization Final Technical Report. Nat Bur Stand (U.S.), Tech Note, 1979,993.
[132] Palik ED Ed. Handbook of Optical Constants, Elsevier, Amsterdam, 1997.
[133] Sennaroglu A. Solid-state Lasers and Applications, 1st edition, CRC Press, Boca Raton, 2017. ISBN 9780367389871.
[134] Zverev PG. The influence of temperature on Raman modes in YVO_4 and $GdVO_4$ crystals. J Phys Conf Ser, 2007, 92, 012073.
[135] Tung JC, Wu TY, Liang HC, Chen YF. Precise measurement of the thermo-optical coefficients of various Nd-doped vanadates with an intracavity self-mode-locked scheme. Laser Phys, 2014, 24, 035804.

8.6 NLO materials

Shilie Pan, Wenqi Jin, Zhihua Yang

8.6.1 Introduction

In 1961, Franken et al. discovered second harmonic generation (SHG) in a quartz crystal, which reflects the response-dependent nonlinearly on the applied electromagnetic field, namely, the nonlinear optical (NLO) phenomena [1]. When a laser beam is transmitted through an NLO medium, it can generate a new beam via frequency conversion such as SHG, difference frequency generation (DFG), or sum-frequency generation (SFG) [2]. Owing to the role in broadening the wavelength range of the laser, NLO materials have been widely used in laser technology, i.e., laser communication, laser guidance, signal processing, laser medical treatment and laser micromachining. In a nonlinear medium, the polarization P depends nonlinearly on the electric field E [3, 4]:

$$P = P_0 + \chi^{(1)}E + \chi^{(2)}E^2 + \chi^{(3)}E^3 + \ldots\ldots \quad (8.6.1)$$

where the nth-order susceptibility $\chi^{(n)}$ ($n = 1, 2, 3, \ldots$) is a (n+1)-rank tensor with $3^{(n+1)}$ elements. In general, NLO phenomena mainly refer to the second-order or third-order nonlinear susceptibility, especially the SHG process, which concerns two photons to produce a third photon ($\omega + \omega \rightarrow 2\omega$) [5]. It should be noticed that the SHG (or higher even-order nonlinear polarization) can only occur in noncentrosymmetric crystals.

According to the application, the spectral region of the second-order NLO materials (NLO materials, for short) can be classified into, (1) ultraviolet (UV)/deep-UV region (200–400 nm/≤200 nm); (2) visible-near infrared (IR) region (400 nm-2 μm); (3) middle and far-IR region (2–20 μm).

During the SHG process, when the fundamental wave and second-harmonic have the same propagation velocity, all second-harmonic waves produced from every position that are passed by the fundamental wave can achieve phase superposition with increased SHG intensity. Thus, a phase-matching wavelength located at the corresponding application spectral region is expected for the targeted crystal, which depends on the bandgap, birefringence and refractive dispersion. Of course, when the birefringence is too small, other methods like quasi-phase matching are also adopted for a periodically poled crystal. In the case of UV/deep-UV inorganic NLO crystals, most of them are alkali/alkaline earth metal borates since β-BaB_2O_4 (β-BBO)[6] and LiB_3O_5 (LBO) [7, 8] identified as NLO crystals; alkali/alkaline earth metal carbonates/phosphates/sulfates can also perform UV/deep-UV properties. Introducing d^0, d^{10} or lone pair metal cation to the systems above can enhance the NLO coefficients, but the application region will be red-shifted to the visible or even to the IR

https://doi.org/10.1515/9783110798890-025

region. In the IR region, niobates, pnictides, chalcogenides are the most commonly researched systems.

This chapter introduces the most widely used and newly developed inorganic NLO crystals with promising applications, which are classified and discussed according to the different systems and the main synthesis methods.

8.6.2 Borates

In the past 30 years, borates have been regarded as state-of-the-art candidates for UV/deep-UV NLO materials, owing to their outstanding intrinsic properties [9, 10]. In borates, the large electronegativity difference between the boron and oxygen atoms is favorable for the transmittance of short-wavelength light [11]. Moreover, boron atoms can be tri- or tetra-coordinated with oxygen atoms, forming $(BO_3)^{3-}$ planar triangles and $(BO_4)^{5-}$ tetrahedra. The π-conjugated $(BO_3)^{3-}$ group with highly anisotropic electron distribution favors large second-order susceptibility and birefringence [12]. The spatial combinations of $(BO_3)^{3-}$ and $(BO_4)^{5-}$ groups contribute to the extremely wide variability of structure types in borate-based compounds. Usually, the B-O spatial architecture in the borates can be in 0D, 1D, 2D and 3D by connecting through oxygen atoms and even other metal cations, such as Be and Al. The first NLO borate described was $KB_5O_8 \cdot 4\,H_2O$ [13] in 1975, but intense research on NLO borate materials began with the advent of low-temperature β-BBO [6, 14] and LBO [7], which are up to now still by far the most frequently used NLO borate materials. A subsequent search for new borate compounds led to the continued discovery of new compounds with improved NLO properties (or in some cases NLO properties that are presumed to be good).

Particularly, many metal borates containing additional halogen anions (i.e., F, Cl, Br or I) have been discovered with novel compositions and crystal structures that could produce the next generation of NLO materials for the UV region and beyond. Generally, metal borates containing additional halogen anions are expressed as $M_xB_yO_zX_n$(M = metal cations, X = halogen anions). In fact, there are two kinds of such compounds but with obvious differences [9]: (i) borates with the halogen atoms exclusively connected to metal cations (regarded as double salt), called "metal borate halides", such as the well-known KBBF [15], (ii) borates in which the halogen atoms are directly bonded to boron atoms (where the halogen acts as a substitute for an oxygen atom). Notably, to date, only fluorooxoborates with B-F bonds were discovered such as BiB_2O_4F [16]. From a chemical point of view, metal borate halides are a kind of double salts containing metal halides and metal borates. It is expected that the introduction of halogen anions into borates can span more diversity in their structures and would be likely to yield new phases with interesting stoichiometries, structures and properties. It has been verified that the proportions of noncentrosymmetric structures increases from borates (35%) to borate fluorides (38%) and especially to fluorooxoborates

(65%) [10]. Particularly functionalized oxyfluoride $[BO_xF_{4-x}]^{(x+1)-}$ $(x = 1, 2, 3)$ chromophores are good units to balance the multiple criteria for deep-UV NLO crystals (birefringence, SHG coefficients and deep-UV cutoff edge), as reflected by the polarizability anisotropy, hyperpolarizability and the highest occupied molecular orbital and lowest unoccupied molecular orbital (HOMO-LUMO) energy gap, respectively [17]. Besides, the introduction of fluorooxoborate modules can reduce the refractive index dispersions, which is good for a short phase-matching wavelength [18, 19]. In addition, the fluorine atom acts as a depolymerizer of the B–O polyanionic architecture to form favorable configurations, which may be beneficial to structural anisotropy.

8.6.2.1 Borates

8.6.2.1.1 LiB_3O_5 (LBO)

LBO crystallizes in the space group $Pna2_1$ with unit cell dimensions $a = 8.4473(7)$, $b = 7.3788(6)$, $c = 5.1359(5)$ Å and $Z = 4$ [20]. The structure is composed by a continuous helix network of $(B_3O_7)^{5-}$ groups along the c axis, which is also the twofold axis in the cell; however, the axis is designated as the y-axis in the optical system (Figure 8.6.1). In 1989, Chen's group reported that the LBO crystal possesses several outstanding NLO properties and constitutes a promising material [7, 21].

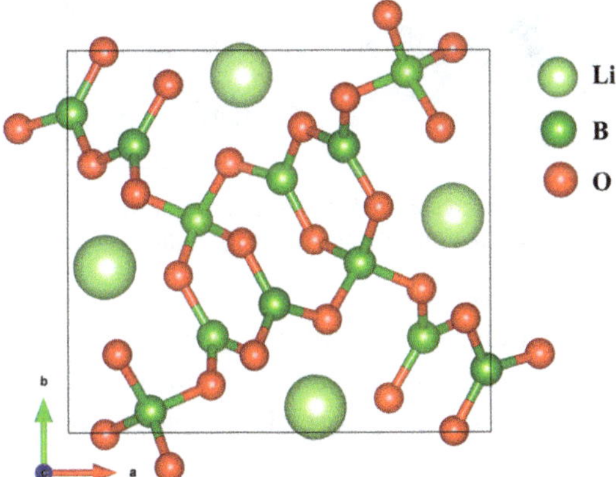

Figure 8.6.1: Structure of LiB_3O_5 viewed along the c axis.

LBO is a negative biaxial crystal with $\Delta n = 0.043–0.040$ @ 532–1064 nm. The SHG coefficients are $d_{31} = \pm 2.34\ (1 \pm 0.08)$ pm V^{-1}, $d_{32} = \pm 2.50\ (1 \pm 0.10)$ pm V^{-1}, $d_{15} \approx d_{32}$, $d_{24} \approx d_{32}$, $d_{33} = \pm 0.04\ (1 \pm 0.12)$ pm V^{-1} [8, 22]. The shortest phase-matching SHG wavelength is about 277 nm for type I and 396 nm for type II.

The first successful growth of small LBO crystals was achieved by the solid-state reaction of B_2O_3 glass covered with LiF powder and reaction at 750 °C for 10 h

[20]. Due to its incongruent melting, large bulk single crystals of LBO have to be grown by flux methods. From early on, a "self-fluxed" melt with excess B_2O_3 has been mainly been used and the top-seeded solution growth technique successfully applied. The composition of the used fluxes varies between Li_2O:B_2O_3 = 1:3.8 to 1:5.6, and slow cooling rates of about 0.3 K d^{-1} are applied to allow the excess B_2O_3 to move away from the growing interface [7, 23–26]. A great problem with the growth of LiB_3O_5 from B_2O_3 fluxes is the rather high viscosity and the resulting low diffusivity of the melt. To reduce melt viscosity, fluxes such as MoO_3 have also been applied. Hu et al. determined that the low-viscosity region suitable for LBO crystal growth in the Li_2O–B_2O_3–MoO_3 system corresponded to a Li_2O: B_2O_3: MoO_3 ratio in the range of 1: (0.8–1.3):(1.7–2.5) and successfully grew LBO crystals with dimensions up to 160 × 150 × 77 mm^3 by seeding in the [011] direction (Figure 8.6.2) [27].

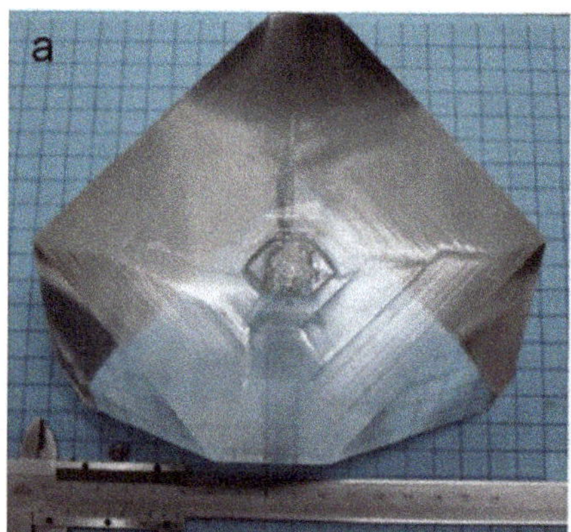

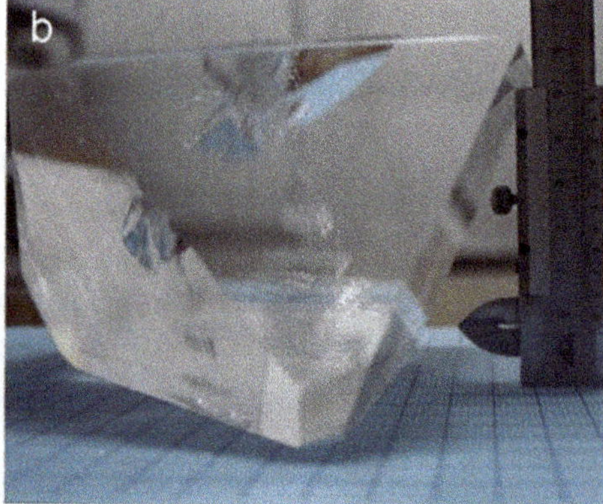

Figure 8.6.2: Top (a) and side (b) view of a LiB_3O_5 single crystal, weight: 1988 g, overall dimensions: 160 × 150 × 77 mm^3. Reproduced with permission from ref. [27]. Copyright 2011 Elsevier.

Because of the outstanding advantages of LBO crystals with respect to their NLO properties mentioned above, the major applications include the following two aspects: the second and third harmonic generation of Nd-based laser systems; optical parametric oscillation and amplification [22].

8.6.2.1.2 β-BaB_2O_4 (β-BBO)

BaB_2O_4 has two known phases, namely α and β. The α-phase, the quenched high-temperature form, was determined by Mighell et al. [28, 29] as $R\bar{3}c$. The β-phase (β-BBO), its low-temperature form, was crystallized in the space group $R3c$ with cell parameters a = 12.532, c = 12.728 Å [30]. Its crystal structure consists of isolated planar anionic $(B_3O_6)^{3-}$ ring groups, and they are parallel to the ab plane. Figure 8.6.3 shows schematically the crystal structure of β-BBO.

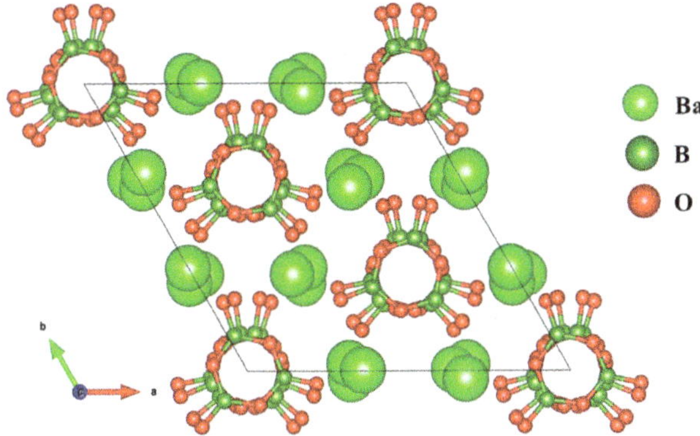

Figure 8.6.3: Structure of β-BBO viewed along the c axis.

The transmittance range of β-BBO is 185–3300 nm. β-BBO is a negative uniaxial crystal with $\Delta n = 0.118 \sim 0.113@546 \sim 1014$ nm [22]. The SHG coefficients of β-BBO are $d_{22} = \pm 1.60\ (1 \pm 0.05)\ \mathrm{pm\,V^{-1}}$, $d_{15} \approx (1 \pm 0.12)\ d_{22}$, $d_{33} \approx 0\ \mathrm{pm\,V^{-1}}$ [6]. $(B_3O_6)^{3-}$ groups, especially the terminal oxygen atoms have sizable contributions to the SHG coefficients [31]. The measurements show that the shortest SHG wavelength output is 204.8 nm [32], while for SFG the wavelength goes down to 189 nm, very close to the absorption edge of the crystal [33].

β-BBO has a melting point of 1095 °C and a reversible but kinetic transition temperature of around 925 °C [6, 28, 29]. Nowadays, the fluxes are usually Na_2O, NaF, and mixtures of both. The phase diagram of the BaB_2O_4–Na_2O system was first studied by Huang and Liang in 1981 [34]. The phase equilibrium diagram in the BaB_2O_4-NaF system was reinvestigated by Bekker et al. in 2009 [35]. β-BBO crystals could be obtained from the solution of 31–45 mol% NaF with a crystallization temperature of 754 °C. High-quality β-BBO crystals with sizes of ∅10 × 16 mm were grown by the traveling solvent zone melting (TSZM) method using Na_2O and B_2O_3 as a flux [36]. Large high-quality β-BBO crystals are also grown by the modified top-seeded solution technique [37]. Figure 8.6.4 shows the commercialized β-BBO.

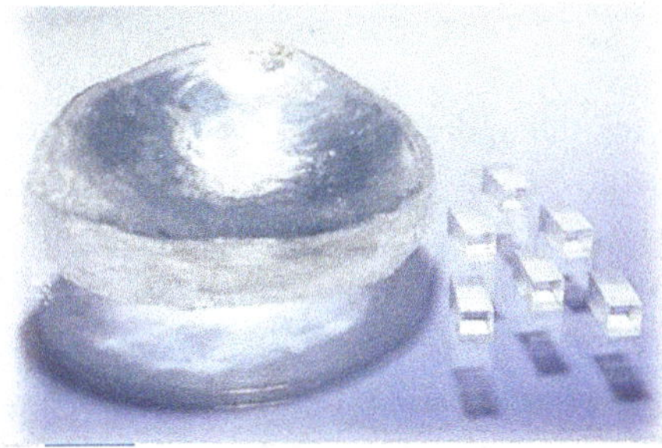

Figure 8.6.4: Single crystal of β-BBO. With courtesy and permission from Castech (https://gb.castech.com/product/106.html).

Because of the outstanding advantages of this crystal for the NLO properties mentioned above, the major applications of β-BBO include three aspects: (1) Fourth and fifth harmonic generation of Nd-based laser systems; (2) SHG of tunable Ti:sapphire lasers; (3) optical parametric oscillation and amplification [22].

8.6.2.1.3 CsB_3O_5 (CBO)

The existence of cesium triborate CBO was first reported by Krogh-Moe [38] in the system of $Cs_2O-B_2O_3$. In 1974, Krogh-Moe refined the crystal structure of CBO and determined that it crystallizes in the orthorhombic space group $P2_12_12_1$ with cell dimensions $a = 6.213(1)$, $b = 8.521(1)$, $c = 9.170(1)$ Å and $Z = 4$ [39]. The structure of CBO is shown in Figure 8.6.5. The structure of CBO can be described as a continuous three-dimensional network of spiral chains formed by $(B_3O_7)^{5-}$ groups and then four oxygen atoms shared by the $(B_3O_7)^{5-}$ units in the same chain or the neighboring chains, with Cs^+ located in the interstices. This boron–oxygen anionic network is very similar to the networks in LBO and $CsLiB_6O_{10}$ (CLBO).

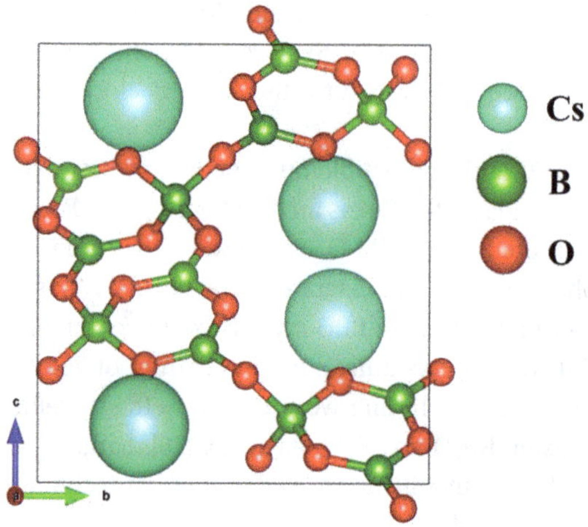

Figure 8.6.5: Structure of CsB_3O_5 viewed along the *a* axis.

The UV and IR absorption edges of CBO are 167 nm and 3400 nm, respectively [40]. CBO is a negative biaxial crystal with $\Delta n = 0.0608–0.0587$ @ 532–1064 nm [41, 42]. The SHG coefficients are $d_{14} = 1.17 \pm 0.11$ pm V^{-1} [41, 43]. The d_{eff} values of the second-order nonlinear coefficient were found to be 1.02 pm V^{-1} (type I) and 0.98 pm V^{-1} (type II) [44].

CBO is a congruently melting compound. Penin et al. [45] presented the complete $Cs_2O-B_2O_3$ phase diagram, confirming that the crystallization of the CBO phase takes place in the composition range from 67 to 81 mol% B_2O_3. Many growth techniques could be used to grow CBO single crystals, including the Czochralski

(CZ) method, the Kyropoulos method and the top-seeded solution growth method. To lower the growth temperature and decrease the volatilization of Cs_2O towards CBO crystals free from scattering centers, many fluxes (NaF, V_2O_5, and MoO_3) have also been applied [22].

The optical properties indicated that CBO is a promising crystal for NLO frequency conversion in the UV region: (1) third harmonic generation (THG, 1064 nm + 532 nm → 355 nm); (2) SFG (1064 nm + 355 nm → 266 nm); (3) sum-frequency generation (SFG, 2000 nm + 213 nm → 193 nm) [22].

8.6.2.1.4 $CsLiB_6O_{10}$ (CLBO)

Cesium lithium borate CLBO was discovered as a new NLO crystal by Sasaki et al. and Mori et al. [46–49]. CLBO belongs to the tetragonal space group $I\bar{4}2d$ with cell dimensions of $a = 10.494(1)$, $c = 8.939(2)$ Å and $Z = 4$ [46]. The structure is composed by eight-coordinated Cs^+ ions and four coordinated Li^+ ions, and a 3D anion network of chains formed from $(B_3O_7)^{5-}$ groups. The Cs^+ and Li^+ ions locate at the alternate sites on the square channel along the c-axis. Figure 8.6.6 shows a projection of CLBO on the (100) plane and (001) plane.

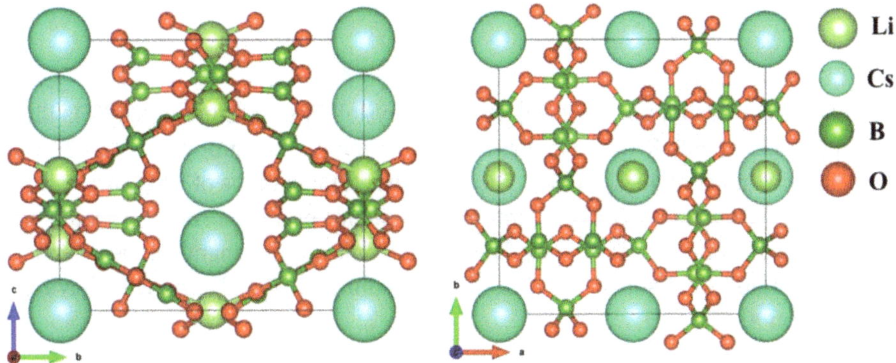

Figure 8.6.6: Structure of $CsLiB_6O_{10}$ viewed along the a and c axis.

The transparency range of CLBO is from 180 to 2750 nm. CLBO is a negative uniaxial crystal. The limits of the type I and type II phase-matching fundamental wavelengths for SHG are 473 and 636 nm, respectively [50]. The values for the coefficient d_{36} at several fundamental wavelengths are as follows: d_{36}(@532 nm) = 0.92 pm V^{-1}, d_{36}(@852 nm) = 0.83 pm V^{-1}, d_{36}(@1064 nm) = 0.74 pm V^{-1} [51].

The CLBO crystal was found to melt congruently at 848 °C. Conventionally, the crystal has been grown from a self-fluxed solution in B_2O_3 by using the top-seeded solution growth method with an a-axis seed crystal. It is much easier to grow when compared to other borates. A crystal dimension of 14 × 11 × 11 cm^3 has been demonstrated within

3 weeks and extra-large NLO crystal dimensions are technically possible [47]. Figure 8.6.7 shows the photograph of a large CLBO crystal produced by CASTECH, Inc.

Figure 8.6.7: Photograph of a large $CsLiB_6O_{10}$ crystal produced by CASTECH, Inc. With courtesy and permission from Castech (https://gb.castech.com/product/CLBO-%E6%99%B6%E4%BD%93-108.html).

The major applications for CLBO are as follows: (1) high-power fourth and fifth harmonic generation of Nd-based laser systems; (2) deep-UV light generation below 200 nm [22].

8.6.2.1.5 $K_2Al_2B_2O_7$ (KABO)

In 1998, KABO was reported [52, 53] as a NLO crystal. The structure of KABO is illustrated in Figure 8.6.8. KABO crystallizes in the trigonal space group *P*321 ($Z = 3$) with $a = 8.5657(9)$ and $c = 8.463(2)$ Å [54]. The basic structural features of KABO are K^+ cations, $(BO_3)^{3-}$ groups and $(AlO_4)^{5-}$ groups. The planes of all the $(BO_3)^{3-}$ groups are approximately parallel to the *c* axis. The whole atomic arrangement can be described as being formed from sheets of $(AlO_4)^{5-}$ tetrahedra and $(BO_3)^{3-}$ triangles having a composition $Al_2(BO_3)_2O$.

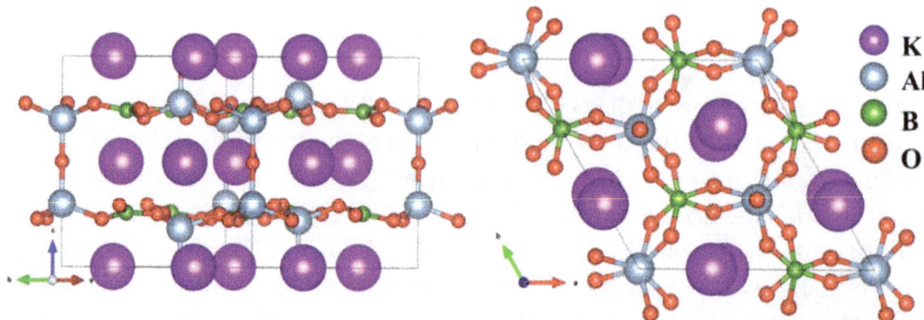

Figure 8.6.8: Crystal structure of $K_2Al_2B_2O_7$.

The UV and IR absorption edges of KABO are 180 nm and 3.6 μm, respectively [55]. KABO is a negative uniaxial crystal with Δn about 0.0704@546 nm. The shortest phase-matching wavelength of type I SHG is 232.5 nm [55, 56]. The measured value of d_{11} is 0.45 pmV^{-1} [22, 57].

KABO decomposes above 900 °C [58], below its melting point of 1109.7 °C, meaning that the crystal growth has to be carried out with flux methods. The best choice to grow KABO is using NaF as the flux [22]. The starting charge was prepared according to the following reaction: $K_2CO_3 + Al_2O_3 + B_2O_3 = K_2Al_2B_2O_7 + CO_2\uparrow$. Stoichiometric amounts of high purity reagents K_2CO_3, Al_2O_3, and B_2O_3 were weighed accurately and mixed thoroughly with the appropriate amount of NaF as a flux in a platinum crucible. Figure 8.6.9 shows the KABO crystal in size of 40 × 6 × 16 mm^3 [59].

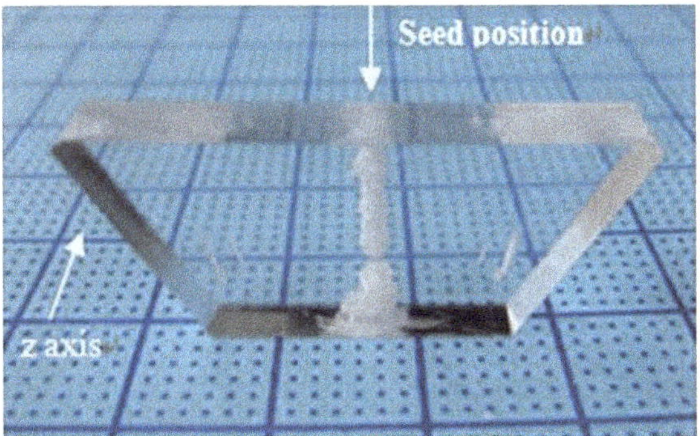

Figure 8.6.9: $K_2Al_2B_2O_7$ crystal in a size of 40 × 6 × 16 mm^3. Reproduced with permission from ref. [59]. Copyright 2011 Elsevier.

Besides the outstanding advantages mentioned above, the crystal has merits of easy growth and good radiation resistance to UV light. The major applications of KABO include two aspects: (1) 266 nm UV generation by fourth harmonic generation of Nd-based laser systems; (2) 193 nm UV generation by SFG [14].

8.6.2.2 Borate halides

8.6.2.2.1 $KBe_2BO_3F_2$ (KBBF)

The space group of KBBF is *R*32, belonging to the uniaxial class, with dimensions of $a = 4.427(4)$, $c = 18.744(9)$ Å [60, 61]. These are quite different from those informed in an earlier report [62], in which the space group of the crystal was described as *C*2. The network of each layer consists of $[Be_2F_2BO_3]$ (Figure 8.6.10). The distance between adjacent layers is rather large, about 6.25 Å, the apparent layering tendency is also the major reason why it is difficult to achieve a larger size along the *c*-axis.

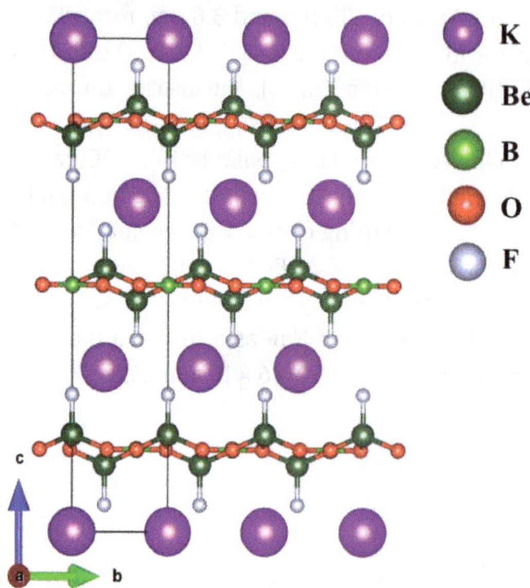

Figure 8.6.10: Crystal structure of $KBe_2BO_3F_2$ viewed along the *a* axis.

The UV and IR absorption edges of KBBF are 147 nm and 3.5 μm, respectively [22]. The birefringence of KBBF is 0.0848@546 nm, and KBBF crystals can produce SHG output down to a wavelength of 161.1 nm [60, 63]. The coefficient d_{11} in a $10 \times 10 \times 1.0\ mm^3$ crystal can be exactly deduced to be $d_{11} = (0.47 \pm 0.01)\ pm\ V^{-1}$ (if d_{36} (KDP) = $0.39\ pm\ V^{-1}$ is adopted) [64].

Crystals of KBBF can now be grown by both flux and hydrothermal methods. The first large KBBF crystal was grown by the former method, which is convenient because the crystal decomposes above (820 ± 3) °C before melting at ≈ 1030 °C [22]. Polycrystalline KBBF was prepared by a normal solid reaction according to the following equation [65]: $6BeO + 3KBF_4 + B_2O_3 \rightarrow 3KBe_2BO_3F_2 + 2BF_3\uparrow$.

Caution: All the operations must be performed in a ventilated system to protect the operators from BeO toxicity.

A strategy called "localized spontaneous nucleation" was developed during the crystal growth process, the key thrust being to control or restrict the nucleation of the crystal at a fixed point. Cyclic temperature variations were used in the early stages to select one single nucleus for further growth [64]. By using the technology above, only one large bulk KBBF crystal possessing a plate-like form can be obtained in most cases. A bulk KBBF crystal with dimensions of $50 \times 40 \times 3.7\ mm^3$ from which a high-quality KBBF bulk crystal can be cut for devices is shown in Figure 8.6.11 [64, 66]. The crystal of KBBF can also be grown by a hydrothermal method [64, 67]. However, the poor SHG efficiency of hydrothermally grown KBBF crystals hinders their application due to the structural defects.

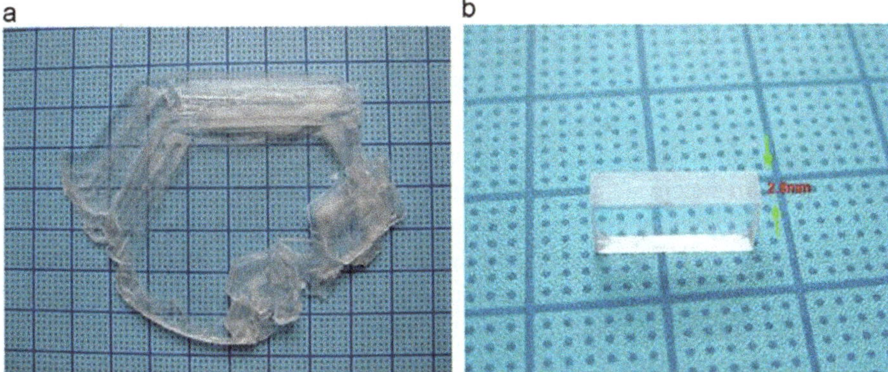

Figure 8.6.11: (a) As-grown $KBe_2BO_3F_2$ crystal with large transparent region. (b) A crystal blank with a thickness of 2.8 mm. Reproduced with permission from ref. [66]. Copyright 2011 Elsevier.

Since the grown KBBF crystal size is not large enough to be cut along the phase-matching direction and the crystal cleaves easily along the *c*-axis, a special prism-coupling technique is adopted [68]. With the optically contacted KBBF-prism-coupled device (KBBF-PCD), the sixth harmonic of Nd-based lasers and fourth harmonic of Ti-sapphire lasers have both been successfully generated. The tunable fourth harmonic generation of Ti:sapphire lasers is useful for the energy-tunable photoemission spectrometer and other deep-UV wavelength spectrometers. A 177.3 nm laser has been achieved, and the output laser of 193.5 nm was realized, which is very useful for 193 nm photolithography [64].

8.6.2.2.2 $RbBe_2(BO_3)F_2$ (RBBF)

Similar to KBBF [53], the space group of the RBBF structure is *R*32, belonging to the uniaxial and rhombohedral system, with unit cell dimensions, $a = 4.4341(9)$, $c = 19.758(5)$ Å and $Z = 4$. [69] The basic building units of RBBF are similar to KBBF, that is, $(BO_3)^{3-}$ and $(BeO_3F)^{5-}$ units form $(Be_2BO_3F_2)_\infty$ infinite layers along the *ab* plane. The distance between neighboring layers is up to 6.59 Å; Rb–F interactions are very weak to bind neighboring layers, resulting in a strong layering tendency along the *c*-axis, which makes it difficult to grow thicker crystals.

RBBF was reported as a promising deep-UV NLO crystal [70]. The cutoff edge of RBBF in the UV and IR region is located at 160 and 3550 nm, respectively. RBBF is a negative uniaxial crystal with a birefringence of 0.0751@546 nm. It is possible to achieve phase-matching SHG wavelength down to 170 nm. Its SHG coefficient is $d_{11} = (0.45\pm0.01)$ pm V^{-1} (d_{36} (KDP) = 0.39 pm V^{-1} as a reference) [69, 71].

Rb and K belong to the same main group in the Periodic Table and RBBF shares the same crystal structure as KBBF, so it also has very similar crystallization properties and growing method. Using the RBBF–PCD device, tunable fourth harmonic generation of Ti:sapphire lasers has been successfully realized. The sixth-harmonic of a Nd-based laser can also be produced using a RBBF-PCD device [64].

8.6.2.2.3 $CsBe_2BO_3F_2$ (CBBF) [72]

The space group of CBBF is also *R*32, belonging to the trigonal and uniaxial system. The lattice parameters are a = 4.4391 and c = 21.125 Å. In the CBBF structure, each $(BO_3)^{3-}$ group connecting two neighboring $(BeO_3F)^{5-}$ groups forms a $(Be_2BO_3F_2)_\infty$ layer along the *ab* plane, which is the same as the frameworks of KBBF and RBBF. Infinite layers of $(Be_2BO_3F_2)_\infty$ link together along the *c*-axis via electrostatic force of Cs–F bonds; thus, the structure displays a tendency, to form layers.

The transparency region of CBBF is 151–3700 nm. CBBF is a negative uniaxial crystal. Its birefringence is 0.0614@546 nm. It is possible to achieve SHG phase matching in CBBF down to the wavelength of 201 nm, which shows that CBBF is also a promising UV NLO crystal, achieving 266 and 213 nm harmonic generation of Nd-based lasers in particular. The SHG coefficient d_{11} is 0.5 pm V^{-1} (d_{36} (KDP) = 0.39 pm V^{-1} as reference).

8.6.2.2.4 $K_3B_6O_{10}Cl$ (KBOC)

KBOC is a structure analogous to perovskite and crystallizes in the rhombohedral space group *R*3*m* [73]. Referring to an ABX_3 format, the formula of KBOC can be represented as $[B_6O_{10}]ClK_3$. The structure is composed of $(B_6O_{10})^{2-}$ units and ClK_6 octahedra or described as two networks (a $[B_6O_{10}]_\infty$ and a ClK_6 net) that are interweaved (Figure 8.6.12), in which $(B_6O_{10})^{2-}$ consists of three $(BO_4)^{5-}$ tetrahedra (T) shared by the oxygen vertex and three $(BO_3)^{3-}$ triangles (Δ) attached to the terminal oxygen atom of these tetrahedra, which can be represented as 6[3Δ+3T] according to the definition given by Burns et al. [74, 75].

The KBOC crystal is transparent in the range of 180–3460 nm. KBOC is a negative uniaxial crystal. The shortest type-I and type-II phase-matching SHG wavelengths are 272 and 374 nm, respectively. The calculated SHG coefficients of KBOC are as follows: d_{22} = −1.033 pm V^{-1}, d_{33} = 0.488 pm V^{-1} and d_{31} = −0.033 pm V^{-1} [73, 76].

KBOC melts incongruently and the growth of KBOC is achieved by using PbF_2 as flux for the top-seeded solution growth method. The most regularly shaped and the high-quality crystal was grown from the [101] seed. The suitable cooling rate is 0.2–0.5 K/day and the suitable rotation rate is about 3–10 rpm. High optical quality KBOC crystals using [101]-oriented seeds with dimensions up to 35 mm × 35 mm × 11 mm have been obtained (Figure 8.6.13) [76]. The first KBOC crystal was grown from a high-temperature solution by using KF-PbO as the flux system (Figure 8.6.13).

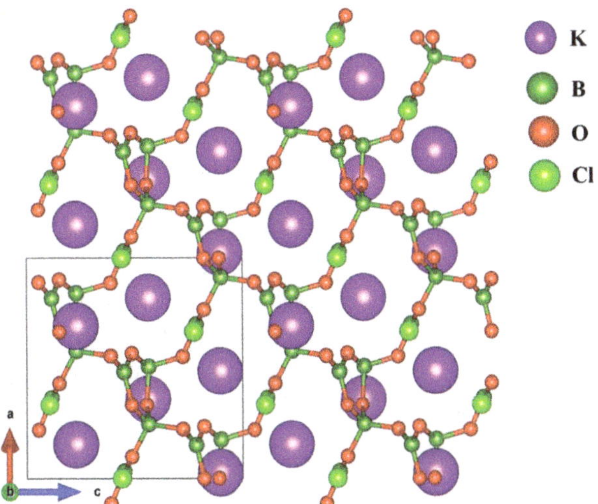

Figure 8.6.12: Crystal structure of $K_3B_6O_{10}Cl$ viewed along the *b* axis.

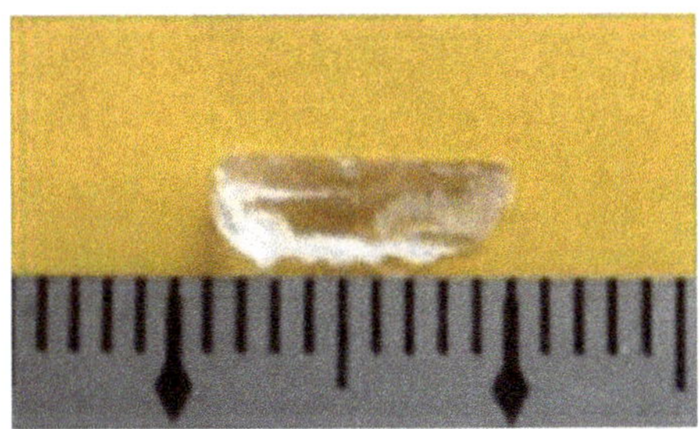

Figure 8.6.13: Photograph of $K_3B_6O_{10}Cl$. Reproduced with permission from ref. [73]. Copyright 2011 American Chemical Society.

UV pulses of 83 fs at 396 nm with a maximum output power of ~220 mW and a highest conversion efficiency of 39.3% are obtained in a 1-mm-long KBOC crystal with a length of 1 mm [77]. Output powers up to 5.9 mW are obtained at a central wavelength of 263 nm, corresponding to a conversion efficiency of 4.5% (second harmonic power, THG process) [78]. The ps green laser was successfully generated from two types of 1064 nm lasers with one being operated at 10 kHz and the other at 10 Hz [79].

8.6.2.2.5 $K_3B_6O_{10}Br$ (KBOB)

KBOB was first reported by Al-Ama et al. in 2006 [80]. KBOB, isostructural with KBOC, crystallizes in space group *R*3*m* with the cell parameters a = 10.1153(8), c = 8.8592(14) Å, Z = 3, V = 785.02(15) Å^3, featuring a pseudo-perovskite structure composed of the $^3_\infty(B_6O_{10})^{2-}$ network and $^3_\infty(BrK_6)^{5+}$ framework with excellent optical properties.

KBOB crystals present a broad transmission of 255–2500 nm with an experimental bandgap of about 4.86 eV. KBOB is a negative uniaxial crystal, with birefringence of 0.046@1064 nm. The shortest type I and type II phase-matching SHG wavelengths are 290 and 400 nm, respectively. The measured SHG coefficients of KBOB are as follows: d_{22}(KBOB) = 0.83 pm V^{-1} = 2.12 d_{36} (KDP), d_{33}(KBOB) = 0.51 pm V^{-1} = 1.32 d_{36} (KDP) [81].

KBOB melts incongruently, meaning that the crystal was grown by the top-seeded solution growth method with a KF-PbO flux system [82, 83]. A raw material molar ratio of $M_{H_3BO_3}/M_{KBr}/M_{KF}/M_{PbO}$ = 6:1:2:1 is suitable to stably grow sizable single crystals. The crystal seed was held in the solution about 20–30 days with a cooling rate in the range of 0.2–1.6 K/day and a seed-rotation rate of 5 rpm. Finally, a large transparent crystal was obtained with size up to 22 mm × 22 mm × 10 mm (Figure 8.6.14) [81].

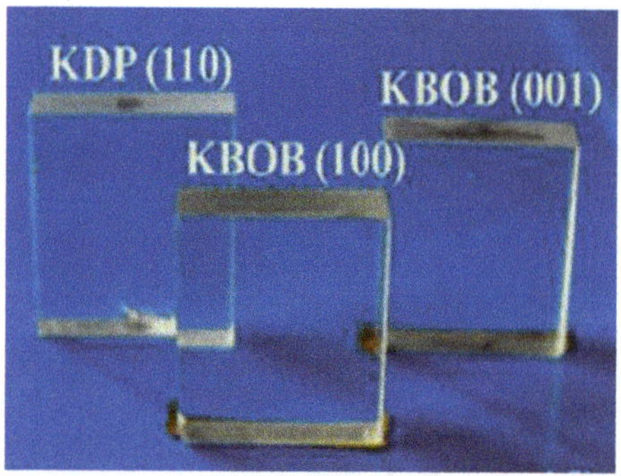

Figure 8.6.14: Photograph of $K_3B_6O_{10}Br$. Reproduced with permission from ref. [81]. Copyright 2014 American Chemical Society.

A 355 nm UV femtosecond laser has been obtained through SHG using a type-I phase-matched nonlinear optical crystal of KBOB [84]. Broadband optical parametric chirped pulse amplification of the KBOB crystal at a central wavelength of 800 nm has been demonstrated, confirming that KBOB crystals represent a potential alternative for efficient, broadband optical parametric chirped pulse amplification near 800 nm [85]. In addition, although KBOB does not allow the realization of 266 nm UV light by direct frequency doubling from the fundamental 1064 nm laser, it is possible to achieve the UV fourth harmonic (266 nm) by frequency sum $\omega + 3\omega$ with a type I configuration.

8.6.2.3 Fluorooxoborates

8.6.2.3.1 $NH_4B_4O_6F$ (ABF)

ABF, the first member of the ABF family, was first reported by Pan et al. in 2017. ABF crystallizes in the orthorhombic system and the polar space group $Pna2_1$ (No. 33) with cell parameters $a = 7.602(2)$, $b = 11.197(3)$, $c = 6.5952(19)$ Å [86]. The basic structure of ABF is shown in Figure 8.6.15. It features a wave-like $^{2}_{\infty}[B_4O_6F]$ layer that is a new anionic structural motif in borates, and the layers are further linked by the NH_4^+ cations through hydrogen bonds, forming the final three-dimensional framework. ABF inherits the favorable structural features of KBBF; that is, all $(BO_3)^{3-}$ groups align almost in the same direction in the $^{2}_{\infty}[B_4O_6F]$ layer, and the dangling bonds of terminal oxygen atoms are eliminated by the $(BO_3F)^{4-}$ groups, i.e., all three oxygen atoms of the $(BO_3)^{3-}$ group are bound to another atom in addition to the boron atom.

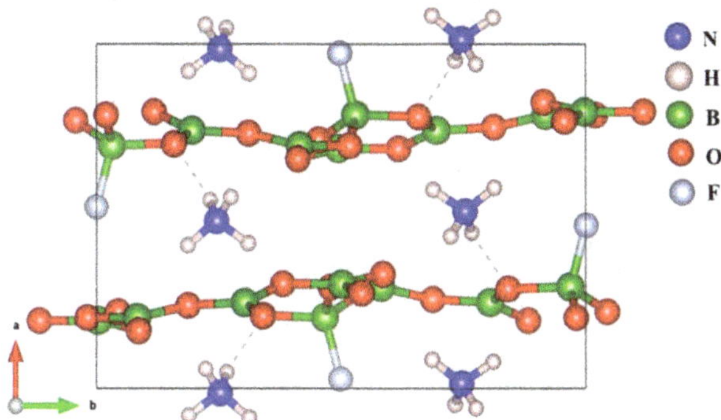

Figure 8.6.15: Crystal structure of $NH_4B_4O_6F$ viewed along the c axis.

The deep-UV cutoff edge of ABF is about 156 nm. ABF is a negative biaxial crystal and its birefringence is 0.1171@1064 nm. ABF can achieve the SHG for both type I and type II in three main directions, and the shortest phase-matching SHG wavelength (λ_{PM}) is down to 158 nm in the *XY* plane calculated via Sellmeier equations. The powder SHG test indicates that the SHG efficiency of ABF is ~3 times that of KDP. The calculated SHG coefficients of ABF are as follows: $d_{31} = 0.02$ pm V^{-1}, $d_{32} = 1.07$ pm V^{-1}, $d_{33} = -1.19$ pm V^{-1} [86].

Single crystals of ABF were grown by a high-temperature method in a closed system. A mixture of B_2O_3 (1.000 g, 14.364 mmol) and NH_4F (0.266 g, 7.182 mmol) were loaded into a tidy quartz tube (∅10 × 20 mm) that was washed with deionized water and dried at high-temperature to remove the impurities, and the tube was flame-sealed under 10^{-3} Pa. The tube was heated to 450 °C for 10 h, and held at this temperature for 24 h, and then cooled to 30 °C with a rate of 1 K/h. The as-grown ABF crystals with the largest dimensions up to 8 mm × 4 mm × 1 mm are shown Figure 8.6.16 [86].

The refractive index measurements were performed using the prism coupling method, and the Sellmeier equations were fitted as:

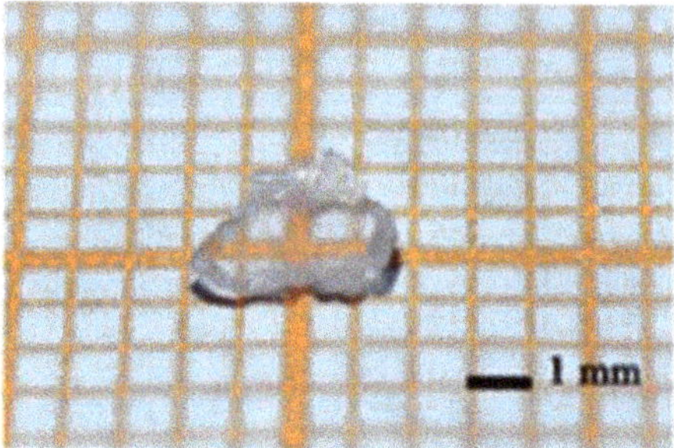

Figure 8.6.16: Crystal plate of ABF (unpolished, thickness ~0.5 mm) for measurement. Reproduced with permission from ref. [86]. Copyright 2017 American Chemical Society.

$$n_x^2 = 2.10333 + \frac{0.00710}{\lambda^2 - 0.01094} - 0.00909 \times \lambda^2$$

$$n_y^2 = 2.45077 + \frac{0.01058}{\lambda^2 - 0.02000} - 0.02517 \times \lambda^2$$

$$n_z^2 = 2.47210 + \frac{0.01139}{\lambda^2 - 0.02130} - 0.02631 \times \lambda^2$$

Besides ABF, NaB_4O_6F [87], RbB_4O_6F [88] and CsB_4O_6F [89] all possess a KBBF-type structure. The ABF family has the potential to produce deep-UV coherent light.

8.6.2.3.2 $MB_5O_7F_3$ (M = Sr, Ca, Mg)

The fluorooxoborate system with the general formula of $MB_5O_7F_3$ was proven as a stable fluorine-terminated framework. $SrB_5O_7F_3$(SBF) was the first-reported structure in this system in 2018 [90]. SBF crystallizes in the orthorhombic system with the noncentrosymmetric space group $Cmc2_1$ with cell parameters a = 10.016, b = 8.654, c = 8.103 Å, Z = 4. The structure features zigzag ${}^2_\infty(B_5O_7F_3)^{2-}$ layers with the Sr^{2+} cation padding the interlayers. Two $(BO_3)^{3-}$ triangles and three $(BO_3F)^{4-}$ tetrahedra constitute the double ring $(B_5O_9F_3)^{6-}$ and further translation yields the ${}^2_\infty(B_5O_7F_3)^{2-}$ layer, then the adjacent layers are stacked along the [010] direction in the -AAAA-sequence (Figure 8.6.17).

SBF has no obvious absorption from 180 to 2600 nm and its deep-UV cutoff edge is lower than 180 nm. The powder SHG test indicates that the SHG efficiency of SBF is about 1.6 times that of KDP. The calculated refractive indices suggest that SBF is a negative biaxial crystal with the birefringence 0.070@1064 nm. The shortest type I phase-matching SHG wavelength corresponds to 180 nm based on the calculated refractive index dispersion. The calculated SHG tensors are as follows: $d_{31} = d_{15} = 0.91$ pm V^{-1}, $d_{32} = d_{24} = -0.37$ pm V^{-1} and $d_{33} = -0.71$ pmV^{-1}.

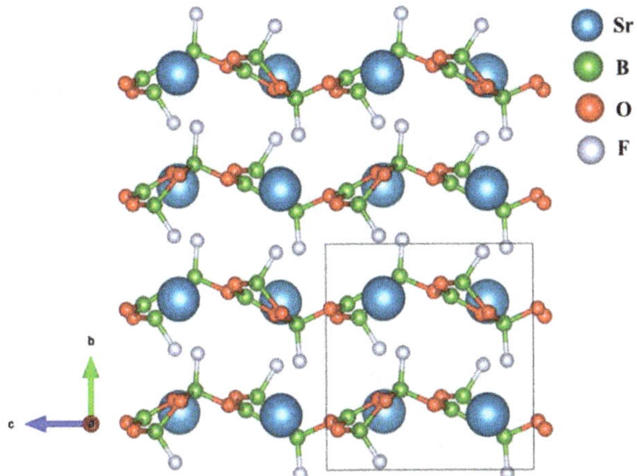

Figure 8.6.17: Crystal structure of $SrB_5O_7F_3$ viewed along the *a*-axis.

Small single crystals of SBF were grown by the high-temperature solution method using a $NaBF_4$ flux in a closed system. Besides SBF, $CaB_5O_7F_3$ [91] and $MgB_5O_7F_3$ [92] also show a relatively large SHG effect with deep-UV phase-matching ability, which indicate the potential of this system to generate deep-UV coherent light.

8.6.3 Phosphates

Phosphates are also one important class of NLO crystals, which can be classified into orthophosphates, pyrophosphates and metaphosphates [93]. In orthophosphates, the anionic group $(PO_4)^{3-}$ is composed of a regular tetrahedral arrangement of four oxygen atoms centered by a phosphorus atom, which can be also called monophosphates. In pyrophosphates and metaphosphates, the $(PO_4)^{3-}$ units are condensed by corner sharing.

Usually, environmentally friendly phosphates are widely regarded as promising UV and deep-UV NLO candidates. Phosphates have the following advantages [94]: (1) short cutoff edge owing to non-π-conjugated $(PO_4)^{3-}$ tetrahedra as the basic structural units; (2) various anionic groups formed by $(PO_4)^{3-}$ via sharing of O atoms like $(P_2O_7)^{4-}$, $(P_3O_{10})^{5-}$, $(P_4O_{12})^{4-}$; (3) diverse structural frameworks from zero-dimensional to three-dimensional, such as clusters, chains, layers and network structures. As the typical representatives of phosphates, KDP [95] and $KTiOPO_4$ (KTP) [96] have an extensive application range, and have been used as benchmarks for the second-order UV or visible NLO materials. In addition to KDP and KTP, further progress has been made in the field of NLO phosphates in the recent years.

It should be noted that the presence of strong SHG response is normally conflicting with short cutoff edge. Hence, achievement of the desired balance between

SHG effect and cutoff edge is a vital criterion to assess the capabilities of NLO materials. The methods to achieve SHG can include aligned groups, introduction of d^0 or d^{10} metal cations, insertion of metal cations with lone pairs or inclusion of multi anion groups/ions, like halides. Further, in analogy to fluorooxoborates, fluorooxophosphates are also explored as NLO crystals.

8.6.3.1 Phosphates

8.6.3.1.1 KH_2PO_4 (KDP), KD_2PO_4 (DKDP) and $NH_4H_2PO_4$ (ADP)

KDP, ADP and their isomorphs were the very first materials to be used and exploited for their NLO and electro-optic properties. KDP crystallizes in the noncentrosymmetric space group $I\bar{4}2d$ with the cell parameters a = 7.444(1), c = 6.967(1) Å (Figure 8.6.18) [97]. Its structure can be seen as an ionic crystal composed by K^+ and $(H_2PO_4)^-$, although the covalent bond and hydrogen bond coexist. All the hydrogen bonds are almost vertical to the c axis. The K atoms are surrounded by eight O atoms to form KO_8 polyhedra, thus, $(PO_4)^{3-}$ groups connect each other via hydrogen and K-O bonds to form 3D framework.

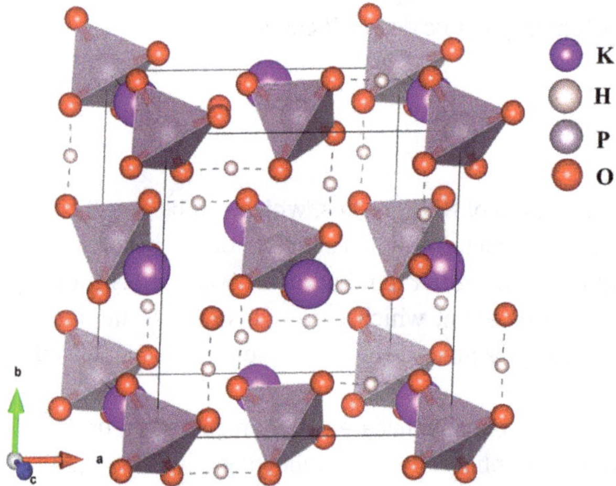

Figure 8.6.18: Crystal structure of KH_2PO_4.

The transparency range of KDP is 174 nm–1.57 µm. KDP is a negative uniaxial crystal. Its birefringence is 0.0418–0.0339@532–1064 nm. Its SHG coefficient is d_{36} = 0.39 pm V^{-1}, which is always used as a benchmark [98, 99].

The crystal growth of KDP uses the aqueous solution growth method, with slow temperature-reduction, and the supersaturation of solution as the driving force for crystal growth [100]. There are two more important parameters affecting the growth rate: impurities and mass transfer, or hydrodynamic conditions. For the traditional growth process of KDP crystals, the supersaturation of the solution is very low, and only the conical surface can grow slowly. If the supersaturation of the solution is further increased, not only the growth rate of the cone will be greatly improved, but even the cylindrical surface will start to grow. This method of simultaneous growth of the conical and cylindrical surfaces of KDP is called the rapid growth method of KDP crystals. Compared to the traditional growth method, the growth rate of the point-seed rapid-growth [101, 102] for the large size can reach up to 10 mm d^{-1}, which is an order of magnitude higher than that of the traditional method (1.5 mm d^{-1}). At present, large KDP crystals have been successfully obtained by long seed crystal free growth (Figure 8.6.19) [103].

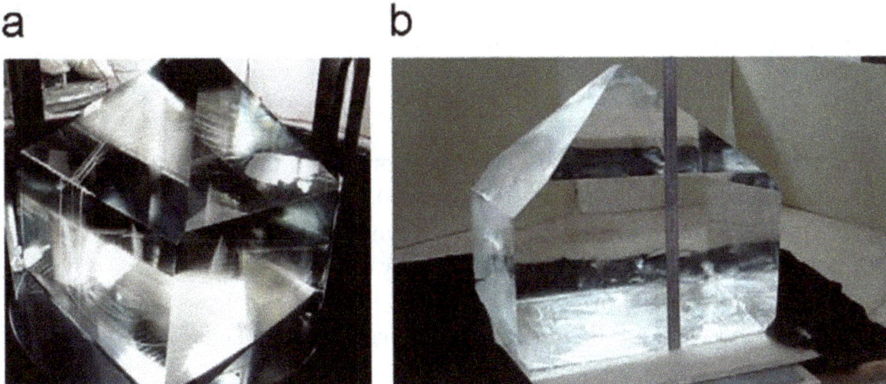

Figure 8.6.19: KH_2PO_4 crystals with sizes 63 × 57 × 55 cm^3 (a) and 57 × 52 × 52 cm^3 (b). Reproduced with permission from ref. [103]. Copyright 2011 Elsevier.

KDP crystals have many advantages, such as wide transmittance band, excellent resistance to laser irradiation damage, high nonlinear conversion efficiency and easy to grow large-diameter single crystals. Therefore, the major applications of the crystal include the following aspects: (1) SHG, THG and fourth harmonic generation can be realized for 1064 nm lasers. (2) KDP can be used as standard reference for the relative frequency doubling coefficient. (3) KDP can be the material for making high-power laser frequency multiplier and parametric oscillator. (4) It can also be used to make laser Q-switches, electro-optic modulators, polarizers, etc.

$K(D_xH_{1-x})_2PO_4$(DKDP) is the isotopic analogue of KDP. There are two phases of DKDP, one of which is the isomorphic compound of KDP, crystallizing in the noncentrosymmetric space group $I\bar{4}2d$ [104]. The other phase is monoclinic. The transparency range of DKDP is 0.2–2.0 μm. DKDP is a negative uniaxial crystal. Its birefringence is 0.0395–0.0373@532–1064 nm. Its SHG coefficient is d_{36} = (0.09 ± 0.04) d_{36} (KDP) = (0.402 ± 0.01) pm V^{-1} [105, 106]. The aqueous solution growth method is adopted to grow DKDP with a slow temperature-reduction. The key to grow large crystals with high quality is to avoid the phase transition of the tetragonal phase and the appearance of the monoclinic phase. Therefore, the determination of the solubility curve and the phase-transition temperature for the two phases is a prerequisite. Because of the benign NLO parameters and obvious size advantage, KDP/DKDP is the only single crystal material to be used for the inertial confinement fusion [107, 108].

ADP is also an important compound of the KDP family with cell parameters a = 7.502 and c = 7.546 Å [109]. It is usually seen as an isomorphic structure of KDP. However, KDP becomes ferroelectric at T_C = 122.7 K while ADP becomes antiferroelectric at T_N = 148.2 K. The transparency range of DKDP is 0.184 ~ 1.5 μm. DKDP is a negative uniaxial crystal. Its birefringence is 0.0460–0.0384@532–1064 nm. Its SHG coefficient is d_{36} = 0.76 pm V^{-1} [110]. ADP is grown from aqueous solution with slow cooling. ADP can be used for electro-optic devices and laser systems to produce SHG, DFG, optical parametric oscillation (OPO), etc.

8.6.3.1.2 $KTiOPO_4$ (KTP)

KTP crystallizes in the space group $Pna2_1$, belonging to the orthorhombic system, with cell parameters a = 12.809, b = 6.420, c = 10.604 Å and Z = 8 [111]. The P and Ti atoms are 4 and 6 coordinated, respectively, while the K atom is 8–9 coordinated. TiO_6 octahedra and $(PO_4)^{3-}$ tetrahedra alternately connect each other in the 3D framework, forming - (PO_4) – (TiO_6) – (PO_4) – (TiO_6) – chain arrays with . . . – O – Ti – O – Ti – . . .bonds. K atoms fill the gaps in the chain network (Figure 8.6.20).

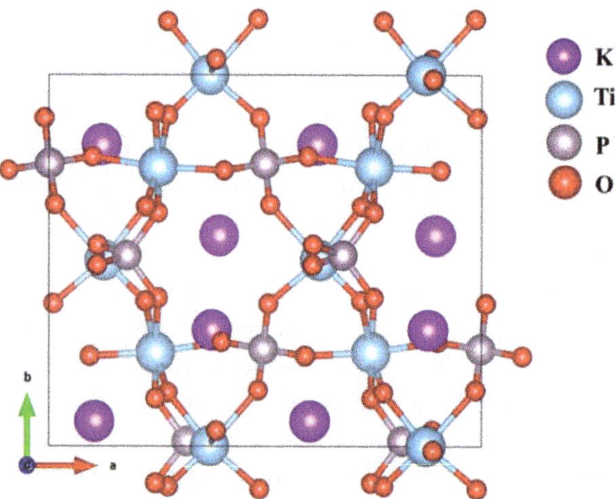

Figure 8.6.20: Crystal structure of $KTiOPO_4$ viewed along the c axis.

KTP is a positive biaxial crystal and transparent from 350 to 4500 nm. Its refractive index dispersion equations are as follows:

$$n_x^2 = 2.10468 + \frac{0.89342}{\lambda^2 - 0.04438} - 0.01036 \times \lambda^2$$

$$n_y^2 = 2.14559 + \frac{0.87629}{\lambda^2 - 0.0485} - 0.01173 \times \lambda^2$$

$$n_z^2 = 1.9446 + \frac{1.3617}{\lambda^2 - 0.047} - 0.1491 \times \lambda^2$$

Its birefringence is 0.1107–0.0920 at 530 nm–1064 nm and the phase-matching range is 984–3400 nm. The SHG coefficients of KTP are d_{31} = 6.5 pm V^{-1}, d_{32} = 5 pm V^{-1}, d_{33} = 13.7 pm V^{-1}, d_{24} = 7.6 pm V^{-1}, d_{15} = 6.1 pm V^{-1} [112–114].

KTP and its isomorphs decompose incongruently below their melting points, and single crystals can be grown only from solutions, at relatively slow rates, e.g. using high-temperature fluxes or hydrothermally [115, 116]. Since the linear growth rate of hydrothermal crystals is low, 0.07–0.15 mm d^{-1}, and the KTP crystal grown hydrothermally exhibits an absorption peak of OH^- at 2.8 μm, the presently leading industrial technique of processing KTP crystals is the top-seeded solution growth method, where more than 1 mm d^{-1} linear growth rates can be achieved. However, the optical uniformity of KTP crystals grown from hydrothermal method is superior to that from high-temperature flux; the conductivity of the former is on the order of 10^{-9}–10^{-11} Ω^{-1} cm^{-1}, 2–3 orders of magnitude lower than the latter [117, 118].

The most important applications of KTP are SHG and OPO applications (laser ranging), especially the OPO applications in recent years. KDP crystals can achieve green light/IR light output by SHG via Nd-doped lasers, and blue light output by SFG. The modulated light in the range of 0.6 to 4.5 μm can be obtained by optical parametric generation (OPG), optical parametric amplification (OPA) and OPO. KTP crystals also can be fabricated for optical switches, optical waveguides and periodically polarized KTP devices.

8.6.3.2 Fluorooxophosphates

The first studies to identify and develop the fluorophosphates as new and ideal NLO material candidates started in 2018 [119]. $(NH_4)_2PO_3F$ crystallizes in the orthorhombic space group *Pna*2$_1$, composed of NH_4^+ cations and non-condensed $(PO_3F)^{2-}$ anionic groups (Figure 8.6.21). $(NH_4)_2PO_3F$ single crystals with centimeter sizes were grown using the solvent-evaporation method. Its deep-UV cutoff edge is shorter than 177 nm, and the oscilloscope traces of SHG signals show that the SHG efficiency of $(NH_4)_2PO_3F$ is about 1 time that of KDP and 0.2 times that of β-BBO in the same particle size range of 150–200 μm at 1064 and 532 nm, respectively. The birefringence for $(NH_4)_2PO_3F$ was measured as larger than 0.03 at 589.3 nm by the immersion method, and the calculated type I phase-matching SHG wavelength is 264 nm.

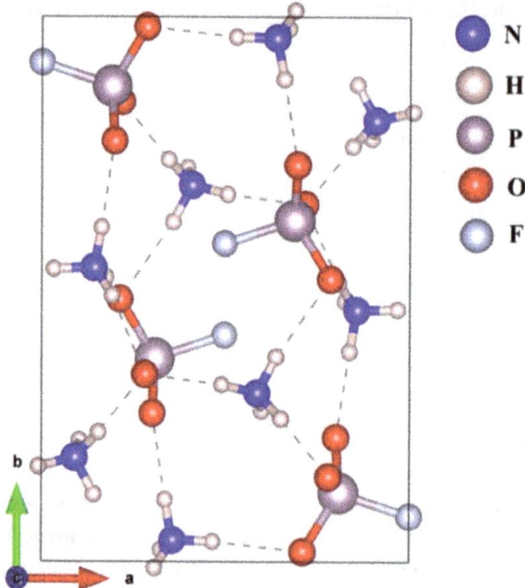

Figure 8.6.21: Crystal structure of $(NH_4)_2PO_3F$ viewed along the *c* axis.

$NaNH_4PO_3F{\cdot}H_2O$ (NNPF) [120] crystallizes in the monoclinic space group *Pn* with $a = 6.0515(6)$, $b = 9.0167(8)$, $c = 4.9402(7)$ Å, $\beta = 90.737(10)$ ° and $V = 269.54$ Å^3. The structure is composed of double-layers that stack along the *b* axis with Na^+ lying in the interlayers (Figure 8.6.22). Uniquely, the $(PO_3F)^{2-}$ tetrahedra within each layer are fastened by H_2O and $(NH_4)^+$ groups via a network of strong hydrogen bond interactions. Since the bases are fastened within the *ab* plane, these tetrahedra cannot move; meanwhile, constrained by the tetrahedral geometry, all the polar P–F bonds are therefore forced to point uniformly to the *c* direction. The

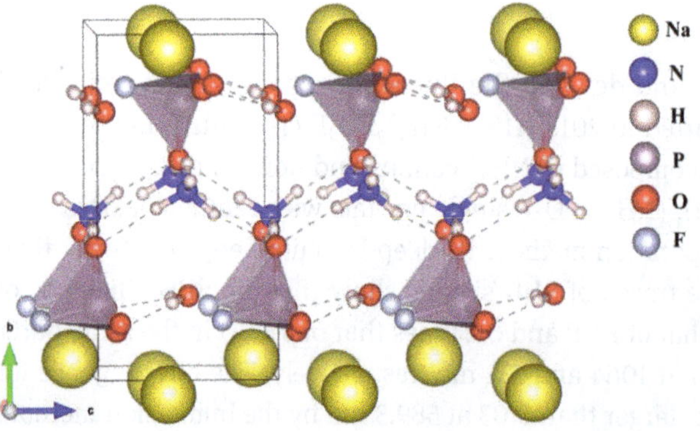

Figure 8.6.22: Crystal structure of $NaNH_4PO_3F{\cdot}H_2O$ viewed along the *a* axis.

powder SHG efficiency of NNPF is about 1.1 × KDP. The birefringence of NNPF was measured as larger than 0.053 at 589.3 nm on a [010] single crystal wafer. The type I phase-matching SHG wavelength is about 194 nm, as estimated from a theoretical calculation. The SHG processes from λ_{ω} ~ 800, 670 and 630 nm to $\lambda_{2\omega}$ ~ 400, 335 and 315 nm, respectively, were obtained by as-synthesized [010] NNPF single crystal wafers.

8.6.4 Carbonates

Carbonates feature π-conjugated $(CO_3)^{2-}$ anionic groups as desirable NLO-active units. The aligned arrangement of the planar triangle configurations is in favor of the generation of large SHG response and birefringence. Recently, researchers introduced alkali fluoride into carbonates and obtained a series of carbonates with good NLO performances. Thus, carbonates with $(CO_3)^{2-}$ groups are becoming one of the research hotspots in the NLO field. It is reported that most of the carbonates crystallize in the hexagonal system (68%), and some crystallize in the monoclinic (12%) and orthorhombic (10%) system [121]. Carbonates also exist in large quantities in natural minerals, which were generated mostly under high-temperature and high-pressure conditions. To date, only a few noncentrosymmetric compounds were discovered. In addition, there are few reports on single crystal growth of carbonates partly owing to their decomposition at high temperature.

8.6.4.1 $LiNaCO_3$ (LNC)

In 1996, based on the binary phase diagram Li_2CO_3 – Na_2CO_3, D'akov et al. [122, 123] successfully grew LNC (a = 14.265, b = 14.261, c = 3.380 Å, α = 88.41, β = 91.70, γ = 119.95 °) with centimeter-size and identified LNC as a potentially attractive crystal for fourth harmonic generation. The crystals were produced by seeded growth from a melt containing 53 mol-% of Li_2CO_3 and 47 mol-% of Na_2CO_3 by using a special treatment at room temperature. The structure of LNC is constructed by a $[Li_2CO_3]$ layer (A) and a $[Na_2CO_3]$ layer (B), the two layers alternate along the c axis as -ABABAB- (Figure 8.6.23).

The UV cutoff edge of the crystal was measured at about 200 nm and its birefringence is around 0.126. Comparison of the SHG efficiency in the LNC crystal with that in the KDP crystal shows that $d_{eff}(\text{LNC}) = (1.20 \pm 0.05) \times d_{eff}(\text{KDP})$. The LNC crystal is slightly biaxial and it is difficult to distinguish n_x and n_y. So, the results of measurement are only for two refractive indexes. The Sellmeier equations for refractive index in UV wavelength are as follows:

$$n_1^2 = 2.26067 + \frac{0.04273}{\lambda^2 - 0.08440} + 0.9156 \times \lambda^2$$

$$n_2^2 = 2.01352 - \frac{0.01919}{\lambda^2 - 0.06402} - 0.26267 \times \lambda^2$$

The Sellmeier equations for refractive index in IR wavelength are as follows:

$$n_1^2 = 2.3500 - \frac{0.08441}{\lambda^2 - 0.02056} - 0.00868 \times \lambda^2$$

$$n_2^2 = 1.9560 + \frac{0.01036}{\lambda^2 - 0.02368} + 0.01052 \times \lambda^2$$

The phase-matching (type I) angles are 20.2 ° for second, 24.70 ° for third and 34.6 ° for fourth-harmonic generation in the plane of the optical axes. These angles are 18 °, 23.4 ° and 33.5 °, respectively, in a perpendicular plane. Further research showed that LNC can realize the second, triple and quadruple-harmonic output based on the Nd-YAG laser.

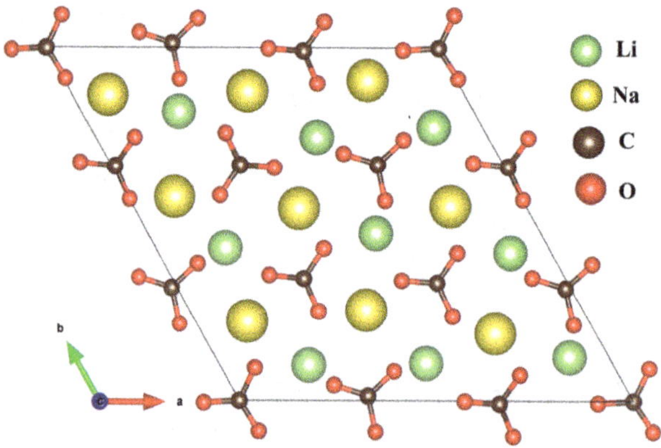

Figure 8.6.23: Crystal structure of $LiNaCO_3$ viewed along the *c* axis.

8.6.4.2 $KSrCO_3F$ and the $ABCO_3F$ family

$KSrCO_3F$, crystallizes in the hexagonal crystal system, space group $P\bar{6}2m$ with the cell parameters a = 5.2598(5) and c = 4.6956(11) Å [124]. The crystal structure can be seen as KO_6F_3 and SrO_6F_2 polyhedra and $(CO_3)^{2-}$ triangular entities, exhibiting the stacking of $[KF]_\infty$ and $[Sr(CO_3)]_\infty$ layers along the c axis with the coplanar alignment of $(CO_3)^{2-}$ triangles (Figure 8.6.24). The $KSrCO_3F$ crystal (001 wafer) shows a UV absorption edge around 195 nm and can transmit well down to 215 nm. The Sellmeier equations of $KSrCO_3F$ are determined by the prism coupling method on a (100) wafer as follows:

$$n_o^2 = 2.28001 + \frac{0.01811}{\lambda^2 - 0.02128} - 0.00244 \times \lambda^2$$

$$n_e^2 = 1.97739 + \frac{0.00888}{\lambda^2 + 0.00175} + 0.00554 \times \lambda^2$$

The birefringence of $KSrCO_3F$ is 0.1145–0.1049 at 0.4502–1.0626 μm and the SHG limit of $KSrCO_3F$ is 200 nm. The SHG coefficient of a $KSrCO_3F$ crystal relative to d_{36} (KDP) is $d_{22} = 1.3 \times d_{36}$ (KDP), i.e., the absolute SHG coefficient of $KSrCO_3F$ crystal is $d_{22} = 0.50$ pm V^{-1} [125].

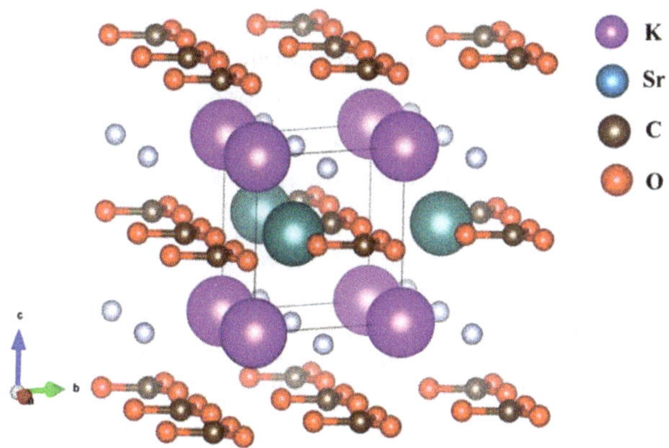

Figure 8.6.24: Crystal structure of $KSrCO_3F$.

Although $KSrCO_3F$ melts congruently, $KSrCO_3F$ crystals were grown by top-seeded solution growth methods using fluxes, due to its volatilization at the melting temperature. KF – Li_2CO_3 was used as a flux to grow $KSrCO_3F$ crystals that can significantly decrease the melting and crystallization temperature to 630 and 591 °C, respectively. The crystal seed was grown by spontaneous crystallization and was introduced into the melt at a rotation rate of 20 rpm at 593 °C in order to reduce surface defects. Then, the temperature was decreased to 591 °C at a cooling rate of 0.4 K min^{-1}. A rotation rate of 20 rpm is proper for the crystal growth quality. The successful single crystal growth of $KSrCO_3F$ suggests that the material is an attractive candidate to generate 266 nm radiation through fourth harmonic generation from 1064 nm photons.

Besides $KSrCO_3F$, the $ABCO_3F$ (A = K, Rb, Cs; B = Mg, Ca, Sr) [124, 126] series all belong to the hexagonal crystal system with the same space group $P\bar{6}2m$, showing the stacking of $[AF]_\infty$ and $[B(CO_3)]_\infty$ layers along the c axis with the coplanar alignment of $(CO_3)^{2-}$ triangles. The absorption edge of all $ABCO_3F$ is below 200 nm. Their SHG efficiency varies from 1.11 × KDP to 4 × KDP due to the different arrangement of the $(CO_3)^{2-}$ group. Therefore, $ABCO_3F$ shows a potential application for UV laser output.

8.6.5 Sulfates

Among the common tetrahedral groups $(BO_4)^{5-}$, $(PO_4)^{3-}$ and $(SO_4)^{2-}$, the sulfate anion has the largest HOMO-LUMO gap, which is beneficial to deep-UV transmission. But the tetrahedral configuration without distortion is not conductive to large optical anisotropy and second-order susceptibility. Even so, a number of researchers are still exploring NLO crystals in this system. Anhydrous alkali metal sulfates always have a wide UV transmission range and high thermal stability but relatively weak SHG effect and small birefringence. When introducing a lone pair, d^0 metal or rare earth cation, the crystal can produce large SHG effect and the UV cutoff edge will redshift to be a visible or near-IR NLO crystal. Most of the sulfates were synthesized by the aqueous solution method.

8.6.5.1 $Li_2SO_4{\cdot}H_2O$ (LS) [127, 128]

LS crystallizes in the monoclinic polar space group $P2_1$ with cell dimensions a = 5.450(3), b = 4.872(3), c = 8.164(4) Å, β = 107.31(3) ° (Figure 8.6.25). LiO_4 tetrahedra connect $(SO_4)^{2-}$ tetrahedra by sharing their common vertex to form a 3D skeleton. The hydrogen bonds (O-H···O) exist among the H_2O molecules and $(SO_4)^{2-}$ groups. $Li_2SO_4{\cdot}H_2O$ has a UV cutoff edge of <200 nm and its birefringence is 0.025–0.024@ 532–1064 nm. The determined Sellmeier equations of LS are as follows:

$$n_1^2 = 2.1575(6) + \frac{0.0088(3)}{\lambda^2 - 0.0114(3)} - 0.0100(3) \times \lambda^2$$

$$n_2^2 = 2.1855(2) + \frac{0.0096(1)}{\lambda^2 - 0.008(1)} - 0.0101(1) \times \lambda^2$$

$$n_3^2 = 2.1164(5) + \frac{0.0078(2)}{\lambda^2 - 0.018(2)} - 0.0098(2) \times \lambda^2$$

The measured SHG coefficients for LS are d_{22} = −0.34(4) pm V^{-1}, d_{21} = 0.12(3) pm V^{-1}, d_{23} = 0.25(6) pm V^{-1}, d_{25} = −0.05(3) pm V^{-1}, d_{14} = −0.06(3) pm V^{-1}, d_{16} = 0.13(2) pm V^{-1}, d_{34} = 0.15(6) pm V^{-1}, d_{36} = −0.06(2) pm V^{-1}, among which the d_{22} is comparable to d_{36} (KDP). The values of d_{eff} for LS range are below 0.13 pm V^{-1} for type I and 0.10 pm V^{-1} for type II phase matching.

Single crystals of LS were grown by a slow evaporation technique at room temperature. Analytical grade LS was used for growing the crystal (Figure 8.6.26) [127]. Using deionized water as a solvent and upon repeated recrystallization, centimeter-sized bright and optically transparent single crystals could be harvested in about

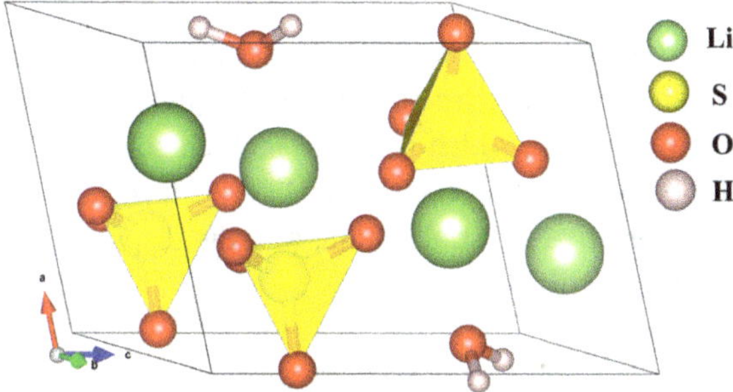

Figure 8.6.25: Crystal structure of $Li_2SO_4{\cdot}H_2O$.

four weeks. Care was taken to minimize the thermal or mechanical disturbances to the solution during growth.

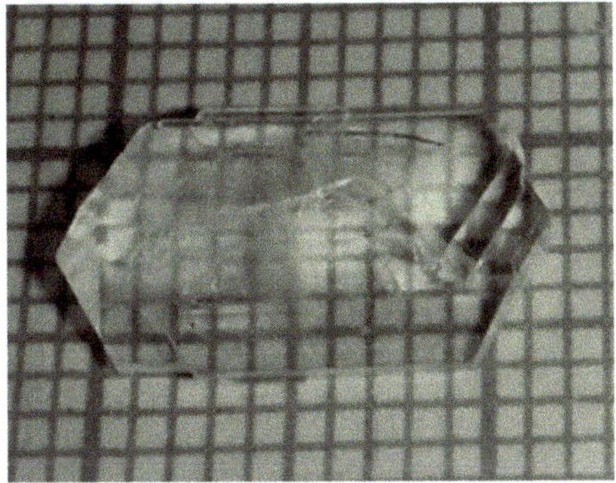

Figure 8.6.26: Photograph of $Li_2SO_4{\cdot}H_2O$. Reproduced with permission from ref. [127]. Copyright 2009 John Wiley and Sons.

8.6.6 Iodates

Iodine ($5s^25p^5$) is a member of the halogen group (seventh main group) and can exhibit the oxidation states −1, +1, +3, +5 and +7. Formally, I^{5+} has a valence electron configuration of $5s^25p^0$. Usually, the I^{5+} can combine with O^{2-} leading to a sterically active asymmetric density on I^{+5}. Common anionic groups include $(IO_3)^-$, $(IO_4)^{3-}$ and the polymerization of these anionic groups to form I_2O_5, $(I_3O_8)^-$, $(I_5O_{14})^{3-}$ and $(I_4O_{11})^{2-}$.

Those anion groups combine with other cations to form iodates. Iodates with stereochemical activity lone pair usually exhibit wide large SHG response and large birefringence as well as high thermal stability. Common iodates are α-$LiIO_3$, α-HIO_3 and KIO_3.

8.6.6.1 α-$LiIO_3$

α-$LiIO_3$ belongs to the space group $P6_3$ [129]. It consists of $(IO_3)^-$ and LiO_6 units. These units form its overall structure through common vertex connections (Figure 8.6.27). Its crystals are grown by thermostatic evaporation, which is due to its high solubility in water but low and negative dissolution temperature coefficient [130]. Its transmission region is 0.3~5.5 μm [131, 132]. It is a negative uniaxial crystal, and the dispersion of refractive index and wavelength can be described as:

$$n_o^2 = 2.083648 + \frac{1.332068\lambda^2}{\lambda^2 - 0.035306} - 0.008525\lambda^2$$

$$n_e^2 = 1.673463 + \frac{1.245229\lambda^2}{\lambda^2 - 0.028224} - 0.003641\lambda^2$$

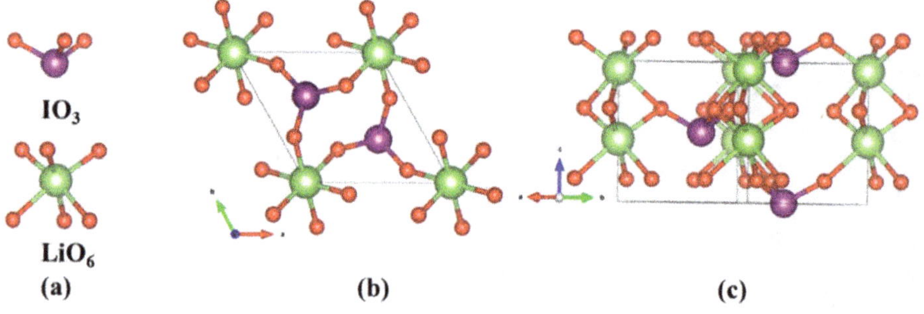

Figure 8.6.27: Crystal structure of α-$LiIO_3$.

Its birefringence is 0.1406 at 1.06 μm. Because of its relatively large birefringence, it can be used to make polarized light mirrors. The SHG coefficient at this wavelength are d_{33} = 7.02 pm V^{-1}, d_{31} = −7.11 pm V^{-1} and d_{15} = −(5.53±0.3) pm V^{-1}. It can be used to make Nd:YAG laser and Cr:Al_2O_3laser devices, respectively, with the sum frequency and difference frequency of green laser and IR laser. Parametric oscillation and up-conversion can also be realized [133–135].

8.6.6.2 α-iodic acid: HIO_3

α-HIO_3 crystallizes in the space group $P2_12_12_1$ [136, 137]. The element I connects six oxygen atoms to form a twisted $(IO_6)^{7-}$ octahedron, and the octahedra are connected by common vertices (Figure 8.6.28). Hydrogen is bonded to an O atom.

The large crystals of α–HIO_3 have been grown by a water solution method. The transmission range of α–HIO_3 is 0.35 ~ 1.6 μm. The principal refractive indices were measured by the minimum deviation method. α-HIO_3 is a negative biaxial crystal. The Sellmeier's equations are:

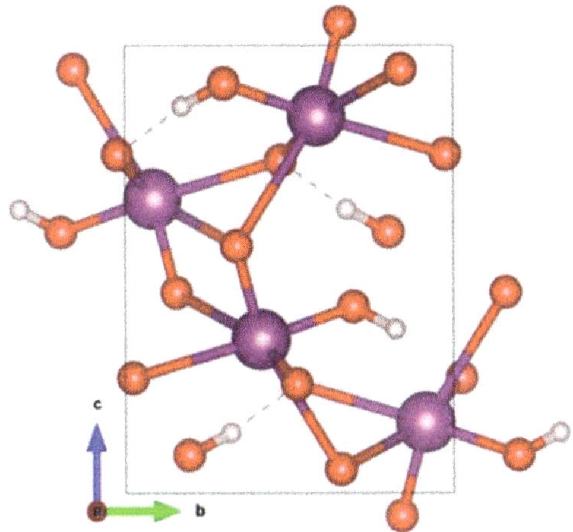

Figure 8.6.28: Crystal structure of α-HIO_3.

$$n_x^2 = 2.5761 + \frac{0.6973\lambda^2}{\lambda^2 - 0.05550736} - 0.0201\lambda^2$$

$$n_y^2 = 2.4701 + \frac{1.2054\lambda^2}{\lambda^2 - 0.05044516} - 0.0152\lambda^2$$

$$n_z^2 = 2.6615 + \frac{1.1316\lambda^2}{\lambda^2 - 0.05202961} - 0.0398\lambda^2$$

The birefringence at 1.064 μm is 0.1371. The SHG response of α-HIO_3 is d_{36} = (4.13±0.56) pm V^{-1}. It is now mostly used for acoustooptic modulators and deflectors [138–140].

8.6.7 Niobates

NLO materials are widely used in solid-state lasers because of their frequency-doubling properties. It is well known that the cations with large distortion including second-order Jahn-Teller (SOJT) effect, such as the d^0 configuration or cations with stereochemically active lone pairs can also contribute to the NLO properties. The octahedrally coordinated d^0 transition metal ions, such as Mo^{+6}, W^{+6}, V^{+5} and Nb^{+5} are typical examples. NLO materials containing these elements have been synthesized recently, and some of them have relatively large NLO coefficients.

8.6.7.1 $LiNbO_3$ (LN)

LN belongs to the trigonal crystal system with space group *R3c* at room temperature with cell parameters of a = 5.1483, c = 13.863 Å and Z = 6 [141]. The LN crystal is approximately hexagonal close packed and built up from $(NbO_6)^{7-}$ distorted octahedra and Li^+ cations (Figure 8.6.29). The positions of the Li and Nb atoms have displacement up or down along the *c*-axis with respect to the octahedra rather than being their centers, resulting in a spontaneous polarization, which is the structural basis for benign NLO properties. The SHG coefficients of LN are d_{33} = 34.4 pm V^{-1}, d_{31} = 5.95 pm V^{-1} and d_{22} = 3.07 pm V^{-1}; d_{eff} = 5.7 pm V^{-1} at 1300 nm. The transmission range of LN is 370–5200 nm and the birefringence is about 0.08@633 nm [142].

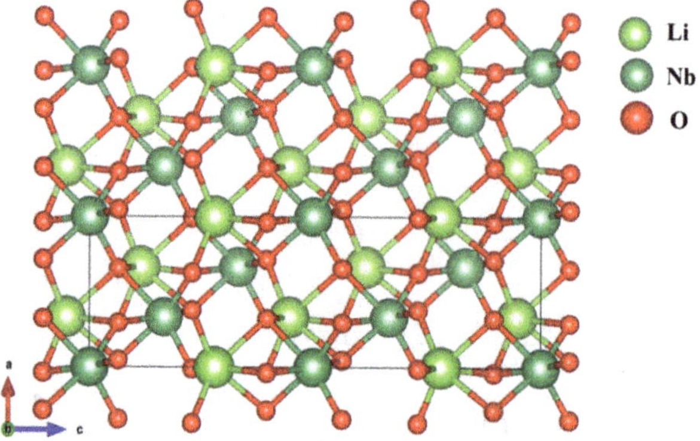

Figure 8.6.29: Crystal structure of $LiNbO_3$ viewed along the *b* axis.

LN melts congruently at 1253 °C and LN crystals were mostly grown from the melt by the Czochralski method. Crystals with a diameter of 75 mm are typically grown at 5 mm h^{-1} with a rotation rate of 9–10 rpm in air when the vertical temperature gradient is 50 K cm^{-1} [143]. The crystal grows with a flat interface under these conditions. A different diameter requires a different set of parameters and their combination may depend to some extent from the apparatus. Congruent LN (CLN) crystal have lots of intrinsic defects with complex chemical properties, so doping and near-stoichiometric ratio growth were used to adjust the defects and performance. Optical damage from the photorefractive effect obstacles the NLO application of LN crystal. To improve the damage resistance, doping with foreign elements like Mg can increase the phase-matching temperature [144]. Mg:LN crystals are suitable for SHG, SFG, OPO, OPA quasi phase-matching and other applications. Besides the photorefractive resistance doping (Mg:LN), the structural defects of LN crystals can also be controlled by single or co-doping of various elements and various valence states to achieve property control, such as doping with Fe to improve the photorefractive effect, or doping with Ti or H to obtain LN crystal waveguides [145]. In addition, the performance of near-stoichiometric LN (SLN) reaches an improvement compared to

CLN. However, owing to the deviation from the co-melting point of solid and liquid of the ratio of SLN, the preparation of SLN always requires a Li-rich melt method, the K_2O flux method or the diffusion method [146]. Figure 8.6.30 shows a commercialized LN crystal.

Figure 8.6.30: Photograph of $LiNbO_3$ crystals. With courtesy and permission from Castech (www.castech.com/product/LiNbO3-153.html).

LN is an excellent multifunctional material and adequate to prepare optical waveguides. Thus, optoelectronic devices based on LN, such as surface acoustic wave filters, light modulators, phase modulators, optical isolators, and electro-optic Q-switches have been widely studied and applied. OPO based on periodically poled LN (PPLN) crystals and quasi-phase matching are suitable for generating continuous wave and tunable coherent radiation in the mid-infrared wavelength range 3–5 μm, which has important applications in atmosphere pollution monitoring, remote sensing, laser radar, spectroscopy analysis and military infrared countermine. Well-established and recent applications, including real-time holography based on polaronic states, THz generation, waveguides, nanostructured devices, etc., open further perspectives for the use of LN. Recently, with the breakthrough in the application fields of the 5th-generation wireless communication (5G), micro-nano photonics, integrated photonics, and quantum optics, lithium niobate crystals have attracted extensive attention again. In 2017, researchers even proposed that LN crystals are as important to photonics as silicon is to electronics [147].

8.6.7.2 $KNbO_3$ (KN)

KN crystal, with large NLO coefficients as discovered by Kurtz and Perry in 1968, belongs to the orthorhombic system with *mm*2 point group and unit cell parameters of $a = 5.697$, $b = 3.971$, $c = 5.721$ Å with $Z = 2$ [148, 149]. Its structure corresponds to a perovskite (ABO_3-type) variant which is built up by distorted $(NbO_6)^{7-}$ octahedra and K atoms; K atoms occupy the A position and Nb atoms occupy the B position (Figure 8.6.31).

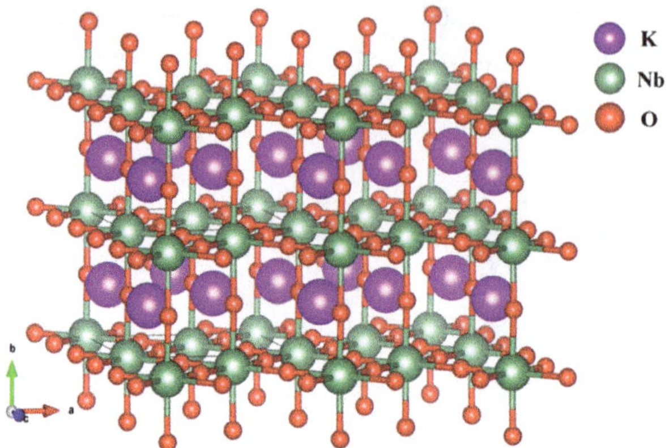

Figure 8.6.31: Crystal structure of $KNbO_3$.

The melting point of KN is about 1039 °C. Three phase tranformations are undergone by this crystal from the melting point to room temperature: cubic (1039 °C) → tetragonal (430 °C) → orthorhombic (200 °C) → rhombohedral (−50 °C) in the case of decreasing temperature [150]. Large transparent KN crystals can be grown by top-seeded solution growth and by the improved Kyropoulos method. The problem in growing KN crystals is that several pieces of KN can be obtained from one batch of starting materials but the first two or three crystals are always blue. Moreover, complex domain structures will form due to the two transformations during the cooling process. Using a prolonged soaking period for the starting materials, or enhanced stirring of the starting materials in solid and molten states during the process of soaking can eliminate the coloration and greatly relieve cracking [150].

KN has a transmission range from UV to near-IR (0.4–4.5 μm). KN is a negative biaxial crystal with a birefringence of 0.1789–0.1323 at 532–1064 nm. The refractive dispersion equations are:

$$n_x^2 = 1 + \frac{3.38361\lambda^2}{\lambda^2 - 0.03448}$$

$$n_y^2 = 1 + \frac{3.79361\lambda^2}{\lambda^2 - 0.03877}$$

$$n_z^2 = 1 + \frac{3.93281\lambda^2}{\lambda^2 - 0.04886}$$

The SHG coefficients of KN are $d_{31} = d_{15} = -11.9$ pm V^{-1}, $d_{32} = d_{24} = 13.7$ pm V^{-1} and $d_{33} = -20.6$ pm V^{-1}. KN is not only a NLO crystal with large coefficients, but also an electro-optic crystal with electro-optic coefficient $\gamma_{42} = 380$ pm V^{-1} [151].

As a multifunctional crystal with large SHG and electro-optic coefficients as well as photorefractive quality factors, excellent piezoelectric properties and benign chemical stability, KN has a relative wide range of applications. KN can be used for low power laser diode frequency doubling in the wavelength range around 860 nm, in the Ti:Sapphire laser in the range of 850–1000 nm and for the Nd:YAG lasers at 1064 nm. In addition, KN can also be used for OPO of Nd:YAG lasers at fundamental or second harmonic wavelengths to generate tunable radiation in the near-IR spectral region of 0.7–3 μm. KN can also be used as electro-optical modulator, electro-optic converter, in photorefractive waveguides and in surface acoustic wave or body wave devices [148]. However, the difficulty in the crystal growth, polarization of multidomain crystals and cold processing of single-domain crystals is not to be ignored; the key issue is to improve the existing KN crystal growth equipment and growth technology to prepare single-domain crystals with sufficient enough size to meet the requirements of large-scale production.

8.6.8 Pnictides

8.6.8.1 GaAs

GaAs belongs to the sphalerite (α-ZnS) structure with point group $3m$ and space group $F\overline{4}3m$ (a = 5.653 Å). The crystal structure is face-centered cubic, in which Ga atoms locate at the vertexes and face-centers and As atoms locate at the 1/4 body diagonal (half of the tetrahedral voids) [152]. GaAs has a relative large nonlinear coefficient (d_{14} = 94 pm V^{-1}), high thermal conductivity (46 W m^{-1} K^{-1}), very wide transparent range (0.9–18 μm) and low absorption loss [153]. With the discovery of periodically polarized LN, the concept of quasi-phase-matching has been applied in nonlinear frequency conversion. Since GaAs belongs to the cubic system, periodically polarized GaAs can be used as quasi-phase-matching material, which has great advantages in the application for phase-matching condition that can be realized in all transmission ranges theoretically without walk-off effect.

Periodically polarized GaAs can be prepared by the wafer bonding method and orientation pattern method (OP-GaAs). Among them, the former can yield large-diameter materials to realize high-power laser output, but optical loss exists at short wavelengths in the orientation reversal interface and the thick monolayer of GaAs is difficult to grow down to 100 μm, so that its application at long wavelengths is limited [154]. OP-GaAs is prepared by hydride vapor phase epitaxy growth based on patterned templates and has been fabricated for OPO devices to output mid and far-IR radiation with high efficiency by several companies and laboratories [155, 156].

8.6.8.2 $ZnGeP_2$ (ZGP)

ZGP is a typical chalcopyrite structure, derived from the III–V zinc blende type by ordered substitution with group IIB (Zn) and group IVA (Ge) atoms on the IIIA (Ga) site (Figure 8.6.32). It belongs to the tetragonal system with $\overline{4}2m$ point group and the cell parameters a = 5.465 and c = 10.711 Å [157, 158]. The introduction of two dissimilar atomic bond lengths (Zn–P and Ge–P) leads to a tetragonal distortion of the cubic zinc blende structure. ZGP has a wide transmission range (0.7–12 μm), large nonlinear coefficient (d_{36} = 75 pm V^{-1}), high laser damage threshold (2090 nm, 10 ns: 100 MW cm^{-2}) and appropriate birefringence (~0.04). Among the mid-infrared NLO crystal materials, ZGP has the good comprehensive performance in the current 3–5 μm laser output [159–165].

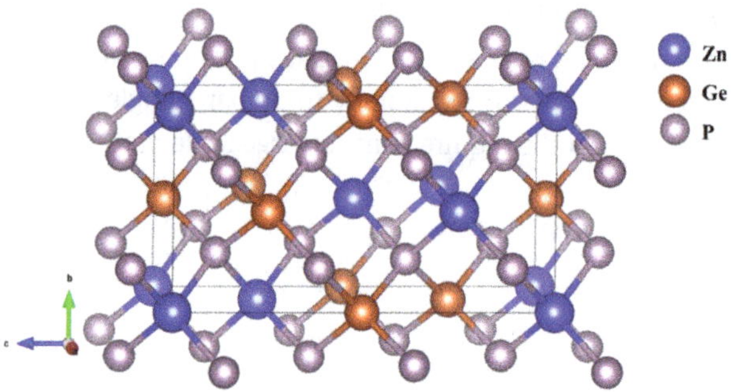

Figure 8.6.32: Crystal structure of $ZnGeP_2$.

Due to the high reaction temperature, complex reaction mechanism and many intermediate products in the synthesis of ZGP, the polycrystalline synthesis is very difficult, and the high steam pressure of Zn and P components makes it easy to damage (by explosion) the ampoule. At present, the reported methods of ZGP polycrystalline synthesis are mainly single temperature zone direct reaction and two temperature zone gas phase transport method. When synthesizing polycrystalline raw materials, precise weighing was carried out according to the ZGP stoichiometric ratio. At the same time, a certain amount of P and Zn in excess is important to compensate for the imbalance of measurement ratio caused by P and Zn vapor pressure in the synthesis process. Currently, the growth method of ZGP is the horizontal gradient freezing approach and the vertical Bridgman method [166, 167]. In 2020, a large size ZGP single crystal of 55 mm × 30 mm × 160 mm was successfully grown by superfine horizontal gradient freezing (0.5–1 K cm^{-1}) and a large aperture (12 mm × 12 mm × 50 mm) ZGP crystal device has been successfully fabricated with an absorption coefficient of only 0.03 cm^{-1} at 2.09 μm, which can be used for high energy and high average power infrared laser output [168]. A large crystal size of ∅50 mm × 140 mm was grown by a vertical Bridgman method and the absorption coefficient at 2.05 μm decreased to 0.02 cm^{-1} after annealing and high-energy electron irradiation (Figure 8.6.33) [169].

The ZGP crystal is employed to realize OPO, DFG, OPA and OPG, and other methods can achieve tunable continuous mid-infrared laser output. Among them, the generation of 3–5 μm mid-IR laser based on ZGP-OPO technology is a research hotspot. Currently, the key technological challenges that limit the use of ZGP-OPO is the laser thermal management, crystal quality and mid-infrared coating technology [170].

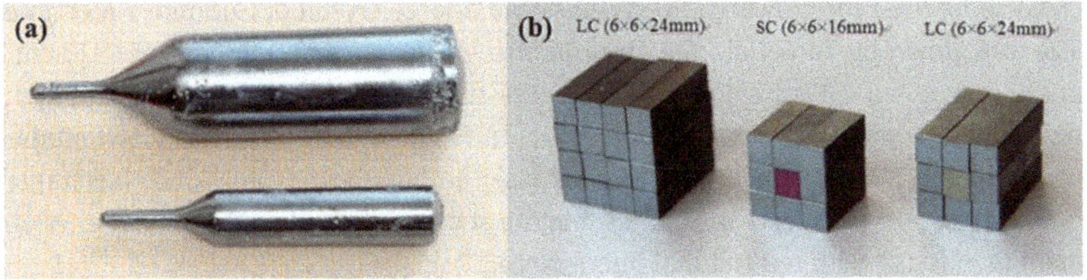

Figure 8.6.33: (a) $ZnGeP_2$. single crystal ingots: large crystal (LC, Ø50×140 mm) and small crystal (SC, Ø20×100 mm). (b) $ZnGeP_2$.–OPO (Optical Parametric Oscillator) devices with size of 6×6×(16–24) mm. Reproduced with permission from ref. [169]. Copyright 2009 Elsevier.

8.6.8.3 $CdSiP_2$ (CSP)

CSP belongs to the chalcopyrite structure, point group $\bar{4}2m$ with cell parameters a = 5.680 and c = 10.431 Å as a negative uniaxial crystal (Figure 8.6.34) [171]. CSP possesses large SHG coefficient d_{14} = 84.5 pm V^{-1}, wide transparency range (0.53–9.5 μm) and high thermal conductivity (13.6 W m^{-1} K^{-1}) [172]. Also, CSP has a large birefringence to realize phase-matching under laser pumping conditions of 1.0, 1.5 and 2.0 μm [171–173].

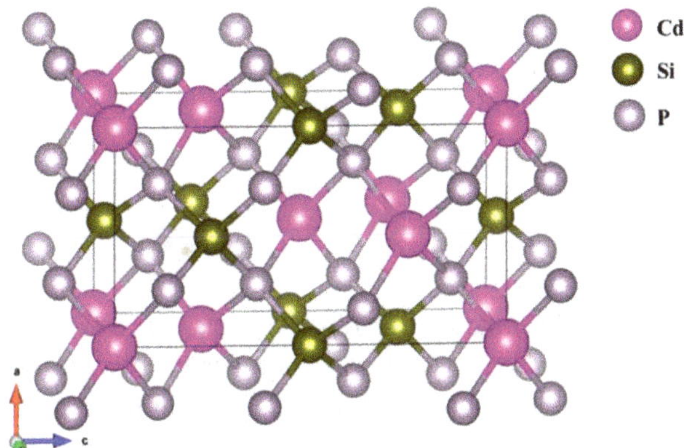

Figure 8.6.34: Crystal structure of $CdSiP_2$.

Polycrystalline samples of CSP are mainly produced by single temperature zone direct reaction and by two temperature zone gas phase transport. When the three elements P, Cd and Si react, Cd and P react first at ~500 °C to generate Cd_3P_2. With the rise of temperature from 600 to 780 °C, the pressure of P in the system increases, and Cd_3P_2 also reacts with free P continuously to further generate CdP_2. Above 1000 °C, CdP_2 initially reacts with elemental Si to generate CSP [174]. In 2010, a high-quality CSP crystal (70 mm × 25 mm × 8 mm) without cracking was successfully grown by the horizontal gradient freezing method (Figure 8.6.35) [175]. The vertical Bridgman method can also be used for growing CSP crystals, and a large crystal of ∅18 mm × 65 mm in size was grown by Tao et al., the absorption coefficients were 0.14 cm^{-1} at 4878 cm^{-1} and 0.06 cm^{-1} at 2500 cm^{-1} [174]. The successful preparation of large size and high-quality CSP crystal laid a foundation for its implementation. Later laser test studies showed that CSP can achieve mid-infrared laser output above 6 μm, which has important application prospects in infrared remote sensing and infrared medical environment monitoring with red external interference [176].

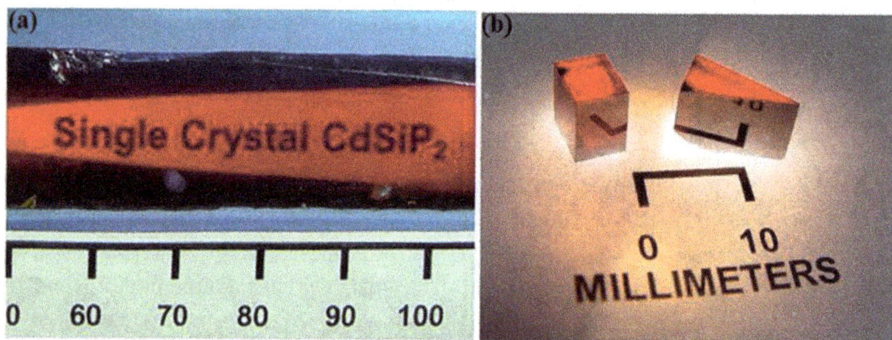

Figure 8.6.35: $CdSiP_2$ (a) Photograph of an oriented and polished $CdSiP_2$ single crystal with "Single Crystal $CdSiP_2$" label viewed in transmission through the crystal; (b) Photograph of oriented, polished $CdSiP_2$ single crystal samples used for material property measurements. Reproduced with permission from ref. [175]. Copyright 2010 Elsevier.

8.6.9 Chalcogenides

8.6.9.1 $AgGaS_2$ (AGS) and $AgGaSe_2$

AGS and $AgGaSe_2$ are already commercially mature reference crystals, and lots of experimental and theoretical explorations for IR NLO crystals focus on this system [177–180]. AGS has a chalcopyrite structure, belonging to the tetragonal system with point group $\bar{4}2m$ (a = 5.758, c = 10.306 Å). In the structure (Figure 8.6.36), Ag, Ga and S atoms form tetrahedra, and the S atom is located in the centers of the tetrahedra while the Ga and Ag atoms occupy the apex position. All the tetrahedra have aligned orientation, which is good for the electric polarizability. AGS has a bandgap of 2.76 eV, a SHG coefficient of d_{36} = 13.4 pm V^{-1} and a transmission range of 0.53–12 μm [181]. The melt temperature oscillation method was used to synthesize the AGS polycrystalline batch.

The synthesized polycrystalline batch was used to grow AGS single crystals using the Bridgman technique. $AgGaSe_2$ is the isostructural compound of AGS. $AgGaSe_2$ has a bandgap of 1.83 eV and a transmission range of 0.7–18 μm [182]. Figure 8.6.37 shows the commercialized AGS crystal. AGS and $AgGaSe_2$ both have a moderate birefringence of 0.0332 and 0.0538, respectively, to realize critical and noncritical phase-matching to

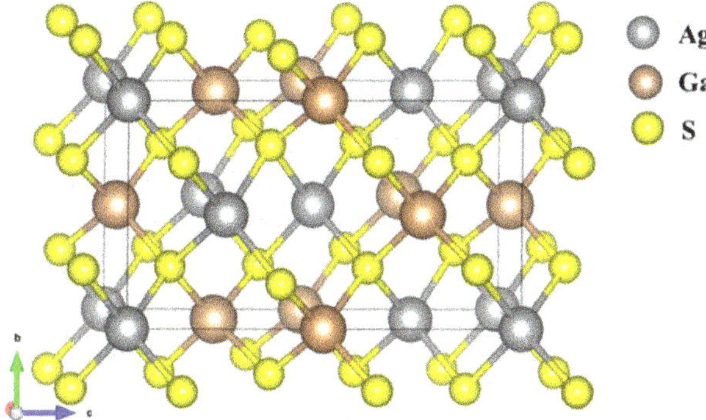

Figure 8.6.36: Crystal structure of $AgGaS_2$.

realize SHG, DFG and OPO lasing [181, 182]. AGS crystals can be pumped and driven by Nd:YAG lasers with a 2.5–10 μm tunable harmonic laser. The double frequency mixing and optical parametric devices made with $AgGaSe_2$ crystals can output continuously adjustable laser light with 3–18 μm wavelength. However, the problems of low thermal conductivity and low damage threshold and anisotropy due to thermal expansion are not suitable for high-energy laser pumping and seriously limit their applications.

Figure 8.6.37: Photograph of $AgGaS_2$. With courtesy and permission from Dientech (https://www.dientech.com/aggas2-crystals-product/).

8.6.9.2 CdSe

CdSe crystals display many structures, such as wurtzite and sphalerite. Among them, wurtzite CdSe crystallizes in space group $P6_3mc$ with cell parameters a = 4.298 and c = 7.015 Å (Figure 8.6.38). CdSe is a good photoconductive material and nonlinear quantum well laser with advantages such as wide transmission range (0.6–20 μm), low optical absorption (<0.01 cm^{-1}, 1–10 μm), no phonon absorption at 8–15 μm and good mechanical properties, showing great potential in the application of middle and far-infrared lasers [183]. However, it is still difficult to prepare high-quality CdSe single crystals in large size due to the following technical difficulties: high melting point (1264 °C), high vapor pressure (1 MPa), segregation precipitation, wrapped twin and honeycomb structure formation during the growth process [184]. At present, there are many methods to grow CdSe crystals, including: (i) high-pressure vertical Bridgman method, (ii) high-pressure vertical zone melting method, (iii) physical gas-transfer method and so on [185, 186]. Recently, high-optical quality single crystals (∅ (50–55) mm × (80–100) mm) were grown by a high-pressure seed-aided Bridgman technique. Some devices ($(6 \times 8)^2 \times (40–50)\ mm^3$) with certain phase-matching angles were fabricated with absorption coefficients ≤ 0.01 cm^{-1} at 10.8 μm [187]. The devices realized 1.05 W power output within the 10.1–10.8 μm far-IR wavelength range. It is necessary to further strengthen the preparation of CdSe single crystals and the study of laser devices.

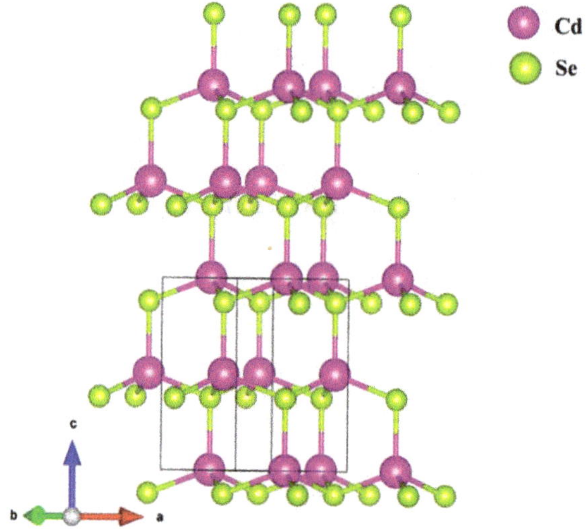

Figure 8.6.38: Crystal structure of CdSe.

8.6.9.3 GaSe

GaSe belongs to the hexagonal system with the space group $P\bar{6}m2$ (Figure 8.6.39). GaSe is a kind of nonlinear infrared optical material with excellent comprehensive properties: large SHG coefficient (d_{22} = 54.4 pm V^{-1}), wide transmission range (0.62–20 μm), relatively large birefringence (Δn is about 0.34) and high thermal conductivity (16.2 W m^{-1} K^{-1}) [188]. However, the layered structure of the material leads to poor mechanical performance, and it cannot be processed according to the designed phase matching, which seriously limits its application in nonlinear frequency conversion devices [189]. The researchers found that the mechanical and optical properties can be effectively improved by doping elements including S, In, Te and

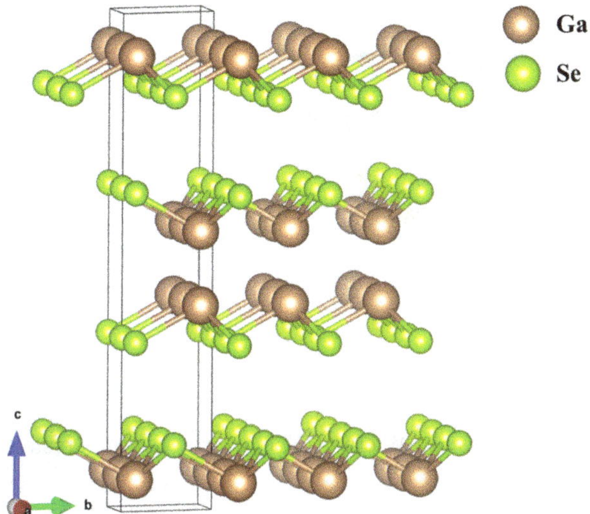

Figure 8.6.39: Crystal structure of GaSe.

so on. Among them, S-doping is the most effective with a proportion as high as 0.413 [190]. However, there are few studies on the properties of doped GaSe crystals as the growth technology is not mature yet, precluding stable performances, high-quality and large-size single crystals.

8.6.9.4 $LiInS_2$

Laser damage threshold is one of the important properties of IR NLO materials. There is a certain correlation between laser damage threshold and bandgap. The commercial IR NLO materials like AGS and $AGaSe_2$ suffer from low laser damage threshold because of their small bandgaps. One way to increase the bandgap is to replace the transition metal elements with alkali metals. $LiInS_2$ was obtained by substituting the transition metal Ag for Li and Ga for In. $LiInS_2$ belongs to the orthorhombic crystal system, with point group *mm*2. The cell parameters of $LiInS_2$ are a = 6.89, b = 8.06 and c = 6.48 Å. The Li^+ cation connects to four S^{2-} anions to form the tetrahedron LiS_4, and the In^+ cation also joins four S atoms to form the tetrahedron InS_4. LiS_4 and InS_4 are connected with each other through shared vertices and accumulate in a direction to form a diamond-like structure with the same arrangement direction. Among them, the LiS_4 tetrahedron forms zigzag chains through common vertex connection, and InS_4 is the same as above (Figure 8.6.40).

In 2000, the Russian researchers Isaenko et al. first reported the crystal growth of $LiInS_2$ and its basic properties [191, 192]. $LiInS_2$ has a large bandgap about 3.57 eV, a relatively wide transmission range (0.35 ~ 12.5 μm), a large SHG coefficient and a

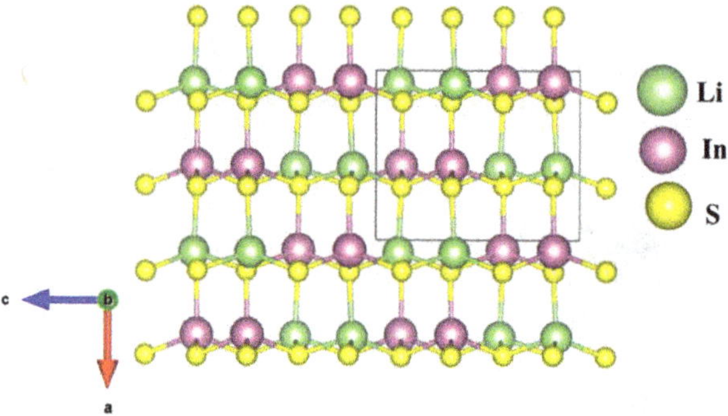

Figure 8.6.40: Crystal structure of $LiInS_2$.

large damage threshold (1064 nm, 10 ns: 1 GW cm^{-2}). The as-grown large crystals are shown in Figure 8.6.41 [193].

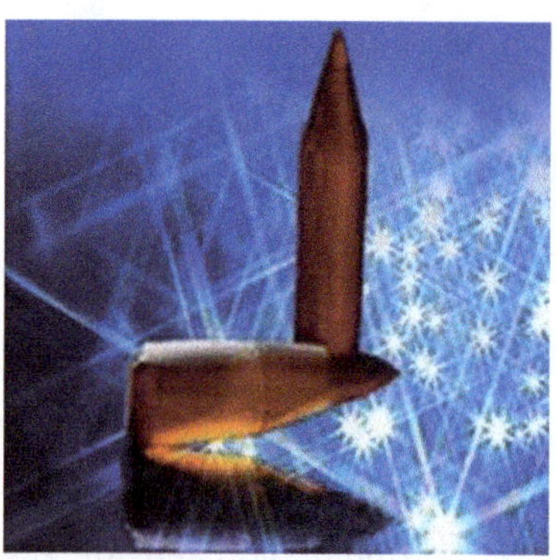

Figure 8.6.41: Photo of large as-grown $LiInS_2$ crystals with a diameter of 16 mm and a length of 50 mm. Reproduced with permission from ref. [193]. Copyright 2014. American Chemical Society.

In 2014, German the researcher Belltler et al. implemented the DFG technology and successfully realized a wide-band tunable laser output of 5–12 μm with a $LiInS_2$ crystal, with single pulse energies exceeding 1 nJ [194].

8.6.9.5 $BaGa_4S_7$ and $BaGa_4Se_7$ [195, 196]

The $BaGa_4S_7$ crystal belongs to the orthorhombic system and point group *mm*2 with cell parameters a = 1.476, b = 0.623 and c = 0.593 nm (Figure 8.6.42). The as-grown $BaGa_4S_7$ and $BaGa_4Se_7$ crystals are shown in Figure 8.6.43.

The UV absorption edge of $BaGa_4S_7$ is 350 nm and the IR absorption edge is 13.7 μm. Its thermal conductivities at 50 °C were 1.45, 1.58 and 1.68 W m^{-1} K^{-1} along the a, b and c-directions, respectively. The laser damage threshold of a single crystal was about 7.1 J cm^{-2} at 2.1 μm, and 26.2 J cm^{-2} at 9.58 μm. In addition, Kato et al. [197] used $BaGa_4S_7$ to achieve a quadruplicated frequency output of 2.65 μm with a CO_2 laser. When the pumped wavelength is about 2.2 μm, the idler noncritical phase-

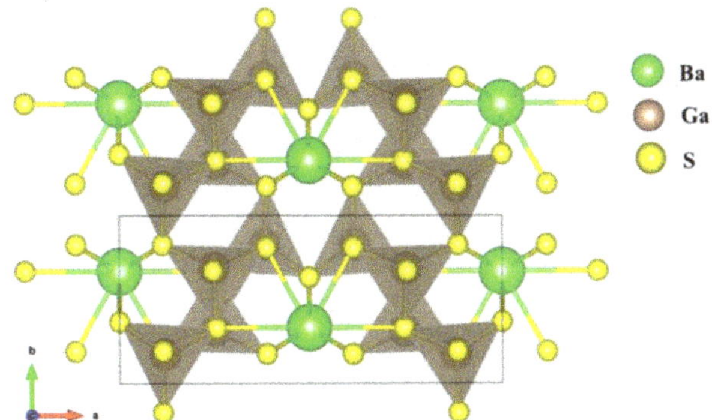

Figure 8.6.42: Crystal structure of $BaGa_4S_7$.

matching OPO wavelength lies between 6 and 10 μm. When the S atom is replaced by Se, the structure of $BaGa_4Se_7$ changes to the monoclinic crystal system, point group *m*, cell parameters $a = 0.763$, $b = 0.651$ and $c = 1.470$ nm. The thermal expansion behavior in $BaGa_4Se_7$ does not exhibit strong anisotropy with $a_a = 9.24 \times 10^{-6}\ K^{-1}$, $a_b = 10.76 \times 10^{-6}\ K^{-1}$ and $a_b = 11.70 \times 10^{-6}\ K^{-1}$ along the three crystallographic axes. The thermal diffusivity/thermal conductivity coefficients measured at 298 K are 0.50(2) $mm^2\ s^{-1}$ / 0.74(3) $W\ m^{-1}\ K^{-1}$, 0.42(3) $mm^2\ s^{-1}$ / 0.64(4) $W\ m^{-1}\ K^{-1}$, 0.38(2) $mm^2\ s^{-1}$ / 0.56(4) $W\ m^{-1}\ K^{-1}$, along the *a*, *b* and *c* crystallographic axis, respectively. Yang et al. [198, 199] realized the tuned laser output of $BaGa_4Se_7$ in the ranges 3–5 μm and 6.4–11 μm by using OPA technology. In 2016 [200], Yuan et al. used the Q-switched Ho:YAG (2.1 μm) as the pumped light source to realize for the first time the OPO laser output in the range 3–5 μm, with an average power of 1.55 W, a light-to-light conversion efficiency of 14.4% and a slope efficiency of 19.9%. In 2017, Xu et al. [201] used a Q-adjusted Nd:YAG (1064 nm) laser to pump, and the $BaGa_4Se_7$ OPO technology was

Figure 8.6.43: (a) $BaGa_4S_7$ crystals with dimension of up to Ø15 × 40 mm, and a polished crystal device with dimension of 5 × 8 × 8 mm. Reproduced with permission from ref 195. Copyright 2014. Elsevier. (b) $BaGa_4Se_7$ crystal with dimensions of Ø20 × 30 mm. Reproduced with permission from ref. [196]. Copyright 2012. Elsevier.

used to output the tuned laser in the range 3.12–5.16 μm, achieving a maximum energy output of 2.56 mJ at 4.11 μm, with a photo-optical conversion efficiency of 4.16%. In 2018, Kolker et al. [202] used a Q-adjusted Nd:$YLiF_4$ (1053 nm, 10 ns, f = 100 Hz) laser to pump and $BaGa_4Se_7$-OPO technology was used to realize the wide band mid-far infrared laser output of 2.6–10.4 μm. The output energy at 3.3 μm and 8 μm was 45 and 14 μJ, respectively.

References

[1] Franken PA, Hill AE, Peters CW, Weinreich G. Phys Rev Lett, 1961, 7, 118–19.
[2] Boyd RW. Nonlinear Optics, 3rd edition, Academic Press, 2008.
[3] Karna SP, Yeates AT. ACS Symp Ser, 1996, 628, 1–22.
[4] Bloembergen N. Nonlinear Optics, W. A. Benjiamin Inc., New York, 1965.
[5] Yang ZH, Pan SL. Mater Chem Front, 2021, 5, 3507–23.
[6] Chen CT, Wu BC, Jiang AD, You GM. Sci Sin, Ser B, 1985, 18, 235–43.
[7] Chen CT, Wu YC, Jiang AD, Wu B, You G, Li R, Lin S. J Opt Soc Am B, 1989, 6, 616–21.
[8] Lin SJ, Sun ZY, Wu BC, Chen CT. J Appl Phys, 1990, 67, 634–38.
[9] Wang Y, Pan SL. Coord Chem Rev, 2016, 323, 15–35.
[10] Mutailipu M, Poeppelmeier KR, Pan SL. Chem Rev, 2021, 121, 1130–202.
[11] Becker P. Adv Mater, 1998, 10, 979–92.
[12] Chen CT, Wang YB, Wu BC, Wu KC, Zeng WL, Yu LH. Nature, 1995, 373, 322–24.
[13] Dewey CF, Cook WR, Hodgson RT, Wynne JJ. Appl Phys Lett, 1975, 26, 714.
[14] Liebertz J, Stähr S. Z Kristallogr, 1983, 165, 91–93.
[15] Chen CT, Xu ZY, Deng DQ, Zhang J, Wong GKL, Wu B, Ye N, Tang D. Appl Phys Lett, 1996, 68, 2930–32.
[16] Li LY, Li GB, Wang YX, Liao FH, Lin JH. Chem Mater, 2005, 17, 4174–80.
[17] Zhang BB, Shi GQ, Yang ZH, Zhang FF, Pan SL. Angew Chem Int Ed, 2017, 56, 3916–19.
[18] Yang ZH, Tudi A, Lei BH, Pan SL. Sci China Mater, 2020, 63, 1480–88.
[19] Cheng HH, Li FM, Yang ZH, Pan SL. Angew Chem Int Ed, 2022, 61, e202115669.
[20] Konig H, Hoppe R. Z Anorg Allg Chem, 1978, 439, 71–79.
[21] Wu BC, Chen N, Chen CT, Deng DQ, Xu ZY. Opt Letters, 1989, 14, 1080–81.
[22] Chen CT, Sasaki T, Li RK, Wu Y, Lin Z, Mori Y, Hu Z, Wang J, Uda S, Yoshimura M, Kaneda Y. Nonlinear Optical Borate Crystals, Wiley-VCH, Weinheim, 2012.
[23] Wu Y, Chen C. Laser Rev, 1991, 19, 15.
[24] Hong H, Zhong W, Lu Z, Zhao T, Hua S. Proc SPIE-Int Soc Opt Eng, 1993, 1863, 184.
[25] Ukatchi T, Lane RJ, Bosenberg WR, Tang CL. J Opt Soc Am B, 1992, 9, 1128.
[26] Markgraf SA, Furukawa Y, Sato M. J Cryst Growth, 1994, 140, 343.
[27] Hu ZG, Zhao Y, Yue YC, Yu XS. J Cryst Growth, 2011, 335, 133–37.
[28] Mighell AD, Perloff A, Block S. Acta Crystallogr, 1966, 20, 819.
[29] Levin EM, McMnrdie HF. J Res Nat Bur Stand, 1949, 42, 131.
[30] Liebertz J, Stähr S. Z Kristallogr, 1983, 165, 91–93.
[31] Lei BH, Pan SL, Yang ZH, Cao C, Singh DJ. Phys Rev Lett, 2020, 125, 187402.
[32] Kato K. IEEE J Quantum Electron, 1986, 22, 1013–14.
[33] Muckenheim W, Lockai P, Burghardt B, Basting D. Appl Phys B, 1988, 45, 259.
[34] Huang QZ, Liang JG. Acta Phys Sin, 1981, 30, 559–64.

[35] Bekker TB, Kokh AE, Kononova NG, Fedorov PP, Kuznetsov SV. Cryst Growth Des, 2009, 9, 4060–63.
[36] Hengel RO, Fischer F. J Cryst Growth, 1991, 114, 656–60.
[37] Nikolaev NA, Andreev YM, Antsygin VD, Bekker TB, Ezhov DM, Kokh AE, Kokh KA, Lanskii GV, Mamrashev AA, Svetlichnyi VA. J Phys: Conf Ser, 2018, 951, 012003.
[38] Krogh-Moe J. Arkiv Kemi, 1958, 12, 247–49.
[39] Krogh-Moe J. Acta Crystallogr, 1974, B30, 1178–80.
[40] Kagebayashi Y, Mori Y, Sasaki T. Bull Mater Sci, 1999, 22, 971–73.
[41] Wu YC. Appl Phys Lett, 1993, 62, 2614.
[42] Kato K. IEEE J Quantum Electron, 1995, 31, 169–71.
[43] Wu YC, Fu PZ, Wang JX, Xu ZY, Zhang L, Kong YF, Chen CT. Opt Letters, 1997, 22, 1840–42.
[44] Ushiyama N, Yoshimura M, Ono R, Kamimura T, Yap YK, Mori Y, Sasaki T. The Advanced Solid-State Lasers Conference, Seattle, Washington, United States, 2001, 622–24. https://doi.org/10.1364/ASSL.2001.WC3
[45] Penin N, Touboul M, Nowogrocki G. J Cryst Growth, 2003, 256, 334–40.
[46] Sasaki T, Mori Y, Kuroda I, Nakajima S, Yamaguchi K, Watanabe S, Nakai S. Acta Crystallogr, 1995, C51, 2222–24.
[47] Mori Y, Kuroda I, Nakajima S, Sasaki T, Nakai S. Appl Phys Lett, 1995, 67, 1818–20.
[48] Mori Y, Kuroda I, Nakajima S, Taguchi A, Sasaki T, Nakai S. J Cryst Growth, 1995, 156, 307–09.
[49] Mori Y, Kuroda I, Nakajima S, Sasaki T, Nakai S. Jpn J Appl Phys, 1995, 34, L296–8.
[50] Umemura N, Yoshida K, Kamimura T, Mori Y, Sasaki T, Kato K. OSA TOPS Adv Solid-State Lasers, 1999, 26, 715–19.
[51] Shoji I, Nakamura H, Ito R, Kondo T, Yoshimura M, Mori Y, Sasaki T. J Opt Soc Am B, 2001, 18, 302–07.
[52] Ye N, Zhang W, Wu B. Chen C Proc SPIE, 1998, 3556, 21.
[53] Hu ZG, Mor Y, Higashiyama T, Yap YK, Kagebayashi Y, Sasaki T. Proc SPIE, 1998, 3556, 156–61.
[54] Hu ZG, Higashiyama T, Yoshimura M, Mori Y, Sasaki T. Z Kristallogr NCS, 1999, 214, 433–34.
[55] Ye N, Zeng W, Jiang J, Wu B, Chen C, Feng B, Zhang X. J Opt Soc Am B, 2000, 17, 764–68.
[56] Umemura N, Ando M, Suzuki K, Takaoka E, Kato K, Hu Z-G, Yoshimura M, Mori Y, Sasaki T. Appl Opt, 2003, 42, 2716–19.
[57] Ye N, Weng Z, Wu B, Chen C. Proc SPIE, 1998, 3556, 21–23.
[58] Zhang CQ, Wang JY, Hu XB, Zhang HJ. J Synth Cryst, 2005, 34, 786–89.
[59] Liu CL, Liu LJ, Zhang X, Wang LR, Wang GL, Chen CT. J Cryst Growth, 2011, 318, 618–20.
[60] Chen CT, Wang GL, Wang XY, Zhu Y, Xu Z, Kanai T, Watanabe S. IEEE J Quantum Electron, 2008, 44, 617–21.
[61] Mei LF, Huang X, Wang YB, Wu Q, Wu BC, Chen CT. Z Kristallogr, 1995, 210, 93–95.
[62] Solov'eva LP, Bakakin VV. Sov Phys Crystallogr, 1970, 15, 802–05.
[63] Chen CT, Xu ZY, Deng DQ, Zhang J, Wong GKL. Appl Phys Lett, 1996, 68, 2930–32.
[64] Chen CT, Wang GL, Wang XY, Xu ZY. Appl Phys B, 2009, 97, 9–25.
[65] Wang JY, Zhang C, Zhang J, Hu X, Jiang M, Chen C, Wu Y, Xu Z. J Mater Res, 2003, 18, 2478–85.
[66] Wang XY, Yan X, Luo SY, Chen CT. J Cryst Growth, 2011, 318, 610–12.
[67] Ye N, Tang DY. J Cryst Growth, 2006, 293, 233.
[68] Chen CT, Lv JH, Wang JY, Xu ZY, Zhang CQ, Liu YG. Chin Phys Lett, 2001, 18, 1081.
[69] Chen CT, Luo SY, Wang XY, Wang G, Wen X, Wu H, Zhang X, Xu Z. J Opt Soc Am B, 2009, 26, 1519–25.

[70] Wen XH. Study on the Growth and Properties of the Nonlinear Optical Crystals: MBBF (M = Na, K, Rb, Cs). Ph. D. Dissertation, Technical Institute of Physics and Chemistry, Chinese Academy of Sciences, 2006. (In Chinese)
[71] Luo SY, Yu JQ, Wang XY, Wen XH, Chen CT. J Synth Cryst, 2010, 39, 28–30.
[72] Huang HW, Chen C, Wang XY, Zhu Y, Wang G, Zhang X, Wang L, Yao J. J Opt Soc Am B, 2011, 28, 2186.
[73] Wu HP, Pan SL, Poepplmeier KR, Li H, Jia D, Chen Z, Fan X, Yang Y, Rondinelli JM, Luo H. J Am Chem Soc, 2011, 133, 7786–90.
[74] Grice JD, Burns PC, Hawthorne FC. Can Mineral, 1999, 37, 731–62.
[75] Burns PC, Grice JD, Hawthorne FC. Can Mineral, 1995, 33, 1131–51.
[76] Wu HP, Yu HW, Yang ZH, Han J, Wu K, Pan SL. J Materiomics, 2015, 1, 221–28.
[77] Zhang NH, Teng H, Huang HD, Tian W-L, Zhu J-F, Wu H-P, Pan SL, Fang S-B, Wei Z-Y. Chin Phys B, 2016, 25, 124024.
[78] Zhang NH, Fang SB, He P, Huang H-D, Zhu J-F, Tian W-L, Wu H-P, Pan SL, Teng H, Wei Z-Y. Chin Phys B, 2017, 26, 064208.
[79] Wu Y, Wu HP, Wang Y, Zhang L, Zhang Q, Meng L, Wang L, He C, Zhang D, Guo H, Lin X, Pan S. Opt Eng, 2018, 57, 066112.
[80] Al-Ama AG, Belokoneva EL, Stefanovich SY, Dimitrova OV, Mochenova NN. Crystallogr Rep, 2006, 51, 225–30.
[81] Zhang M, Su X, Pan SL, Wang Z, Zhang H, Yang Z, Zhang B, Dong L, Wang Y, Zhang F, Yang Y. J Phys Chem C, 2014, 118, 11849–56.
[82] Zhang M, Pan SL, Fan XY, Zhou ZX, Poeppelmeier KR, Yang Y. CrystEngComm, 2011, 13, 2899.
[83] Fan X, Zhang M, Pan SL, Yang Y, Zhao WW. Mater Lett, 2012, 68, 374–77.
[84] Zhang L, Zhang M, Wang LR, Chen X, Jiao J, Pan S, Tang G, Zhang X, Chen Z, Zhang X. Optic Mater, 2020, 107, 110088.
[85] Liu XD, Xu L, Pan SL, Liang XY. Laser Phys Lett, 2017, 14, 095403.
[86] Shi GQ, Wang Y, Zhang FF, Zhang B, Yang Z, Hou X, Pan S, Poeppelmeier KR. J Am Chem Soc, 2017, 139, 10645–48.
[87] Zhang ZZ, Wang Y, Zhang BB, Yang ZH, Pan SL. Angew Chem, Int Ed, 2018, 57, 6577–81.
[88] Wang Y, Zhang BB, Yang ZH, Pan SL. Angew Chem, Int Ed, 2018, 57, 2150–54.
[89] Wang XF, Wang Y, Zhang BB, Zhang FF, Yang ZH, Pan SL. Angew Chem, Int Ed, 2017, 56, 14119–23.
[90] Mutailipu M, Zhang M, Zhang B, Wang L, Yang Z, Zhou X, Pan S. Angew Chem, Int Ed, 2018, 57, 6095–99.
[91] Zhang ZZ, Wang Y, Zhang BB, Yang ZH, Pan SL. Inorg Chem, 2018, 57, 4820–23.
[92] Xia M, Li FM, Mutailipu M, Han SJ, Yang ZH, Pan SL. Angew Chem, Int Ed, 2021, 60, 14650–56.
[93] Corbridge DEC. Phosphorus. An Outline of Its Chemistry, Biochemistry and Technology, 2nd edition, Elsevier Scientific Publishing Company, Amsterdam, Oxford, New York, 1980.
[94] Dan JH, Mei DJ, Wu YD, Lin ZS. Coord Chem Rev, 2021, 431, 213692.
[95] Yoreo JJD, Burnham AK, Whitman PK. Int Mater Rev, 2002, 47, 113–52.
[96] Wang J, Wei J, Liu Y, Yin X, Hu X, Shao Z, Jiang M. Prog Cryst Growth Charact Mater, 2000, 40, 3–15.
[97] Itoh K, Uchimoto M. Ferroelectrics, 1998, 217, 155–62.
[98] Dmitriev VG, Gurzadyan, Nikogosyan DN. Handbook of nonlinear optics crystals. 2rd Edition, Springer, 1997.
[99] Shoji I, Kondo T, Kitamoto A, Shirane M, Ito R. JOSA B, 1997, 14, 2268–94.
[100] He Y, Zeng J, Wu D, Su G, Yan M. J Cryst Growth, 1996, 169, 196–98.

[101] Qi HJ, Shao JD, Wu FL, Wang B, Chen DY. J Synth Cryst, 2020, 49, 1004–09.
[102] Zhang LY, Wang SH, Liu H, Xu LY, Li XL, Sun X, Wang B. J Synth Cryst, 2021, 50, 724–31.
[103] Zhuang X, Ye L, Zheng G, Su G, He Y, Lin X, Xu Z. J Cryst Growth, 2011, 318, 700–02.
[104] Sliker TR, Burlage SR. J Appl Phys, 1963, 34, 1837–40.
[105] Sonin AS, Vasilevskaya AS. Electrooptic Crystals, (Russian translation), Izd Mir, Moscow, 1971.
[106] Roberts DA. IEEE J Quant Electr, 1992, 28, 2057–74.
[107] Su GB, Zeng JB, He YP, Li ZD, Huang BR, Jiang RH. J Chin Ceram Soc, 1997, 25, 717–19.
[108] Shao JD, Dai YP, Xu Q. Optics and Precision Engineering, 2016, 24, 2889–95.
[109] Haussühl S. Z Kristallogr, 1964, 120, 401–14.
[110] Eimerl D. Ferroelectrics, 1987, 72, 95–139.
[111] Voloshina IV, Gerr RG, Antipin MYu, Tsirel'son VG, Pavlova NI, Struchkov YuT, Ozerov RP, Rez IS. Kristallografiya, 1985, 30, 668–76.
[112] Fan TY, Huang CE, Hu BQ, Eckardt RC, Fan YX, Byer RL, Feigelson RS. Appl Opt, 1987, 26, 2390–94.
[113] Bierlein JD, Vanherzeele H. J Opt Soc Am B, 1989, 6, 622–33.
[114] Liao H, Shen H, Zheng Z, Lian T, Zhou Y, Huang C, Zeng R, Yu G. Opt Laser Technol, 1988, 20, 103–04.
[115] Wang JY, Wei JQ, Liu YG, Yin Xin, Hu XB, Shao ZS, Jiang MH. Prog. Crystal Growth and Charact, 2000, 40, 3–15.
[116] Vanherzeele H, Bierlein JD, Zumsteg FC. Appl Opt, 1988, 27, 3314–16.
[117] Shen HY, Zhou YP, Lin WX, Zeng ZD, Zeng RR, Yu GF, Huang CH, Jiang AD, Jia SQ, Shen DZ. IEEE J Quant Electr, 1992, 28, 48–51.
[118] Shen DZ. J Synth Cryst, 2001, 30, 28–35.
[119] Zhang BB, Han GP, Wang Y, Chen XL, Yang ZH, Pan SL. Chem Mater, 2018, 30, 5397–403.
[120] Lu J, Yue JN, Xiong L, Zhang WK, Chen L, Wu LM. J Am Chem Soc, 2019, 141, 8093–97.
[121] Liu YC, Shen YG, Zhao SG, Luo JH. Coord Chem Rev, 2020, 407, 213152.
[122] D'yakov VA, Laptiiiskava TV, Prvalkiii VI. Int Soc Opt Phot, 1999, 3734, 415–19.
[123] D'yakov VA, Ebbers CA, Pchelkin MV, Pryalkin VI. J Russ Laser Res, 1996, 17, 489–94.
[124] Zou GH, Ye N, Huang L, Lin XS. J Am Chem Soc, 2011, 133, 20001–07.
[125] Zhang W, Halasyamani PS. CrystEngComm, 2017, 19, 4742–48.
[126] Tran TT, He JG, Rondinelli JM, Halasyamani PS. J Am Chem Soc, 2015, 137, 10504–07.
[127] Priya R, Krishnan S, Justin RC, Jerome DS. Cryst Res Technol, 2009, 44, 1272–76.
[128] Bayarjargal L. Cryst Res Technol, 2008, 43, 1138–42.
[129] Rosenzweig A, Morosin B. Acta Crystallogr, 1966, 20, 758–61.
[130] Unezawa T, Ninomiya Y, Tatuoka S. J Appl Crystallogr, 1970, 3, 417.
[131] Gurzadian GG, Dmitriev VG, Nikogosian DN. Handbook of Nonlinear Optical Crystals, Springer, Heidelberg, 1997.
[132] Nath G, Mehmanesch H, Gsänger M. Appl Phys Lett, 1970, 17, 286–88.
[133] Kabelka VI, Piskarskas AS, Stabinis AY, Sher LR. Sov J Quant Electron, 1975, 5, 255–56.
[134] Volosov V, Karpenko SG, Kornienko NE, Strizhevskii VL. Sov J Quant Electron, 1975, 4, 1090.
[135] Deserno U, Nath G. Phys Lett, 1969, 30, 483–84.
[136] Kurtz SK, Perry TT. Appl Phys Lett, 1968, 12, 186–88.
[137] Pinnowb DA, Dixon RW. Appl Phys Lett, 1968, 13, 156–58.
[138] Naito H, Inaba H. Opto-electronics, 1972, 4, 335–37.
[139] Kovrigin AI, Nikles PV. JETP Lett, 1971, 13, 313.
[140] Andrews RA. IEEE J, 1970, 6, 68–80.
[141] Abrahams SC, Hamilton WC, Reddy JM. J Phys Chem Soilds, 1966, 27, 1013–18.

[142] Günter P, Huignard JP. Photorefraction Materials and Their Application, Springer-Verlag, Berlin, Heidelberg, 1988.
[143] Kimura S, Kitamura K. J Ceram Soc, 1993, 101, 22–37.
[144] Bridenbaugh PM. J Cryst Growth, 1973, 19, 45–52.
[145] Sun J, Hao YX, Zhang L, Xu JJ, Zhu SN. J Synth Cryst, 2020, 49, 947–64.
[146] Xu JY, Lu BL, Xia ZR. J Synth Cryst, 2003, 23, 626–30.
[147] Burrows L. Now entering, lithium niobate valley [EB/OL]. https://www.Seas.Harvard.edu/news/2017/12/now-entering-lithium-niobate-valley. 2017. Accessed december 19th, 2021.
[148] Shen DZ. J Synth Cryst, 2002, 31, 192–200.
[149] Katz L, Megaw HD. Acta Crystallogr, 1967, 22, 639–48.
[150] Shen DZ. Prog Crystal Growth Charact, 1990, 20, 161–74.
[151] Dmitriev VG, Gurzadyan GG, Nikogosyan DN. Handbook of Nonlinear Optical Crystals, Springer-Verlag, Berlin, Heidelberg, 1991.
[152] Cooper AS. Acta Crystallogr, 1962, 15, 578–82.
[153] Schunemann PG, Pomeranz LA, Setzler SD, Jones CW, Budni PA. CW mid-IR OPO based on OP-GaAs. Conference on lasers and electro-optics: Europe & international quantum electronics conference, Munich, Germany, 2013.
[154] Jian W, Cheng HJ, Gao YZ. J Synth Cryst, 2020, 49, 1397–442.
[155] Vodopyanov KL, Levi O, Kuo PS, Pinguet TJ, Harris JS, Fejer MM, Gerard B, Becouarn L, Lallier E. Opt Letters, 2004, 29, 1912–14.
[156] Peterson RD, Whelan D, Bliss D, Lynch C. Proceedings of SPIE, 2009, 7197, 09.
[157] Buehler E, Wernick JH, Wiley JD. J Electron Mater, 1973, 2, 445–54.
[158] Zhang JQ, Zhao BJ, Zhu SF, Chen B-J. J Synth Cryst, 2013, 42, 392–96.
[159] Zhu CQ, Lei ZT, Yang CH. J Synth Cryst, 2012, 41, 160.
[160] Wu HX, Ni YB, Geng L, Mao M-S, Wang Z-Y, Cheng G-C, Yang L. J Synth Cryst, 2007, 36, 507–11.
[161] Bliss DF, Harris M, Horrigan J, Higgins WM, Armington AF, Adamski JA. J Cryst Growth, 1994, 137, 145–49.
[162] Zhang G, Tao X, Wang S, Shi Q, Ruan H, Chen L. J Cryst Growth, 2012, 352, 67–71.
[163] Zhang G, Tao X, Wang S, Liu G, Shi Q, Jiang M. J Cryst Growth, 2011, 318, 717–20.
[164] Lei Z, Okunev AO, Zhu C, Verozubova GA, Ma T, Yang AC. J Cryst Growth, 2016, 450, 34–38.
[165] Zhong K, Li JS, Xu DG, Wang JL, Wang Z, Wang P, Yao JQ. Optoelectron Lett, 2010, 6, 179–82.
[166] Schunemann PG, Zawilski KTL, Pomeranz A, Creeden DJ, Budni PA. J Opt Soc Am B, 2016, 33, D36.
[167] Zawilski KT, Schunemann PG, Setzler SD, Pollak TM. J Cryst Growth, 2008, 310, 1891–96.
[168] Yuan ZR, Dou YW, Chen Y, Fang P, Yin WL, Kang B. J Synth Cryst, 2020, 49, 8.
[169] Lei Z, Zhu C, Xu C, Yao B, Yang C. J Cryst Growth, 2014, 389, 23–29.
[170] Su N, Zhang M, Reng G, Liu QX, Xing JG. Opt Technol, 2013, 39, 359.
[171] Wu SL, Zhao BJ, Zhu SF, He Z-Y. J Synth Cryst, 2014, 43, 492–96.
[172] Yang H, Zhu SF, Zhao BJ, Chen B-J. J Synth Cryst, 2012, 41, 11–14.
[173] Kumar SC, Jelínek M, Baudisch M, Zawilski KT, Schunemann PG, Kubeček V, Biegert J, Ebrahim-Zadeh M. Opt Express, 2012, 20, 15703–09.
[174] Zhang GD, Ruan HP, Zhang X, Wang SP, Tao XT. CrystEngComm, 2013, 15, 4255–60.
[175] Zawilski KT, Schunemann PG, Pollak TC, Zelmon DE, Fernelius NC, Hopkins FK. J Cryst Growth, 2010, 312, 1127–32.
[176] Kumar SC, Agnesi A, Dallocchio P, Pirzio F, Reali G, Zawilski KT, Schunemann PG, Ebrahim-Zadeh M. Opt Lett, 2011, 36, 3236–38.
[177] Cai WB, Abudurusuli, Xie CW, Tikhonov E, Li JJ, Pan SL, Yang ZH. Adv Funct Mater, 2022, 32, 2200231–39.

[178] Zhao BJ, Zhu SF, Li ZH, Yu FL, Zhu XH, Gao DY. Chin Sci Technol, 2001, 46, 1132.
[179] Li G, Chu Y, Zhou Z. Chem Mater, 2018, 30, 602–06.
[180] Hanna DC, Rampal VV, Smith RC. Opt Commun, 1973, 8, 151–53.
[181] Boyd G, Kasper H, McFee J. IEEE J Quantum Electron, 1971, 7, 563–73.
[182] Singh NB, Hopkins RH, Feichtner JD. J Mater Sci, 1986, 21, 837–41.
[183] Wu HX, Huang F, Ni YB, Wang ZY, Chen L, Cheng GC. Chin J Quant Elect, 2010, 27, 711.
[184] Zeng TX, Zhao BJ, Zhu SF, He ZY, Chen BJ, Lu DZ. J Synth Cryst, 2009, 38, 1068–72.
[185] Kolesnikov NN, James RB, Berzigiarova NS, Kulakov MP. Proc SPIE, 2003, 4784, 93–104.
[186] Jie WQ. Principle and Technology of Crystal Growth, Science Press, Beijing, 2010.
[187] Ni YB, Han WM, Wu HX, Mao M, Huang C, Wang Z, Jiang P. J Synth Cryst, 2020, 49, 1488–89.
[188] Anis MK. J Cryst Growth, 1981, 55, 465–69.
[189] Petrov V, Panyutin VL, Tyazhev A, Marchev G, Zagumennyi A, Rotermund F, Noack F, Miyata K, Iskhakova L, Zerrouk A. Laser Phys, 2011, 21, 774–81.
[190] Tikhomirov AA, Lanski GV. Proc SPIE, 2006, 6258, 625809.
[191] Isaenko L, Yelisseyev A, Lobanov S, Petrov V, Rotermund F, Zondy JJ, Knippels GHM. Mater Sci Semicond Proc, 2001, 4, 665–68.
[192] Isaenko L, Vasilyeva I, Yelisseyev A, Lobanov S, Malakhov V, Dovlitova L, Zondy JJ, Kavun I. J Cryst Growth, 2000, 218, 313–22.
[193] Wang S, Gao Z, Zhang X, Zhang X, Li C, Dong C, Lu Q, Zhao M, Tao X. Cryst Growth Des, 2014, 14, 5957–61.
[194] Beutler M, Rimke I, Büttner E, Petrov V, Isaenko L. Opt Lett, 2014, 39, 4353–55.
[195] Guo YF, Zhou YQ, Lin XS, Chen WD, Ye N. Opt Mater, 2014, 36, 2007–11.
[196] Yao JY, Yin WL, Feng KX, Li X, Mei D, Lu Q, Ni Y, Zhang Z, Hu Z, Wu Y. J Cryst Growth, 2012, 346, 1–4.
[197] Kato K, Okamoto TT, Mikami V, Petrov V, Badikov D, Badikov V. Proc SPIE, 2013, 8604, 860416.
[198] Yang F, Yao JY, Hu HY, Feng K, Yin W-L, Li F-Q, Yang J, Du S-F, Peng Q-J, Zhang J-Y, Cui D-F, Wu Y-C, Chen C-T, Xu Z-Y. Opt Lett, 2013, 38, 3903–05.
[199] Yang F, Yao JY, Xu HY, Zhang -F-F, Zhai N-X, Lin Z-H, Zong N, Peng Q-J, Zhang J-Y, Cui D-F, Wu Y-C, Chen C, Xu Z-Y. IEEE Phot Techn Lett, 2015, 27, 1100–03.
[200] Yuan JH, Li C, Yao BQ, Yao J-Y, Duan X-M, Li -Y-Y, Shen Y-J, Wu Y-C, Cui Z, Dai T-Y. Opt Express, 2016, 24, 6083–87.
[201] Xu WT, Wang YY, Xu DG, Li C, Yao J-Y, Yan C, He Y-X, Nie M-T, Wu Y-C, Yao J-Q. Appl Phys B, 2017, 123, 8.
[202] Kolker DB, Kostyukiva NY, Boyko AA, Badikov VV, Badikov DV, Shadrintseva AG, Tretyakova NN, Zenov KG, Karapuzikov AA, Zondy J-J. J Phys Commun, 2018, 2, 035039.

8.7 Excimer, mercury and sodium dischargers

Manfred Salvermoser

8.7.1 Introduction

Electromagnetic waves coming from the sun, a thermal, incandescent light source directly driven by nuclear fusion, constitute the main energy source sustaining almost all forms of life on earth. The sun's emission has a continuous and broad color range, with about 40% of the emitted electromagnetic energy being within the visible spectrum with wavelengths (λ) between 400 and 800 nm. The sun's luminous efficacy is 93 lm/W [1]. By considering the nuclear power to electrical energy conversion efficiency of modern nuclear power plants, today's best solid-state LED light source with a luminous efficacy of about 170 lm/W, reaches only 0.4·170 lm/W = 68 lm/W. In case one wishes to employ the photons energy content $\varepsilon(\lambda) = \frac{h \cdot c}{\lambda}$, $(h = 6 \cdot 62607015 \cdot 10^{-34} Js, c = 299792458 \frac{m}{s})$ to photochemically initiate reactions, the photons energy must exceed a certain threshold value.

Typical binding energies in molecules fall into the range of 3 to 10 eV (1 eV = $1.602 \cdot 10^{-19}$ J), while visible light only has photon energies between 1.6 and 3.2 eV. In order to access specific photochemical reactions, energy-efficient, narrowband light sources with emission wavelengths sufficiently short (such that the photons carry more energy than the molecules binding energy we wish to overcome) are necessary. An efficient way to generate the desired photons is achieved by means of gas discharges.

8.7.2 Gas discharges

8.7.2.1 Thermal and non-thermal plasma

In a plasma, gas particles are mixed with free electrons and ions. Both charged species behave like gases as well. As a result, a plasma is electrically conductive. Plasmas can be generated by simply heating matter to the point that atoms melt. The thermal energies necessary are of order $k_B \cdot T > 1$ eV, with $k_B = 1.38064852 \times 10^{-23}$ J/K being Boltzmann's constant. This results in temperatures exceeding 12,000 K. Thermal plasmas are characterized by thermal equilibrium between electrons, ions and neutral particles, leading to high degrees of ionization. Well-known examples for thermal plasmas are our sun and other stars. Electrons, ions and neutral gas particles couple thermally together through elastic collisions. Since ions and atoms have essentially the same mass, they couple strongly together. Electrons, however, due to their small mass, lose only a small fraction of their kinetic energy during an elastic collision with a heavy collision partner. They therefore couple weakly with ions and gas particles. As a result, it is possible to create plasmas with hot, very energetic

https://doi.org/10.1515/9783110798890-026

electrons having an electron energy distribution function (EEDF) characterized by a high electron temperature T_e with $k_B \cdot T_e \sim eV$, embedded within a cold gas environment consisting of neutral particles and ions being at room temperature ($T_n = T_i \sim 300$ K). Plasmas with $T_e \gg T_i = T_n$ are non-thermal plasmas [2]. They are characterized by very low degrees of ionization (≪1). Non-thermal plasmas can be created by means of electric discharges in gases.

8.7.2.2 Low-pressure glow discharges

8.7.2.2.1 Low-pressure sodium and mercury discharges

Figure 8.7.1 shows a cold cathode discharge tube filled with a rare gas (RG), e.g. argon, at a pressure of $p_{Ar} = 4$ mbar ($T = 293$ K, $n_{Ar} = 10^{17} cm^{-3}$), a setup typical for low-pressure mercury (Hg) and sodium (Na) lamps.

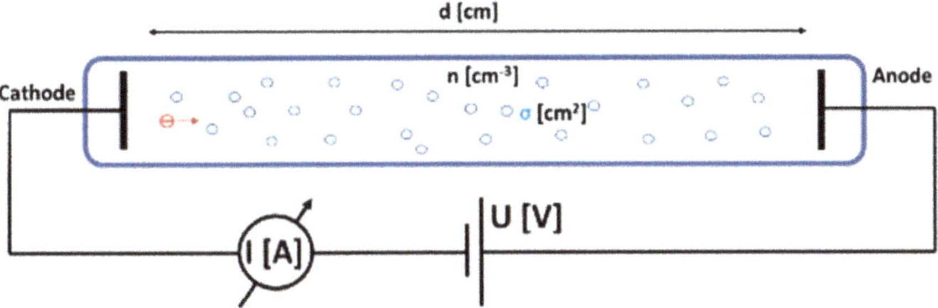

Figure 8.7.1: Simplified sketch of a cold cathode discharge tube.

A voltage U ([U] = 1 V) is applied between electrodes located at a distance d ([d] = 1 cm) from each other, resulting in an electric field E = U/d being able to move charged particles between electrodes. At low voltages, E is just strong enough to remove all the seed electrons that were created by cosmic rays within the tube, resulting in extremely small currents (<1 nA). At sufficiently strong electric fields, these seed electrons may multiply by means of Townsend avalanches. But without seed electrons, there is no current. These discharges are called dark discharges or Townsend discharges. They are used in Geiger counters. The amplification factor of a seed electron traveling a distance d between cathode and anode is $\sim\exp(\alpha \cdot d)$, with α being the first Townsend coefficient describing the number of ionizing collisions an electron encounters per unit length ([α] = 1/cm) [2]. Here, $\alpha \sim \sigma_{ion} \cdot n$, where n [cm^{-3}] is the gas density. σ_{ion} [cm^2] is the ionization cross section, and strongly depends on the average electron energy, which, in turn, is proportional to the electron temperature T_e. T_e itself depends on the reduced electric field $E/N = \frac{U}{n \cdot d}$, which is the electric field E divided by gas density n. E/N's unit is 1 Townsend: $[E/N] = 10^{-17}\ V \cdot cm^2 = 1$ Td. E/N basically determines the average kinetic energy an electron can gain between two collisions with gas particles, which is proportional to the electric field E times the mean free path of the drifting electron (~1/n). For argon, the first Townsend coefficient is [2]:

$$\alpha\left(n, \frac{U}{n \cdot d}\right) = n \cdot \sigma_{ion}\left(\frac{E}{N}\right) = n \cdot \sigma_0 \cdot \exp\left(\frac{-B}{U/(n \cdot d)}\right),$$

with $\sigma_0 = 3.64 \cdot 10^{-16} cm^2$ and $B = 545Td$.

Increasing E further will lead to electrical breakdown of the gas, meaning that the discharge becomes self-sustaining. Note that each ionizing collision produces one new electron-ion pair. The number of ions generated within a Townsend avalanche is exactly the number of electrons minus the seed electron. If slowly drifting ions hit the cathode, some of them are able to knock out a secondary electron. The probability of this happening is described by the second Townsend coefficient γ, a material constant of the cathode, typically being between 10^{-3} to 10^{-2} [2]. Using special heated electrode materials with low work functions, γ values approaching 1 are possible. Breakdown conditions are met, as soon as each electron traveling from anode to cathode produces enough ions within its own avalanche, so that if all of these ions reach the cathode, at least one new electron is released:

$$\gamma \cdot \left(\exp\left(\alpha\left(n, \frac{U_{bd}}{n \cdot d}\right) \cdot d\right) - 1\right) = 1$$

Now, the discharge does not need externally generated seed electrons to sustain its electric current flow. Solving this equation for the breakdown voltage U_{bd} as a function of $n \cdot d$, and γ results in Paschen's curve [2]:

$$U_{bd}(n \cdot d\gamma) = \frac{B \cdot n \cdot d}{\ln\left(\frac{\sigma_0 \cdot n \cdot d}{\ln(1+\gamma^{-1})}\right)}$$

For argon, $U_{bd}(n \cdot d, \gamma)$ is shown in Figure 8.7.2. [2].

$U_{bd}(n \cdot d, \gamma)$ has a distinct minimum. It describes the minimum voltage necessary to induce breakdown within a gas of density n, which will happen at a well-defined gap width d_{min}. At these parameters, ideal conditions for electron multiplication exist. As soon as breakdown occurs, space charge effects drive the formation of a small thin region on front of the cathode called cathode fall. Size ($d_{CF} \sim 1/n$, typically $d_{CF} < cm$) and voltage drop ($U_{CF} \sim 200$ V) of the cathode fall are characterized by the minimum of Paschen's curve [2]. All the ionization necessary to sustain the discharge occurs there. Note that if γ grows towards 1, U_{CF} as well as d_{CF} becomes progressively smaller. In the rest of the discharge, i.e. in the positive column, electrons drift in a quasi-constant electric field E that is too small to cause ionization. Discharges with a well-defined cathode fall region and a positive column are called glow discharges. Within the positive column [2], electrons can only collide elastically with minimal energy loss, or inelastically with trace atom admixtures, for which excitation energies are significantly below the excitation thresholds of the main rare gas component.

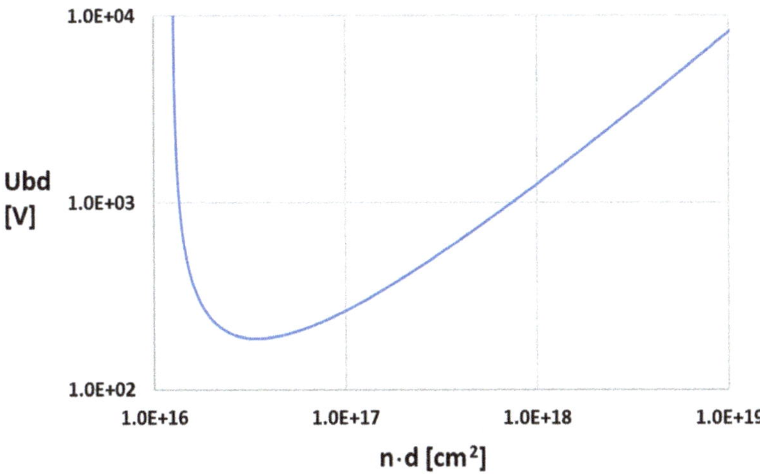

Figure 8.7.2: Paschen's curve for Argon.

As a result, within the positive column, highly efficient excitation of atoms emitting narrow band radiation is possible. This is realized within low-pressure Na and Hg discharge lamps. The reduced electric field in the positive column is about 1 Td with a typical current density ~0.1 A/cm^2 and low electron density ($n_e < 10^{13} cm^{-3}$). The length of the positive column can be from several cm to ~2 m, which results in positive column voltages of ~100 V. In order to minimize cathode-fall voltage, the lamps' tungsten electrodes are often coated with a mixture of BaO, SrO, and CaO, which have low work functions. These coatings result in a high secondary electron emissivity, such that $U_{CF} < 20$ V is achievable [2]. Note that within the discharge's cathode fall region, the ion current heats up a small cathode area of about 1 mm^2 size that is responsible for supplying all the discharge current.

8.7.2.2.1.1 Low-pressure sodium lamps (LPS)

Addition of about 1000 ppm of Na atoms to a low-pressure rare gas glow discharge will lead to excitation of the first two exited states of the Na atom radiating at $\lambda =$ 588.99 nm $\left([\mathrm{Ne}]3p^1\left(^2P_{3/2}\right) \rightarrow [\mathrm{Ne}]3s^1\left(^2S_{1/2}\right)\right)$ and $\lambda = 589.60$ nm $\left([\mathrm{Ne}]3p^1\left(^2P_{1/2}\right) \rightarrow [\mathrm{Ne}]\ 3s^1\left(^2S_{1/2}\right)\right)$ respectively. LPS reach a luminous efficiency close to 200 lm/W [3], while experimental devices can exceed 300 lm/W [4]. For comparison, the best commercially available LED lamps available today reach about 170 lm/W [5]. The reason for the LPS's high efficiency, besides the glow discharges' positive columns efficiently coupling the electron energy with inelastic collisions involving Na atoms, is due to the fact that the two yellow Na lines are very close to the peak of the phototopic human eye sensitivity located at 555 nm. Optimum efficiency is obtained at a Na partial pressure of only 0.4 Pa, which occurs above molten Na at a temperature of 260 °C (see Figure 8.7.3) [6].

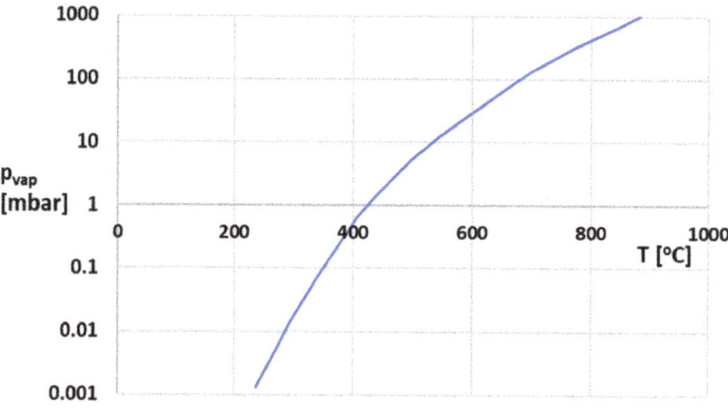

Figure 8.7.3: Vapor pressure of Na as a function of temperature.

Any deviation from the optimal Na partial pressure decreases the efficiency. Na vapor is chemically extremely aggressive. Glass and quartz are quickly stained brown by Na, leading to lamp failure. Standard soda lime glass tubes must be coated with a thin (about 0.02 mm) protective layer of Na-resistant borate glass [3]. Since Na vapor is opaque to its own resonance radiation, the product of Na density times discharge tube diameter must be kept small. LPS lamps are most efficient at low wattage per volume (~0.5 W/cm^3). In order to reach the lamps operating temperature of 260 °C at such low power loads and large surface areas, the lamps are designed and operated inside a vacuum jacket coated with an IR reflector, providing thermal isolation.

Lamps are available up to about 180 W electrical input power with operational lifetimes reaching up to 18,000 h. LPS lamps are designed primarily for use in remote general lighting applications, where good efficiency and long life are desired with no color rendering requirements. These lamps are ideally suited for areas where night sky observations are performed, since the narrow-band 589 nm emission can be filtered out easily. LPS lamps are well suited for remote areas where energy costs are high, including parking lots, roadway, bridges and tunnel lighting, security lighting, area floodlighting, railway crossings and airport lighting. Today, LPS lamps are gradually being replaced by LEDs.

8.7.2.2.1.2 Low-pressure mercury lamps (LPM)

Adding about 2000 ppm of Hg atoms to a low-pressure RG glow discharge results in efficient population of the $6s^16p^1(^3P_1)$ and $6s^16p^1$ $(^1P_1)$ states of the Hg atom, inducing emission lines at $\lambda = 253.652$ nm $(6s^16p^1(^3P_1) \rightarrow 6s^2(^1S_0))$ and $\lambda = 184.950$ nm $(6s^16p^1(^1P_1) \rightarrow 6s^2(^1S_0))$ ground-state transitions. The natural radiative lifetime of the 185 nm transition is 1.3 ns, while the natural radiative lifetime of the 254 nm emission in the ultraviolet C (UVC) spectral region is 108 ns [7]. The optimal vapor pressure for highest UVC yields is ~0.0092 mbar (a), corresponding to an elementary Hg source temperature of 42 °C (see Figure 8.7.4) [9].

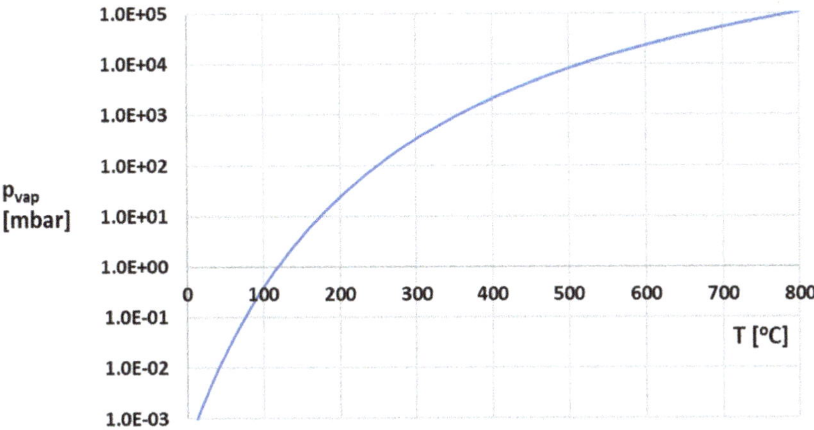

Figure 8.7.4: Vapor pressure of Hg as a function of temperature.

The electron temperature T_e is about 1 eV and the electron density n_e is of the order of $2 \cdot 10^{11} cm^{-3}$ [7, 8]. Under ideal conditions, the LPM's conversion efficiency of electric input energy into the lamp to 254 nm radiation emission coming out of the plasma is ~56%, while the conversion efficiency into 185 nm radiation is solely ~9.4% [7, 8]. Electric input power per length is limited to below 1 W/cm, in order to not exceed the optimal lamps' surface temperature of 42 °C. An electric field of about 1 V/cm results in a reduced field of about 1 Td, limiting the electric current density to about 100 mA/cm^2. Higher input power leads to higher Hg densities, enhancing radiation trapping and diminishing the UVC yields [10]. Significantly higher input power levels are possible if amalgams of Hg and other metals, such as bismuth, indium, lead or tin are used as the Hg source. The ideal Hg partial pressures are now achievable at lamp surface temperatures of about 100 °C, allowing input power levels of several W/cm lamp length [11]. Due to radiation trapping, it takes about $\tau_{eff} = 10$ µs [7, 10] until a 254 nm photon actually leaves the lamp. Even in amalgam low-pressure Hg lamps, first the 254 nm yield and then the 185 nm yield drop significantly, once current densities of about 400 mA/cm^2 and electric input power densities of 0.4 W/cm^3 are exceeded. UV lamp discharge vessels are made from quartz glass having a high transmission for 254 nm radiation.

An important application of LPMs is the inactivation of microorganisms by UV radiation [12, 13]. LPM lamps in UV-transmitting quartz envelopes led to the first UV disinfection installation for drinking water in Marseilles, France, in 1910. The main commercial application of the low-pressure Hg discharges is, in general, lighting by means of luminescent materials, resulting in fluorescent lamps. The lamp body, made out of inexpensive Na silicate or lead silicate glasses, is coated on the inside with a blend of phosphors efficiently converting the plasma's 254 nm UVC emission into visible, mostly white light, resulting in efficiencies of about 100 lm/W [14]. Typical lamp lifetimes are longer than 10,000 h. Until today, fluorescent lamps illuminate commercial and office buildings. They are gradually being replaced by retrofit LED lamps.

8.7.2.3 Discharges in dense gases

8.7.2.3.1 Arc discharges: medium and high-pressure mercury and sodium discharges

Due to self-absorption of the radiating species, low-pressure glow discharge lamps are long and thin. Due to lateral diffusion of electrons, the plasma fills the discharge tube (with typical outer diameters <5 cm) seemingly homogenously. For an electron drifting a distance x in a gas of density n under the influence of an electric field $E = U/d$, the lateral diffusion distance R(x) decreases with rising gas density n: $R(x) \sim \sqrt{\frac{d \cdot x}{U \cdot n}} \sim \frac{1}{\sqrt{n}}$. Note that the lateral electron movement through the lamp resembles a random walk [15]. Once a stable glow discharge with electron/ion-producing cathode fall and positive column is established, the discharge current can be easily increased by just enlarging the surface area where cathode fall interacts with the cathode surface. No increase in applied voltage is necessary. Once all the cathode's surface is touched by the cathode fall region, the voltage must rise in order to increase the current, since electrons must be pushed through the gas faster and more electrons have to be produced by more ionization. At one point, the current densities at the cathode surface get high enough (~10^2 to 10^4 A/cm^2), such that the cathode heats up considerably. Electrons can now be easily supplied by thermionic emission from very hot cathode spots. Now the cathode fall's voltage drops significantly [2]. Once the cathode fall voltage drops below 10 V, the discharge transitions from a glow into an arc discharge. A higher current is now achievable with decreasing voltage, since high electron current densities lead to frequent ionization events of already excited atoms. Increasing the current further, the plasma's electron, ion and atom-gas components drift further towards thermal equilibrium ($T_e = T_i = T_g$), until a thermal arc is established [2]. Rising pressure speeds up the glow to arc transition, since decreasing lateral diffusion leads to increasing power density within the discharge column. Arc discharges are realized in medium and high-pressure Hg and Na lamps operating at temperatures of several 100 °C and Na and Hg partial pressures exceeding several 100 mbar. Electric input power within these discharges can reach up to several 100 W/cm arc length [16]. High gas pressure leads to pressure broadening and Doppler broadening as well as resonance imprisonment of the dominant emission lines. High excitation rates of all transitions within the dominant gas component lead to a broad, multiline emission spectrum [3]. In the case of Na, satellite bands induced by molecular effects (Na_2) result in a broad-band, quasi-continuous light output in the visible spectral region, which made high-pressure sodium lamps (HPS) well suited for general lighting applications [3]. Many technical applications like curing of surface coatings and inks as well as air, water and surface disinfection rely on Hg arc lamps [17].

8.7.2.3.2 Dielectric barrier discharges and excimers

Since lateral diffusion decreases with pressure, discharges in dense, atmospheric pressure gases ($n \cong 10^{19} cm^{-3}$) are not homogenous. The discharge current prefers to flow in many fine filaments, being distributed seemingly randomly over the electrode surface. The diameter of these microdischarge filaments is well below 1 mm. During formation, power densities within these microdischarges can exceed 100 kW/cm^3. As a result, a microdischarge can transition from glow to arc within a very short time frame (<1 µs). Unfortunately, the dominant microdischarge that first successfully transitions from glow to arc will short circuit the whole discharge gap and claim all the available discharge current from the power source for itself. For this reason, glow discharges in dense gases are only possible for short periods of time. Nevertheless, cold, non-thermal glow discharge plasmas can be realized in dense gases (p ~ bar(a) at T ~ 300 K) favoring high three-body collision rates (~n^2) necessary for molecular formation processes. In order to suppress arc formation, one must make sure that the discharge voltage is applied only for a short time (~1 ns to 100 ns), such that ions cannot travel a significant distance in the electric field. The easiest way to achieve this is to cover one or both electrodes with non-conducting dielectric layers. During the discharge, the insulators accumulate charges on their surface. As a result, an electric field within the discharge's gas space opposing the external driving field is established. This leads to rapid quenching of the discharge filaments before they can reach the arc phase. This type of discharge is called silent discharges or dielectric barrier discharges (DBD) [18]. DBDs are ideally suited to generate large volumes of non-thermal plasmas in dense, atmospheric pressure gases. They offer very favorable conditions for molecule formation out of excited atoms as needed for excimer or ozone generation [19]. The microdischarges themselves, due to their short lifetime, are characterized as transient glow discharges [18].

8.7.2.3.2.1 Excimers

The term "excimer" is the abbreviation for excited dimer, describing diatomic molecules that are solely stable in an excited state. The ground state of an excimer correspond to two separated atoms that do not show chemical bonding (often they correspond to strongly repulsive electron configurations). This leads to rapid dissociation of the excimer as soon as the transition from the excited to the ground state occurs.

8.7.2.3.2.1.1 Rare gas excimers

A simplified potential diagram of the xenon excimer is shown in Figure 8.7.5.

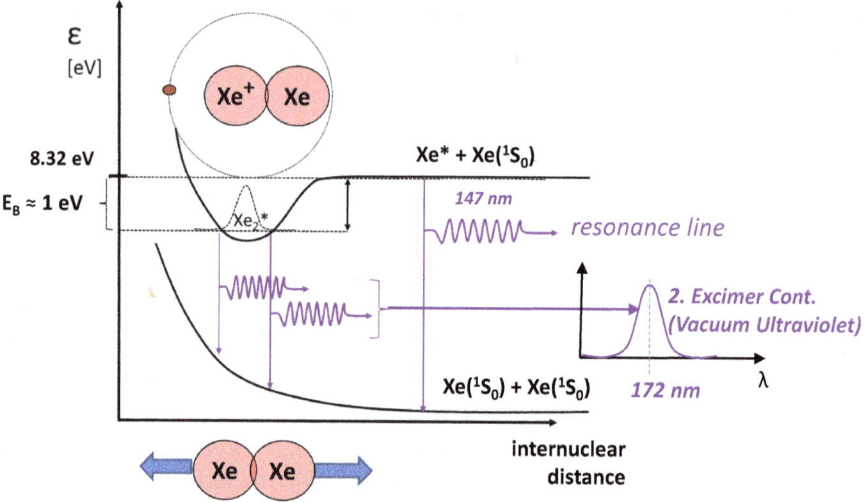

Figure 8.7.5: Simplified potential diagram of the xenon excimer Xe_2^* molecule. The excited Xe^* atom can be in the $Xe(^3P_1)$, $Xe(^3P_2)$, $Xe(^1P_1)$ or $Xe(^3P_0)$ state.

Due to the Franck-Condon principle, transitions appear as vertical lines within the potential diagram [20, 21]. In order to form an excimer molecule, we first need an excited atom. This is typically achieved by an inelastic collision between an electron with kinetic energy exceeding the atoms excitation energy threshold and an atom (e.g. a RG atom) itself: $e^- + RG \rightarrow e^- + RG^*$

Now the excited atom has to collide with two ground-state atoms:

$$RG^* + 2RG \rightarrow RG_2^* + RG$$

The third collision partner is necessary to carry away the binding energy of the excimer molecule, thus allowing for conservation of impulse and energy. Since three-body collision rates are proportional to the particle density squared, a dense gas at modest to low temperature (p > 300 mbar (a) at 298 K) is necessary to efficiently generate excimers.

Typical binding energies of rare gas excimer molecules are of the order of 1 eV only. Cold RG-excimer molecules emit predominantly within the second excimer continuum (see Figure 8.7.6) with its full width at half maximum (FWHM) being determined by the binding distance of Xe_2^* in the lowest vibrational state as well as the repulsive nature of the two ground-state atoms. Due to the excimer's binding energy as well as the repulsive ground state, the photon energy (7.22 eV for Xe_2^* 172 nm emission) is significantly lower than the energy of the lowest lying excited state

(8.32 eV for Xe*) of individual atoms. As a result, excimers do not suffer from resonance absorption, leaving the RG medium transparent to the excimer's emission. That is why efficient conversion of electrical pumping power into excimer radiation even at extreme power densities of up to 10^6 W/cm^3 is possible. The high achievable pumping power density combined with the lack of a ground state makes excimer systems suitable for laser action [22]. Due to the low binding energy of the RG_2^* excimer molecule, a large part of the RG* excitation energy can be converted into second continuum emission. More than 50% of the plasma's input energy can be converted into the second continuum [23–26]. This leads to the possibility of achieving high wall-plug-to-VUV emission efficiencies within Xe-excimer lamps exceeding 30% [27, 28]. At low pressures (<50 mbar), the RG-atoms resonance line dominates the emission spectrum. From 100 mbar on, the second continuum starts to emerge, dominating the emission from $p > 300$ mbar on. Table 8.7.1 shows the wavelength of the resonance line as well as the second continuum for the rare gases [29].

Table 8.7.1: Wavelength of resonance and second continuum emission of rare gases.

Rare gas	Resonance line [nm]	Second ontinuum (band center) [nm]
He	60	70–80
Ne	74.4	82
Ar	106.7	126
Kr	123.6	146
Xe	146.4	172

Note that all the rare gas excimers emit in the vacuum of ultraviolet spectral range (VUV) below 180 nm. The smaller the rare gas atom, the shorter the wavelength of the resonance line and the second continuum. A typical emission spectrum of the Xe excimer is shown in Figure 8.7.6. [25].

Only the xenon excimer can be transmitted through high-purity synthetic quartz windows. Krypton and argon excimers can be transmitted through brittle CaF_2, MgF_2, or LiF windows. Helium and neon excimer lamps can only be realized with windowless systems.

8.7.2.3.2.1.2 Rare gas halogen excimers and halogen dimers

Excimer molecules can also form between rare gases and halogens X. Typical gas mixtures consist of several 100 mbar RG with several mbar of X_2 halogen. The formation of these dimers depends on fast harpoon reactions,

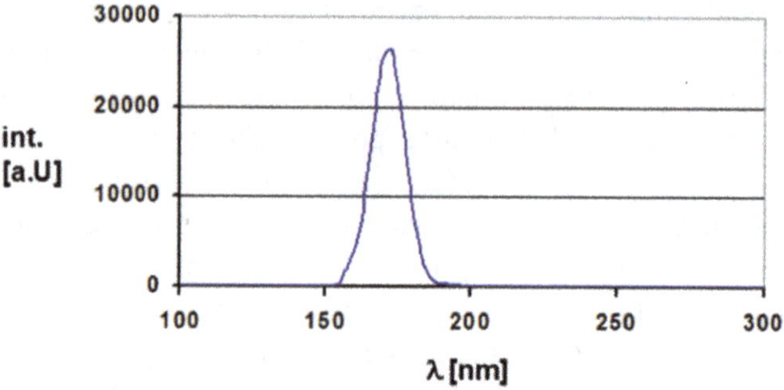

Figure 8.7.6: Typical spectra of the second continuum of the Xe_2* excimer at a pressure of 1.3 bar(a).

$$RG^*_{(2)} + X_2 \rightarrow RGX + X + (RG)$$

dissociative attachment reactions between slow electrons and X_2 molecules,

$$e^- + X_2 \rightarrow X^- + X$$

and subsequent recombination of X^- ions with RG^+ ions or molecules [30]:

$$RG^+ + ^- + RG \rightarrow RGX + RG \text{ or } RG_2^+ + X^- \rightarrow RGX + RG$$

Due to the strong electronegativity of the halogen, the binding of these dimers has a significant ionic character, $RG^+(^2P_{3/2}) + X^-(^1S_0) \rightarrow RGX\ (B_{1/2})$, resulting in a rather deep potential well with a binding energy of several eV.

Due to the strongly bound upper level and a quite flat ground-state RGX $(X_{1/2})$ potential curve consisting of a rare gas atom $RG(^1S_0)$ and a halogen atom $X(^2P_{3/2})$, the RGX* excimer's $B_{1/2} \rightarrow X_{1/2}$ transitions have small FWHM of a few nm. Radiative lifetimes of the RGX* $B_{1/2} \rightarrow X_{1/2}$ transitions are typically of order of 10 ns. Since the stimulated emission cross-section is ~1/FWHM, these dimers are excellent candidates for excimer lasers [31]. An important application is the ArF* 193 nm excimer laser used in photolithography of semiconductors. Due to the high binding energy of the RGX* dimer, a large part of the RG*excitation energy gets lost during the RGX* formation process. As a result, the RGX excimer lamp's wall-plug-UV-emission efficiencies are typically below 10% [32].

Finally, also halogen dimers (X_2*) are possible. Table 8.7.2 shows an overview over all the rare gas, rare gas halide and halide excimers. An upper limit for efficiency $\eta \leq \frac{h\nu}{\varepsilon(RG^*)}$ is also given.

8.7.2.3.2.1.3 Efficiency of excimer systems

Highest efficiencies are only achievable with RG excimer systems at low pumping power densities resulting in degrees of ionization below 10^{-7}. This implies small

Table 8.7.2: Overview of all available RG_2^*, RGX^* and X_2^* excimer emission wavelengths.

Excimer	λ [nm]
He_2^*	60
Ne_2^*	80
Ar_2^*	128
Kr_2^*	145
Xe_2^*	172

RG_2^*:
FWHM: 10nm
Efficiency ≤ 87%

Excimer	λ [nm]
NeF*	108
ArF*	193
KrF*	248
XeF*	353

Excimer	λ [nm]
ArCl*	175
KrCl*	222
XeCl*	308
KrBr*	206
XeBr*	282
KrI*	185
XeI*	253

RG-Halide*:
FWHM: 3nm
Efficiency ≤ 30%

Excimer	λ [nm]
F_2^*	157
Cl_2^*	258
Br_2^*	290
I_2^*	343

$Halide_2^*$:
FWHM: 3nm
Efficiency ≤ 30%

electron densities $n_e < 3 \cdot 10^{12}\ cm^{-3}$. Only then inelastic, ionizing collisions between electrons and excimer molecules can be avoided [33]. Unfortunately, DBDs in dense gases have high electron densities within the filaments. In order to achieve high efficiency, the electric pumping power must be distributed homogenously across the discharge volume. Since in dense (~300 mbar) Xe a typical microdischarge lasts about 100 ns, diffuse, quasi homogenous discharges can be generated by applying fast, steep high voltage pulse with $\frac{\partial U}{\partial t} > \frac{1kV}{100ns} = 10^{10}\,\frac{V}{s}$. On the time scale of microdischarge development, ignition conditions are reached everywhere within the whole discharge gap simultaneously. Streamers start and form everywhere within the gap and no streamer can gain an advantage over another [23]. Realizing such fast pulsed HV power supplies for large, high-power lamps is quite difficult. The lamp's electric input capacitance increases with the lamp size. As a result, especially for big lamps, it is necessary to drive large electric currents in a short time. Latest developments in fast silicon carbide-based HV solid state switches can be expected to lead to efficient and powerful power supplies also for large excimer lamps. Pumping xenon excimer lamps with cost-efficient sinusoidal HV power supplies in the 10 kHz frequency range (with $\frac{\partial U}{\partial t} < 10^8\,\frac{V}{s}$) leads to filamentary discharges and efficiencies well below 20%.

8.7.3 Outlook: applications of excimer lamps

Excimer-based lamps are at least 10 times less expensive than UV or VUV lasers. They can be instantly turned on and off, and their output power level can be easily modulated from 0 to 100%. By adjusting the gas mixtures, excimer lamps are able to deliver photons with energies ranging from 3 to more than 10 eV, with wall-plug efficiencies from several % (RGX*) and exceeding 30% (RG_2^*). Combining these features, excimer lamps are well suited for almost all well-known UV and/or VUV-assisted photo processes [34–36]. Excimer lamps are commercially available from well-known light-source producers like Heraeus Noblelight (Germany), Philips (The Netherlands) or Ushio Inc. (Japan) [29]. Commercial excimer lamps predominately use quartz envelopes and gas mixtures that emit wavelengths above the cut off of synthetic quartz at 160 nm. Figure 8.7.7 shows the photon energies of some RG_2^* and RGX^* excimers as well as the binding energies of selected bonds within molecules. Due to energy conservation, a photon can only be absorbed and thereby photochemically break a bond if the photon's energy $h\nu$ exceeds the binding energy E_B: $h\nu = \frac{hc}{\lambda} > E_B$. This represents a key lock principle for photons being able to trigger specific chemical reactions.

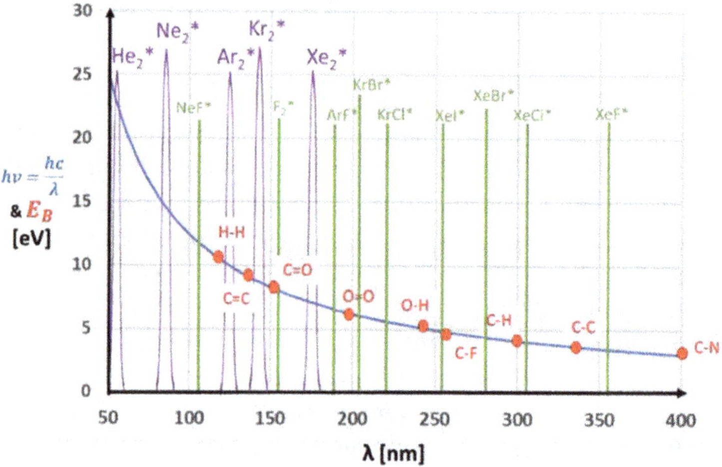

Figure 8.7.7: Comparison of several rare gas and rare gas halide excimer photon energies with binding energies of covalent bonds between selected atoms.

Using the proper UV radiation sources with suitable wavelengths and an appropriate UV emission intensity, one can selectively switch on and off the desired photochemical reactions at will, enabling the synthesis of new coatings with customized surface layers. Right now, we are just at the very start of this exciting development. In the near future, the combination of efficient (30% wall-plug efficiency) xenon lamps with special phosphor coatings that convert 172 nm VUV excimer photons into longer wavelength light will enable the production of Hg-free UVC lamps with wall-plug efficiencies of up to 20% [37–39]. These lamps will combine all the advantages of present xenon excimer lamps (instant on/off, 100% dimmable, high efficiency) with the advantages of modern LPM lamps (high output power, cost-effective) while being able to deliver the desired UV wavelength, tunable by the phosphor used. Many available wavelengths are significantly shorter than the theoretical lower limit of modern (Al, Ga)N-based deep-UV LEDs of 210 nm [40].

References

[1] Tannous C. Light production metrics of radiation sources. Eur J Phys, 2014, 35, 045006. DOI: https://doi.org/10.1088/0143-0807/35/4/045006.

[2] Raizer Y. In: Allen JE. (Ed.) Gas Discharge Physics, Springer Verlag, Berlin, 1997, ISBN: 978-3-642-64760-4. DOI: https://dx.doi.org/10.1007/978-3-642-61247-3.

[3] Waymouth JF. Electric discharge lamps. In: Kusko A. (Ed.). Monographs in Modern Electrical Technology. MIT Press, Cambridge, MA, 1971. ISBN 978-0-262-23048-3.

[4] Pirani M. Z Tech Phys, 1930, 11, 482.

[5] Pattison PM, Hansen M, Tsao JY. LED lighting efficacy: Status and directions. C R Physique, 2018, 19, 134–45. DOI: https://doi.org/10.1016/j.crhy.2017.10.013.

[6] Rodebush WH, Walters EG. The vapor pressure and vapor density of sodium. J Am Chem Soc, 1930, 52, 2654–65. DOI: https://doi.org/10.1021/ja01370a011.

[7] Winkler RB, Wilhelm J, Winkler R. Kinetics of the Ar-Hg plasma of fluorescent lamp discharges, I. Model – basic equations – Hg partial pressure variation. Ann Phys, 1983, 495, 90–118. DOI: https://doi.org/10.1002/andp.19834950203.

[8] Winkler RB, Wilhelm J, Winkler R. Kinetics of the Ar-Hg plasma of fluorescent lamp discharges, II. Ar partial pressure and discharge current variation. Ann Phys, 1983, 495, 119–39. DOI: https://doi.org/10.1002/andp.19834950204.

[9] Huber M, Laesecke A, Friend D. The Vapor Pressure of Hg, NIST Interagency/Internal Report (NISTIR 6643), National Institute of Standards and Technology, Gaithersburg, MD, 2006. DOI: https://doi.org/10.6028/NIST.IR.6643.

[10] Anderson JB, Maya J, Grossman MW, Lagushenko R, Waymouth JF. Monte Carlo treatment of resonance-radiation imprisonment in fluorescent lamps. Phys Rev A, 1985, 31, 2968–75. DOI: https://doi.org/10.1103/PhysRevA.31.2968.

[11] Schmalwieser AW, Wright H, Cabaj A, Heath M, Mackay E, Schauberger G. Aging of low-pressure amalgam lamps and UV dose delivery. J Env Eng Sci, 2014, 9, 113–24. DOI: http://dx.doi.org/10.1680/jees.13.00009.

[12] Downes A, Blunt TP. The Influence of light upon the development of bacteria. Nature, 1877, 16, 218. DOI: https://doi.org/10.1038/016218a0.

[13] Gates FL. A study of the bactericidal action of ultra violet light: III. The absorption of ultra violet light by bacteria. J Gen Physiol, 1930, 14, 31–42. DOI: https://doi.org/10.1085/jgp.14.1.31.

[14] Gruber W, Schliplage M, Zachau M. Patent EP1306885.

[15] Pearson K. The problem of the random walk. Nature, 1905, 72, 294. DOI: https://doi.org/10.1038/072294b0.

[16] Drakakis E, Karabourniotis D, Zissis G. Atomic and excitation temperatures in medium- and high-pressure Hg arc lamps. J Light Vis Env, 2013, 37, 166–70. DOI: https://doi.org/10.2150/jlve.IEIJ130000514.

[17] Heering W. UV sources – basics, properties and applications. IUVA News, 2004, 6, 7–13. https://uvsolutionsmag.com/stories/pdf/archives/060401Heering_2004.pdf.

[18] Kogelschatz U. Silent discharges for the generation of ultraviolet and vacuum ultraviolet excimer radiation. Pure Appl Chem, 1990, 62, 1667–74. DOI: https://doi.org/10.1351/pac199062091667. 

[19] Eliasson B, Hirth M, Kogelschatz U. Ozone synthesis from oxygen in dielectric barrier discharges. J Phys D: Appl Phys, 1987, 20, 1421–37. DOI: https://doi.org/10.1088/0022-3727/20/11/010.

[20] Franck J, Dymond EG. Elementary processes of photochemical reactions. Trans Faraday Soc, 1926, 21, 536–42. DOI: https://doi.org/10.1039/TF9262100536.

[21] Condon E. A theory of intensity distribution in band systems. Phys Rev, 1926, 28, 1182–201. DOI: https://doi.org/10.1103/PhysRev.28.1182.

[22] McCusker M. The rare gas excimers. In: Rhodes C. (Ed.) Excimer Lasers, Topics Appl Phys, vol. 30, Springer, Berlin, 1984. ISBN 978-3-662-11716-3.

[23] Carman RJ, Mildren RP. Computer modelling of a short-pulse excited dielectric barrier discharge xenon excimer lamp (λ~172nm). J Phys D: Appl Phys, 2002, 36, 19–33. DOI: https://doi.org/10.1088/0022-3727/36/1/304.

[24] Beleznai S, Mihajlik G, Agod A, Maros I, Juhasz R, Nemeth Z, Jakab L, Richter P. High-efficiency dielectric barrier Xe discharge lamp: Theoretical and experimental investigations. J Phys D: Appl Phys, 2006, 39, 3777–87. DOI: https://doi.org/10.1088/0022-3727/39/17/012.

[25] Salvermoser M, Murnick DE. Efficient, stable, corona discharge 172nm xenon excimer light source. J Appl Phys, 2003, 94, 3722–31. DOI: https://doi.org/10.1063/1.1598300.

[26] Salvermoser M, Murnick DE. High-efficiency, high-power, stable 172 nm xenon excimer light source. Appl Phys Lett, 2003, 83, 1932–34. DOI: https://doi.org/10.1063/1.1605798.

[27] Vollkommer F, Hitzschke L. Durchbruch bei der effizienten Erzeugung von Excimer-Strahlung. Phys Bl, 1997, 53, 887–89. DOI: https://doi.org/10.1002/phbl.19970530912.

[28] Vollkommer F, Hitzschke L. US Patent 5,604,410, 1997.

[29] Sosnin E, Tarasenko V, Lomaev M. UV and VUV Excilamps, LAP LAMBERT Academic Publishing, ISBN: 978-3-659-21756-2.

[30] Salvermoser M, Murnick DE. Stable high brightness radio frequency driven micro-discharge lamps at 193 (ArF^*) and 157 nm (F^*_2). J Phys D: Appl Phys, 2003, 37, 180–84. DOI: https://doi.org/10.1088/0022-3727/37/2/006.

[31] Brau C. Rare gas halogen excimers. In: Rhodes C. (Ed.) Excimer Lasers, Topics Appl Phys, vol. 30, Springer, Berlin, 1984. ISBN 978-3-662-11716-3.

[32] Salvermoser M, Murnick DE, Wieser J, Ulrich A. Energy Flow and excimer yields in continuous wave rare gas-halogen systems. J Appl Phys, 2000, 88, 453–59. DOI: https://doi.org/10.1063/1.373680.

[33] Eliasson B, Kogelschatz U. Modeling and applications of silent discharge plasmas. IEEE Trans Plasma Sci, 1991, 19, 309–23. DOI: https://doi.org/10.1109/27.106829.

[34] Boyd IW, Zhang JY. New large area ultraviolet lamp sources and their applications. Nucl Instr Meth Phys B, 1997, 121, 349–56. DOI: https://doi.org/10.1016/S0168-583X(96)00538-1.

[35] Kogelschatz U, Eliasson B, Egli W. Dielectric-barrier discharges. principle and applications. J Phys IV France Colloque, 1997, 07(C4), 47–66. DOI: https://doi.org/10.1051/jp4:1997405.

[36] Oppenländer T. Hg-free sources of VUV/UV radiation: Application of modern excimer lamps (excilamps) for water and air treatment. J Environ Eng Sci, 2007, 6, 253–64. DOI: https://doi.org/10.1139/s06-059.

[37] Jüstel T, Huppertz P, Mayr W, Wiechert DU. Temperature dependent spectra of YPO_4:Me(Me = Ce, Pr, Nd, Bi). J Lumin, 2004, 106, 225–33. DOI: https://doi.org/10.1016/j.jlumin.2003.10.004.

[38] Broxtermann M, Den Engelsen D, Fern GR, Harris P, Ireland TG, Jüstel T, Silver J. Cathodoluminescence and Photoluminescence of $YPO_4:Pr^{3+}$, $Y_2SiO_5:Pr^{3+}$, $YBO_3:Pr^{3+}$ and $YPO_4:Bi^{3+}$. ECS J Solid State Sci Technol, 2017, 6, R47–52. DOI: http://dx.doi.org/10.1149/2.0051704jss.

[39] Broxtermann M, Dierkes T, Funke LM, Salvermoser M, Laube M, Natemeyer S, Braun N, Hansen MR, Jüstel T. An UV-C/B emitting Xe excimer discharge lamp comprising $BaZrSi_3O_9$ – A lamp performance and phosphor degradation analysis. J Lumin, 2018, 200, 1–8. DOI: https://doi.org/10.1016/j.jlumin.2018.03.043.
[40] Taniyasu Y, Kasu M, Makimoto T. An aluminum nitride light-emitting diode with a wavelength of 210 nanometers. Nature, 2006, 441, 325–28. DOI: https://doi.org/10.1038/nature04760.

8.8 Inorganic scintillators

Florian Baur, Thomas Jüstel

Scintillators are materials that emit UV radiation or visible light upon excitation with high-energy radiation, i.e. ionizing radiation. They are an important class of materials, since direct detection of high-energy radiation is difficult by technical means and not possible at all for a human being.

8.8.1 Introduction and history

Scintillators have been used as radiation detectors ever since the discovery of X-rays by Röntgen in 1895. A screen coated with barium tetracyanidoplatinate(II) ($Ba[Pt(CN)_4]$) showed luminescence if placed close to a discharge tube [1]. This first discovered scintillator showed only weak luminescence upon irradiation, which led to intense research activity and already in 1896 to the discovery of the highly efficient scintillators ZnS (introduced independently by Crookes, Henry, and Pupin) and $CaWO_4$ "by Edison" [2]. In 1897, Braun invented the cathode ray tube, in which the energy of accelerated electrons is converted into visible radiation. The mechanism of action of the so-called cathodoluminescence is similar to that of the conversion of high-energy radiation [3]. In 1948, Hofstaedter developed Tl^+-activated scintillators, namely NaI:Tl, which lead to further research activity in the class of alkali halides, both with and without activators [4]. In 1973, $Bi_4Ge_3O_{12}$ (BGO) was discovered, which sparked a search for more complex scintillators, resulting in the discoveries of $Lu_2SiO_5:Ce^{3+}$ (LSO) in 1992 and $(Lu,Y)_2SiO_5:Ce^{3+}$ (LYSO) in 2004 [5]. As an alternative to scintillators, direct detectors were developed in the 1970s, i.e. semiconductors such as lithium-doped Si or Ge, where high-energy radiation generates electron-hole pairs that can be detected as a current. In gamma-spectroscopy, Ge-semiconductor detectors are widely employed, while silicon drift detectors (SDD) are used for the detection of low-energy X-rays in X-ray fluorescence spectroscopy. Regardless, scintillators are dominating many areas, since the semiconductor detectors require cooling to prevent leak currents [6]. New, fast scintillators based on valence-core transitions such as BaF_2 were developed in the 1980s and 1990s [7].

8.8.2 Physics of scintillators

As described by Klintenberg et al. and Nikl, the mechanism of luminescence upon excitation by high-energy radiation can be separated into three stages [4, 8]: *conversion*, *transport* and *luminescence*. Figure 8.8.1 depicts the process schematically. In the first stage, conversion, the incident photons of ionizing radiation, the primary radiation,

https://doi.org/10.1515/9783110798890-027

transfer their energy to the scintillator. This occurs mainly via the photoelectric effect and Compton scattering, while at photon energies higher than 1.022 MeV, pair production, i.e. the generation of a positron and an electron (mass = 0.511 MeV/c^2), can occur. In the two former cases, the energy of the incident photon is fully (photoelectric effect) or partly (Compton scattering) transferred to an atom of the scintillator. As the energy of the primary photons is higher than the binding energy of inner shell electrons, an electron from an inner shell of the atom is removed. This "hot" electron carries some energy with it and leaves a hole behind. Subsequently, an electron from an outer shell can immediately recombine with the hole, which releases energy. Thereby, either a photon is generated (secondary X-ray) or another electron is removed from the atom (secondary electron). Electrons created via pair production are another source of secondary electrons.

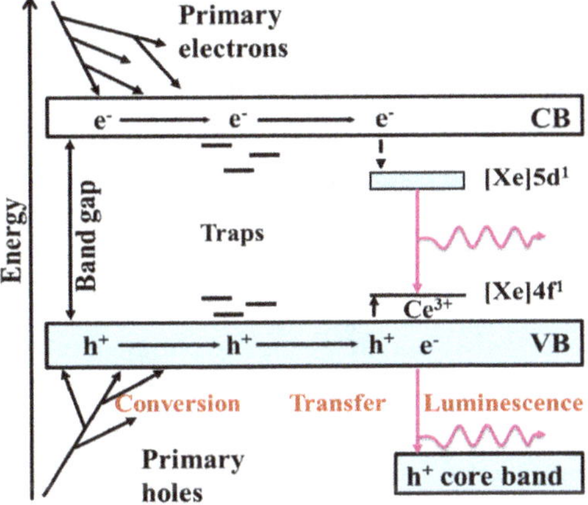

Figure 8.8.1: Simplified sketch to illustrate the three steps of the energy flow in a scintillator.

The secondary X-rays and electrons are scattered at atoms, transfer their energy and generate additional secondary X-rays, electrons and holes. This process of the generation of secondary X-rays and secondary electrons continues until the remaining energy is too low for further electron-hole-pair production. Typically, two to seven times the band gap energy is required to remove an electron from an inner shell. This stage has a duration of less than 1 ps and will result in a specific number of electron-hole pairs. The number of pairs is roughly proportional to the energy deposited in the scintillator by the incident primary radiation.

During the second stage, the "hot" electrons and holes thermalize and can become trapped at defects or self-trapped by the crystal structure or form excitons. During this stage, non-radiative relaxation can occur, and the probability of this event greatly influences the efficiency of the scintillator. A high defect density will increase the probability of non-radiative relaxation, as these additional energy states shorten the distance between the excited state and the ground state. If the electron-hole pairs reach a luminescence center, the center will be excited. Additionally, the luminescence centers can be excited directly by “hot” electrons moving through the crystal. This transport stage has a duration between 1 and 10 ps and determines the length of the rise-time of the scintillator in which the emission intensity increases along with the increasing number of excited luminescence centers.

In stage three, the excited luminescence centers return radiatively to the ground state. Figure 8.8.1 shows Ce^{3+} as an example, but various kinds of luminescence centers are used in scintillators [4]: Self-trapped excitons are found in undoped alkali halides like NaI or CsI, where a hole localizes on an atom and traps a spatially diffuse electron. The respective radiative transition is either a spin-forbidden triplet-singlet transition (decay time in µs range) or a spin-allowed singlet-singlet transition (decay time in ns range). Valence-core luminescence occurs in scintillators where the energy gap between valence band and core band is less than the principal band gap. The radiative transition is an electron moving from the valence band to a hole in the core band. These transitions are fast (less than 1 ns), but since only few holes can be created in the core band, and the light yield is low (2000 photons/MeV).

In self-activated scintillators, the luminescent species are a building unit of the crystal. Examples are $CaWO_4$ or $Bi_4Ge_3O_{12}$ where the radiative transitions occur within the WO_4^{2-} tetrahedra – i.e. charge-transfer luminescence – or the Bi^{3+} cation – i.e. a 6p-6s transition. The general mechanism is the same for scintillators comprising an activator, e.g. $Lu_2SiO_5:Ce^{3+}$ or $NaI:Tl^+$. A hole is trapped at the activator ion with an excited electron being localized at the same ion. The radiative return of the electron to the ground state is responsible for the emission of a photon.

The time-scale of the luminescence stage depends on the rate of the involved transition. Electron-hole recombination, exciton emission or valence-band core-band (cross-luminescence) transitions are very fast at a scale of 1 ns or less. Cross-luminescence has, for example, been observed in binary halides such as BaF_2 or CsBr, with a reported decay time of 0.07 ns for the latter [7]. Scintillators whose luminescence is based on 5d-4f transitions (e.g. Ce^{3+}, Pr^{3+}, Eu^{2+}) exhibit decay times between 20 ns (Pr^{3+}) and 1 µs (Eu^{2+}). Despite being quantum-mechanically allowed, the transitions rates are limited by the poor spatial overlap between 5d and 4f orbitals and in case of Eu^{2+} additionally by the partly spin-forbidden nature of the transition [4]. Quantum-mechanically forbidden transitions such as 4f-4f (e.g. Eu^{3+}, Tb^{3+}) have longer decay times up to several milliseconds and are mostly used for static imaging where fast reaction times are of no concern. Since even the faster radiative transitions occur on a time scale that are 1–2 magnitudes longer than that of the first two stages, they constitute the step that determines the overall response time of the scintillator.

The mechanism of action in case of charged particle excitation, like the He^{2+} nuclei of α-radiation, is basically the same as in the case of ionizing radiation. The charged particles interact with the atoms mostly via Coulomb scattering and generate secondary electrons. This is followed by steps comparable to those described above.

The main difference is that the secondary electrons are generated in a much smaller volume as the charged particles deposit their energy in a closer radius than the photons of ionizing radiation [5]. This can become significant if the activator concentration or the defect density shows a gradient within the scintillator. In many cases, impurities such as activator ions tend to accumulate closer to the surface of a crystal. This observation also means that increasing the size of the scintillator crystal does not necessarily increase the sensitivity, as the range of the incident particles is limited.

8.8.3 Application areas of scintillators

Scintillators are used in a wide range of applications. These can be categorized, depending on whether the energy of the individual incident photons must be determined or solely their combined intensity. As already discussed, the total number of visible light photons that is generated in the scintillator is proportional to the total energy that has been deposited. The relation is expressed by the light yield of a scintillator. By counting the number of emitted photons per unit of time, the amount of energy that was deposited within that time frame can be calculated.

If it is assured that solely one high-energy photon interacts with the scintillator within that time frame, the calculated amount of energy is equal to the energy of the incident photon. The photon output caused by an individual incident high-energy photon is called "pulse". The number of times a specific pulse-height was detected can be determined with a pulse-height analyzer (PHA) and can be plotted in a histogram. A simplified histogram is shown in Figure 8.8.2. Ideally, an incident photon of a specific energy would always generate the same pulse-height and only a line would emerge in the histogram. Since there is a certain probability that more or less photons are emitted per absorbed high-energy photon, the ideal line broadens to a peak. The broadness can be quantified as the full-width at half-maximum (FWHM) and is a measure for the energy resolution of the scintillator.

If two pulses overlap in time, the PHA will detect one high-intensity pulse instead of two lower-intensity pulses. This means that the rate of incident photons has to be low enough to ensure that solely one incident photon interacts with the scintillator within its decay time. Otherwise, it is not possible to distinguish between one incident photon with energy E and two simultaneously incident photons of energy 1/2 E. Scintillators with a shorter decay time can handle a higher rate of incident photons.

In the next application area to consider, it is assured that solely one type of high-energy photons, i.e. with a specific, known energy, interacts with the scintillator. Thus, the deposited energy divided by the energy of a single incident photon yields the total number of incident photons. In this case, more than one photon can be detected simultaneously and the rate of incident photons is only limited by the saturation threshold of the scintillator.

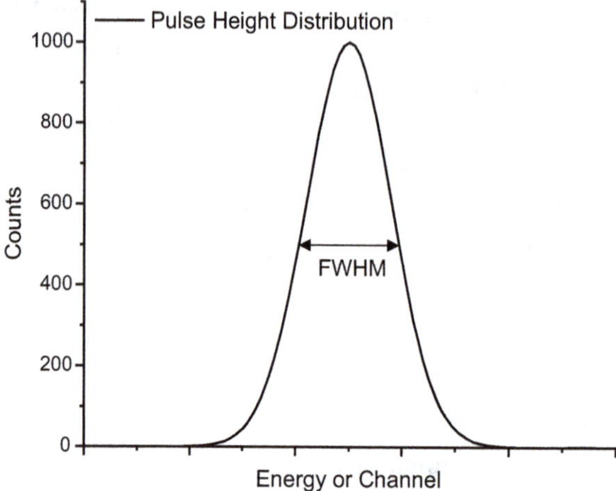

Figure 8.8.2: Simplified pulse-height distribution histogram.

Thirdly, if neither the number nor the energy of incident photons is controlled or known, the measurement only yields the total amount of energy deposited in the scintillator. All three cases have their specific application areas.

To detect the emitted visible photons, a suitable detector, usually a photomultiplier tube (PMT), is attached to the scintillator. To accurately determine the deposited energy, it is mandatory that a large fraction of all emitted photons reaches the detector, since otherwise the amount of energy calculated from it will be too low. In most cases, this is ensured by using a single crystal, as it exhibits almost no scattering and even photons that are generated several millimeters or more inside the scintillator can reach the surface. To guide the emitted photons towards the detector, the scintillator is usually encased in a highly reflective material, such as optical grade PTFE.

Examples of the first case, i.e. where the energy of the incident photons has to be determined, comprise various analytical methods:

- Energy-dispersive X-ray spectroscopy (EDX)
- X-ray fluorescence (XRF)
- Gamma spectroscopy
- High-energy physics

EDX and XRF can be used to identify elements based on the characteristic X-rays that are emitted from the atom after excitation by either electrons (EDX) or X-rays (XRF). The energy resolution lies between 150 and 300 eV with measurement ranges of approximately 1 to 50 keV. Gamma spectroscopy relies on the same method of analysis; however, it is mainly applied to identify radioactive specimens and, therefore, is used in measurement ranges of up to several MeV. The scintillator has to be stable towards high-energy radiation and should exhibit a short decay time to allow higher intensities of incident X-ray or gamma photons without coinciding absorption. In high-energy physics, scintillators are frequently used to detect low-intensity radiation with very high energies. Due to the low intensities, a high light yield of the scintillator and a high efficiency in the respective energy range are required. In astrophysics, the *Fermi Gamma-ray Space Telescope* (FGST) makes use of scintillators for its anticoincidence shield that detects charged particles, which occur with a very high flux and would disturb the gamma ray detection. The actual gamma rays are detected by silicon strip detectors behind the anticoincidence shield.

Detection of X-ray and gamma-ray intensities, instead of their energy, is important for applications such as X-ray diffraction (XRD). Here, X-rays of energies in the lower keV range are diffracted in a crystalline sample and their intensities are recorded in dependence of the measurement angle. The energy of the X-rays is known and close to monochromatic; therefore, the photon output of the scintillator is directly proportional to the intensity of the incident X-rays.

In oil well logging, gamma-rays are used to investigate the well-bore environment. A radioactive source is lowered into the well-bore, i.e. the hole that has been drilled into the ground, and the intensity of the scattered gamma-rays is detected. The dampening of the gamma-rays allows conclusions to be drawn as to the physical properties of the surroundings, especially their density. For this purpose, NaI:Tl is commonly employed as a scintillation detector [9]; however, semiconductor detectors are also in widespread use [10].

Another important field of application for scintillators is imaging. Computer tomography (CT) makes use of scintillators to detect X-rays that have passed through a patient. Depending on the absorption cross-section of the various types of biomatter, the intensity of the transmitted X-rays is decreased. By plotting these intensities over spatial position, an image is created. Scintillators used for CT must have a large absorption cross-section for X-rays, radiation damage resistance and a relatively short decay time, to allow for fast measurements and high counting rates. Commonly used scintillators for CT are: $CdWO_4$, $Gd_2O_2S:Pr^{3+},Ce^{3+},F^-$, $(Y,Gd)_2O_3:Eu^{3+}$ and "GE Gemstone", the latter being a Ce^{3+} activated garnet in the form of a transparent ceramic [11]. X-ray scanners in non-medical fields, such as baggage security checks at airports or to screen moving trucks, place similar demands on the scintillator. Here, the combination of one of the aforementioned scintillators with a CCD detector is used to quickly generate images.

For positron emission tomography (PET), a radioactive tracer that decays by emitting a positron is injected into a patient. The positron interacts within less than 10^{-9} s with an electron in an annihilation reaction, whereupon two 511 keV photons are generated that move in opposite directions. If both photons are detected, it is possible to reconstruct their point of origin in the patient's body. To achieve this, spatial information has to be retained. To that end, several scintillation detectors are arranged in a ring around the patient. The scintillator has to exhibit various properties such as [12]:

- high density for a large absorption cross-section
- short decay time for good time-resolution and high rates of incident photons
- good energy resolution for specific detection of 511 keV photons

As its properties agree well with these conditions, $(Lu,Y)_2SiO_5{:}Ce^{3+}$ is commonly used as a scintillator for PET.

Furthermore, scintillators can be employed in dosimetry, where the dose of radiation is measured. The dose is the amount of energy of ionizing radiation absorbed per unit mass. This value is proportional to the number of photons emitted by the scintillator if the light yield is constant for the energy and rate of incident photons. The region in which this condition is fulfilled determines the range of photon energies and rates for which the scintillator can be used in dosimetry.

8.8.4 Established scintillators and their properties

A luminescent material has to fulfil a range of criteria to be applicable as a scintillator. These include [4, 8, 13]:

Large absorption cross-section

Only the incident photons that are absorbed by the scintillator can be detected. The larger the fraction of absorbed photons, the higher the efficiency of the scintillator. A high density and large atomic number Z of the elements increase the absorption cross-section. By increasing the volume of the scintillator crystal, the absorption can also be increased.

Short decay time

After a high-energy photon has been absorbed, a certain amount of visible photons is emitted by the scintillator. During this time span, the scintillator is not ready to discriminate another high-energy photon, as the detection events would overlap. In that case, it is not possible to distinguish between the simultaneous absorption of two lower-energy photons and the absorption of one high-energy photon. A short decay time is therefore crucial for good time resolution and to allow accurate detection of high rates of incident photons.

Low cost

Scintillators often comprise rare earth elements such as lutetium or europium, which can be costly. The growth of large single crystals also factors into the cost of production.

High light yield

The energy of the absorbed incident photons is converted into visible photons by the scintillator. As discussed in the previous section, there are loss processes and, therefore, the number of emitted photons per absorbed energy differs between various scintillators. It is expressed in photons per unit energy, usually as photons per MeV. A high light yield is especially crucial for detection of low-energy X-rays.

Chemical and physical stability

Commonly used scintillators such as NaI are hygroscopic and have to be encapsulated to prevent degradation. Scintillators of the garnet or silicate type exhibit a very high stability. Resistance against radiation damage is important if the scintillator is used in an application with high fluxes of ionizing radiation, such as computer tomography (CT).

Emission wavelength

As the photons that are emitted by the scintillator must be detected by a photomultiplier tube (PMT) or charge-coupled device CCD, their wavelength should be close to the maximum sensitivity of the respective photodetector. For PMTs, this is usually in the blue to green spectral range.

Energy resolution

The number of emitted photons is proportional to the energy deposited in the scintillator by the incident high-energy photons as expressed by the light yield. This relation can be used to determine the energy of the incident photons. However, some deviation from this proportionality will invariably occur, which results in a broadening of the peaks on an intensity vs. incident photon energy scale. The resolution is expressed as the full-width at half maximum of the peak divided by the peak energy. Typical energy resolutions are 8–9% for $(Lu,Y)_2SiO_5:Ce^{3+}$ or 6% for NaI:Tl scintillators.

Refractive index

To detect the photons emitted by the scintillator, they must be outcoupled from the scintillator and into the photodetector. To realize high efficiencies for this process, the refractive index of the scintillator must closely match that of the photodetector's surface. For glass, this would amount to a refractive index of around 1.5.

During the decades of development, several materials have been found that meet the aforementioned requirements and are used in various application areas. The following Table 8.8.1 lists important parameters of established scintillators. Figure 8.8.3 shows a kg scale single crystal of $Y_3Al_5O_{12}:Ce^{3+}$ (YAG:Ce). Scintillators are cut from such large boules with size and geometry according to the demands of the respective application.

Figure 8.8.3: Photographic image of a single crystal of $Y_3Al_5O_{12}:Ce^{3+}$. (Photograph: Dr. Patrick Pues).

Table 8.8.1: Selected scintillators and their properties [14].

	NaI:Tl	CsI:Tl	BGO	YAG:Ce	CWO	LYSO:Ce	LSO:Ce	GOS:Tb	$LaBr_3$:Ce
Density (g/cm^{-3})	3.67	4.51	7.13	4.57	7.9	7.1	7.4	7.3	5.3
Hygroscopic	Yes	Yes	No	No	No	No	No	No	Yes
Emission wavelength (nm)	415	550	480	550	490	420	420	550	358
Decay time (ns)	230	900	300	65	5000	35	40	600000	35
Light yield (phot. MeV^{-1})	38000	52000	2500	8000	28000	33000	30000	50000	61000

BGO = $Bi_4Ge_3O_{12}$.
YAG:Ce = $Y_3Al_5O_{12}:Ce^{3+}$.
CWO = $CdWO_4$.
LYSO:Ce = $(Lu,Y)_2SiO_5:Ce^{3+}$.
LSO:Ce = $Lu_2SiO_5:Ce^{3+}$.
GOS:Tb = $Gd_2O_2S:Tb^{3+}$.

8.8.5 Outlook

Due to the increasing demand on quantity and quality of scintillators by many applications, there is still tremendous research activity in the field of scintillators. Especially the requirement for high-quality single crystals causes increased production costs due to the need for specialized equipment and experienced personnel. An alternative to single inorganic crystals are scintillators in the form of transparent ceramics. Using ceramics poses several advantages over single crystals, such as faster and less expensive production, large size and unconventional geometries are easier to obtain, and multilayer materials can be manufactured [15]. General Electric (GE) has developed a transparent ceramic on the basis of Ce^{3+}-activated garnets, which is used as a scintillation detector in various types of medical equipment [16].

Furthermore, plastic scintillators, comprised of organic polymers, have been under investigation for some decades. They usually exhibit a very short decay time, are relatively simple to manufacture and are highly versatile, e.g. large volume detectors are relatively easy to produce. Their disadvantage is mainly the low absorption cross section for high-energy radiation and their low radiation damage resistance. The low absorption cross-section can be overcome to some extent by using sensitizers [17].

Liquid scintillators offer similar advantages and disadvantages as the plastic scintillators, and additionally allow very close contact between the emitting species and the scintillator, as the investigated sample can be dispersed or dissolved in the liquid [18]. For that reason they are often employed in alpha-particle or beta-spectroscopy.

References

[1] Röntgen WC. Über eine neue Art von Strahlen. Sb Phys Med Ge Würzburg, 1895, 137, 132.
[2] Stürmer W. Zur Geschichte der Röntgen-Leuchtstoffe. RöFo, 1962, 97, 514–19.
[3] Ozawa L, Itoh M. Cathode ray tube phosphors. Chem Rev, 2003, 103, 3835–56. https://doi.org/10.1021/cr0203490.
[4] Derenzo SE, Weber MJ, Bourret-Courchesne E, Klintenberg MK. The quest for the ideal inorganic scintillator. Nucl Instrum Methods Phys Res A, 2003, 505, 111–17. https://doi.org/10.1016/S0168-9002(03)01031-3.
[5] Yanagida T. Inorganic scintillating materials and scintillation detectors. Proc Jpn Acad Ser B Phys Biol Sci, 2018, 94, 75–97. https://doi.org/10.2183/pjab.94.007.
[6] Lowe BG, Sareen RA. Semiconductor X-Ray Detectors, CRC Press, Boca Raton, FL, 2014.
[7] Kubota S, Ruan J-Z, Itoh M, Hashimoto S, Sakuragi S. A new type of luminescence mechanism in large band-gap insulators: Proposal for fast scintillation materials. Nucl Instrum Methods Phys Res A, 1990, 289, 253–60. https://doi.org/10.1016/0168-9002(90)90267-A.
[8] Nikl M. Scintillation detectors for X-rays. Meas Sci Technol, 2006, 17, R37–R54. https://doi.org/10.1088/0957-0233/17/4/R01.
[9] Williams J, Johanning J. US20040178346A1, 2004.
[10] Gul S, Shiriyev J, Singhal V, Erge O, Temizel C. Advanced materials and sensors in well logging, drilling, and completion operations. In: Temizel C, Sari MM, Canbaz CH, Saputelli LA, Torsaeter O. (Eds.) Sustainable Materials for Oil and Gas Applications, Gulf Professional Publishing, Amsterdam, 2021, pp. 93–123.
[11] Shefer E, Altman A, Behling R, Goshen R, Gregorian L, Roterman Y, Uman I, Wainer N, Yagil Y, Zarchin O. State of the art of CT detectors and sources: A literature review. Curr Radiol Rep, 2013, 1, 76–91. https://doi.org/10.1007/s40134-012-0006-4.

[12] Nassalski A, Kapusta M, Batsch T, Wolski D, Mockel D, Enghardt W, Moszynski M. Comparative study of scintillators for PET/CT detectors. In: Yu B (Ed.) 2005 IEEE Nuclear Science Symposium Conference Record, IEEE, Piscataway, NJ, 2005, pp. 2823–29.
[13] Weber MJ. Inorganic scintillators: Today and tomorrow. J Lumin, 2002, 100, 35–45. https://doi.org/10.1016/S0022-2313(02)00423-4.
[14] Lecoq P. Development of new scintillators for medical applications. Nucl Instrum Methods Phys Res A, 2016, 809, 130–39. https://doi.org/10.1016/j.nima.2015.08.041.
[15] Xiao Z, Yu S, Li Y, Ruan S, Kong LB, Huang Q, Huang Z, Zhou K, Su H, Yao Z, Que W, Liu Y, Zhang T, Wang J, Liu P, Shen D, Allix M, Zhang J, Tang D. Materials development and potential applications of transparent ceramics: A review. Mater Sci Eng B, 2020, 139, 100518. https://doi.org/10.1016/j.mser.2019.100518.
[16] Zhang D, Li X, Liu B. Objective characterization of GE discovery CT750 HD scanner: Gemstone spectral imaging mode. Med Phys, 2011, 38, 1178–88. https://doi.org/10.1118/1.3551999.
[17] Hajagos TJ, Liu C, Cherepy NJ, Pei Q. High-Z sensitized plastic scintillators: A review. Adv Mater, 2018, 30, e1706956. https://doi.org/10.1002/adma.201706956.
[18] Beaulieu L, Beddar S. Review of plastic and liquid scintillation dosimetry for photon, electron, and proton therapy. Phys Med Biol, 2016, 61, R305–R343. https://doi.org/10.1088/0031-9155/61/20/R305.

Subject index

https://doi.org/10.1515/9783110798890-028

Formula index

https://doi.org/10.1515/9783110798890-029

www.ingramcontent.com/pod-product-compliance
Lightning Source LLC
Chambersburg PA
CBHW060956210326
41598CB00031B/4842